T0213285

Lecture Notes in Computer Science 11522

Commenced Publication in 1973
Founding and Former Series Editors:
Gerhard Goos, Juris Hartmanis, and Jan van Leeuwen

More information about this series at http://www.springer.com/series/7407

Susanna Donatelli · Stefan Haar (Eds.)

Application and Theory of Petri Nets and Concurrency

40th International Conference, PETRI NETS 2019
Aachen, Germany, June 23–28, 2019
Proceedings

 Springer

Editors
Susanna Donatelli ⓘ
Università di Torino
Turin, Italy

Stefan Haar ⓘ
Inria and LSV, CNRS & ENS Paris-Saclay
Cachan Cedex, France

ISSN 0302-9743   ISSN 1611-3349 (electronic)
Lecture Notes in Computer Science
ISBN 978-3-030-21570-5   ISBN 978-3-030-21571-2 (eBook)
https://doi.org/10.1007/978-3-030-21571-2

LNCS Sublibrary: SL1 – Theoretical Computer Science and General Issues

This Springer imprint is published by the registered company Springer Nature Switzerland AG
The registered company address is: Gewerbestrasse 11, 6330 Cham, Switzerland

Foreword

Welcome to the proceedings of the 40th International Conference on Application and Theory of Petri Nets and Concurrency. We were very proud to be able to host the 40th instance of the conference in the beautiful city of Aachen. The conference took place in the Tivoli football stadium close to the city center of Aachen. This beautiful venue provided a unique atmosphere with great views and excellent conference facilities. In the same week, the First International Conference on Process Mining (ICPM 2019) and the 19th International Conference on Application of Concurrency to System Design (ACSD 2019) as well as several workshops and tutorials took place at the same location.

I would like to thank the members of the Process and Data Science (PADS) group at RWTH Aachen University for helping to organize this event. Thanks to all authors, presenters, reviewers, Program Committee members, and participants. Special thanks go to the program chairs of Petri Nets 2019: Susanna Donatelli and Stefan Haar. The three conferences were supported by Celonis, Deloitte, ProcessGold, SAP, myInvenio, Fluxicon, AcceleraLabs, Minit, PuzzleData, PAFnow, Software AG, StereoLOGIC, BrightCape, Logpickr, Mehrwerk, QPR, KPMG, LanaLabs, Wintec, RWTH, Fraunhofer FIT, Gartner, the Alexander von Humboldt Foundation, the Deutsche Forschungsgemeinschaft, FernUniversität Hagen, and Springer.

A special feature of this year's conference was the special Friday afternoon session "Retrospective and Perspective of Petri Net Research" celebrating 40 years of Petri Net conferences organized by Jörg Desel. This was a wonderful opportunity to reflect on the history of our field and to celebrate the great scientific achievements of our community.

June 2019   Wil van der Aalst

Preface

This volume constitutes the proceedings of the 40th International Conference on Application and Theory of Petri Nets and Concurrency (Petri Nets 2019). This series of conferences serves as an annual meeting place to discuss progress in the field of Petri nets and related models of concurrency. These conferences provide a forum for researchers to present and discuss both applications and theoretical developments in this area. Novel tools and substantial enhancements to existing tools can also be presented.

Petri Nets 2019 was colocated with the 19th Application of Concurrency to System Design Conference (ACSD 2019). Both were organized by the Process and Data Science (PADS) group at RWTH Aachen University, Aachen, Germany. The conferences were co-organized with the First International Conference on Process Mining (ICPM 2019). The conference took place in the Tivoli business and event area in Aachen, during June 23–28, 2019.

This year, 41 papers were submitted to Petri Nets 2019 by authors from 19 different countries. Each paper was reviewed by three reviewers. The discussion phase and final selection process by the Program Committee (PC) were supported by the EasyChair conference system. From 33 regular papers and eight tool papers, the PC selected 23 papers for presentation: 20 regular papers and three tool papers. After the conference, some of these authors were invited to submit an extended version of their contribution for consideration in a special issue of a journal.

We thank the PC members and other reviewers for their careful and timely evaluation of the submissions and the fruitful constructive discussions that resulted in the final selection of papers. The Springer LNCS team (notably Anna Kramer, Ingrid Haas and Alfred Hofmann) provided excellent and welcome support in the preparation of this volume.

We are also grateful to the invited speakers for their contributions:

- Luca Cardelli, University of Oxford, who delivered the Distinguished Carl Adam Petri Lecture (CAP) on "Programmable Molecular Networks"
- Dirk Fahland, Eindhoven University of Technology, on "Describing, Discovering, and Understanding Multi-Dimensional Processes"; and
- Philippas Tsigas, Chalmers University of Technology, on "Lock-Free Data Sharing in Concurrent Software Systems"

Alongside *ACSD 2019*, the following workshops were also colocated with the conference:

- Workshop on Petri Nets and Software Engineering (PNSE 2019)
- Workshop on Algorithms and Theories for the Analysis of Event Data (ATAED 2019)

Other colocated events included the Petri Net Course and the tutorials on Model Checking for Petri Nets - from Algorithms to Technology (Karsten Wolf) and on Model-Based Software Engineering for/with Petri Nets (Ekkart Kindler).

We hope you enjoy reading the contributions in this LNCS volume.

June 2019

Susanna Donatelli
Stefan Haar

Organization

Steering Committee

W. van der Aalst	RWTH Aachen University, Germany
J. Kleijn	Leiden University, The Netherlands
L. Pomello	Università degli Studi di Milano-Bicocca, Italy
G. Ciardo	Iowa State University, USA
F. Kordon	Sorbonne University, France
W. Reisig	Humboldt-Universität zu Berlin, Germany
J. Desel	FernUniversität in Hagen, Germany
M. Koutny (Chair)	Newcastle University, UK
G. Rozenberg	Leiden University, The Netherlands
S. Donatelli	Università di Torino, Italy
L. M. Kristensen	Western Norway University of Applied Sciences, Norway
M. Silva	University of Zaragoza, Spain
S. Haddad	Ecole Normale Supérieure Paris-Saclay, France
C. Lin	Tsinghua University, China
A. Valmari	University of Jyväskylä, Finland
K. Hiraishi	JAIST, Japan
W. Penczek	Institute of Computer Science PAS, Poland
A. Yakovlev	Newcastle University, UK

Program Committee

Elvio Amparore	Università di Torino, Italy
Ekkart Kindler	Technical University of Denmark
Paolo Baldan	Università di Padova, Italy
Jetty Kleijn	Leiden University, The Netherlands
Didier Buchs	University of Geneva, Switzerland
Michael Köhler-Bußsmeier	University of Applied Science Hamburg, Germany
Robert Lorenz	University of Augsburg, Germany
Jose Manuel Colom	University of Zaragoza, Spain
Roland Meyer	TU Braunschweig, Germany
Isabel Demongodin	LSIS, UMR CNRS 7296, France
Lukasz Mikulski	Nicolaus Copernicus University, Torun, Poland
Susanna Donatelli (Co-chair)	Università di Torino, Italy
Andrew Miner	Iowa State University, USA
Javier Esparza	Technical University of Munich, Germany
G. Michele Pinna	University of Cagliari, Italy

Dirk Fahland	Eindhoven University, The Netherlands
Pascal Poizat	Université Paris Nanterre and LIP6, France
Gilles Geeraerts	Université Libre de Bruxelles, Belgium
Pierre-Alain Reynier	Aix-Marseille Université France
Stefan Haar (Co-chair)	Inria and LSV, CNRS and ENS Paris-Saclay, France
Olivier Roux	LS2N/Ecole Centrale de Nantes, France
Henri Hansen	Tampere University of Technology, Finland
Arnaud Sangnier	IRIF, University of Paris Diderot, CNRS, France
Petr Jancar	Palacky University of Olomouc, Czech Republic
Irina Virbitskaite	A.P. Ershov Institute, Russia
Ryszard Janicki	McMaster University, Canada
Matthias Weidlich	Humboldt-Universität zu Berlin, Germany
Gabriel Juhas	Slovak University of Technology Bratislava, Slovakia
Karsten Wolf	Universität Rostock, Germany
Anna Kalenkova	RWTH Aachen, Germany

Additional Reviewers

Kamila Barylska	Pieter Kwantes
Benoit Delahaye	Rabah Ammour
Marcin Piatkowski	Alfons Laarman
Claudio Antares Mezzina	Sebastian Muskalla
Tiziana Cimoli	Prakash Saivasan
Leonardo Brenner	Peter Chini
Johannes Metzger	Philipp J. Meyer
Jorge Julvez	Nataliya Gribovskaya
Natalia Garanina	Thomas Brihaye
Damien Morard	Anna Gogolinska
Stefan Klikovits	Petr Osička
Luca Bernardinello	Dimitri Racordon
Hanifa Boucheneb	

Sponsors

Deloitte.

Contents

Models with Extensions

Models

Describing Behavior of Processes with Many-to-Many Interactions

Dirk Fahland$^{(\boxtimes)}$ (iD)

Eindhoven University of Technology, Eindhoven, The Netherlands
d.fahland@tue.nl

Abstract. Processes are a key application area for formal models of concurrency. The core concepts of Petri nets have been adopted in research and industrial practice to describe and analyze the behavior of processes where each instance is executed in isolation. Unaddressed challenges arise when instances of processes may interact with each other in a one-to-many or many-to-many fashion. So far, behavioral models for describing such behavior either also include an explicit data model of the processes to describe many-to-many interactions, or cannot provide precise operational semantics.

In this paper, we study the problem in detail through a fundamental example and evolve a few existing concepts from net theory towards many-to-many interactions. Specifically, we show that three concepts are required to provide an operational, true concurrency semantics to describe the behavior of processes with many-to-many interactions: unbounded dynamic synchronization of transitions, cardinality constraints limiting the size of the synchronization, and history-based correlation of token identities. The resulting formal model is orthogonal to all existing data modeling techniques, and thus allows to study the behavior of such processes in isolation, and to combine the model with existing and future data modeling techniques.

Keywords: Multi-instance processes · Many-to-many interactions · Modeling · True-concurrency semantics · Petri nets

1 Introduction

Processes are a key application area for formal models of concurrency, specifically Petri nets, as their precise semantics allows both describing and reasoning about process behavior [1]. The basic semantic concepts of Petri nets, locality of transitions which synchronize by "passing" tokens, are at the core of industrial process modeling languages [21] designed to describe the execution of a process in a *process instance* which is isolated from all other instances. At the same time, processes behavior in practice is often not truly isolated in single process instances, but instances are subject to interaction with other instances, data objects, or other processes. Modeling and analyzing such processes has been the

© Springer Nature Switzerland AG 2019
S. Donatelli and S. Haar (Eds.): PETRI NETS 2019, LNCS 11522, pp. 3–24, 2019.
https://doi.org/10.1007/978-3-030-21571-2_1

focus of numerous works such as proclets [3], artifact-centric modeling [6], UML-based models [5], BPMN extensions [16], DB-nets [19], object-centric declarative modeling [2], and relational process structures [27].

The existing body of work can be considered in two major groups. Artifact-centric modeling that can address one-to-many and many-to-many relations first defines a relational data model; process behavior is then defined by a control-flow model for each entity in the data model, and logical constraints and conditions that synchronize the steps in one entity based on data values in the data model of another entity [5,6]. DB-nets adopt this principle to Petri net theory through Coloured Petri nets [19]. Numerous decidability results and verification techniques are available for these types of models, such as [5,7,17]. However, the behavior described by these models cannot be derived easily by a modeler through visual analysis as the interaction of one entity with other entities depend on complex data conditions not shown in the visual model, inhibiting their application in practice [22]. While object-centric declarative models [2,12] make the dependency between behavior and data visually explicit, declarative constraints themselves are challenging to interpret. Proclets do explicitly describe interactions between instances of multiple processes [3] but do not provide sufficient semantic concepts to describe many-to-many interactions [10]. Relational process structures [26,27] turn the observations of [10] into a model with implemented operational semantics for describing many-to-many interactions through so-called coordination processes. However, the language requires numerous syntactic concepts, and no formal semantics is available, prohibiting analysis, which also applies to data-aware BPMN extensions [16].

In this paper, we investigate semantic concepts that are required to provide formal semantics for a modeling language that is able to describe many-to-many interactions. The language shall bear a minimal number of syntactic and semantic concepts building on established concepts from Petri net theory. Our hope is that such a minimal, yet maximally net-affine language allows to project or build richer modeling languages on top of our proposed language, while allowing to apply or evolve existing Petri net analysis techniques for analyzing behavior with many-to-many interactions.

In Sect. 2, we study a basic example of many-to-many interactions through the formal model of Proclets and analyze the core challenges that arise in describing the behavior of such processes. In Sect. 3, we show that these challenges can be overcome by a paradigm shift in describing many-to-many relations. Just as many-to-many relations in a data model have to be reified into its own entity, we show that many-to-many interactions require reifying the message exchange into its own entity. Based on this insight, we then propose in Sect. 4 *synchronous proclets* as a formal model that extends [3] with *dynamic unbounded synchronization* of transitions. We provide a formal true concurrency semantics for our model. We then show in Sect. 5 how the semantics of relations between entities can be realized on the level transition occurrences as *cardinality* and *correlation* constraints over pairs of instance identifiers to fully describe many-to-many interactions in an operational semantics. We discuss some implications of our work in Sect. 6.

2   Multi-dimensional Dynamics - A Simple Example

The running example for this paper describes a very simple logistics process. After an order has been created, it gets fulfilled by sending packages to the customer. Typically, not all products are available in the same warehouse, resulting the order to be split into *multiple* packages. Packages are transported to the customer through a delivery process, where multiple packages are loaded for one delivery tour; these packages may originate from *multiple* orders, resulting in a many-to-many relationship between orders and deliveries. Packages are then delivered one by one, being either successfully delivered or the package could not be delivered, leading either to a retry in a new delivery or in considering the package as undeliverable. The customer is billed only after all deliveries of all packages in the order concluded.

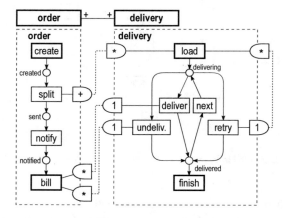

Fig. 1. Proclet model describing asynchronous message exchange between multiple instances

The proclet model [3] in Fig. 1 describes this process. The behavior of order and delivery instances are described in their respective *proclet*, initial and final transitions describe the creation and termination of instances. Instances of order and delivery *interact* by exchanging messages via channels (dashed lines); the cardinality inscriptions at the ports indicate how many messages are sent or received in the occurrence of the transition in one instance. Figure 2 shows a *partially-ordered run* that satisfies the model of Fig. 1: order17 gets split into two packages 1 and 3, while order18 requires just a single package 2. Packages 1 and 2 are loaded into delivery23 where package 2 requires a retry with delivery24 where it is joined by package 3.

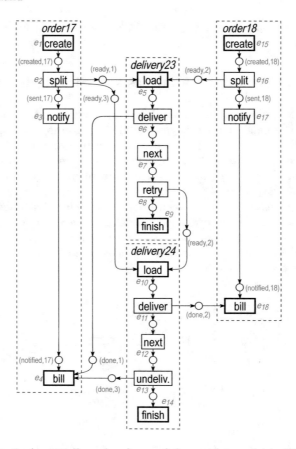

Fig. 2. A partially-ordered run of the proclet model in Fig. 1

The proclet semantics [3] however is not defined on instance identifiers and hence also allows undesired behaviors in many-to-many interactions, such as the run in Fig. 3: First, package 1 gets duplicated and then both delivered and gets a retry with delivery24. Second, package 3 originates in order17 but gets billed for order18. Third, package 2 is created and loaded but then disappears in the run.

3   Reifying Behavior of Relations into Conversations

The core problem of the proclet model in Fig. 1 is that order and delivery are in a many-to-many relation that is not explicitly described. We know from data modeling that implementing a many-to-many relationship in a relational data model requires to reify the relationship into its own entity, which then also results in the well-known Second Normal Form (2NF). By the same reasoning, we reify

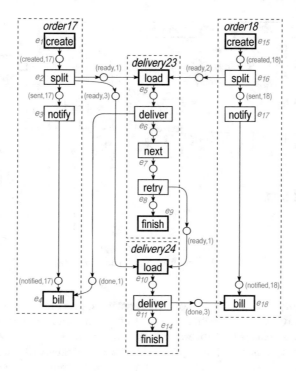

Fig. 3. Another partially-ordered run of the proclet model in Fig. 1 describing undesired behavior

the structural many-to-many relation in Fig. 1 into a package entity which has one-to-many relations to order and delivery.

The first central idea of this paper is that we also reify the *behavioral* relation between order and delivery, i.e., the channels, into its own sub-model. The resulting composition, shown in Fig. 4, contains a proclet for the package entity. Where the model in Fig. 1 described interaction through asynchronous message exchange, the model in Fig. 4 uses *synchronization of transitions* along the indicated channels. Hence, we call this model a *synchronous* proclet model. Crucially, *multiple* instances of a package may synchronize with an order instance in a single transition occurrence. Figure 5 shows a distributed run of the synchronous proclet system of Fig. 4, where the run of each instance is shown separately and the dashed lines between events indicate synchronization. For example, events e_2, e_2', and e_2'' in order17, package1, and package3 occur together at once in a single *synchronized event* which also causes the creation of the package1, and package3 instances. The run in Fig. 5 is identical to the run in Fig. 2 except for the condition loaded in the different package instances.

In line with the idea of a relational model in 2NF, the synchronization channels in the model in Fig. 4 only contain one-to-many, and one-to-one cardinalities. We consider a model such as the one in Fig. 4 to be in *behavioral second normal*

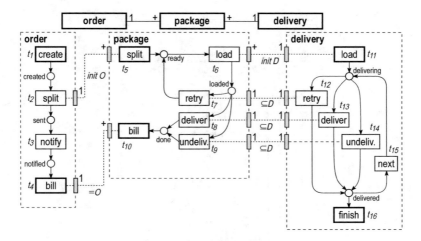

Fig. 4. Synchronous Proclet model

form: During any interaction between two entities, one entity is always uniquely identified, by having only one instance participate in the interaction. Specifically, a **package** always relates and interacts with one **delivery** and one **order**. However, that relation is dynamic as **package2** first relates to **delivery23** and later to **delivery24**. To ensure that "the right" instances remain synchronized, we propose using *correlation* constraints: the annotation init O and =O shall ensure that only package instances created by an order also synchronize in the bill step.

4   Dynamic Unbounded Synchronization

In the following, we develop the required formal concepts for the model proposed in Sect. 3. We first define dynamic synchronization of transitions in a true-concurrency semantics. In Sect. 5 we constrain synchronization through cardinality and correlation constraints.

4.1   Notation on Nets

A *net* $N = (P, T, F)$ consists of a set P of *places*, a set T of *transitions*, $P \cap T = \emptyset$, arcs $F \subseteq (P \times T) \cup (T \times P)$. We call $X_N = P \cup T$ the *nodes* of N. We write $^\bullet t$ and $t^\bullet$ for the set of pre-places and post-places of $t \in T_N$ along the arcs F, respectively; pre- and post-transitions of a place are defined correspondingly. We write $N_1 \cap N_2$, $N_1 \cup N_2$, and $N_1 \subseteq N_2$ for intersection, union, and subset of nets, respectively, which is defined element-wise on the sets of nodes and arcs of N_1 and N_2, and we write $\emptyset$ for the empty net.

A *labeled* net $N = (P, T, F, \ell)$ additionally defines a *labeling* $\ell : P \cup T \to \Sigma$ assigning each node $x \in X_N$ a label $\ell(x) \in \Sigma$; w.l.o.g, we assume for any two

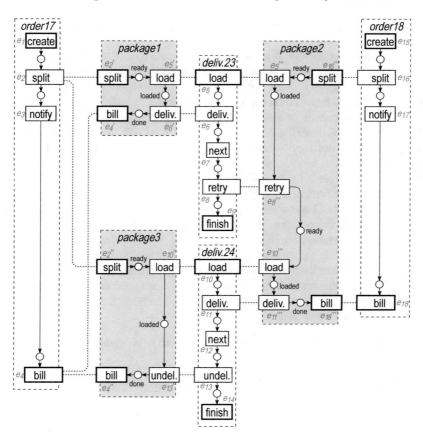

Fig. 5. Synchronization of multiple partially ordered runs corresponding to the synchronous proclet model of Fig. 4

labeled nets N_1, N_2 that $\ell_1(x) = \ell_2(x)$ for any node $x \in X_1 \cap X_2$. We use labels to synchronize occurrences of different transitions with the same label.

For a node $x \in X_N$ and a fresh node $x^* \notin X_N$ we write $N[x/x^*]$ for the net obtained by replacing in N simultaneously all occurrences of x by x^*.

An *occurrence net* $\pi = (B, E, G)$ is a net (B, E, G) where each place $b \in B$ is called a condition, each $e \in E$ is called an event, the transitive closure G^+ is acyclic, the set $past(x) = \{y \mid (y, x) \in G^+\}$ is finite for each $x \in B \cup E$, and each $b \in B$ has at most one pre-event and at most one post-event, i.e., $|^\bullet b| \leq 1$ and $|b^\bullet| \leq 1$. We will consider *labeled* occurrence nets $\pi = (B, E, G, \lambda)$, where each condition (event) is labeled with a *set* of labels of the from (x, id) where x refers to a place (transition) of another net, and id is an instance identifier. The behavior of any net N (with an initial marking m_0) can be described as a set of occurrence nets $R(N, m_0)$ called the runs of N.

4.2 Entities, Instances, and Synchronous Proclet System

A classical process model N_E describes the processing of a single entity E, e.g., an order, a delivery. This can be formalized as a single labeled net $N_E = (P_E, T_E, F_E, \ell_E)$. The behavior of *one instance* of E, e.g., a concrete order, then follows from consuming and producing "black" tokens in N_E which defines a firing sequence or a run of N_E [1].

Our aim is to describe the behavior of *multiple entities* and *multiples instances* of these entities *together in one* run as discussed in Sect. 2. We adopt Petri nets with token identities [11,24,25] to distinguish different instances of an entity E. Let $\mathcal{I}$ be an infinite set of instance *identifiers*. Each $id \in \mathcal{I}$ is a unique identifier. By distributing $\mathcal{I}$ (as tokens) over the places of N_E, we describe the state of instance id. The state of N_E (for several instances) is then a distribution of *multiple* tokens from $\mathcal{I}$ over the places of N_E.

To describe the interplay of multiple entities, we adopt concepts of *proclets* [3, 4] and *open nets* [14]. A system describes the behavior of each entity in its own net; the nets of all entities are *composed* along channels. Where the earlier works use asynchronous channels for composition, we use *synchronous channels* which are connecting pairs of transitions, as motivated in Sect. 3. This gives rise to the notion of a *synchronous proclet system*:

Definition 1 (Synchronous Proclet System). *A synchronous proclet system* $S = (\{N_1, \ldots, N_n\}, m_\nu, C)$ *defines*

1. *a set of labeled nets* $N_i = (P_i, T_i, F_i, \ell_i), i = 1, \ldots, n$, *each called a* proclet *of S,*
2. *an* initial marking $m_\nu : \bigcup_{i=1}^{n} P_i \to \mathbb{N}^{\mathcal{I}}$ *assigning each place p a multiset $m(p)$ of identifier tokens such that proclets have disjoint sets of identifiers,* $\forall 1 \leq i < j \leq n, \forall p_i \in P_i, p_j \in P_j : m_\nu(p_i) \cap m_\nu(p_j) = \emptyset$, *and*
3. *a set C of* channels *where each channel $(t_i, t_j) \in C$ is a pair of identically-labeled transitions from two different proclets:* $\ell(t_i) = \ell(t_j)$, $t_i \in T_i, t_j \in T_j, 1 \leq i \neq j \leq n$.

Figure 4 shows a synchronous proclet system for entities order, package, and delivery with several channels, for example, channel (t_2, t_5) connects two split-labeled transitions in order and package. Each proclet in Fig. 4 has a transition without pre-places and the initial marking is *empty*. This allows creating an unbounded number of new instances of any of the three entities in a run.

4.3 Intuitive Semantics for Synchronous Proclet Systems

A single proclet N_i describes the behavior of a single entity E_i. We assume each instance of E_i is identified by an identifier id. A distribution of id tokens over the places in N_i describes the current state of this instance. The instance id advances by an *occurrence* of an enabled transition of N_i in that instance id: Any transition $t \in N_E$ is *enabled* in instance id when each pre-place of t contains an id token; firing t in instance id then consumes and produces id tokens as usual. A *new instance* of E_i can be created by generating a new identifier id_ν as proposed for ν-nets [25]. We limit the creation of new identifiers to transitions without pre-set. Such an "initial" transition t_{init} is always enabled (as it has no pre-places); t_{init} may occur in instance $id_\nu \in \mathcal{I}$ only if id_ν is a fresh identifier never seen before; its occurrence then produces one id_ν token on each post-place of t_{init}. For example, in the run in Fig. 5, we see three occurrence of t_5 (split) in proclet package, each occurrence creates a different token package1, package2, package3 $\in \mathcal{I}$ on place ready describing the creation of three package instances.

In the entire proclet system, a local transition that is not connected via any channel, such as t_3 (notify) in order in Fig. 4, always occurs on its own. However, for transitions that are connected to each other via a channel, such as t_2 and t_5 (split), their occurrences *may synchronize*.

The modality "may synchronize" is important in the context of true concurrency semantics. Considering the partially-ordered run in Fig. 5, we can see two occurrence of t_2 (split) in instances order17 and order18, and three occurrences of t_5 in package1–package3. Bearing in mind that all instances are concurrent to each other, we may not enforce that one occurrence of t_2, say, in order17 *must* synchronize with *all* occurrences of t_5 in package1–package3. If we did, we would silently introduce a notion of global state and a global coordination mechanism which knows all order instances in that state, i.e., at a particular point in time. Rather, by synchronizing on *non-deterministically chosen subsets of possible occurrences*, we can express local knowledge. The occurrence of t_2 in order17 synchronizes with occurrences of t_5 in package1 and package2 in Fig. 5 because they happen to be "close to each other"—because package1 and package2 are created for order17. In other words, this non-determinism on synchronizing transition occurrences allows to abstract from a rather complex data-driven mechanism describing *why* occurrences synchronize while preserving *that* occurrences synchronize. While this very broad modality also leads to undesired behaviors intermittently, the notion of channel will allow us to rule out those undesired behaviors through a *local mechanism* only.

4.4   Partial Order Semantics for Synchronous Proclet Systems

In the following, we capture these principles in a true concurrency semantics of runs by an inductive definition over labeled occurrence nets. Specifically, we adopt the ideas proposed for plain nets [8,23] to our setting of multiple, synchronizing instances:

A *run* describes a partial order of transition occurrences which we represent as a special *labeled*, acyclic net π as shown in Fig. 2. A place b in π is called *condition*; its label $\lambda(b) = (p, id)$ describing the presence of a token id in a place p. A transition e in π is called an *event*; its label $\lambda(e) = t$ describes the occurrence of t. For example, the event e_2 in the run in Fig. 2 describes the occurrence of a transition which consumes token 17 from place created produces token 17 on sent and tokens 1 and 3 on place ready; e_2 is un-ordered, or *concurrent*, to e_{16}.

We construct such runs inductively. The initial state of a proclet system is a set of *initial conditions* representing the initial marking m_ν. The run in

Fig. 2 has no initial conditions. To extend a run, we term the *occurrence* of a transition t in an instance id. An occurrence o of t is again small net with a single event e labeled with (t, id); e has in its pre-set the conditions required to enabled t in instance id: for each pre-place p of t, there is a pre-condition of e with label (p, id); the post-set of e is defined correspondingly. Each occurrence of t describes the enabling condition for t and the effect of t in that particular instance id. Figure 6 (top) shows two occurrences of t_2 (split) in instance order17 and in instance order18, and three occurrences of t_5 (split) in instances package1-package3.

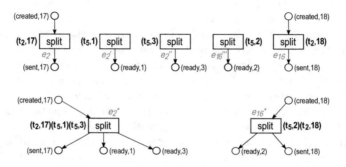

Fig. 6. Occurrences of transitions t_2 and t_5 in different instances (top), and synchronized occurrences of t_2 and t_5 in different instances (bottom)

Intuitively, a run is obtained by repeatedly appending occurrences (of enabled transitions) to the run. The *maximal* conditions of a run π describe which places hold which tokens, that is, the current marking of the system. For instance, the prefix π_1 shown in Fig. 7 has reached the marking where only place created holds the tokens 17 and 18. An occurrence o of a transition t is then *enabled* in a run π if the pre-conditions of o also occur in the maximal condition of π. For instance, the occurrence of t_2 (split) in order17 shown in Fig. 6 (top) is enabled in π_1 in Fig. 7. Also the occurrences of t_5 in package1-package3 and of t_2 in order18 are enabled π_1.

In a classical net, we could now append the occurrence of t_2 in order17 to π_1. In a synchronous proclet system, occurrences of enabled transitions that are connected via a channel may synchronize. For example, we may synchronize the occurrences of t_2 in order17 with the occurrences of t_5 in package1 and package3 which we express as a *synchronized occurrence*. The synchronized occurrence unifies the events in all individual occurrences into a single event e^* and otherwise preserves all pre- and post-conditions; we label e^* with the multiset of transitions occurring together as shown in Fig. 6 (bottom). The synchronized occurrence is then appended to the run. For example, we obtain run π_2 in Fig. 7 by appending the synchronized occurrence of t_2, t_5, t_5 in order17, package1, package3 to run π_1, where synchronization is shown in Fig. 7 through a dashed line. In run π_2

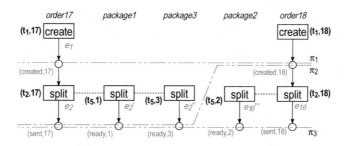

Fig. 7. Prefixes of the run of Fig. 5

the synchronized occurrence of t_2, t_5 in order17 and package2 (shown in Fig. 6 (bottom)) is enabled; appending it yields run π_3.

4.5 Formal Semantics for Dynamic Synchronization

We formalize the semantics of proclet systems laid out in Sect. 4.4 by an inductive definition. For the remainder of this section, let $S = (\{N_1, \ldots, N_n\}, m_\nu, C)$ be a proclet system, and let $T_S = \bigcup_{i=1}^{n} T_i$ be all transitions in S.

An *occurrence* of a transition t in an instance id is an occurrence net o with a single event e describing the occurrence of t; the labeling of o describes an injective homomorphism between the pre-places of t and the pre-conditions of e, and between the post-conditions and post-places, respectively. Technically, the event's label is a singleton set $\{(t, id)\}$ and a condition's label is a singleton set $\{(p, id)\}$; using sets of labels will allow us later to establish an injective homomorphism from synchronized occurrences of transitions to the transitions in the system definition.

Definition 2 (Occurrence of a transition). *Let $t \in T_S$ be a transition. Let $id \in \mathcal{I}$ be an identifier.*

An occurrence of t in id is a labeled net $o = (B_{pre} \uplus B_{post}, \{e_t\}, G, \lambda)$ with

1. *for each $p \in {}^\bullet t$ exists a condition $b_p \in B_{pre} : \lambda(b_p) = \{(p, id)\}(b_p, e_t) \in G$*
2. *for each $p \in t^\bullet$ exists a condition $b_p \in B_{post} : \lambda(b_p) = \{(p, id)\}, (e_t, b_p) \in G$*
3. *$\lambda(e_t) = \{(t, id)\}$.*

The labeling ℓ_i canonically lifts to o: for each node $x \in B_{pre} \cup B_{post} \cup \{e_t\}$ referring to $\lambda(x) = \{(y, id)\}$ set $\ell(x) := \ell_i(y)$.

We call $pre_o := B_{pre}$ the precondition of o, we call $con_o := B_{post} \cup \{e_t\}$ the contribution of o, and write $e_o := e_t$ for the event of o. Let $O(t, id)$ be the set of all occurrences of t in id.

Figure 6 shows among others occurrences of t_2 in order1 and of t_5 in package2. Each transition (also in the same instance) may occur multiple times in a run, see for instance t_6 (load) in package2 in Fig. 5. To make our task of composing runs from transition occurrences easier, and not having to reason about the identities of events or conditions, we therefore consider the set $O(t, id)$ of all occurrences of t in id. Technically, it contains an infinite supply of isomorphic copies of *the* occurrence of t in id. When composing runs and when synchronizing occurrences, we will simply take a suitable copy.

We now turn to synchronizing transition occurrences. Although we will later only synchronize transition occurrences along channels, we propose here a simpler and more general definition. Where an occurrence of a transition is parameterized by the transition t and the instance id, a *synchronized occurrence* is parameterized by a label a and a finite *set* $I \subseteq \mathcal{I}$ of instances. For each instance $id \in I$, one transition with label a participates in the synchronized occurrences. We first define, given a and I, the sets of occurrences of transitions that may be synchronized. In the spirit of high-level nets which also parameterize transition occurrences through variables, we call this set of occurrences an *occurrence binding* to a and I. The synchronized occurrence is then a canonical composition of all occurrences in the binding.

Definition 3 (Occurrence binding). *Let $a \in \Sigma$ be a label. Let $I \subseteq \mathcal{I}$ be a nonempty finite set of identifiers.*

A set O_I of occurrences is an occurrence binding *for a and I (in system S) iff for each $id \in I$ exists exactly one occurrence $o_{id} \in O(t, id)$ of some $t \in T_S$ with $o_{id} \in O_I$. We write $O(a, I)$ for the set of all occurrence bindings for a and I.*

In Fig. 6, the occurrences $o_1 \in O(t_5, 1), o_3 \in O(t_5, 3), o_{17} \in O(t_2, 17)$ form an occurrence binding $\{o_1, o_3, o_{17}\}$ for split and $I = \{1, 3, 17\}$. In this example, there is no other occurrence binding for this label and set of instances (up to isomorphism of the occurrences themselves). Without loss of generality, we may assume that any two occurrences $o, o' \in O_I$ in an occurrence binding $O_I \in O(a, I)$ are pair-wise disjoint, i.e., $o \cap o' = \emptyset$.

We obtain a *synchronized occurrence* of instances I at a by composing the occurrences in the binding for a and I along the events labeled a. For example, the synchronized occurrence of $\{1, 3, 17\}$ at split is shown in Fig. 6 (bottom left). To aid the composition, we use (1) simultaneous replacement $o[e/e^*]$ of an event by a new event $e*$, as defined in Sect. 4.1, and (2) we lift the union of nets in Sect. 4.1 to union of occurrences $o_1 \cup o_2$ by defining $\lambda_1 \cup \lambda_2(x) := \lambda_1(x) \cup \lambda_2(x)$ for each $x \in X_1 \cup X_2$. The formal definition of a synchronized occurrence reads as follows.

Definition 4 (Synchronized occurrence of a label). *Let $a \in \Sigma$, let $I \subseteq \mathcal{I}$ be a nonempty finite set of identifiers, and let $O_I \in O(a, I)$ be an occurrence binding for a and I in system S.*

The synchronized occurrence *of instances I at label a is the net*

$$\tilde{o} = \bigcup_{o_{id} \in O_I} (o_{id}[e_o/e^*])$$

obtained by replacing the event e_o in each occurrence o_{id} be a fresh event $e^* \notin X_i$, for all $i = 1, \ldots, n$ which unifies the occurrences along the event, and compos-

ing all occurrences by union. We write $\tilde{O}(a, I)$ for the set of all synchronized occurrences of I at a.

Figure 6 (bottom left) shows the synchronized occurrence of order17, package1, and package3 at split. Note that the event of this occurrence is labeled with the set $\{(t_2, 17), (t_5, 1), (t_5, 3)\}$ resulting from the synchronizing composition of events labeled with $(t_2, 17)$, $(t_5, 1)$, and $(t_5, 3)$. Note that this synchronized occurrence describes the *instantiation* of two new packages package1, package2 by order17.

Labeling event e^* with a set $\lambda(e^*)$ does not allow us to describe auto-concurrency of a transition, i.e., a synchronization of two occurrences of (t, id) when ${}^\bullet t$ is marked with two or more id tokens. This case is excluded by our definition of occurrence binding (Definition 3) allowing for each $id \in I$ exactly one occurrence; this limitation could be overcome by using multisets.

The labeling $\ell(.)$ which we lifted to occurrences of a transition (Definition 2) also lifts to synchronized occurrences, as the synchronization in Definition 4 only merges events $e_1, \ldots, e_n$ carrying the same $\ell(e_1) = \ldots = \ell(e_n) = a$. Also the precondition, contribution, and the event of an occurrence lift to synchronized occurrences written $pre_{\tilde{o}}$, $con_{\tilde{o}}$, and $e_{\tilde{o}}$, respectively.

Each synchronized occurrence structurally preserves its constituting occurrences through the labeling $\lambda(.)$. Let $o = (B, E, G, \lambda)$ be a labeled occurrence net. Let $I \subseteq \mathcal{I}$. We write $o|_I$ for the restriction of o to those nodes labeled with an $id \in I$: $o|_I = (B \cap Z, E \cap Z, G|_{Z \times Z}, \lambda|_I)$ with $Z = \{x \in X_o | (y, id) \in \lambda(x), id \in I\}$ and $\lambda|_I(x) := \{(y, id) \in \lambda(x) | id \in I\}$. Restricting any synchronized occurrence to a single identifier results in the occurrence of the corresponding transition in that instance.

Lemma 1. *Let $a \in \Sigma$, let $I \subseteq \mathcal{I}$ be a nonempty finite set of identifiers, and let $O_I \in O(a, I)$ be an occurrence binding for a and I in system S. For each $\tilde{o} \in \tilde{O}(a, I)$ holds: for each $id \in I$, $o := \tilde{o}|_{\{id\}} \in O(t, id)$ where $\lambda(e_o) = \{(t, id)\}$.*

Note that occurrence bindings and synchronized occurrences also apply to singleton sets of identifiers. In that case any "synchronized" occurrence $\tilde{o} \in \tilde{O}(a, \{id\})$ is an occurrence $\tilde{o} \in O(t, id)$. In case the system has only one transition t_a labeled with a, synchronized occurrences and transition occurrences coincide: $\tilde{O}(a, \{id\}) = O(t, id)$; see for instance t_3 (notify) in Fig. 4. This allows us to consider from now on only synchronized transition occurrences.

We may now give a formal inductive definition of the partially-ordered runs of a system with dynamic unbounded synchronization. This definition still does not consider specific semantics of channels of a proclet system which we discuss in Sect. 5.

Definition 5 (Runs with dynamic unbounded synchronization). *Let $S = (\{N_1, \ldots, N_n\}, m_\nu, C)$ be a proclet system. The set of runs of S with dynamic unbounded synchronization is the smallest set $R(S)$ such that*

1. *The initial run $\pi_0 = (B, \emptyset, \emptyset, \lambda)$ that provides for each id token on a place p in m_ν a corresponding condition, i.e., $|\{b \in B | \lambda(b) = \{(p, id)\}\}| = m_\nu(p)(id)$, is in $R(S)$.*

2. *Let $\pi \in R(S)$, let $a \in \Sigma$, and $I \subseteq \mathcal{I}$ be finite. Let $\tilde{o} \in \tilde{O}(a, I)$ be a synchronized occurrence of I at a.*
   (a) *$\tilde{o}$ is* enabled *in π iff exactly the preconditions of $\tilde{o}$ occur at the end of π, i.e., $pre_{\tilde{o}} \subseteq \max \pi := \{b \in B_\pi | b^\bullet = \emptyset\}$ and, w.l.o.g, $con_{\tilde{o}} \cap X_\pi = \emptyset$*
   (b) *if $\tilde{o}$ is enabled in π then appending $\tilde{o}$ to π is a run of S, formally $\pi \cup \tilde{o} \in R(S)$*

The run of Fig. 5 is a run of the proclet system in Fig. 4 according to Definition 5. The wrong run in Fig. 3 is *not* a run of that proclet system: by events e_6 instance package1 moves to state done, hence events e_7 and e_8 cannot be synchronized occurrences in instance package1 that lead to state ready.

The following lemma states that the labeling λ of the runs of S defines a local injective homomorphism to the syntax of S in the same way as the labeling in the runs of a classical net N defines a local injective homomorphism the syntax of N [9]. The lemma follows straight from the inductive definition and from Lemma 1.

Lemma 2. *Let $S = (\{N_1, \ldots, N_n\}, m_\nu, C)$ be a proclet system, let $\pi \in R(S)$. Then π is an occurrence net where*

1. *for each $b \in B_\pi$: $\lambda(b) = \{(p, id)\}$ for some $p \in P_i, 1 \leq i \leq n$*
2. *for each $e \in E_\pi$: $\lambda(e) = \{(t_1, id_1), \ldots, (t_k, id_k)$ with $t_j \in T_i, 1 \leq i \leq n$ for each $1 \leq j \leq k$, and $\ell(t_j) = \ell(t_y)$ for all $j, y = 1, \ldots, k$*
3. *$|\{b \in B | \lambda(b) = \{(p, id)\}| = m_\nu(p)(id)$, for each $p \in P_i, 1 \leq i \leq n$*
4. *for each instance id occurring in π and the run $\pi|_{id} = (B', E', G', \lambda')$ of instance id holds: for each event $e \in E'$ with $\lambda(e) = \{(t, id)\}, t \in N_i, 1 \leq i \leq n$, λ' defines an injective homomorphism from $\{e\} \cup {}^\bullet e \cup e^\bullet$ to $\{t\} \cup {}^\bullet t \cup t^\bullet$ in N_i.*

5 Relational Synchronization

While the runs defined in Sect. 4 provide an operational formal semantics for proclet systems, Definition 5 is not restrictive enough to correctly model the intended behaviors. It, for instance, allows two order instances to synchronize with a single package in an occurrence of split, e.g., synchronizing o_{17}, o_{18}, o_2 in Fig. 6. Likewise, Definition 5 allows that a package instance created by order17 does not synchronize with order17 but with order18 at bill. In this section, we rule out such behavior by constraining the occurrence bindings via the channels. We first define *cardinality constraints* and *correlation constraints* for synchronization at channels, and then provide the semantics of both constraints.

5.1 Cardinality and Correlation Constraints

We adopt the notion of cardinality constraints known from data modeling, and applied in relational process structures [27], to channels (t_i, t_j) between two proclets N_i and N_j, see Definition 1. Each channel constraint specifies how many occurrences of t_i may synchronize with how many occurrences of t_j in one synchronized occurrence. As each occurrence of t_i and t_j is related to a specific instance, we thus constrain which instances of N_i and N_j may synchronize in one step. A *cardinality constraint* specifies for each transition of a channel a lower bound l and an upper bound u between 0 and ∞. We will later formalize that in any synchronized occurrence, the number of occurrences of the transition has to be between these two bounds. For example, according to the channel constraint for (t_2, t_5) in Fig. 4, exactly one instance of order synchronizes with one or more instances of package at any occurrence of split.

To ensure consistency of synchronization over multiple steps, we adopt the concept of *correlation identifiers*. Correlation in message-based interaction between processes [20, 21] is achieved by specifying a particular attribute of a message as a *correlation attribute a*. A process instance R receiving a message m from an unknown sender instance S *initializes* a local *correlation key* $k := m.a$ with the value of a in m. To send a response to the unknown sender instance S, R creates a message m_2 where attribute $m_2.a := k$ holds the value of k. If R later only wants to receive a response from S (and no other sender instance), R will only accept a message m_3 where $m_3.a = k$. This is called *matching of correlation keys*. This concept can be extended to multi-instance interaction, using local data for correlation, instead of dedicated correlation keys [16].

For the synchronous interaction model proposed in this paper, we define correlation over synchronous channels instead of messages. A channel c_{init} can be labeled to *initialize* a correlation set S, meaning all instances which synchronize at a step over c_{init} are in S. Another channel c_{match} can be labeled to *match* a previously initialized correlation set S, meaning the instances synchronizing at a step over c_{match} have to be either a subset of S or equal to S. For example, according to the correlation constraints at channels (t_2, t_5) (split) and (t_4, t_{10}) (bill), exactly the package instances which were created at split by an order instance must synchronize at bill with the same order instance. In contrast, the package instances synchronizing at a deliver step with a delivery instance only have to be a *subset* of the package instances loaded into the delivery.

Definition 6 (Channel constraints). *Let* $S = (\{N_1, \ldots, N_n\}, m_\nu, C)$ *be a synchronous proclet system.*

1. *A cardinality constraint for C is a function card which specifies for each channel $c = (t, t') \in C$ a lower and an upper bound for each transition t and t' in the channel card$(c) = ((l, u) : (l', u'))$ with $0 \leq l \leq u \leq \infty$, $0 \leq l' \leq u' \leq \infty$.*
2. *A correlation constraint $K = (C_{init}, C_{match}^{\subseteq}, C_{match}^{=})$ for C specifies a set of initializing channels $C_{init} \subset C$, a set of partially matching channels $C_{match}^{\subseteq}$, and a set of fully matching channels $C_{match}^{=}$, where all sets of channels are pair-wise disjoint.*

Given a cardinality constraint card and a set corr of correlation constraints for C, we call $S = (\{N_1, \ldots, N_n\}, m_\nu, C, card, corr)$ a constrained synchronous proclet system.

Channel constraints are visualized as shown in Fig. 4. Cardinality constraints are indicated at the ends of the edges indicating channels, we use the standard abbreviations of ? for $(0,1)$, 1 for $(1,1)$, $*$ for $(0,\infty)$, and $+$ for $(1,\infty)$. The channels in Fig. 4 are constrained by $1 : +$ and $1 : 1$ cardinality constraints. Correlation constraints are annotated in the middle of a channel, marking initialization with $\mathsf{init}K$, partial matching with $\subseteq \mathsf{K}$, and full matching with $= \mathsf{K}$. The proclet system in Fig. 4 has two correlation constraints O and D. Note that in general, a channel may be part of different correlation constraints, initializing in one constraint while matching in another constraint.

5.2 Semantics of Cardinality Constraints

A cardinality constraint of a channel (t, t') restricts the number of occurrences of t and t' synchronizing in a step. We formalize this by restricting the *occurrence bindings* of a label (Definition 3) to adhere to the channel constraints of all transitions involved. For any set O of transition occurrences, let $O[t] = \{o \in O \mid \ell(e_o) = t\}$ be the set all occurrences of transition t in O. Further, we write $O[t_1, t_2] = \{(o_1, o_2) \mid o_1 \in O[t_1], o_2 \in O[t_2]\}$ for the *relation of occurrences* between t_1 and t_2. By Lemma 1, we may also write $\tilde{O}[t_1, t_2] = O[t_1, t_2]$ for the synchronized occurrence $\tilde{O}$ of binding O. Writing $inst(o) = id$ for the instance of an occurrence $o \in O(t, id)$, we obtain $inst(O[t_1, t_2]) = \{(inst(o_1), inst(o_2)) \mid (o_1, o_2) \in O[t_1, t_2]\}$.

If we interpret a channel (t_1, t_2) between two proclets N_1 and N_2 as a *relation* between two transitions in two different proclets, then $O[t_1, t_2]$ are the "records" of this relation that we can observe in O. Assuming there is a relational data model that underlies the proclet system and provides relational tables for the entities E_1 and E_2 described by N_1 and N_2, then $inst(O[t_1, t_2])$ are the "records" of the relationship between E_1 and E_2. In this spirit, the cardinality constraint only allows occurrence bindings where the constraints on this relation is satisfied.

Definition 7 (Occurrence binding satisfies cardinality constraint). *Let* $S = (\{N_1, \ldots, N_n\}, m_\nu, C, card, corr)$ *be a constrained proclet system. Let* $O_I \in O(a, I)$ *be an occurrence binding for instances* $I \subseteq \mathcal{I}$ *at* $a \in \Sigma$.

The occurrence binding O_I *satisfies the cardinality constraint* $card(t_1, t_2) = ((l_1, u_1) : (l_2, u_2))$ *of channel* $(t_1, t_2) \in C$ *iff if* $\ell(t_1) = a$, *then* $l_1 \leq |O_I[t_1]| \leq u_1$ *and* $l_2 \leq |O_I[t_2]| \leq u_2$. *We then also say that the synchronized occurrence* $\tilde{O}_I$ *of* O_I *is an* occurrence *of channel* (t_1, t_2).

O_I *satisfies the cardinality constraints of* S *iff* O_I *satisfies the cardinality constraint of each channel of* S. *We then also say that the synchronized occurrence* $\tilde{O}_I$ *satisfies the cardinality constraints of* S.

For example, considering the occurrences $o_1, o_2, o_3, o_{17}, o_{17}$ in Fig. 6, the occurrence binding $\{o_{17}, o_1, o_3\}$ satisfies the $1 : +$ cardinality constraint of (t_2, t_5) in Fig. 4, whereas the occurrence binding $\{o_{17}, o_{18}, o_3\}$ does not satisfy this constraint. A run of S can only be extended with a synchronized occurrence $\tilde{O}$ if the occurrence binding O satisfies the cardinality constraints of S.

Following the reasoning of Sect. 3, $1 : +$ and $1 : 1$ cardinality constraints have the most natural interpretation from an operational perspective as in any occurrence of the channel, the "1" side can take a local "coordinating" role for synchronization with the "+" side.

5.3   Semantics of Correlation Constraints

Correlation between different transition occurrences is a *behavioral property*. Thus, we will not extend the notion of state of a proclet system to hold values of correlation properties which can be initialized and matched. Rather, we give a behavioral definition over the history of the run.

A correlation constraint may be initialized multiple times in a run, each time with the relation $inst(\tilde{O}[t, t'])$ of instances involved in the synchronized occurrence $\tilde{O}$. For example, consider the synchronized occurrence $\tilde{O}_{\mathsf{split},17}$ of $(t_2, 17), (t_5, 1), (t_5, 3)$ at split (synchronization of e_2, e'_2, e''_2) in the run of Fig. 8. According to Fig. 4, $\tilde{O}_{\mathsf{split},17}$ initializes the correlation constraint O with $inst(\tilde{O}_{\mathsf{split},17}[t_2, t_5]) = \{(17, 1), (17, 3)\}$. Likewise, the synchronized occurrence $\tilde{O}_{\mathsf{split},18}$ of $(t_2, 18), (t_5, 2)$ at split (synchronization of e_{16}, e'_{16}) initializes the constraint O with $inst(\tilde{O}_{\mathsf{split},18}[t_2, t_5]) = \{(18, 2)\}$.

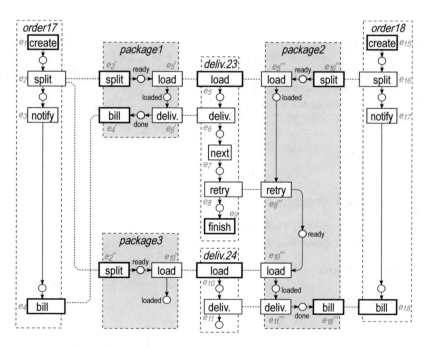

Fig. 8. The synchronized occurrence of order17 and package1 at bill violates the matching constraint = O of Fig. 4

While initialization can occur arbitrarily, matching is constrained: any synchronized occurrence involving channel (t_4, t_{10}) at bill has to match an initialization of O that occurred before. For instance, the synchronized occurrence $\tilde{O}_{\text{bill},17}$ of $(t_4, 17), (t_{10}, 1)$ at bill (synchronization of e_4, e'_4) in Fig. 8 satisfies the cardinality constraint $1 : +$ of channel (t_4, t_{10}) and involves the relation $inst(\tilde{O}_{\text{bill},17}[t_4, t_{10}]) = \{(17, 1)\}$. This occurrence violates the matching constraint $= \mathsf{O}$ of channel (t_4, t_{10}) because the initialization of cardinality constraint O that precedes $\tilde{O}_{\text{bill},17}$ (at e_4, e'_4) is the synchronized occurrence $\tilde{O}_{\text{split},17}$ (at e_2, e'_2, e''_2), and $inst(\tilde{O}_{\text{bill},17}[t_4, t_{10}]) = \{(17, 1)\} \neq inst(\tilde{O}_{\text{split},17}[t_2, t_5]) = \{(17, 1), (17, 3)\}$. In contrast, the synchronized occurrence $\tilde{O}'_{\text{bill},17}$ shown in Fig. 5 satisfies the correlation constraint $= \mathsf{O}$.

Definition 8 (Occurrence satisfies correlation constraint in a run). *Let $S = (\{N_1, \ldots, N_n\}, m_\nu, C, card, corr)$ be a constrained proclet system. Let π be a labeled occurrence net. Let $\tilde{o}_m \in \tilde{O}(a_m, I_m)$ be a synchronized occurrence of instances I_m at a_m such that $pre_{\tilde{o}_m} \subseteq \max \pi$.*

$\tilde{o}_m$ satisfies correlation constraint $(X_{init}, X_{match}^{\subseteq}, X_{match}^{=})$ in π iff

1. *if $\tilde{o}_m$ is an occurrence of a channel $(t_m, t'_m) \in X_{match}^{\subseteq}$, then there exists a synchronized occurrence $\tilde{o}_i \in \tilde{O}(a_i, I_i)$ of an initializing channel $(t_i, t'_i) \in X_{init}$ such that*
   (a) *the initializing occurrence $\tilde{o}_i$ is in π, i.e., $\tilde{o}_i \subseteq \pi$,*
   (b) *and precedes the matching occurrence $\tilde{o}_m$, i.e., for the event $e_{\tilde{o}_i}$ of $\tilde{o}_i$ exists a path in π to some $b \in pre_{\tilde{o}_m} : (e_{\tilde{o}_m}, b) \in G^+$, and*
   (c) *the relation of instances involved in $\tilde{o}_m$ matches the relation of instances involved in $\tilde{o}_i$, i.e., $id(\tilde{o}_m[t_m, t'_m]) \subseteq id(\tilde{o}_i[t_i, t'_i]))$*
2. *if $\tilde{o}_m$ is an occurrence of a channel $(t_m, t'_m) \in X_{match}^{=}$, then additionally $id(\tilde{o}_m[t_m, t'_m]) = id(\tilde{o}_i[t_i, t'_i]))$ has to hold.*

$\tilde{o}_m$ satisfies the correlation constraints of S iff $\tilde{o}_m$ satisfies each correlation constraint in corr.

5.4   Runs of a Constrained Proclet System

We can now easily extend Definition 5 to limit the runs of a proclet system to those allowed by the cardinality and correlation constraints.

Definition 9 (Runs of a constrained proclet system). *Let $S = (\{N_1, \ldots, N_n\}, m_\nu, C, card, corr)$ be a constrained proclet system. The set of runs of S is the smallest set $R(S)$ such that*

1. *The initial run $\pi_0 \in R(S)$ as in Definition 5.*
2. *Let $\pi \in R(S)$, let $a \in \Sigma$, and $I \subseteq \mathcal{I}$ be finite. Let $\tilde{o} \in \tilde{O}(a, I)$ be a synchronized occurrence of I at a.*
   (a) *$\tilde{o}$ is enabled in π iff*
      i. *exactly the preconditions of $\tilde{o}$ occur at the end of π, i.e., $pre_{\tilde{o}} \subseteq \max \pi$,*

   ii. õ satisfies the cardinality constraints card of S (Definition 7), and
   iii. õ satisfies the cardinality constraints corr of S in π (Definition 8).
(b) if õ is enabled in π then appending õ to π is a run of S, formally π ∪ õ ∈ R(S).

The occurrence net of Fig. 5 is a run of the constrained proclet system in Fig. 4 (assuming all events connected by dashed lines, such as e_2, e'_2, e''_2 are synchronized into a single event). The occurrence net of Fig. 8 is not a run of that system. However, the system of Fig. 4 cannot ensure termination of delivery instances: finish should only occur when all packages have been handled. The proclet system of Fig. 9 ensures correct termination through an extended package life-cycle and an additional channel. The system also illustrates that a proclet may interact with more than two other proclets, in this case a return process that must be completed for each undelivered package prior to billing.

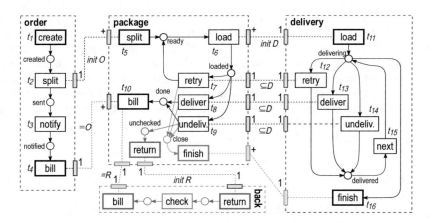

Fig. 9. Extension of the proclet system of Fig. 4

6   Conclusion

We have shown that by reifying a behavioral many-to-many relation into its own entity with a behavioral model, the behavior of processes with many-to-many interactions can be described both visually clear and formally precise. The foundational concept is that of *dynamic unbounded synchronization*, which can be seen as a generalization of the synchronization in artifact-choreographies with 1-to-1 relationships proposed by Lohmann and Wolf [15]. We have shown that this basic behavioral model can be extended orthogonally with cardinality and correlation constraints that limit the allowed synchronization much in the same way as guards in Coloured Petri Nets limit firing modes. We believe that this restriction to purely behavioral concepts allows adoption and integration with other more data-aware modeling techniques such as [19, 27].

The only "higher-level" concept required by our model are case identity tokens as in ν-nets, and we only use equality of token identities, and subsets of pairs of token identities. This suggests that verification results on ν-nets [18] may be lifted to our model. Though, we suspect undecidability to arise in the general case for correlation constraints testing for equality.

Further, the structure of a "reified" process model in behavioral second normal form (having only 1-to-1 and 1-to-many synchronizations) allows some further reflection on the structure of such processes. The asynchronous message exchange between two processes is replaced by an entity explicitly describing the interaction. This entity such as the package in our running example, is a "passive" (data) object, as changes to its state are due to activities in the processes operating on it, making the processes "active" entities. This leads for many, but possibly not all use cases to a bipartite structure of entities. Two "active" processes never synchronize directly as each transition describes a task that is manipulating a "passive" object; two "passive" objects never synchronize directly as they require an "active" process to trigger the necessary state change. In this understanding, an asynchronous message channel is passive object, synchronizing with sender and receiver, and each instance of the channel is a message. Lohmann and Wolf [13] have shown that this interpretation allow for formulating new types of research questions: given a set of objects, synthesize the active processes synchronizing them; given a set of processes, synthesize the passive objects realizing their synchronization.

References

1. van der Aalst, W.M.P.: The application of petri nets to workflow management. J. Circ. Syst. Comput. **8**(1), 21–66 (1998)
2. van der Aalst, W.M.P., Artale, A., Montali, M., Tritini, S.: Object-centric behavioral constraints: integrating data and declarative process modelling. In: Proceedings of the 30th International Workshop on Description Logics, Montpellier. CEUR Workshop Proceedings, vol. 1879. CEUR-WS.org (2017)
3. van der Aalst, W.M.P., Barthelmess, P., Ellis, C.A., Wainer, J.: Proclets: a framework for lightweight interacting workflow processes. Int. J. Cooperative Inf. Syst. **10**(4), 443–481 (2001)
4. van der Aalst, W.M.P., Mans, R.S., Russell, N.C.: Workflow support using proclets: divide, interact, and conquer. IEEE Data Eng. Bull. **32**(3), 16–22 (2009)
5. Calvanese, D., Montali, M., Estañol, M., Teniente, E.: Verifiable UML artifact-centric business process models. In: CIKM 2014, pp. 1289–1298. ACM (2014)
6. Cohn, D., Hull, R.: Business artifacts: a data-centric approach to modeling business operations and processes. IEEE Data Eng. Bull. **32**(3), 3–9 (2009)

7. Damaggio, E., Deutsch, A., Hull, R., Vianu, V.: Automatic verification of data-centric business processes. In: Rinderle-Ma, S., Toumani, F., Wolf, K. (eds.) BPM 2011. LNCS, vol. 6896, pp. 3–16. Springer, Heidelberg (2011). https://doi.org/10.1007/978-3-642-23059-2_3
8. Desel, J., Erwin, T.: Hybrid specifications: looking at workflows from a run-time perspective. Comput. Syst. Sci. Eng. 15(5), 291–302 (2000)
9. Engelfriet, J.: Branching processes of petri nets. Acta Inf. 28(6), 575–591 (1991)
10. Fahland, D., de Leoni, M., van Dongen, B.F., van der Aalst, W.M.: Many-to-many: some observations on interactions in artifact choreographies. In: Eichhorn, D., Koschmider, A., Zhang, H. (eds.) ZEUS 2011. CEUR Workshop Proceedings, vol. 705, pp. 9–15. CEUR-WS.org (2011)
11. van Hee, K.M., Sidorova, N., Voorhoeve, M., van der Werf, J.M.E.M.: Generation of database transactions with petri nets. Fundam. Inform. 93(1–3), 171–184 (2009)
12. Li, G., de Carvalho, R.M., van der Aalst, W.M.P.: Automatic discovery of object-centric behavioral constraint models. In: Abramowicz, W. (ed.) BIS 2017. LNBIP, vol. 288, pp. 43–58. Springer, Cham (2017). https://doi.org/10.1007/978-3-319-59336-4_4
13. Lohmann, N.: Compliance by design for artifact-centric business processes. Inf. Syst. 38(4), 606–618 (2013)
14. Lohmann, N., Massuthe, P., Wolf, K.: Operating guidelines for finite-state services. In: Kleijn, J., Yakovlev, A. (eds.) ICATPN 2007. LNCS, vol. 4546, pp. 321–341. Springer, Heidelberg (2007). https://doi.org/10.1007/978-3-540-73094-1_20
15. Lohmann, N., Wolf, K.: Artifact-centric choreographies. In: Maglio, P.P., Weske, M., Yang, J., Fantinato, M. (eds.) ICSOC 2010. LNCS, vol. 6470, pp. 32–46. Springer, Heidelberg (2010). https://doi.org/10.1007/978-3-642-17358-5_3
16. Meyer, A., Pufahl, L., Batoulis, K., Fahland, D., Weske, M.: Automating data exchange in process choreographies. Inf. Syst. 53, 296–329 (2015)
17. Montali, M., Calvanese, D.: Soundness of data-aware, case-centric processes. STTT 18(5), 535–558 (2016)
18. Montali, M., Rivkin, A.: Model checking petri nets with names using data-centric dynamic systems. Formal Asp. Comput. 28(4), 615–641 (2016)
19. Montali, M., Rivkin, A.: DB-Nets: on the marriage of colored petri nets and relational databases. In: Koutny, M., Kleijn, J., Penczek, W. (eds.) Transactions on Petri Nets and Other Models of Concurrency XII. LNCS, vol. 10470, pp. 91–118. Springer, Heidelberg (2017). https://doi.org/10.1007/978-3-662-55862-1_5
20. OASIS: Web Services Business Process Execution Language, Version 2.0, April 2007. http://docs.oasis-open.org/wsbpel/2.0/wsbpel-v2.0.html
21. OMG: Business Process Model and Notation (BPMN), Version 2.0, January 2011. http://www.omg.org/spec/BPMN/2.0/
22. Reijers, H.A., et al.: Evaluating data-centric process approaches: does the human factor factor in? Softw. Syst. Model. 16(3), 649–662 (2017)
23. Reisig, W.: Understanding Petri Nets - Modeling Techniques, Analysis Methods, Case Studies. Springer, Heidelberg (2013). https://doi.org/10.1007/978-3-642-33278-4
24. Rosa-Velardo, F., Alonso, O.M., de Frutos-Escrig, D.: Mobile synchronizing petri nets: a choreographic approach for coordination in ubiquitous systems. Electr. Notes Theor. Comput. Sci. 150(1), 103–126 (2006)
25. Rosa-Velardo, F., de Frutos-Escrig, D.: Name creation vs. replication in petri net systems. Fundam. Inform. 88(3), 329–356 (2008)

26. Steinau, S., Andrews, K., Reichert, M.: Modeling process interactions with coordination processes. In: Panetto, H., Debruyne, C., Proper, H., Ardagna, C., Roman, D., Meersman, R. (eds.) OTM 2018, Part I. LNCS, vol. 11229. Springer, Cham (2018)
27. Steinau, S., Andrews, K., Reichert, M.: The relational process structure. In: Krogstie, J., Reijers, H.A. (eds.) CAiSE 2018. LNCS, vol. 10816, pp. 53–67. Springer, Cham (2018). https://doi.org/10.1007/978-3-319-91563-0_4

Modal Open Petri Nets

Vitali Schneider and Walter Vogler[✉]

Institut für Informatik, University of Augsburg, Augsburg, Germany
walter.vogler@informatik.uni-augsburg.de

Abstract. Open nets have an interface of input and output places for modelling asynchronous communication; these places serve as channels when open nets are composed. We study a variant that inherits modalities from Larsen's modal transition systems. Instantiating a framework for open nets we have developed in the past, we present a refinement preorder in the spirit of modal refinement. The preorder supports modular reasoning since it is a precongruence, and we justify it by a coarsest-precongruence result. We compare our approach to the one of Haddad et al., which considers a restricted class of nets and a stricter refinement. Our studies are conducted in an extended class of nets, which additionally have transition labels for synchronous communication.

1 Introduction

On an abstract level, concurrent systems can be specified and developed with the well-known labelled transition systems (LTS). The labels of such an LTS are the actions of the system, including the *hidden* action τ. To combine components to larger systems according to synchronous communication, parallel composition $\|$ merges equally-labelled transitions of two components; one might also hide such labels. Furthermore, a relation for stepwise refinement is needed that supports modular reasoning: if one refines a component of a parallel composition, then this should result in a refinement of the overall system. A refinement relation with this property is called a *precongruence* w.r.t. $\|$.

Such a precongruence can be defined as inclusion of the LTS-languages or some other trace-based semantics, or it can be some kind of bisimilarity, see [7] for an overview. These refinement relations can easily be transferred to (labelled) Petri nets, cf. e.g. [14, 21], with precongruence results for an analogous parallel composition. Advantages of Petri nets are that they are distributed by nature as are concurrent systems, and that they can give a finite representation for infinite state systems.

Bisimilarity allows one, in particular, to refine an LTS to a parallel composition with new hidden transitions resulting from communication. Such a composition can be a step forward to an implementation. But bisimilarity, being an

Research support provided by the DFG (German Research Foundation) under grant no. VO 615/12-2.

© Springer Nature Switzerland AG 2019
S. Donatelli and S. Haar (Eds.): PETRI NETS 2019, LNCS 11522, pp. 25–46, 2019.
https://doi.org/10.1007/978-3-030-21571-2_2

equivalence, does not offer much leeway and is, thus, in general not so appropriate for refinement.

Modal transition systems (MTS) are a ground breaking improvement towards *loose specifications* [11]. An MTS is an LTS with two kinds of transitions: *must*-transitions are required, while *may*-transitions are allowed, but only optional. A *modal refinement relation* can be described as an alternating simulation: each must-transition of the *specification* has to be simulated by an equally labelled must-transition of the refinement – possibly using additional hidden must-transitions; analogously, a may-transition of the *refinement* has to be allowed in the specification by a number of may-transitions. Also modalities and modal refinement can be transferred to Petri nets, see the discussion of modal Petri nets (MPN) below.

Petri nets are particularly well suited for modelling asynchronous communication, where the sender of a message does not have to wait for the receiver. If the order of messages on the same channel is not relevant, one can simply connect sending transitions via a (channel) place to receiving transitions instead of merging transitions as in the synchronous case. For such a setting, we model systems as so called *open* nets,[1] which have an interface consisting of two disjoint sets of special input and output places.

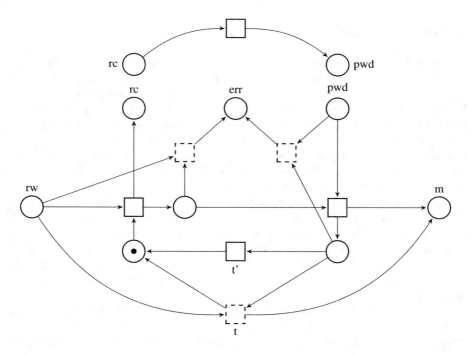

Fig. 1. Modal open nets *ATM* and password provider

---

[1] The term open in this sense presumably stems from [1].

As an example, consider the lower net *ATM* in Fig. 1; this is actually a modal open net (MON), where the dashed boxes denote may-transitions while the others denote must-transitions. The places *rw* and *pwd* are input places (no ingoing arcs from the net), the other named places are output places. If a user of the modelled system puts a token onto *rw*, this requests to withdraw some money. Initially, the system requests additional credentials by putting a token onto *rc*. After receiving a password via *pwd*, money is handed out on *m*. Now the system can decide to repeat this, but it optionally can do without asking for a password. In the latter case, a password is needed for the request after. The system can also be implemented with some ability to detect errors: an error message can be sent via place *err* if a withdrawal request is sent while the previous request is being handled or if a password is sent already when the previous interaction has just been finished.

The upper net describes a piece of software providing the user's password whenever it is required. Asynchronous composition ⊕ merges interface places with the same name, removing the name from the new interface. In the example, after installing the software and saving her password, the user deals with the composed system and does not have to enter her password again.

An early paper considering the composition of place-bordered nets as the above (without modalities) is [16], where nets are built composing deterministic nets. It is shown how to check (Petri net) liveness for such compositions. In [20], refining a transition *t* means to replace *t* (via place merging) by a so-called daughter-net whose border (interface) consists of the places incident to *t*. Results are given for which daughter-nets the replacement preserves behaviour like liveness and boundedness. Results on liveness for asynchronous compositions of more general nets can be found in [18]. A compositional semantics building Petri-net processes by place merging is presented in [10].

Extending [20], a general framework is suggested in [21] how to transfer semantics and refinement relations from a synchronous setting to open nets in a "sensible way". This is worked out e.g. in [19], which is one of a number of papers on open nets and operating guidelines like [12]. It is also applied in [2] in a setting with labelled open Petri nets; it is shown that some variants of bisimilarity are congruences for one operator that combines ‖ and ⊕. By a "sensible" refinement relation we mean a relation that accepts a refinement unless there is a formal reason against this. One general formal requirement is that the relation supports modular reasoning by being a precongruence. Additionally, one chooses some behavioural requirements; in [19] for example, it is required that a refinement step does not introduce a deadlock. Two trace sets are defined, and inclusion of these is shown to be sensible in the above sense, since it is the *coarsest* precongruence with the latter requirement.

In the present paper, we demonstrate that the above framework works also in a setting with modalities and alternating simulation. Although we are mainly interested in MON, we conduct our study for the larger class of *labelled MON* (ℓMON), which also have action-labelled transitions and include MON and MPN. The main contribution is a kind of modal refinement, which we show to be a precongruence for ‖ and the coarsest precongruence for ⊕ respecting modal refinement on MPN.

MPN have been introduced in [6] in combination with a modal language as in [15]. The main issue is to decide weak determinism (a variant of determinacy as in [13]) and – for weakly deterministic, but possibly unbounded MPN – modal-language inclusion. Asynchronous composition is only defined to build large MPN, which are often unbounded due to the channel places. There are no precongruence results.

The net class and the composition from [6] are further studied in [8], which is very close to the present paper. There, the nets are called modal asynchronous I/O-Petri nets (MAIOPN or, as we write here, MAP); they are a different representation for a restricted class of MON: the nets are actually MPN and the interface places are left implicit. The places only become explicit during the composition $\oplus_{HH}$. The refinement relation is modal refinement and is shown to be a precongruence. Due to the special interests in [8], a MAP may have so-called *internal* actions, showing how the generated channel places are accessed. They are not really visible but still taken into account in modal refinement. As a consequence, a MAP can only be refined by another one, if the latter has the same channel places; one cannot refine a monolithic specification by an asynchronous composition. This is noted in [6,8], so for stepwise refinement it is suggested to hide the internal actions at the end of composition. With this modification, one can translate MAP into MON such that composition is preserved, i.e. MAP can be seen as a sub-setting of our setting.

We show that our modal refinement is coarser than the one on MAP, i.e. it is better from our perspective. One difference concerns a typical feature of asynchronous communication: if two messages are sent on different channels one after the other, there can be overtaking such that the environment cannot observe the order of sending. Hence, this order should not matter for the refinement relation. This is indeed the case in our approach, but it does matter in [8]. We also give an alternative proof for the MAP precongruence result, which we believe to be simpler conceptually. This paper revises and generalizes [17].

Section 2 introduces MTS and transfers parallel composition and hiding to MPN; the latter is also done in [8], but the actual net variant there is more complicated and an MTS (!) variant with additional Petri net places is used. Section 3 defines asynchronous composition and our refinement relation, pointing out that it preserves some liveness notion. The coarsest-precongruence for $\oplus$ and precongruence for the MTS operators are shown. Section 4 compares ours to the MAP-approach. The paper ends with a sketch how to restrict our approach to bounded nets in Sect. 5 and with some conclusions. We thank Alexander Knapp and Ayleen Schinko for supporting us with the figures, and the reviewers for their helpful comments.

2   Preliminaries

This section provides some basic notation for modal transition systems and modal Petri nets. Refinement and basic operations such as parallel composition, relabelling and hiding are transferred from MTS to MPN. The same holds for the precongruence results provided by Hüttel and Larsen in [9].

Most of the structures in this paper have an action *alphabet*, usually denoted by Σ. There is one hidden or invisible action τ, which is never in an alphabet. We denote $\Sigma \cup \{\tau\}$ by Σ^τ; a and α often stand for a typical action in Σ and Σ^τ resp. $\mathbb{N}$ denotes the set of natural numbers including zero.

2.1   Modal Transition Systems

In the introduction, we have already explained that MTS [11] have required must- and optional may-transitions. The condition $\longrightarrow \subseteq \dashrightarrow$ below reflects that every required transition should also be allowed.

Definition 1 (MTS). A *modal transition system (MTS)* is a tuple $Q = (S, \Sigma, \dashrightarrow, \longrightarrow, s^0)$ where S is a set of *states* containing the *initial state* s_0; Σ is an alphabet,

- $\dashrightarrow \subseteq S \times \Sigma^\tau \times S$ is the set of *may-transitions*, and
- $\longrightarrow \subseteq S \times \Sigma^\tau \times S$ is the set of *must-transitions* satisfying $\longrightarrow \subseteq \dashrightarrow$. $\Diamond$

We add the name of the MTS as an index to the components when needed or use e.g. S_i for the state set of Q_i etc., and similarly for nets later on. We write $s \stackrel{\alpha}{\dashrightarrow} s'$ for $(s, \alpha, s') \in \dashrightarrow$, and extend this to words $w \in (\Sigma^\tau)^*$: $s \stackrel{w}{\dashrightarrow} s'$ means that there is a sequence $s \stackrel{\alpha_1}{\dashrightarrow} s_1 \stackrel{\alpha_2}{\dashrightarrow} s_2 \ldots s_{n-1} \stackrel{\alpha_n}{\dashrightarrow} s'$ with $w = \alpha_1 \ldots \alpha_n$. Let $\widehat{w}$ be obtained from w by *removing all* τs. With this, we define the weak may-transition $s \stackrel{w}{\Longrightarrow} s'$ as $\exists v \in (\Sigma^\tau)^* : \widehat{v} = w \wedge s \stackrel{v}{\dashrightarrow} s'$. We have the analogous notations for must-transitions, writing $\Longrightarrow$ for $=\Rightarrow$. A state s is *reachable* in Q when $s^0 \stackrel{w}{\dashrightarrow} s$ for some $w \in \Sigma^\tau$.

The following defines the standard (weak) modal refinement for MTS as explained in the introduction.

Definition 2 (MTS refinement). Let Q_1 and Q_2 be two MTS over the same alphabet. We say that Q_1 is a *(modal)refinement* of Q_2, written $Q_1 \sqsubseteq_{MTS} Q_2$, if there exists an *MTS-relation* $\mathfrak{R} \subseteq S_1 \times S_2$ with $(s_1^0, s_2^0) \in \mathfrak{R}$ such that for every $(s_1, s_2) \in \mathfrak{R}$:

- $s_2 \stackrel{\alpha}{\longrightarrow} s_2' \;\Rightarrow\; s_1 \stackrel{\widehat{\alpha}}{\Longrightarrow} s_1' \wedge (s_1', s_2') \in \mathfrak{R}$ and
- $s_1 \stackrel{\alpha}{\dashrightarrow} s_1' \;\Rightarrow\; s_2 \stackrel{\widehat{\alpha}}{\Longrightarrow} s_2' \wedge (s_1', s_2') \in \mathfrak{R}$. $\Diamond$

Note that, for two *implementations* (MTS with coinciding may- and must-transitions), MTS-relations and weak bisimulations [13] are the same. Next we define the operations of relabelling, hiding, parallel composition and parallel composition with hiding.

Definition 3 (MTS relabelling, hiding). A *relabelling function* for an alphabet Σ (and for MTS and MPN below with this alphabet) is a surjective function $f : \Sigma \to \Sigma'$; additionally, we set $f(\tau) = \tau$. The respective *relabelling* of an MTS Q is denoted by $Q[f]$ and obtained from Q by replacing Σ with Σ' and each action α of a transition with $f(\alpha)$.

Similarly, for an alphabet H, the *hiding* of H in Q, denoted by Q/H, is obtained from Q by replacing Σ with $\Sigma \setminus H$ and each action $a \in H$ of a transition with τ. ◇

The idea of parallel composition is that two systems synchronize on common (visible) actions and perform all other actions independently.

Definition 4 (MTS parallel composition). The *MTS parallel composition* of two MTS Q_1 and Q_2 is defined as the MTS $Q_1 \| Q_2 = (S_1 \times S_2, \Sigma_1 \cup \Sigma_2, \dashrightarrow, \longrightarrow, (s_1^0, s_2^0))$ with

$$\longrightarrow = \{((s_1, s_2), \alpha, (s_1', s_2)) \mid s_1 \xrightarrow{\alpha}_1 s_1' \ \wedge \ \alpha \notin \Sigma_2\}$$
$$\cup \ \{((s_1, s_2), \alpha, (s_1, s_2')) \mid s_2 \xrightarrow{\alpha}_2 s_2' \ \wedge \ \alpha \notin \Sigma_1\}$$
$$\cup \ \{((s_1, s_2), a, (s_1', s_2')) \mid s_1 \xrightarrow{a}_1 s_1' \ \wedge \ s_2 \xrightarrow{a}_2 s_2' \ \wedge \ a \in \Sigma_1 \cap \Sigma_2\}$$

and $\dashrightarrow$ is defined analogously. ◇

Note that two equally labelled must-transitions synchronize to a must-transition, and the same for may-transitions. In effect, a must- and a may-transition synchronize to a may-transition, because the must-transition has an *underlying* may-transition. Finally, we define a variant of parallel composition where the synchronized actions are hidden.

Definition 5 (MTS parallel composition with hiding). The *parallel composition with hiding* of MTS Q_1 and Q_2 is the MTS $Q_1 \Uparrow Q_2 = (Q_1 \| Q_2)/H$ with $H = \Sigma_1 \cap \Sigma_2$. ◇

In [9], there is a parametric precongruence result for modal refinement (in a version for MTS without an initial state), which can be instantiated to obtain the following result. The details have been worked out in [17].

Theorem 6. *For relabelling, hiding, parallel composition and parallel composition with hiding, $\sqsubseteq_{MTS}$ is a precongruence, i.e.: for MTS Q_1, Q_2 and R with $Q_1 \sqsubseteq_{MTS} Q_2$, a relabelling function f for Q_1 (and thus for Q_2), and an alphabet H we have:*

$$Q_1[f] \sqsubseteq_{MTS} Q_2[f], \quad Q_1/H \sqsubseteq_{MTS} Q_2/H,$$
$$Q_1 \| R \sqsubseteq_{MTS} Q_2 \| R, \quad Q_1 \Uparrow R \sqsubseteq_{MTS} Q_2 \Uparrow R$$

Since (Petri net) liveness is an issue in the related literature, we define a corresponding property on MTS in such a way that it is preserved under refinement. An action a is action live in an MTS if it surely remains possible whatever happens. Formally:

Definition 7 (action live). For an MTS Q, $a \in \Sigma$ is *action live* in Q if, for each reachable state s, $s \xRightarrow{wa} s'$ for some word $w \in \Sigma^\tau$.

Proposition 8. *For MTS Q_1 and Q_2 with $Q_1 \sqsubseteq_{MTS} Q_2$, $a \in \Sigma_1$ is action live in Q_2 implies a is action live in Q_1.*

Proof. The assumptions imply that there is a suitable MTS-relation $\mathfrak{R}$. A reachable state s_1 of Q_1 is reached by a sequence of may-transitions. Each of these is matched by a small path in Q_2 according to $\mathfrak{R}$; stringed together, these paths reach some s_2 with $(s_1, s_2) \in \mathfrak{R}$. By assumption for a, there is some w and s_2' with $s_2 \stackrel{wa}{\Longrightarrow} s_2'$. In turn, the respective must-transitions are matched in Q_1, implying $s_1 \stackrel{wa}{\Longrightarrow} s_1'$. $\qquad\qquad\square$

For implementations, action liveness directly corresponds to Petri net liveness. In Definition 7, s is reached by may-transitions and a is performed along a sequence of must-transitions. To see that this is the right choice of modalities, think of a variant where only states s reachable by must-transitions are considered. If Q consists of states s^0 and s with $s^0 \stackrel{a}{\longrightarrow} s^0$ and $s^0 \stackrel{\tau}{\dashrightarrow} s$, then a would be action live in Q, but not in a refinement having the τ-transition as a must. Vice versa, think of a variant where it suffices that a is performed along a sequence of may-transitions. If Q consists of state s^0 with $s^0 \stackrel{a}{\dashrightarrow} s^0$, then a would be action live in Q, but not in a refinement having no transition.

2.2 Modalities for Petri Nets

Also for Petri nets, one can distinguish between must- and may-transitions. Additionally, one can label transitions with actions, which form an interface for synchronous communication (MPN). Alternatively, one can distinguish specific input and output places, and these form an interface for asynchronous communication (MON). Our focus lies on the latter, but we need also MPN for our approach, and we even need a combination for the envisaged coarsest precongruence result. For generality, we start from this combination. Note that all transitions are may-transitions, their set is denoted by T as usual. We also treat infinite nets, but observe the assumption in the paragraph after the following definition.

Definition 9 (ℓMON). A *labelled modal open net* (ℓMON) is a tuple

$$N = (P, I, O, \Sigma, T, T^\square, W, m^0, l)$$

where P and T are disjoint sets of *places* and *(may-)transitions*, and $T^\square \subseteq T$ is the set of *must-transitions*; $W : (P \times T) \cup (T \times P) \to \mathbb{N}$ is the set of *weighted arcs*; m^0 is the *initial marking*, where a *marking* is a mapping $m : P \to \mathbb{N}$.

Furthermore, $I \subseteq P$ and $O \subseteq P$ are disjoint sets of *input* and *output places*, which are empty under the initial marking. Finally, Σ is an alphabet disjoint from I and O, and $l : T \to \Sigma^\tau$ is the *labelling*; τ-labels are omitted in figures.

A *modal open net* (MON) is an ℓMON where Σ is empty, cf. Fig. 1; we will often omit Σ and l, which maps all transitions to τ. A *modal Petri net* (MPN) is an ℓMON where I and O are empty and often omitted. $\qquad\diamond$

We call $F = \{(x, y) \mid W(x, y) \neq 0\}$ the *flow relation* of N. For an $x \in P \,\dot\cup\, T$, we call the sets ${}^\bullet x = \{y \mid (y, x) \in F\}$ the *preset* and $x^\bullet = \{y \mid (x, y) \in F\}$ the *postset* of x. At some stage, we will need that transitions have finite presets, so we *assume* this throughout.

The behaviour of an ℓMON N is given by the occurrence rule. A transition $t \in T$ is *enabled* at a marking m, if $\forall p \in {}^\bullet t : W(p, t) \leq m(p)$. When t is enabled at m, it can *occur* or *fire*, changing the marking to m' with $m'(p) = m(p) - W(p, t) + W(t, p)$; we write $m \overset{t}{\dashrightarrow} m'$, or $m \overset{t}{\longrightarrow} m'$ if t is a must-transition. Furthermore, the same notation is used for transition labels, i.e. we also write $m \overset{l(t)}{\dashrightarrow} m'$ or $m \overset{l(t)}{\longrightarrow} m'$.

The latter notations in fact define the may- and must-transitions of an MTS *associated* to N: its alphabet is Σ, m^0 the initial state, and the reachable markings are the states. With this view, the other MTS notations like $\overset{w}{\dashrightarrow}$ and $\overset{w}{\Longrightarrow}$ for words carry over to ℓMON. Whenever $m \overset{w}{\dashrightarrow} m'$ or $m = \overset{w}{\Rightarrow} m'$, there exists an *underlying* transition sequence, a *firing sequence* leading from m to m'.

2.3   MPN: Refinement and Operators

First, we will concentrate on MPN. With the concept of an associated MTS, MPN refinement can be defined according to Definition 2, i.e. the MPN-relation below is just an MTS-relation between the associated MTS:

Definition 10 (MPN refinement). For MPN N_1 and N_2 over the same alphabet, we say that N_1 is a *refinement* of N_2, written $N_1 \sqsubseteq_{MPN} N_2$, if there is an *MPN-relation* $\mathfrak{R}$ between the reachable markings of N_1 and N_2 with $(m_1^0, m_2^0) \in \mathfrak{R}$ such that for every $(m_1, m_2) \in \mathfrak{R}$:

- $m_2 \overset{\alpha}{\longrightarrow} m_2' \;\Rightarrow\; m_1 \overset{\widehat{\alpha}}{\Longrightarrow} m_1' \,\wedge\, (m_1', m_2') \in \mathfrak{R}$ and
- $m_1 \overset{\alpha}{\dashrightarrow} m_1' \;\Rightarrow\; m_2 = \overset{\widehat{\alpha}}{\Rightarrow} m_2' \,\wedge\, (m_1', m_2') \in \mathfrak{R}$. $\Diamond$

In some cases, we might use MPN-relations that include unreachable markings. This can make arguments easier, e.g. we do not have to prove reachability. Strictly, we would have to remove all pairs containing an unreachable marking.

Next, we define the operations for MTS also for MPN. The essential point for parallel composition is that, for a common label a, each a-labelled transition in the first and each a-labelled transition in the second MPN are merged to a new transition, which inherits both presets and both postsets. This implies the lemma after the definition. Note that we identify isomorphic structures; hence, we can e.g. assume place sets to be disjoint in this definition:

Definition 11 (MPN operators). Let N_1 and N_2 be MPN, where w.l.o.g. the place sets are disjoint. Then, we define their *parallel composition* to be

$$N_1 \| N_2 = (P, \Sigma, T, T^\square, W, m^0, l)$$

where P and Σ are the componentwise unions, i.e. $P = P_1 \cup P_2$ etc.

- $T = \{(t_1, \tau) \mid t_1 \in T_1 \ \wedge \ l_1(t_1) \notin \Sigma_2\} \cup \{(\tau, t_2) \mid t_2 \in T_2 \ \wedge \ l_2(t_2) \notin \Sigma_1\}$
$\cup \ \{(t_1, t_2) \mid t_1 \in T_1 \ \wedge \ t_2 \in T_2 \ \wedge \ l_1(t_1) = l_2(t_2) \in \Sigma_1 \cap \Sigma_2\},$

- $T^{\square}$ is defined analogously,

- $\forall p \in P, \ (t_1, t_2) \in T: \quad W(p, (t_1, t_2)) = \begin{cases} W_1(p, t_1) \text{ if } p \in P_1 \ \wedge \ t_1 \in T_1 \\ W_2(p, t_2) \text{ if } p \in P_2 \ \wedge \ t_2 \in T_2 \\ 0 \qquad\qquad \text{otherwise} \end{cases}$

  $W((t_1, t_2), p)$ is defined analogously,

- $\forall p \in P : m^0(p) = \begin{cases} m_1^0(p) \text{ if } p \in P_1 \\ m_2^0(p) \text{ if } p \in P_2, \end{cases}$

- $\forall (t_1, t_2) \in T : l(t_1, t_2) = \begin{cases} l_1(t_1) \text{ if } t_1 \in T_1 \\ l_2(t_2) \text{ if } t_2 \in T_2. \end{cases}$

With this, we define relabelling, hiding and parallel composition with hiding word by word as in Definitions 3 and 5. $\diamond$

Note that in the last item above, in case of a merged transition, $l_1(t_1)$ and $l_2(t_2)$ coincide. For the next lemma, note that markings of $N_1 \| N_2$ can be written (m_1, m_2), where m_1 is a marking of N_1 and m_2 one of N_2.

Lemma 12. *Let N_1 and N_2 be two MPN. If t_1 and t_2 are a-labelled transitions of N_1 and N_2 resp., then $(m_1, m_2) \xrightarrow{(t_1, t_2)} (m_1', m_2')$ if and only if $m_1 \xrightarrow{t_1} m_1'$ and $m_2 \xrightarrow{t_2} m_2'$. If t_1 is a transition of N_1 with $l_1(t_1) \notin \Sigma_2$, then $(m_1, m_2) \xrightarrow{(t_1, \tau)} (m_1', m_2)$ if and only if $m_1 \xrightarrow{t_1} m_1'$, and analogously for N_2. The same statements hold for must-transitions.*

This lemma implies that the MTS associated to $N_1 \| N_2$ is the parallel composition of the two MTS associated to $N_1 \| N_2$. Similar statements hold for the other three operators defined above. Hence, we obtain the following corollary to Theorem 6.

Corollary 13. *W.r.t. the above operators for MPN, $\sqsubseteq_{MPN}$ is a precongruence.*

Observe that, in the same way, the notion of action liveness and its preservation under refinement carry over to MPN and $\sqsubseteq_{MPN}$. We close with a technical operation and lemma, which will be important in the next section. The operation contracts special τ-must-transitions by merging the only place in the preset with the only place in the postset. This is illustrated in Fig. 2.

Definition 14 (τ-contraction). Let N be an MPN and A a set of τ-labelled must-transitions with the following properties:

- for each $t_i \in A$: $^{\bullet}t_i = \{p_i\}$, $t_i^{\bullet} = \{p_i'\}$ and $W(p_i, t) = W(t, p_i') = 1$; furthermore, $^{\bullet}p_i' = p_i^{\bullet} = \{t_i\}$ and $m^0(p_i) = m^0(p_i') = 0$;

- all these places are different.

Then, the τ-contraction $N[A]$ is obtained from N by removing the transitions $t_i \in A$ and the associated places p_i', and changing the values $W(p_i, t)$ for the remaining transitions t from 0 to $W(p_i', t)$. $\Diamond$

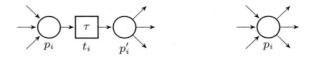

Fig. 2. Transformation from MPN N to MPN $N[A]$

Lemma 15. *Let N be an MPN and $A \subseteq T_N^{\square}$ as in Definition 14, then $N \sqsubseteq_{MPN} N[A]$ and $N[A] \sqsubseteq_{MPN} N$.*

Proof. Let m and m' be markings of N and $N[A]$, resp., that are identical on the common places except that, for each of the p_i, $m'(p_i) = m(p_i) + m(p_i')$; then we denote m' by $[m]$. Now consider the relation $\mathfrak{R} = \{(m, [m]) \mid m$ reachable in $N\}$. This relation proves the first claim, and its reverse proves the second claim.

First, consider some $m \xrightarrow{t_i} m'$ with $t_i \in A$. This can be matched by firing no transition: $(m', [m]) \in \mathfrak{R}$ since t_i does not change $m(p_i) + m(p_i')$, i.e. $[m'] = [m]$. It remains to deal with each transition t of $N[A]$, and we restrict ourselves to must-transitions, since the case of may-transitions is similar. So second, consider $m \xrightarrow{t} m'$ in N. From each $p_i' \in {}^{\bullet}t$, t removes $W(p_i', t)$ tokens, which are present in and removed from p_i under $[m]$; for the other places, t removes the same number of tokens in both nets, and it adds the same number of tokens to each place in both nets.

Third, consider $[m] \xrightarrow{t} [m']$ in $N[A]$. Here, some tokens on some $p_i' \in {}^{\bullet}t$ might be missing in N. This is remedied by first firing invisibly some t_i: for each of the (finitely many!) $p_i' \in {}^{\bullet}t$, we fire the invisible must-transitions t_i until p_i is empty. Now t removes the same number of tokens from $p_i' \in {}^{\bullet}t$ in N as it removes from p_i in $N[A]$, and it removes the same number of tokens in both nets from each other place; then, it adds the same number of tokens in both nets to each place. $\square$

3   Asynchronous Communication

While the definition of composing nets according to asynchronous communication, i.e. by merging places, should be pretty clear, the question is how to define a refinement framework that deals with the interface places in a suitable way. The idea of [19,21] is to make visible how an environment interacts via these places. The environment observes that it puts a token onto an input place, but not when the token is taken, and vice versa for output places. Thus, for each input (output) place a, we add an arc from (to) a new a-labelled transition and compare the resulting MPNs with $\sqsubseteq_{MPN}$. In the following definition, we assume that a^- and a^+ are fresh in the sense that they are not in $P \cup T$.

Definition 16 (ℓMON wrapper). The ℓMON *wrapper* of an ℓMON N is the MPN $wrap(N)$ (also denoted here by N_w). N_w is obtained from N by renaming $a \in I$ ($a \in O$) to the fresh a^- (a^+) – defining P_w; these inherit the arcs and initial marking from a – defining m_w^0. We set $\Sigma_w = \Sigma \mathbin{\dot\cup} I \mathbin{\dot\cup} O$ and add a-labelled transitions a to T and $T^\square$ for all $a \in I \cup O$ – defining also T_w, $T_w^\square$ and l_w. The modified W is extended on the new pairs involving some new transition a by $W_w(a, a^-) = 1$ for $a \in I$, $W_w(a^+, a) = 1$ for $a \in O$ and 0 otherwise – defining W_w. See Fig. 3 for the example N_2 and $wrap(N_2)$.

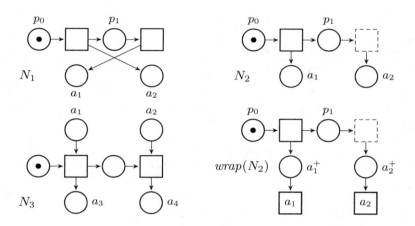

Fig. 3. Three MON N_1, N_2, N_3 and $wrap(N_2)$

Definition 17 (ℓMON refinement). Let N_1 and N_2 be two ℓMON with the same alphabet as well as input and output places. We say that N_1 is a *refinement* of N_2, written $N_1 \sqsubseteq_{\ell MON} N_2$, if $wrap(N_1) \sqsubseteq_{MPN} wrap(N_2)$. ◇

This definition extends Definition 10: since an MPN N is identical to $wrap(N)$, $\sqsubseteq_{\ell MON}$ and $\sqsubseteq_{MPN}$ coincide if applied to two MPN.

As a first example, we show that $N_1 \sqsubseteq_{\ell MON} N_2$ for the MON N_1 and N_2 in Fig. 3 by showing a suitable MPN-relation $\mathfrak{R}$. We write markings as a formal sum: e.g. if p_1 and a_1 have one token each, we write $p_1 + a_1$, and we write this also for a marking of $wrap(N_2)$ although the place has changed its name to a_1^+ there; 0 is the empty marking. With this, $\mathfrak{R} = \{(p_0, p_0), (p_1 + a_2, p_1 + a_1), (a_2, p_1), (p_1, a_1), (a_1 + a_2, a_1 + a_2), (a_1, a_1), (a_2, a_2), (0, 0)\}$.

An interesting pair is $(p_1 + a_2, p_1 + a_1)$: the only enabled must-transition on the specification side is a_1; although the token on a_1 is produced after the token on a_2 in N_1, this can be matched using the second τ-transition, which is a must transition. Thus, the pair (a_2, p_1) is reached. On the refinement side, a_2 is enabled, which can be matched with the second τ-transition in N_1; it is sufficient that this is a may-transition. Additionally, the two second τ-transitions match each other as required.

Here, the specification produces two messages in some order while the refinement produces them the other way round. We justify this intuitively after the definition of asynchronous composition.

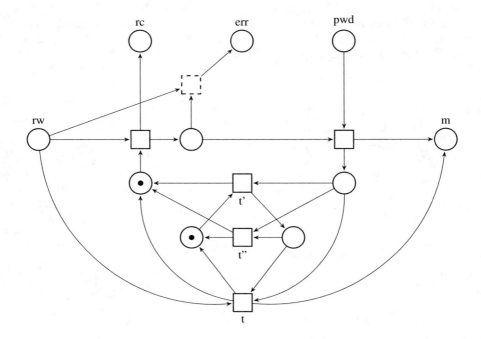

Fig. 4. Refinement ATM' of ATM

As another example, Fig. 4 shows a refinement ATM' of the lower MON ATM in Fig. 1. Here, the optional top right may-transition is omitted, possibly because we assume that the password provider in Fig. 1 is used and we expect no errors due to premature passwords. Furthermore, the optional shortcut t is now a must-transition. But there is a difference that makes it less obvious that ATM' really is a refinement of ATM: if the shortcut is used in ATM, it can only be used again the next but one time; in ATM', it can only be used one time later. Let us prove the refinement.

The two MON have the same places except for the lower two places of ATM'; obviously, these together will always have one token. For each reachable marking m of $wrap(ATM)$, we denote by $m+l$ $(m+r)$ the same marking of $wrap(ATM')$ with an additional token on the left (right) additional place. The MPN-relation $\mathfrak{R}$ for $wrap(ATM)$ and $wrap(ATM')$ consists of all pairs $(m+l, m)$ and $(m+r, m)$ for such reachable m, obviously relating the initial markings.

The new visible must-transitions of $wrap(ATM)$ and the upper must-transitions of ATM have the same labels and effects in $wrap(ATM')$. Furthermore, $m \xrightarrow{\tau} m'$ due to t' in $wrap(ATM)$ if and only if $m + l \xrightarrow{\tau} m' + r$ due to t' in $wrap(ATM')$ if and only if $m + r \xrightarrow{\tau} m' + l$ due to t'' in $wrap(ATM')$. Thus, all must- and their underlying may-transitions (except t) are matched appropriately.

The may-transition of $wrap(ATM')$ is matched by itself in $wrap(ATM)$. Finally, t can only fire under a marking $m + r$ in $wrap(ATM')$ resulting in $m + r \dashrightarrow^{\tau} m' + l$. This is matched in $wrap(ATM)$ by $m \dashrightarrow^{\tau} m'$ due to t.

We call some $a \in \Sigma \cup I \cup O$ *action live* in an ℓMON N if it is action live in $wrap(N)$. This is the case for each $a \in I$, whereas for $a \in O$ it means: N can always put another token onto a provided that sufficiently many tokens are provided on the input places. Obviously, action liveness is also preserved under $\sqsubseteq_{\ell MON}$.

We now come to the most important operator of this paper; it merges common interface places, modelling asynchronous communication.

Definition 18 (ℓMON asynchronous composition). Two ℓMON N_1 and N_2 are called *(async-)composable* whenever $(\Sigma_1 \cup I_1 \cup O_1) \cap (\Sigma_2 \cup I_2 \cup O_2) = (I_1 \cap O_1) \cup (I_2 \cap O_2) =: asc(N_1, N_2)$. We can further assume that the four place and transition sets are pairwise disjoint except for $asc(N_1, N_2)$.

The *(asynchronous) composition* of such ℓMON is the ℓMON $N_1 \oplus N_2 = (P, I, O, \Sigma, T, T^\square, W, m^0, l)$ where P, Σ, T and $T^\square$ are the componentwise unions. The interface places are $I = (I_1 \cup I_2) \setminus asc(N_1, N_2)$ and $O = (O_1 \cup O_2) \setminus asc(N_1, N_2)$. For $i = 1, 2$, marking m^0 coincides on $p \in P_i$ with m_i^0, and l coincides on $t \in T_i$ with l_i. Finally,

$$W(p,t) = \begin{cases} W_1(p,t) \text{ if } p \in P_1 \wedge t \in T_1 \\ W_2(p,t) \text{ if } p \in P_2 \wedge t \in T_2 \\ 0 \qquad\quad \text{otherwise} \end{cases} \quad — W(t,p) \text{ is defined analogously.} \qquad \Diamond$$

Composability ensures that $N_1 \oplus N_2$ is well-defined. In particular, it ensures that synchronous and asynchronous channels do not get confused. One could also think of a variant that combines parallel and asynchronous composition where the components also synchronize on common actions – actions we have forbidden here. We observe that the composition $\oplus$ is commutative and associative up to isomorphism for pairwise composable components. Note that for three ℓMON N_1, N_2 and N_3 with some $a \in I_1 \cap O_2 \cap I_3$, $(N_1 \oplus N_2) \oplus N_3$ and $N_1 \oplus (N_2 \oplus N_3)$ might be well-defined, but would have different behaviour in general: N_2 communicates on a with N_1 in one and with N_3 in the other composition. So also N_1 and N_3 have to be composable.

Let us reconsider the nets N_1 and N_2 in Fig. 3. As a potential argument against reordering messages, one might come up with N_3, which looks like it is sensitive to the order in which tokens arrive on a_1 and a_2. In $N_2 \oplus N_3$, the first token arrives on a_1, and a token on a_3 can be produced immediately; in $N_1 \oplus N_3$, the first token arrives on a_2. But using the second τ-transition of N_1, a token can be put onto a_3 before marking a_4, so N_3 cannot "see" the reordering. Thus, this reordering *should* be allowed in a refinement step. To prove that $\sqsubseteq_{\ell MON}$ is indeed a precongruence w.r.t. $\oplus$, we need another lemma.

Lemma 19. *Let N_1 and N_2 be two composable ℓMON and $A = \{(a,a) \mid a \in asc(N_1, N_2)\}$, then $(wrap(N_1) \Uparrow wrap(N_2))[A]$ and $wrap(N_1 \oplus N_2)$ are isomorphic.*

Proof. By composability, $asc(N_1, N_2)$ is the set of the common actions of $wrap(N_1)$ and $wrap(N_2)$. For $wrap(N_1) \Uparrow wrap(N_2)$, the unique a-labelled transition a in one net is merged with the unique a-labelled transition a in the other if $a \in asc(N_1, N_2)$, and then all these a-labels are hidden. Hence, A is a set as required in Definition 14 and $[A]$ merges a^+ and a^- into one place, which we may call a again; cf. Fig. 2. Now the only difference is that transitions are pairs, where always one component is τ. Removing these components results in $wrap(N_1 \oplus N_2)$.   □

Theorem 20. *The refinement relation $\sqsubseteq_{\ell MON}$ is a precongruence for $\oplus$, i.e. for three ℓMON N_1, N_2 and N_3 where N_2 is composable with N_3 and $N_1 \sqsubseteq_{\ell MON} N_2$, also N_1 is composable with N_3 and $N_1 \oplus N_3 \sqsubseteq_{\ell MON} N_2 \oplus N_3$.*

Proof. Composability only depends on the interfaces, so the first claim is obvious. Let $A = \{(a,a) \mid a \in asc(N_1, N_3)\} = \{(a,a) \mid a \in asc(N_2, N_3)\}$.

By definition of $\sqsubseteq_{\ell MON}$ and the precongruence properties of $\sqsubseteq_{MPN}$, we have

– $wrap(N_1) \Uparrow wrap(N_3) \sqsubseteq_{MPN} wrap(N_2) \Uparrow wrap(N_3)$.

Now by Lemmas 19 and 15,

– $wrap(N_1 \oplus N_3) \sqsubseteq_{MPN} (wrap(N_1) \Uparrow wrap(N_3))[A]$
  $\sqsubseteq_{MPN} wrap(N_1) \Uparrow wrap(N_3)$ and
– $wrap(N_2) \Uparrow wrap(N_3) \sqsubseteq_{MPN} (wrap(N_2) \Uparrow wrap(N_3))[A]$
  $\sqsubseteq_{MPN} wrap(N_2 \oplus N_3)$.

Thus, $wrap(N_1 \oplus N_3) \sqsubseteq_{MPN} wrap(N_2 \oplus N_3)$ and $N_1 \oplus N_3 \sqsubseteq_{\ell MON} N_2 \oplus N_3$.   □

It might seem that the *wrap*-based definition of our refinement relation is somewhat arbitrary; our next aim is to show its optimality by proving a coarsest-precongruence result. The starting point is that modal refinement is accepted for MTS, so its translation $\sqsubseteq_{MPN}$ to MPN is in a sense just right. Hence, the optimal refinement relation $\sqsubseteq$ on ℓMON should respect this: if $N_1 \sqsubseteq N_2$ for MPN N_1 and N_2, then also $N_1 \sqsubseteq_{MPN} N_2$. Furthermore, $\sqsubseteq$ should be a precongruence w.r.t. $\oplus$. To be optimal, it should allow all refinements consistent with these two requirements, so $\sqsubseteq$ should be the coarsest MPN-respecting precongruence w.r.t. $\oplus$; such a coarsest precongruence always exists.

In principle, this coarsest precongruence could be finer than $\sqsubseteq_{MPN}$ for MPN but – being $\sqsubseteq_{\ell MON}$ – it actually coincides with $\sqsubseteq_{MPN}$ for MPN, which is even more pleasing.

One could question why a precongruence for all ℓMON is needed. Our proof below also supports another argument with the same starting point to answer this. In this argument, we call an ℓMON N_o an *observer* of a MON N if it has the same interface places as N but with input and output interchanged; thus,

$N \oplus N_o$ is an MPN. Hence, N_o interacts with N on the complete asynchronous interface of the latter, and it can make its observations visible on its synchronous interface. Now one could alternatively aim for some $\sqsubseteq$ on MON that is the coarsest precongruence w.r.t. $\oplus$ such that $N \sqsubseteq N'$ implies, for all observers N_o of N (i.e. also of N'), that $N \oplus N_o \sqsubseteq_{MPN} N' \oplus N_o$. Again, this $\sqsubseteq$ is $\sqsubseteq_{\ell MON}$ (restricted to MON). In fact, for all MON N and N', $N \sqsubseteq_{\ell MON} N'$ if and only if $N \oplus N_o \sqsubseteq_{MPN} N' \oplus N_o$ for all observers N_o of N, as we show more generally in the next proposition.

Definition 21. A relation $\sqsubseteq$ on ℓMON is called *MPN-respecting* if it implies $\sqsubseteq_{MPN}$ on MPN. An ℓMON N_o is an *observer* of an ℓMON N, if N_o and N are composable, $I_o = O$ and $O_o = I$. $\qquad\qquad\qquad\qquad\qquad\qquad\qquad\qquad\qquad\Diamond$

Proposition 22. *Let N and N' be ℓMON with the same alphabet as well as input and output places. Then $N \sqsubseteq_{\ell MON} N'$ if and only if $N \oplus N_o \sqsubseteq_{MPN} N' \oplus N_o$ for all observers N_o of N.*

Proof. For the "if"-direction, we construct a specific N_o: it has, for each $a \in I \cup O$, an empty place a and an a'-labelled transition, where a' is a fresh action, i.e. $a' \notin \Sigma \cup I \cup O$. (These fresh actions are needed, since an interface place is not allowed to be an action as well.) The only arcs have weight one and connect each $a \in I_o$ to the a'-labelled transition and, for $a \in O_o$, the a'-labelled transition to the place a. Further, let f be the relabelling that maps each a' to a and is the identity on Σ.

Now, $N \oplus N_o$ is isomorphic to $wrap(N)$ except that it has labels a' instead of a and, so, $(N \oplus N_o)[f]$ is isomorphic to $wrap(N)$. Thus, $N \oplus N_o \sqsubseteq_{MPN} N' \oplus N_o$ implies $(N \oplus N_o)[f] \sqsubseteq_{MPN} (N' \oplus N_o)[f]$, which implies $N \sqsubseteq_{\ell MON} N'$.

The "only if"-direction follows from Theorem 20 and the observation after Definition 17.

Theorem 23. *Relation $\sqsubseteq_{\ell MON}$ is the coarsest MPN-respecting precongruence w.r.t. $\oplus$ on ℓMON.*

Proof. Let $\sqsubseteq$ be the coarsest MPN-respecting precongruence w.r.t. $\oplus$ on ℓMON. Due to Theorem 20 and the observation after Definition 17, $\sqsubseteq_{\ell MON}$ is an MPN-respecting precongruence as well. Thus, it is contained in $\sqsubseteq$ by the definition of the latter.

This definition also gives us that $N \sqsubseteq N'$ implies $N \oplus N_o \sqsubseteq N' \oplus N_o$ and $N \oplus N_o \sqsubseteq_{MPN} N' \oplus N_o$ for all observers N_o of N. The latter implies $N \sqsubseteq_{\ell MON} N'$ by Proposition 22, showing that $\sqsubseteq$ is contained in $\sqsubseteq_{\ell MON}$ as well. $\qquad\qquad\square$

We close with a quick look at the operators that we have only defined for MPN so far. The following definition extends Definition 11 to ℓMON; in particular, par-composability holds automatically for MPN.

Definition 24 (further ℓMON operators). Two ℓMON N_1 and N_2 are called *par-composable* whenever $(\Sigma_1 \cup I_1 \cup O_1) \cap (\Sigma_2 \cup I_2 \cup O_2) = \Sigma_1 \cap \Sigma_2$. We can further assume that the place sets are disjoint. Then, we define their *parallel composition* $N_1 \| N_2$ as in Definition 11, letting also I and O be the componentwise unions.

For an ℓMON N, a *relabelling function* f is defined as in Definition 3 except that, additionally, we require that Σ' and $I \cup O$ be disjoint. With this, *relabelling* $N[f]$ and *hiding* N/H are defined word by word as in Definition 3. Similarly, the *parallel composition with hiding* $N_1 \Uparrow N_2$ of two par-composable ℓMON N_1 and N_2 is defined as before as $(N_1 \| N_2)/H$ with $H = \Sigma_1 \cap \Sigma_2$. $\Diamond$

Also the operations $\|$ and $\Uparrow$ are commutative and associative up to isomorphism for pairwise par-composable components.

Theorem 25. *The relation $\sqsubseteq_{\ell MON}$ is a precongruence for $\|$ and $\Uparrow$ on ℓMON, i.e. for three ℓMON N_1, N_2 and N_3 where N_2 is par-composable with N_3 and $N_1 \sqsubseteq_{\ell MON} N_2$, also N_1 is par-composable with N_3, $N_1 \| N_3 \sqsubseteq_{\ell MON} N_2 \| N_3$ and $N_1 \Uparrow N_3 \sqsubseteq_{\ell MON} N_2 \Uparrow N_3$. The relation is also a precongruence for relabelling and hiding.*

Proof. For parallel composition, observe that $wrap(N_i) \| wrap(N_3)$ and $wrap(N_i \| N_3)$ are isomorphic for $i = 1, 2$, since $wrap$ adds the same transitions to the same places for both systems, and these new transitions are also not synchronized in the first system. By definition of $\sqsubseteq_{\ell MON}$, we have $wrap(N_1) \sqsubseteq_{MPN} wrap(N_2)$, which implies $wrap(N_1) \| wrap(N_3) \sqsubseteq_{MPN} wrap(N_2) \| wrap(N_3)$ by Corollary 13. The above observation gives $wrap(N_1 \| N_3) \sqsubseteq_{MPN} wrap(N_2 \| N_3)$ and we are done.

For $N_1 \sqsubseteq_{\ell MON} N_2$ and a suitable relabelling function f, we extend f to f_{IO}, which additionally is the identity on $I_1 \cup O_1$; f_{IO} is a relabelling function for each $wrap(N_i)$. Now we only have to observe that $wrap(N_i)[f_{IO}])$ and $wrap(N_i[f])$ are isomorphic. With this, we are done as above, using again the definition of $\sqsubseteq_{\ell MON}$ and Corollary 13.

The case of hiding is easier, and the case of $\Uparrow$ is implied. $\Box$

4   Modal Asynchronous I/O-Petri Nets (MAP)

For comparison, we have a closer look at MAP, which are MPN where the visible actions are subdivided into input, output and internal actions [8]. An input action a indicates that an a-labelled transition takes a token from the (only implicit) place a, and analogously for an output. For composition, a common action a must always be an input of one and an output of the other component. A new place a is created and connected to a-labelled transitions as explained above; it represents an internal channel of the overall system. The label a is changed to *internal* actions $a^\triangleright$ on the output and $^\triangleright a$ on the input side.

The main issue in [8] is to decide the property *message consuming* (and a variation thereof): a net is message consuming w.r.t. internal channel a if, whenever there is a token on a, it is possible to perform a must-$^\triangleright a$, possibly preceded by output, internal or hidden must-transitions. This is regarded as a quality criterion for communication since no message in a channel will necessarily be ignored. Message consuming is preserved under composition and refinement, and to achieve this, internal actions must be visible.

Message consuming can be too strict: possibly, a message on a can only be processed sensibly if another message on b is received first. Also, a message consumption is certainly not so relevant if it has no effect for the environment. So we will not pursue this issue here. But note that the main idea in [19] is similar in spirit: there, the aim is to construct systems that only stop when a final marking (from a predefined set) is reached where all channel places are empty. Thus, the system will not stop while a message is pending. In contrast to the MAP approach, this property is not checked for the components; the aim is only to achieve it in the final system where it really is essential.

Since it is argued in [8] that, for stepwise refinement, the internal transitions should be hidden in the end, we will do so immediately in our comparison. To avoid a partitioning of Σ, we present MAP as MPN where the visible actions have the form $a^\triangleright$ or $^\triangleright a$ and, for no a, we have $a^\triangleright$ *and* $^\triangleright a$ in Σ. The refinement is simply $\sqsubseteq_{MPN}$. MAP are *composable* if their alphabets are disjoint. In the composition $\oplus_{HH}$, whenever some $a^\triangleright$ is in one and $^\triangleright a$ in the other alphabet, a new place a is created together with an arc from each $a^\triangleright$-labelled transition and an arc to each $^\triangleright a$-labelled transition as sketched in Fig. 5 on the left; the respective transitions are hidden.

Essentially, we could produce the same net by adding a place a and the resp. connections to each of the two MAP first and then apply $\oplus$. The first part gives us a function that embeds MAP into MON.

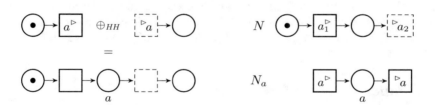

Fig. 5. Composition $\oplus_{HH}$, MAP N and the MPN N_a

Definition 26. The function *map2mon* maps each MAP N to a MON by adding for each $a^\triangleright$ and $^\triangleright a$ in Σ a new empty place a together with a weight-1 arc from each $a^\triangleright$-labelled transition or an arc to each $^\triangleright a$-labelled transition resp. All actions are hidden.

For a symbol a, N_a denotes an MPN (not a MAP!) as shown in Fig. 5. For a set A, we denote the disjoint union of the N_a with $a \in A$ by N_A. $\Diamond$

For the MAP N in Fig. 5, *map2mon*(N) is the MON N_2 in Fig. 3. Clearly, any *map2mon*(N) is a MON with the restriction (violated by the MON in Fig. 1) that each transition is only connected to at most one interface place, and then with an arc of weight one. From each such MON, we can shear off the interface places and reconstruct the respective transition labels for a corresponding MAP. The following theorem states that MAP is a proper sub-setting of our MON-setting with a stricter refinement.

Theorem 27. *Function map2mon embeds MAP into MON in the sense that it is injective but not surjective and, for all composable MAP N_1 and N_2, map2mon(N_1) and map2mon(N_2) are composable and*

$$map2mon(N_1 \oplus_{HH} N_2) = map2mon(N_1) \oplus map2mon(N_2) \ .$$

If we have $N_1 \sqsubseteq_{MPN} N_2$ instead, then map2mon(N_1) $\sqsubseteq_{\ell MON}$ map2mon(N_2), but not vice versa.

Proof. The first sentence should be clear. For the refinement, let $A = \{a \mid a^\triangleright$ or $^\triangleright a$ in $\Sigma\}$ and let f rename $a^\triangleright$ (!) to a if $^\triangleright a \in \Sigma$ and $^\triangleright a$ (!) to a if $a^\triangleright \in \Sigma$. Now, $(N_i \Uparrow N_A)[f]$ is isomorphic to $wrap(map2mon(N_i))$. Since $(N_1 \Uparrow N_A)[f] \sqsubseteq_{MPN} (N_2 \Uparrow N_A)[f]$ by the MPN-precongruence results, we are done.

Finally, we prove that the implication is strict. Consider the MAP N in Fig. 5 with $map2mon(N) = N_2$ in Fig. 3, and the similar MAP N' with $map2mon(N') = N_1$ in Fig. 3. We have $N_1 \sqsubseteq_{\ell MON} N_2$, but $N \sqsubseteq_{MPN} N'$ fails due to the reordering. □

In [8], it is shown that $\sqsubseteq_{MPN}$ on MAP is a precongruence for $\oplus_{HH}$ with a (not so difficult) proof that goes into the details of the definition of $\sqsubseteq_{MPN}$. One can also prove this from general precongruence results on MPN. Let N_1, N_2 and N_3 be MAP such that $N_1 \sqsubseteq_{MPN} N_2$ and N_1 and N_3 are composable. Let A be the set of those a where $^\triangleright a$ is in one of Σ_1 and Σ_3 while $a^\triangleright$ is in the other. Then, $N_i \Uparrow N_A \Uparrow N_3$ is isomorphic to $N_i \oplus_{HH} N_3$ and, with $N_1 \Uparrow N_A \Uparrow N_3 \sqsubseteq_{MPN} N_2 \Uparrow N_A \Uparrow N_3$, we are done.

5   Bounded Modal Open Nets

To argue that $N_1 \sqsubseteq_{\ell MON} N_2$ for MON N_1 and N_2, we have to exhibit an MPN-relation for $wrap(N_1)$ and $wrap(N_2)$. The problem is that the latter usually have infinitely many reachable markings, since arbitrarily many tokens can be put on each input place. One solution to this problem is to work with finite nets and to require that the final system (a closed MON as defined below) is *b-bounded* for some fixed bound b, i.e. that all reachable markings are b-bounded, assigning at most b tokens to each place. We sketch below how to modify MPN-relations for a setting where more than b tokens on a place are considered to be an error.

It can also be helpful to observe that a $wrap(N)$ is a special MPN, where each visible action a appears just once, and on a must-transition. From the position of such a transition, we can read off whether a is an input or an output action and whether the incident place (still denoted by a below) was an input or output place originally without having specific components in the MPN-tuple. We give here a first observation only, calling an MPN a *special MPN (sMPN)* if it is $wrap(N)$ for some MON N.

Proposition 28. *Let $\mathfrak{R}$ be an MPN-relation for sMPN N_1 and N_2, and $(m_1, m_2) \in \mathfrak{R}$. Writing $m + i$ for a marking m with an additional token on input place i, also $\mathfrak{R} \cup \{(m_1 + i, m_2 + i)\}$ is an MPN-relation for N_1 and N_2.*

Proof. We check the two conditions for $(m_1 + i, m_2 + i)$.

— Let $m_2 + i \xrightarrow{\alpha} m_2''$. By $(m_1, m_2) \in \mathfrak{R}$ and $m_2 \xrightarrow{i} m_2 + i$, there is some $(m_1', m_2 + i) \in \mathfrak{R}$ with $m_1 \xRightarrow{} m_1'$; we can assume that the underlying firing sequence starts with $m_1 \xrightarrow{i} m_1 + i$, since the i-transition does not remove any token; thus, $m_1 + i \Longrightarrow m_1'$. Furthermore, $m_1' \xRightarrow{\alpha} m_1''$ with $(m_1'', m_2'') \in \mathfrak{R}$. Hence, $m_1 + i \xRightarrow{\alpha} m_1''$ matches $m_2 + i \xrightarrow{\alpha} m_2''$.

— Let $m_1 + i \xdashrightarrow{\alpha} m_1''$. By $(m_1, m_2) \in \mathfrak{R}$ and $m_1 \xdashrightarrow{i} m_1 + i$, there is some $(m_1 + i, m_2') \in \mathfrak{R}$ with $m_2 \xRightarrow{i} m_2'$; we can again assume that the underlying firing sequence starts with $m_2 \xdashrightarrow{i} m_2 + i$, so that $m_2 + i \Longrightarrow m_2'$. Furthermore, $m_2' \xRightarrow{\alpha} m_2''$ with $(m_1'', m_2'') \in \mathfrak{R}$. Hence, $m_2 + i \xRightarrow{\alpha} m_2''$ matches $m_1 + i \xdashrightarrow{\alpha} m_1''$. $\qquad\square$

This observation shows that $m_1 \xrightarrow{i} m_1 + i$ can always be matched by $m_2 \xrightarrow{i} m_2 + i$ and vice versa; no other pair than $(m_1 + i, m_2 + i)$ is needed for this in $\mathfrak{R}$. This can help to prove or disprove $N_1 \sqsubseteq_{MPN} N_2$.

Often, it is desirable that systems are finite state and channels have a finite capacity. The final systems in such a setting can be modelled by finite b-bounded Petri nets; for the rest of this section, we fix some arbitrary positive bound b.

Definition 29. A MON is *closed* if it has no input or output places. A marking m of a MON or an sMPN that is not b-bounded is called an *error*; then, a marking m' with $m' \Longrightarrow m$ is called *illegal*. $\qquad\qquad\diamond$

A closed MON describes a final system, which usually arises as the composition of a system with the final user. We consider a setting where such a closed MON is required to be b-bounded. Note that a closed MON N coincides with $wrap(N)$.

If such a MON is built with a system component N', a marking of $wrap(N')$ that is not b-bounded is an *error*; it cannot occur in the final system and subsequent behaviour is irrelevant. In fact, this already holds for an illegal marking m', since nothing can prevent $wrap(N')$ to move autonomously from m' to an error. Note that the occurrence of a transition t can only lead from a legal to an illegal marking if t is an input.

Interface automata (IA) [5] form a similar setting (with synchronous communication), where an "unexpected" input leads to an error. While IA are a kind of LTS, there is quite some literature on combinations with modalities, see [4] for an advanced approach called modal interface automata (MIA). Similarly to transferring refinement and precongruence results from MTS to MPN, one can transfer these with some care from MIA to sMPN. The refinement definition looks as follows; note that any behaviour is better than an error, so an illegal marking does not have to be matched.

Definition 30 (sMPN-b-refinement). For sMPN N_1 and N_2 with the same input and output actions, we say that N_1 is an *sMPN-b-refinement* of N_2, written $N_1 \sqsubseteq_{sMPN}^b N_2$, if there is an *sMPN-$b$-relation* $\mathfrak{R}$ between the reachable markings of N_1 and N_2 with $(m_1^0, m_2^0) \in \mathfrak{R}$ such that for every $(m_1, m_2) \in \mathfrak{R}$ where m_2 is legal:

- m_1 is legal,
- $m_2 \xrightarrow{\alpha} m_2' \;\Rightarrow\; m_1 \xRightarrow{\widehat{\alpha}} m_1' \;\wedge\; (m_1', m_2') \in \mathfrak{R}$,
- $m_1 \dashrightarrow^{\alpha} m_1' \;\Rightarrow\; m_2 =\!\!\xRightarrow{\widehat{\alpha}} m_2' \;\wedge\; (m_1', m_2') \in \mathfrak{R}$. $\qquad\qquad\qquad \Diamond$

For closed MON N_1 and N_2, $wrap(N_1) \sqsubseteq_{sMPN}^b wrap(N_2)$ simply means that m_1^0 must be legal if m_2^0 is. Intuitively, this means: if the system specification composed with the user is b-bounded, then the refinement composed with the user is b-bounded as well.

Some details are simpler for sMPN than for MIA. Here, m' is illegal if $m' =\!\!\Rightarrow m$ for some error m. For MIA, also outputs must be considered for the transition sequence; we can ignore these here, since output transitions only remove tokens. Furthermore, for the matching of transitions as in Definition 30, inputs and outputs are treated differently for MIA: in case of an input, the matching transition sequence must *start* with the respective input. Here, this does not matter; if the only visible transition in a firing sequence is an input, we can just as well move it to the front since it does not remove tokens.

Additionally, MIA have so-called disjunctive must-transitions [4] for defining conjunction on MIA. It is not at all clear to us how a conjunction for Petri nets ("real" Petri nets with concurrency) could look like. Compared to a setting without conjunction, disjunctive must-transitions make [4] unnecessarily difficult to read. Therefore, we plan to work out a self-contained presentation of the b-bounded setting. There, it will be worthwhile to explicitly accompany each sMPN by a modified reachability graph, where – as in MIA – all illegal markings are merged into a special error state.

As a final remark, we point out that our b-bounded setting is *optimistic* like IA and MIA. An sMPN might have behaviour that leads to an error. As long as an error cannot be reached autonomously (i.e. the initial marking is illegal), there is an environment such that the composition is error-free; e.g. the environment may simply not provide any inputs. In fact, if the respective MON has some input place, reachable errors are unavoidable. A *pessimistic* approach as in [3] forbids components where errors are reachable, it cannot be applied here.

6   Conclusion

In [6,8], Petri nets were augmented with may- and must-modalities and modal refinement for stepwise design, and they were used for modelling asynchronous communication via merging implicit interface places (MAP). We have here applied a much older framework for nets with interface places [21], see also [19], and developed an according refinement relation for nets with modalities and explicit interface places (MON). We have justified this relation with a coarsest-precongruence result. Our studies were carried out in a larger setting with modal nets having interface places for asynchronous as well as action-labelled transitions for synchronous communication.

Details of the MAP-approach are related to checking so-called *message consumption*, a property that holds if, intuitively speaking, each message can eventually be received, i.e. removed from the channel. For stepwise refinement, it is more appropriate to abstract from these details, as also suggested in [6,8]. With this abstraction, it turned out that MAP is a subsetting of MON with a stricter refinement relation. With an example, we have shown that some reordering of messages leads to a rejection as a refinement in the MAP-approach, although it is intuitively acceptable for asynchronous communication (and in the MON-approach).

To show that one MON refines another, a suitable alternating simulation has to be exhibited. These simulations have special properties (compared to the general alternating simulations used for MAP), which could help to find one or prove that none exists. We have given one such property here and will look into this issue in the future.

Often, it is desirable that the components are finite-state and channels have a finite capacity. We have given a rough sketch how this can be integrated into the MON-approach and plan to work this out in detail. Furthermore, also motivated by the idea of message consumption, we think about integrating final markings as in [19] such that a system can only stop when all channels are empty.

References

1. Baldan, P., Corradini, A., Ehrig, H., Heckel, R.: Compositional modeling of reactive systems using open nets. In: Larsen, K.G., Nielsen, M. (eds.) CONCUR 2001. LNCS, vol. 2154, pp. 502–518. Springer, Heidelberg (2001). https://doi.org/10.1007/3-540-44685-0_34
2. Baldan, P., Corradini, A., Ehrig, H., Heckel, R., König, B.: Bisimilarity and behaviour-preserving reconfigurations of open Petri Nets. Log. Methods Comput. Sci. 4(4) (2008). https://doi.org/10.2168/LMCS-4(4:3)2008
3. Bauer, S.S., Mayer, P., Schroeder, A., Hennicker, R.: On weak modal compatibility, refinement, and the MIO workbench. In: Esparza, J., Majumdar, R. (eds.) TACAS 2010. LNCS, vol. 6015, pp. 175–189. Springer, Heidelberg (2010). https://doi.org/10.1007/978-3-642-12002-2_15
4. Bujtor, F., Fendrich, S., Lüttgen, G., Vogler, W.: Nondeterministic modal interfaces. Theoret. Comput. Sci. **642**, 24–53 (2016)
5. de Alfaro, L., Henzinger, T.A.: Interface-based design. In: Broy, M., Grünbauer, J., Harel, D., Hoare, T. (eds.) Engineering Theories of Software Intensive Systems. NSS, vol. 195, pp. 83–104. Springer, Dordrecht (2005). https://doi.org/10.1007/1-4020-3532-2_3
6. Elhog-Benzina, D., Haddad, S., Hennicker, R.: Refinement and asynchronous composition of modal petri nets. In: Jensen, K., Donatelli, S., Kleijn, J. (eds.) Transactions on Petri Nets and Other Models of Concurrency V. LNCS, vol. 6900, pp. 96–120. Springer, Heidelberg (2012). https://doi.org/10.1007/978-3-642-29072-5_4

7. Glabbeek, R.J.: The linear time — branching time spectrum II. In: Best, E. (ed.) CONCUR 1993. LNCS, vol. 715, pp. 66–81. Springer, Heidelberg (1993). https://doi.org/10.1007/3-540-57208-2_6

8. Haddad, S., Hennicker, R., Møller, M.H.: Specification of asynchronous component systems with modal I/O-petri nets. In: Abadi, M., Lluch Lafuente, A. (eds.) TGC 2013. LNCS, vol. 8358, pp. 219–234. Springer, Cham (2014). https://doi.org/10.1007/978-3-319-05119-2_13

9. Hüttel, H., Larsen, K.G.: The use of static constructs in a model process logic. In: Meyer, A.R., Taitslin, M.A. (eds.) Logic at Botik 1989. LNCS, vol. 363, pp. 163–180. Springer, Heidelberg (1989). https://doi.org/10.1007/3-540-51237-3_14

10. Kindler, E.: A compositional partial order semantics for Petri net components. In: Azéma, P., Balbo, G. (eds.) ICATPN 1997. LNCS, vol. 1248, pp. 235–252. Springer, Heidelberg (1997). https://doi.org/10.1007/3-540-63139-9_39

11. Larsen, K.G., Thomsen, B.: A modal process logic. In: Logic in Computer Science 1988, pp. 203–210. IEEE (1988)

12. Massuthe, P., Reisig, W., Schmidt, K.: An operating guideline approach to the SOA. Ann. Math. Comput. Teleinformatics **1**, 35–43 (2005)

13. Milner, R.: Communication and Concurrency. Prentice-Hall, Inc., Upper Saddle River (1989)

14. Pomello, L.: Some equivalence notions for concurrent systems. An overview. In: Rozenberg, G. (ed.) APN 1985. LNCS, vol. 222, pp. 381–400. Springer, Heidelberg (1986). https://doi.org/10.1007/BFb0016222

15. Raclet, J.B.: Residual for component specifications. Electr. Notes Theor. Comput. Sci. **215**, 93–110 (2008)

16. Reisig, W.: Deterministic buffer synchronization of sequential processes. Acta Inf. **18**, 117–134 (1982)

17. Schneider, V.: A better semantics for asynchronously communicating Petri nets. M.Sc. Thesis, Universität Augsburg (2017)

18. Souissi, Y.: On liveness preservation by composition of nets via a set of places. In: Rozenberg, G. (ed.) ICATPN 1990. LNCS, vol. 524, pp. 277–295. Springer, Heidelberg (1991). https://doi.org/10.1007/BFb0019979

19. Stahl, C., Vogler, W.: A trace-based service semantics guaranteeing deadlock freedom. Acta Inf. **49**, 69–103 (2012)

20. Vogler, W.: Behaviour preserving refinements of Petri nets. In: Tinhofer, G., Schmidt, G. (eds.) WG 1986. LNCS, vol. 246, pp. 82–93. Springer, Heidelberg (1987). https://doi.org/10.1007/3-540-17218-1_51

21. Vogler, W.: Modular Construction and Partial Order Semantics of Petri Nets. LNCS, vol. 625. Springer, Heidelberg (1992). https://doi.org/10.1007/3-540-55767-9

Stochastic Evaluation of Large Interdependent Composed Models Through Kronecker Algebra and Exponential Sums

Giulio Masetti[1,3]($\boxtimes$), Leonardo Robol[2,3], Silvano Chiaradonna[3], and Felicita Di Giandomenico[3]

[1] Department of Computer Science, Largo B. Pontecorvo 3, 56125 Pisa, Italy
[2] Department of Mathematics, Largo B. Pontecorvo 1, 56125 Pisa, Italy
[3] Institute of Science and Technology "A. Faedo", 56124 Pisa, Italy
giulio.masetti@isti.cnr.it

Abstract. The KAES methodology for efficient evaluation of dependability-related properties is proposed. KAES targets systems representable by Stochastic Petri Nets-based models, composed by a large number of submodels where interconnections are managed through synchronization at action level. The core of KAES is a new numerical solution of the underlying CTMC process, based on powerful mathematical techniques, including Kronecker algebra, Tensor Trains and Exponential Sums. Specifically, advancing on existing literature, KAES addresses efficient evaluation of the Mean-Time-To-Absorption in CTMC with absorbing states, exploiting the basic idea to further pursue the symbolic representation of the elements involved in the evaluation process, so to better cope with the problem of state explosion. As a result, computation efficiency is improved, especially when the submodels are loosely interconnected and have small number of states. An instrumental case study is adopted, to show the feasibility of KAES, in particular from memory consumption point of view.

Keywords: Stochastic Petri Nets · Stochastic Automata Networks · Markov chains · Mean Time To Absorption · Kronecker algebra · Exponential sums · Tensor Train

1 Introduction

Stochastic modeling and analysis is a popular approach to assess a variety of non-functional system properties, depending on the specific application domain the system is employed in.

Given the increasing complexity and sophistication of modern and future contexts where cyber systems are called to operate, their modeling and analysis becomes on one side more and more relevant to pursue, and on the other side

© Springer Nature Switzerland AG 2019
S. Donatelli and S. Haar (Eds.): PETRI NETS 2019, LNCS 11522, pp. 47–66, 2019.
https://doi.org/10.1007/978-3-030-21571-2_3

more and more difficult to achieve (especially when high accuracy of analysis outcomes is requested due to criticality concerns). Modularity and composition are widely recognized as foundational principles to manage system complexity and largeness when applying model-based analysis. Sub-models, tailored to represent specific system components at the desired level of abstraction, are first defined, then composed to derive the overall model, representative of the totality of the system under analysis. However, in order to be effective and scalable, such compositional approach needs to be efficient not only at modeling level, but also at model evaluation level. This topic has been addressed by a plethora of studies. When dependability, performance and performability related measures are of interest, a variety of modeling and solution approaches and automated supporting tools have been proposed, typically adopting high level modeling formalisms (among which the Stochastic Petri Nets family is a major category) and either simulation-based or analytical solution techniques [14, 27].

In this paper, we focus on state-based analytical numerical evaluation and propose a new approach to address the problem of the state explosion in the quantitative assessment of dependability and performability related indicators of large, interconnected systems modeled using Stochastic Petri Net (SPN). The reference picture is an overall system model, resulting from the composition of a set of relatively small models (e.g. expressed through the Superposed GSPN (SGSPN) formalism [13]), each one representing individual system component(s) at a desired level of abstraction, then composed through transition-based synchronization.

Specifically, the paper addresses the solution of the Continuous Time Markov Chain (CTMC) underlying the SPN, whose evolution represents the behavior of systems under analysis at a reasonable level of detail. The focus here is on those CTMCs that present absorbing states, and on the evaluation of the Mean Time To Absorption (MTTA), i.e., the expected time needed to arrive into an irreversible state. To the best of our knowledge, this kind of CTMCs has received low attention in past studies in terms of efficient solutions when dealing with large interconnected systems. However, addressing this context is relevant, since it is meaningful in a variety of modeling scenarios of the system under analysis. For example, depending on the performance or dependability measures under analysis, absorbing states represent system conditions directly involved in the computation of the measure, such as:

- in a reliability model [27], absorbing states can be those representing the system failure,
- in a security model, absorbing states are those representing the fact that a certain level of confidentiality has been violated or a part of the system is under the attacker's control,
- in a safety model, absorbing states are states where the system is considered unsafe.

Resorting to well known symbolic representation of the CTMC to gain in efficiency, the new approach, called Kronecker Algebra Exponential Sums (KAES), advances on existing solutions by exploiting powerful mathematical technologies such as Kronecker algebra [6], Tensor Trains [24] and Exponential sums [3].

The rest of the paper is organized as follows. In Sect. 2, related work is discussed. In Sect. 3 an overview of the proposed contribution is presented. Basic concepts and model design principles are introduced in Sect. 4. A detailed description of the MTTA is offered in Sect. 5. Then, the proposed KAES method is described in Sect. 6. In order to demonstrate the benefits of the new method, KAES has been implemented in the MATLAB evaluation environment and applied to a case study, detailed in Sect. 7. Obtained numerical results, discussed in Sect. 8, show the feasibility of KAES when the MTTA is evaluated, at increasing the size of the system under analysis, while standard numerical approaches fail due to the state-explosion problem. Finally, in Sect. 9 conclusions are drawn and future work is briefly discussed.

2   Related Work

It is well known that state-space analysis of discrete event systems has to cope with the problem of state space largeness, which in many cases makes unaffordable the analysis of realistic systems. Therefore, many studies have appeared in the literature, all attempting to alleviate the state space explosion problem.

Among them, a well established strategy consists in promoting state space reduction through a symbolic representation of the CTMC. Proposals in this direction were already formulated a few decades ago (e.g., in [12,25]), and there was active research for several years, as documented in the survey in [6]. The overall system model, resulting from composition of a number of system component models, is typically expressed through a SPN-like formalism (e.g., Generalized SPN (GSPN) [8,16]). The component models are orchestrated by the synchronization of a distinguished set of transitions, called synchronization transitions, that implement interdependencies among components. Such "high-level" model is then automatically translated into a "lower level" representation (such as in a Stochastic Automata Network (SAN) [4]). Moreover, the implicit representation of the CTMC is not obtained through constructing the Infinitesimal generator matrix (Q), but through a symbolic representation of Q, the Descriptor matrix $(\tilde{Q})$ that is the sum of two parts: one is the composition of the independent behaviors of the component automata (all the transitions of each submodel are not synchronized with transitions of other submodels), called here Local matrix (R), and the other one takes into account only the interdependencies, typically called Synchronization matrix (W) [12]. The matrix-vector product, a key mathematical operation common to all numerical methods, is then performed through the descriptor matrix-vector product, as in the shuffle, slice and split algorithms [10].

In these studies, since an irreducible CTMC [27] is assumed, it is required that the *reachability graphs* of all component models are fully connected [16]. Notice that "there is no requirement on the number of input and output arcs for synchronization transitions" [13].

Research on how to manipulate symbolically $\tilde{Q}$ in order to efficiently extract information needed to generate the relevant part of the Reachable state-space

($\mathcal{RS}$) of the system model, as well as fast implementation of the descriptor matrix-vector product, has been the subject of many investigations in the last twenty five years. A concise survey can be found in [6].

Although the relevant benefits obtained from the symbolic representation and manipulation of $\tilde{Q}$, when the state-space becomes so large that even storing in memory vectors of size $|\mathcal{RS}|$ is unfeasible, symbolic representations of the vectors, called descriptor vectors, would be desirable. This is the research area where we concentrate in this paper. To the best of our knowledge, only two other papers address symbolic representation of descriptor vectors: Kressner et al. [17] and Buchholz et al. [5]. In [17] the same symbolic vector representation as in KAES, i.e., the *Tensor Train* (TT) format [24], is employed together with standard numerical solvers, such as Alternating Minimal ENergy (AMEN) [11], for the evaluation of the steady-state probability vector, meaningful when the Markov chain is irreducible and finite. In [5] a different representation, the Hierarchical Tucker Decomposition, is employed again for the evaluation of the steady-state probability vector in the irreducible context. However, these solutions cannot be easily generalized to address wider measures of interest, such as the evaluation of transient properties, or adapt to analyze Markov chains with absorbing states, which is the target of KAES.

Finally, although not relevant for the developments in this paper, but for completeness on the literature on efficient management of the generated state space, we recall that an alternative approach to the symbolic representation and manipulation of the $\tilde{Q}$ is to exploit a symbolic state-space exploration with multi-valued decision diagrams (MDDs) [2,7].

3   Overview of the Novel Contribution

As already introduced, the contribution offered by the KAES approach is an efficient solution to evaluate MTTA when the CTMC is large and has absorbing states, working on symbolic representation of the descriptor vectors. First of all, the KAES approach builds upon the following assumptions, which are also common to most of the research studies from the literature review:

- the state space generated from each submodel has to be bounded;
- the marking dependencies of synchronization transition rates have strict rules (see Sect. 4);
- the Descriptor matrix $\tilde{Q}$ is obtained in two consecutive steps, deriving: (i) first the matrix R, that describes the CTMC generated from each submodel when all the transitions of the submodel are not synchronized with the transitions of the other submodels; (ii) then the matrix W, that describes only the interactions among the CTMC generated from the submodels when the synchronized transitions are considered.

In this paper, in order to ease the notation, no instantaneous transition is considered, even if both instantaneous local and synchronization transitions can be tackled, as shown in [6]. The logical view and reasoning behind the contribution offered in this paper is now outlined:

– The standard representation of the vectors involved in the computations would require a storage exponential in the number of interconnected systems. To overcome this difficulty, a compressed representation is employed. Under suitable assumptions, this only requires a storage linear in the number of interconnected systems. A vector or matrix which can be compressed in this format is said to have low tensor train rank.
– Unfortunately, arithmetic operations performed using this representation degrade the low tensor train rank property, which can be restored by recompression.
– The evaluation of the MTTA is recast into solving a linear system with a modified descriptor matrix $\tilde{Q} - S$, where S is a rank 1 correction – efficiently representable in TT form. Linear system solvers are available in the TT format, but are ineffective for the problem under consideration.
– Therefore, a new splitting of $\tilde{Q}$ as

$$\tilde{Q} = Q_1 + Q_2, \tag{1}$$

is considered, where Q_1 is represented in terms of Kronecker sums and Q_2 in TT form.
– The inverse of Q_1 can be easily applied to a vector (in TT form) using *exponential sums* [3], since the exponential of Kronecker sums is the Kronecker product of exponentials. This property is exploited to efficiently solve the linear system through an iterative method.
– The way the MTTA is computed guarantees a conservative assessment.

4   System Architecture and Model Design

The systems category we address comprises n components $C_1,\ldots,C_n$. These components are interconnected, according to a specific topology that depends on the application domain the system operates in. Such interconnections, also called dependencies, allow inter-operability among system components, but they also represent formidable vehicles through which potential malfunctions or attacks propagate, possibly leading to cascading or escalating failure effects. The analysis of such systems needs to account for the impact of error/failure propagation due to dependences, especially when focusing on dependability-critical systems. This requires cautiousness in building models for such systems, to properly master the resulting complexity, both at model representation and model solution levels.

   At the current stage of development, we target loosely interconnected systems. Although this might appear a significant limitation of the proposed approach, loosely interconnection is actually encountered in realistic contexts, such as the electric infrastructure where grid topologies of hundreds of buses have number of dependencies around 2–3 on average. On the other side, we aim at alleviating the problem of state explosion in analytical modeling, that the KAES approach fulfills at some extent.

Exploiting the modular modeling approach of the SGSPN formalism [13], each system component C_i is modeled through a GSPN extended with synchronization transitions, and the model that corresponds to C_i is called Mi. The overall SGSPN model, called M^{sync}, is a set of submodels M_i which interact only through synchronized transitions.

To fix the notation, a GSPN [1] can be defined as an 8-tuple

$$M = (P, T, I, O, H, pri, w, m_{init}),$$

where P is the set of *places* and T is the set of (timed and immediate) *transitions* with $P \cap T = \emptyset$. The functions $I\colon P \times T \to \mathbf{N}$, $O\colon T \times P \to \mathbf{N}$ and $H\colon P \times T \to \mathbf{N}$ are respectively the input, output and inhibition functions that map arcs (p, t) or (t, p) onto *multiplicity* values. In the graphical representation, the multiplicity is written as a number next to the arc (when grater than 1). The function $pri\colon T \to \mathbf{N}$ specifies the priority level associated to each transition, that is 0 for timed transitions and a value greater than 0 for immediate transitions. The weight function $w\colon T \to \mathbf{R}^+$ assigns rates to timed transitions and weights to immediate transitions. A marking m of M is a function $m\colon P \to \mathbf{N}$. A place p has n *tokens* if $m(p) = n$. The initial marking of the GSPN is denoted by m_{init}. GSPN formalism considered in this paper is extended to allow marking-dependent rates and weights, and marking-dependent multiplicities of arcs. Transition t is *enabled* in a marking m, written $m \xrightarrow{t}$, if t has concession (to fire), i.e., $m(p) \geq I(p, t)$ and $m(p) < H(p, t)$, and if no other transition t' exists that has concession in m, with $pri(t') > pri(t)$. The firing delay, i.e., the time that must elapse before the enabled transition can fire, is an exponentially distributed random variable for timed transitions and is zero for immediate transitions. Firing of a transition t enabled in a marking m yielding a new marking m' is denoted by $m \xrightarrow{t} m'$, with $m'(p) = m(p) - I(p, t) + O(t, p)$. The set of markings that are reachable from m_{init} (reachability set) is denoted by $\mathcal{RS}$. A GSPN is called *bounded* if for all $p \in P$ and $m \in \mathcal{RS}$ the value of $m(p)$ is bounded. A GSPN is called *structurally bounded* if it is bounded for every initial marking [22]. Following the reasoning briefly outlined in [6], in order to guarantee that every M_i will have a finite state-space, in this paper all the component submodels M_i will be assumed structurally bounded.

In this paper, the standard definition of synchronized transitions is restricted to timed transitions.

Definition 1 (Synchronization transitions). *Let be T^{sync} and T_i the sets of transitions defined respectively in M^{sync} and M_i. Let $\mathcal{ST} \subseteq T^{sync}$ the set of synchronization transitions of M^{sync}. A timed transition t is a* synchronization *(or superposed) transition, i.e., $t \in \mathcal{ST}$, if there is an occurrence of t in two or more submodels, i.e., $t \in T_{i_1} \cap \ldots \cap T_{i_k}$, with $k \geq 2$. A synchronized transition t is enabled in a marking of M^{sync} if all the occurrences of t within submodels are enabled in the same marking restricted to the submodels. Formally, calling*

m a marking of M^{sync} and m_i its projection on M_i, $m \xrightarrow{t}$ if $m_i \xrightarrow{t}$ for all i such that $t \in T_i$. In the overall model M^{sync}, all the occurrences of t are enabled at the same time and a unique exponentially distributed firing delay is defined for all them, thus all of them fire at the same instant of time. The overall SGSPN model M^{sync} is equivalent to the whole GSPN model M^{sys} obtained joining all the submodels M_i where all the occurrences of t are merged into one transition, also named t. Firing of t in M^{sys} corresponds to the firing of all the occurrences of t within the submodels, i.e., formally

$$m \xrightarrow{t} m' \iff m_i \xrightarrow{t} m'_i \text{ for all } i \text{ such that } t \in T_i.$$

All the transitions t that are not synchronization transitions, i.e., those for which there exists a unique i such that $t \in T_i$, are called *local transitions*.

Allowing general marking-dependent rates and weights for the design of M_i can lead to inconsistent components models and this issue is strictly related to the granularity of the model and the tensor algebra of choice (see [4,6,9]). In this paper, as in [8], rates and weights of all the local transitions that belong to M_i and multiplicities of the corresponding arcs are allowed to depend on the marking of M_i, whereas rates and weights of the synchronization transitions and multiplicities of the corresponding arcs should be constant.

As described in [6], the system model SGSPN can be translated into a SAN and then the state space of M^{sync} called $\mathcal{RS}$, is not fully explored, and the CTMC associated to M^{sync} is not assembled. Instead of working with Q, the SAN provides an implicit representation, called descriptor matrix $\tilde{Q}$, of Q. In particular, calling $\mathcal{RS}^{(i)}$ the state-space of M_i when each occurrence of the synchronization transitions is considered local and $N_i = |\mathcal{RS}^{(i)}|$, $\tilde{Q}$ is defined as

$$\tilde{Q} = R + W + \Delta, \tag{2}$$

i.e., the sum of local contributions, called R, and synchronization contributions, called W, where

$$R = \bigoplus_{i=1}^{n} R^{(i)}, \tag{3}$$

$$W = \sum_{t \in \mathcal{ST}} \bigotimes_{i=1}^{n} W^{(t,i)}, \tag{4}$$

$R^{(i)}$ and $W^{(t,i)}$ are $N_i \times N_i$ matrices, the diagonal matrix Δ is defined as $\Delta = -\text{diag}((R+W)e)$ and the operators $\oplus$ and $\otimes$ are the *Kronecker sum* and *Kronecker product*, respectively. The matrices $R^{(i)}$ and $W^{(t,i)}$ are assembled exploring $RS^{(i)}$. Specifically, $R^{(t,i)} = l_t \tilde{W}^{(t,i)}$ where l_t is the constant rate associated to t, equal in every M_i, and $\tilde{W}^{(t,i)}$ is a $\{0,1\}$-matrix defined as follows:

$$\tilde{W}^{(t,i)}_{m_i, m'_i} = \begin{cases} 1 & \text{if } t \text{ is enabled in } m_i \text{ inside } M_i \text{ and } m_i \xrightarrow{t} m'_i, \\ 0 & \text{otherwise.} \end{cases} \tag{5}$$

In particular, if the transition t has no effect on the component M_i, we have $\tilde{W}^{(t,i)} = I$. The potential state-space of M^{sync}, called PS, is defined as

$$\mathcal{PS} = \mathcal{PS}^{(1)} \times \cdots \times \mathcal{RS}^{(n)},$$

and $|\mathcal{PS}| = N_1 \cdot \ldots \cdot N_n$ will be indicated as N in the following. Using this notation, R, W and Δ are $N \times N$ matrices.

Performance, dependability and performability properties can be defined in terms of *reward structures* [26, 27] at the level of the SGSPN model. These reward structures are automatically translated to reward structures at the Stochastic Activity Network (SAN) level and represented by symbolic reward structures at the CTMC level.

5   Mean Time To Absorption

For simplicity, in the rest of the paper it is assumed that, fixed m_{init}, there exists a unique[1] absorbing state in $\mathcal{PS}$ that is the last of the chain defined by $\tilde{Q}$. This is not restrictive because the problem can be always reduced to this situation by collapsing all the absorbing states of M_i into a single one and reordering the CTMC of M_i so that the absorbing state has index N_i. This guarantees, as a consequence of the lexicographic ordering defined by the Kronecker product, that the last state of $\mathcal{PS}$ is absorbing and corresponds to the last state of $\mathcal{RS}$, where all the component models are in their absorbing state. Thus, in the following N will indicate the absorbing state of $\mathcal{PS}$.

Calling $X(\tau) \in \mathcal{PS}$ the stochastic process defined by $\tilde{Q}$, the MTTA is defined as the expected time for transitioning into the absorbing state, which can be formalized as

$$MTTA = \int_0^\infty \mathbb{P}\{X(\tau) \neq N\} \, d\tau. \tag{6}$$

Given the unique absorbing state assumption, $\tilde{Q}$ can be replaced by $\hat{Q}$, the submatrix of $\tilde{Q}$ obtained by removing the last row and column (as shown in [27]), that is

$$\tilde{Q} = \left[\begin{array}{c|c} \hat{Q} & \begin{array}{c} v_1 \\ \vdots \\ v_{N-1} \end{array} \\ \hline 0 \ldots 0 & 0 \end{array} \right] \tag{7}$$

Then the MTTA can be expressed as

$$MTTA = -\hat{\pi}_0^T \hat{Q}^{-1} \mathbb{1}, \tag{8}$$

where $\hat{\pi}_0$ contains the first $N - 1$ entries of π_0, and therefore the problem has been recast into the solution of a linear system.

---

[1] Notice that this assumption does not imply that $\tilde{Q}$ has an unique row of zeros, as for the case of the stochastic process defined by Q.

6 The KAES Approach

Targeting the efficient evaluation of the MTTA as in (8), the KAES approach develops solutions to treat both the descriptor matrix and the descriptor vector in a symbolic representation. Specifically, KAES is an iterative method, and relies on the following steps:

- A compressed representation scheme for the descriptor vector $\tilde{V}$ is devised by using *tensor trains*. This representation will be used throughout the iterations, and is described in Sect. 6.1.
- The linear system (8) is solved by a Neumann iteration obtained by splitting the descriptor matrix $\tilde{Q}$ as in (1), and analyzed in Sect. 6.2.
- The core of the iteration is the inversion of Q_1, which can be efficiently performed in the compressed format using exponential sums; this technique is described in Sect. 6.3.
- Some further remarks on the efficient computation of the Neumann iteration are reported in Sect. 6.4.

6.1 Symbolic Representation of the Descriptor Vector

As already discussed when presenting the related work, studies on the symbolic representation of the descriptor matrix in the Kronecker algebra are already well consolidated.

Concerning the descriptor vector, a few approaches have recently appeared on compact representations, as already reviewed in Sect. 2, but in the context of irreducible CTMC. Here, we exploit the Tensor Train (TT)-representation as in [17], since it is a convenient low-rank tensor format, but addressing CTMC with absorbing states.

We refer the reader to [24] for an overview of the philosophy and the theory of TT tensors, including an accurate description of the truncation procedure.

In a nutshell, a TT-representation of a tensor $\mathcal{X}$ can be given by a tuple $(G_1, \ldots, G_n)$ of arrays, where G_1 and G_n are matrices (so they have two indices), and G_j for $j = 2, \ldots, n-1$ are order 3 tensors (that is, arrays with 3 indices) such that

$$\mathcal{X}(i_1, \ldots, i_n) = G_1(i_1, :)G_2(:, i_2, :) \ldots G_{n-1}(:, i_{n-1}, :)G_n(:, i_n),$$

where we have used the MATLAB notation : to denote "slices" of the tensors, and the products are the usual matrix-matrix or matrix-vector products. More precisely, given an array with two indices $G(\alpha, \beta)$, we define $G(:, \beta)$ as the column vector with entry in position α equal to $G(\alpha, \beta)$, and $G(\alpha, :)$ is a row vector with entry in position β equal to $G(\alpha, \beta)$. Similarly, given an array with three indices $G(\alpha, \beta, \gamma)$, we define $G(:, \beta, :)$ as the matrix whose entry (α, γ) is equal to $G(\alpha, \beta, \gamma)$.

The G_j, often called *carriage*, are tensors of dimension $\nu_{j-1} \times N_j \times \nu_j$, where we fix $\nu_0 = \nu_n = 1$ (and thus G_1 and G_n are matrices).

The vector $(\nu_0, \ldots, \nu_n)$ is called the TT-rank of the tensor $\mathcal{X}$. In our context, the initial probability vector π_0 and vector $\mathbb{1}$ can be easily expressed in the Kronecker form

$$\pi_0 = \pi_0^{(1)} \otimes \ldots \otimes \pi_0^{(n)}, \tag{9}$$

$$\mathbb{1} = \mathbb{1}^{(1)} \otimes \ldots \otimes \mathbb{1}^{(n)}, \tag{10}$$

and in TT-format as:

$$\pi_0(i_1, \ldots, i_n) = \pi_0^{(1)}(i_1) \cdot \ldots \cdot \pi_0^{(n)}(i_n),$$

$$\mathbb{1}(i_1, \ldots, i_n) = \mathbb{1}^{(1)}(i_1) \cdot \ldots \cdot \mathbb{1}^{(n)}(i_n).$$

Similarly, also the auxiliary vectors necessary to perform the iterative computation of KAES are expressed in TT-format.

The matrix Q and the other auxiliary matrices used in the following have low TT-ranks (and so are expressed in TT-format) when the CTMC is obtained from a loosely interconnected system model, as discussed in Sect. 4. We refer the reader to [20] for further details on the justification for the presence of such low-rank structures.

TT-format representation is convenient, since it employs $\mathcal{O}(N_{\max} \cdot n \cdot \nu_{\text{eff}}^2)$ flops for each matrix-vector product, instead of the generally larger $\mathcal{O}(N_{\max}^n)$ flops of the corresponding standard representation, where $N_{\max} = \max\{N_1, \ldots, N_n\}$ and ν_{eff} is the *effective rank*[2].

When two tensors are added or other matrix operations are performed, the result is still represented in the TT format, but usually with a suboptimal value of the ranks ν_j. For this reason, it is advisable to recompress the result using a rounding procedure, available in the TT-format, that has a complexity $\mathcal{O}(N_{\max} \cdot n \cdot \nu_{\text{eff}}^2 + n \cdot \nu_{\text{eff}}^4)$. When the rank r is low, this number is still very small compared to the number of states, which is $N_{\max}^n$.

Although this unavoidably leads to rounding errors, the accuracy can be chosen by the user. Note that, differently from the floating point arithmetic, the trade-off between the rounding error parameter and the required number of correct digits is more complex to devise, since the computational effort is not an increasing function of the accuracy level.

Often, in the following, TT-tensor will be treated as first-order objects, assuming that the arithmetic on these objects has been overloaded. When this happens, it is assumed that truncation is performed after each operation, to restore an optimal representation of the data.

---

[2] The effective ranks have been obtained through the **erank** function provided by the TT-toolbox [24].

6.2 Matrix Splitting and Neumann Expansion

In order to exploit the low-rank format, it is necessary to avoid the extraction of the submatrix $\hat{Q}$, since it cannot be directly expressed in the language of Kronecker algebra. Therefore, an auxiliary rank 1 matrix S that satisfies

$$\hat{\pi}_0^T \hat{Q}^{-1} \mathbb{1} = \pi_0^T (\tilde{\hat{Q}} - S)^{-1} \mathbb{1}, \tag{11}$$

where $\mathbb{1}$ is the vector of all ones of appropriate dimension, is defined as

$$S = (\tilde{Q}u)u^T - uu^T, \qquad u \in \mathbb{C}^N, \qquad u_j = \begin{cases} 0 & \text{if } j < N \\ 1 & \text{if } j = N \end{cases},$$

where $N = |\mathcal{PS}|$ is the dimension of $\tilde{Q}$. If $\tilde{Q}$ has a low TT-rank, the same holds for $\tilde{Q} - S$, and therefore it can be expected that exploiting an existing TT-enabled system solver to compute the MTTA would maintain the TT-ranks low.

The solvers AMEN [11] and DMRG [23], used in [17] where $\tilde{Q}$ is irreducible, have been tested to solve Eq. (11). Unfortunately, there was not always convergence, thus making the measure of interest not assessable in many cases.

For this reason, a different approach has been designed to compute the MTTA. The idea is to make use of the so-called *Neumann expansion*:

$$(I - M)^{-1} = \sum_{j=0}^{\infty} M^j, \tag{12}$$

valid for each matrix M that has spectral radius[3] $\rho(M) < 1$.

The crucial point in KAES is the definition of the splitting of Eq. (1) such that $M = -Q_1^{-1}(Q_2 - S)$ verifies the necessary condition for the Neumann expansion applicability and promotes fast evaluation of Q_1^{-1}. This is done in two steps: first a diagonal matrix Δ' is chosen such that $\Delta' \leq \Delta$, and $\Delta' = \Delta_1' \oplus \ldots \oplus \Delta_n'$ and then Q_1, Q_2 are defined as $Q_1 = \Delta' + R$, and $Q_2 = W + \Delta - \Delta'$. From the definition of Q_1 and Q_2 follows that

$$(\tilde{Q} - S)^{-1} = (I + Q_1^{-1}(Q_2 - S))^{-1} Q_1^{-1}, \tag{13}$$

and it is possible to prove [20] that $\rho(M) < 1$. Using (12) one can approximate the row vector $y = \pi_0^T (\tilde{Q} - S)^{-1}$ by truncating the infinite sum to k terms:

$$y_k = \sum_{j=0}^{k} (-1)^j \pi_0^T (Q_1^{-1}(Q_2 - S))^j Q_1^{-1}. \tag{14}$$

and then compute

$$MTTA = -y_k \cdot \mathbb{1} + \mathcal{O}(\rho(M)^{k+1}) \tag{15}$$

---

[3] The spectral radius is defined as the maximum of the moduli of the eigenvalues.

with a straightforward dot product. The notation $\mathcal{O}(\rho(M)^{k+1})$ is used to indicate that the error is bounded by a constant times $\rho(M)^{k+1}$. The choice of Δ' can be tuned to choose a trade-off between the speed of convergence and the memory consumption, determined by the rank growth in the iterations.

Notice that, defining $z_{k+1} = Q_1^{-1}(Q_2 - S)z_k$ and $z_0 = Q_1^{-1}(\mathbb{1} - e_N^T Q_1^{-1} \mathbb{1} \cdot e_N)$, it is possible to re-write Eq. (15) as $MTTA = -\pi_0^T \cdot z_k + \mathcal{O}(\rho(M)^{k+1})$, where $z_{k+1} \geq z_k$ for all $k = 0, 1, \ldots$ because $e_N^T z_0 = 0$ and both Q_1^{-1} and Q_2 are non-negative matrices. This means that the MTTA can be computed in a conservative way, being the approximation $-\pi_0^T \cdot z_k$ a lower bound.

In this paper, a variation of (14) is employed; this modification yields a method with quadratic convergence, overcoming difficulties encountered when $\rho(M)$ gets close to 1. It is based on refactoring $(I - M)^{-1}$ as

$$(I - M)^{-1} = (I + M)(I + M^2) \cdots (I + M^{2^k}) \cdots$$

The downside is that this variation requires to store powers of the matrix $Q_1^{-1}(Q_2 - S)$ in place of just results of matrix vector products and system solves. This has higher memory requirements – but all these matrices are stored in the TT-format, ensuring linear memory storage in the number of subsystems when the TT-ranks (measuring the level of interaction between components) are low.

6.3   Inversion Through Exponential Sums

The main ingredient for implementing KAES is to efficiently evaluate the action of the inverse of Q_1 on a TT-vector and on a TT-matrix. To this aim, in this paper a well-known *exponential sums* construction is adopted. This construction has been used in a variety of contexts (see, for instance, [15,18,19] and the references therein), often being rediscovered by different authors. The construction is built upon a few important observations. The first one is that in Sect. 6.2 all the addends are expressed as Kronecker sums, namely

$$Q_1 = Q_1^{(1)} \oplus \ldots \oplus Q_1^{(n)}. \tag{16}$$

Thus, a very important property of the standard splitting in Eq. (2) is maintained in the new splitting: all the Kronecker products belong to only one of the splitters, i.e., Q_2, and the Kronecker sums to the other one, namely Q_1.

The second consideration is that, given a TT-tensor $\mathcal{X}$, it is possible to efficiently evaluate the product $\mathcal{Y} = (M_1 \otimes \ldots \otimes M_n)\mathcal{X}$, as this can be performed in $\mathcal{O}(n)$ flops, assuming a low TT-rank for $\mathcal{X}$. Moreover, the result is still a TT-tensor with the same rank.

All the Kronecker products are in Q_2 and the assumption of dealing with loosely interconnected components implies that there are only a few non-identity matrices in W, and then in Q_2. Thus, in this setting $\mathcal{X}$ is the sum of a few terms with TT-rank 1, and consequently has low TT-rank.

The third observation is that, from Eq. (16) follows that

$$e^{Q_1} = e^{Q_1^{(1)}} \otimes \ldots \otimes e^{Q_1^{(n)}}. \tag{17}$$

This can be easily proved using the addends defining the Kronecker sum in Eq. (16) commute, and that $e^{A+B} = e^A e^B$ whenever $AB = BA$. Then, the conclusion follows by $(I \otimes A)(B \otimes I) = A \otimes B$.

Taking this remarks into account, let to consider the approximated expansion

$$\frac{1}{x} \approx \sum_{j=1}^{\ell} \alpha_j e^{-\beta_j x}, \tag{18}$$

which can be obtained truncating the expansion of $1/x$ to ℓ terms; the error in the approximation on $[1, \infty]$ performed when truncating to ℓ terms can be controlled with a-priori estimates. Several constructions are available, we refer the reader to [3] which provides the optimal result, and can guarantee an error term that converges to zero exponentially in ℓ. According to the construction in [20], one can choose the decomposition Q_1 in a way that the eigenvalues of Q_1 are the ones of R shifted to be in the left half of the complex plane. For simplicity, here the case where the eigenvalues of R are real is considered[4]—the general case can be handled with minimal modifications [20].

In particular, the spectrum of Q_1 is contained in $(-\infty, \sigma_{\min}]$, and the action of the inverse can be approximated, applying Eqs. (17) and (18), as

$$\mathcal{Q}_1^{-1} \approx \sum_{j=1}^{\ell} \alpha_j e^{-\beta_j Q_1^{(1)}} \otimes \ldots \otimes e^{-\beta_j Q_1^{(n)}} \tag{19}$$

where $\sigma_{\min}$ is the eigenvalue with minimum modulus of Q_1 and α_j, β_j are computed working on $-\frac{1}{\sigma_{\min}} Q_1$, that has eigenvalues enclosed in $[1, \infty)$.

Since Q_1 is a Kronecker sum, the computation of its eigenvalues can be performed almost for free; in fact, if $Q_1 = Q_1^{(1)} \oplus \ldots \oplus Q_1^{(n)}$ and we denote by $\sigma(Q_1)$ its spectrum,

$$\sigma(Q_1) = \left\{ \sigma_{i_1}^{(1)} + \ldots + \sigma_{i_n}^{(n)} \mid \sigma_{i_k}^{(k)} \in \sigma\left(Q_1^{(k)} \right) \right\}$$

In particular, computing the minimum and maximum eigenvalue just requires to compute the extreme eigenvalues of each factor $Q_1^{(k)}$.

Consequently, the action of the right-side expression in Eq. (19) is cheap to evaluate, being the sum of l actions of Kronecker products.

6.4   Efficient Computation of the Neumann Iterations

In computing MTTA, one has to evaluate $-\pi_0^T (Q - S)^{-1} \mathbb{1}$. To accomplish this, it is possible to either evaluate $\pi_0^T (Q - S)^{-1}$ and then compute the dot product with $\mathbb{1}$, or to compute $(Q - S)^{-1} \mathbb{1}$ instead, and take the dot product with π_0.

It can be seen that the former strategy is more convenient. In fact, the graph with M^T as adjacency matrix is a subgraph of the one induced by $\tilde{Q}^T$. In particular, states in $\mathcal{PS} \backslash \mathcal{RS}$ have no impact on the evaluation of MTTA because they

---

[4] This assumption is verified in the cases considered in the numerical experiments.

correspond to zero entries in π_0, and these entries will remain zero in $\pi_0^T M^k$ for any $k > 0$. This guarantees that this part of the chain has no effect on the computation: there is no need to have an explicit algorithm to detect the reachable states as in [6], because these are implicitly ignored.

Moreover, this strategy is seen to provide lower TT-ranks during the Neumann iterations, compared to computing $(Q - S)^{-1} \mathbb{1}$ first.

This choice has another beneficial effect: the addends in the series (14) are non-negative, and therefore the MTTA is approximated from below—and at every step the partial result is effectively a lower bound [20].

7   Case Study

To illustrate the effectiveness of the proposed approach, we consider a complex computer system composed by n interconnected components $C_1, \ldots, C_n$, properly functioning at time 0. Each properly functioning component C_i fails after an exponentially distributed time with rate λ_i. With probability p the failed component C_i can be repaired and restarted as properly functioning after an exponentially distributed time with rate μ_i. Instead, with probability $1 - p$ the failure of C_i propagates instantaneously to all the components directly interconnected to it. In this case, all the failed components cannot be repaired. The list of the $\bar{d}_i$ indexes of the components where the failure of C_i can propagate is $\bar{D}_i = \{h_1, h_2, \ldots, h_{\bar{d}_i}\}$. The list of the d_i indexes of the components whose failure can propagate to C_i is $D_i = \{j_1, j_2, \ldots, j_{d_i}\}$. The topology of interactions among components is given by the $n \times n$ adjacency matrix $\mathcal{T} = [\mathcal{T}_{i,j}]$, where $\mathcal{T}_{i,j} = 1$ if $j \in \bar{D}_i$, else $\mathcal{T}_{i,j} = 0$. hus, $\mathcal{T}$ defines an oriented graph that represents how the n components depend on each other and how they are connected to form the overall system. Although different topologies $\mathcal{T}$ can be defined, for example when different access rights to components are defined for different types of service or customers, for the sake of simplicity only one topology $\mathcal{T}$ is considered in the following.

7.1   Model of the Case Study

The SGSPN model representing the overall system of the case study is obtained defining a submodel for each single component C_i, with $i = 1, \ldots, n$, and composing all such submodels through a transition-synchronization approach, as described in Sect. 4. The model of the component C_i is depicted in Fig. 1. The places On_i (initialized with one token), $Down_i$ and F_i are local to the model and represent the states where, respectively, C_i works properly (one token in On_i), is under repair (one token in $Down_i$), and is failed and cannot be repaired. The transitions $TDown_i$ and TOn_i are local to the model and represent respectively the exponentially distributed time with rate $p_i \lambda_i$ to the occurrence of a failure, when the failed component can be repaired, and the exponentially distributed time with rate μ_i after which the component returns to operate properly. The transitions $TFail_i$ and $TFail_{j_k}$ with $k = 1, \ldots, d_i$ are synchronization transitions used to synchronize the models representing each component of the system, i.e., to propagate the failure that affects C_i to its neighbors with probability $1 - p_i$. $TFail_i$ represents the exponentially distributed time with rate $(1 - p_i)\lambda_i$ to the occurrence of a failure on C_i, that instantaneously propagates to C_h, with $h \in \bar{D}_i$ (without the possibility to repair the failed components). $TFail_i$ is replicated in the models of C_i and C_h, for each $h \in \bar{D}_i$. In each C_h model, it exists a transition $TFail_{j_k}$ with $j_k = i$, synchronized with $TFail_i$, that propagates the failure occurred in C_i. The transitions $TFail_{j_k}$ for each $k = 1, \ldots, d_i$ in Fig. 1 represent the time to the occurrence of a failure on C_{j_k} that instantaneously propagates to C_i (without the possibility to repair the failed components). Each transition $TFail_{j_k}$ is replicated in the models of C_i and C_h with $j_k \in D_h$. In each model C_h exists a transition $TFail_h$ with $j_k = h$, synchronized with $TFail_{j_k}$, that represents the occurrence of the failure in C_h that propagates to C_i.

In absence of immediate transitions, a synchronized transition is enabled when it, and all the transitions synchronized with it, have concession. As shown in Fig. 1, the transition $TFail_i$ has concession when one token is in the place On_i. All the transitions $TFail_{j_k}$, for $k = 1, \ldots, d_i$, have always concession, being the multiplicity of each input arc equal to the number of tokens in the corresponding input place On_i and $Down_i$, as shown in Fig. 1. Thus, $TFail_i$ is enabled when there is one token in On_i. The firing of $TFail_i$ occurs simultaneously in the model of C_i where $TFail_i$ removes the token from On_i and adds one token to F_i (the component is failed and cannot be repaired), and in the model of C_h, for each $h \in \bar{D}_i$, where, as shown in Fig. 1 replacing i with h, $TFail_{j_k}$, with $j_k = i$, removes one token from On_h and $Down_h$ (if any) and adds one token to F_h (the failure of C_i propagated to C_h that cannot be repaired).

On the model the following reward structure is considered

$$r = \begin{cases} 1 & \text{if } \#\mathrm{F}_i = 1 \text{ for all } i, \\ 0 & \text{otherwise,} \end{cases}$$

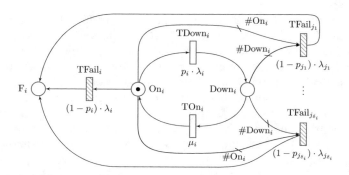

Fig. 1. Model of the component C_i of the case study. The shaded transitions are synchronization transitions.

so that the mean time to system failure τ_F corresponds to the cumulative measure defined by r on the interval of time $[0, \infty)$.

The model depicted in Fig. 1 can be classified as a reliability model [27] and produces a CTMC with a unique absorbing state. The model of Fig. 1 is structurally bounded, being a stochastic finite state machine and the corresponding stochastic automaton is characterized by:

$$R^{(i)} = \begin{pmatrix} 0 & p_i\lambda_i & 0 \\ \mu_i & 0 & 0 \\ 0 & 0 & 0 \end{pmatrix}, \quad W^{(t,i)} = \begin{cases} \begin{pmatrix} 0 & 0 & (1-p_i)\lambda_i \\ 0 & 0 & 0 \\ 0 & 0 & 0 \end{pmatrix} & \text{if } j = i \\ \begin{pmatrix} 0 & 0 & 1 \\ 0 & 0 & 1 \\ 0 & 0 & 1 \end{pmatrix} & \begin{array}{l} \text{if } j \neq i \text{ and} \\ \mathcal{T}(i,j) = 1 \end{array} \\ \begin{pmatrix} 1 & 0 & 0 \\ 0 & 1 & 0 \\ 0 & 0 & 1 \end{pmatrix} & \text{otherwise} \end{cases} \tag{20}$$

The mean time to system failure τ_F is then evaluated through the MTTA, because of the correspondence between the unique absorbing state of the CTMC and the system failure state.

8   Evaluation Results

In this section, details on how the case study described in Sect. 7 has been evaluated through KAES are discussed, and the obtained results in terms of time and memory consumption are presented in Table 1. The analysis is carried out for different numbers of components $n = 10, 20, \ldots, 50$ and generating random topologies of interactions following a predefined template. Results are obtained implementing[5] the case study SAN, i.e., Eq. (20), and the KAES method in MATLAB [21].

As a form of validation of KAES, it has been verified that for $n \leq 10$ the values obtained for the mean time to system failure τ_F with KAES coincide with the values obtained with the standard technique (full exploration of $|\mathcal{RS}|$ followed by the linear system of Eq. (8) solution). However, since here the analysis focuses on assessing the efficiency of the proposed method, the results obtained for τ_F are out of the scope of this paper, and then are not shown.

In order to demonstrate the ability of KAES in improving on current limitations suffered by standard techniques, analyzed scenarios are characterized by: (1) both $|\mathcal{PS}|$ and $|\mathcal{RS}|$ are large; (2) the model parameters define a *stiff* [27] CTMC. In particular, $|\mathcal{PS}| = 3^n$, being $|\mathcal{RS}^{(i)}| = 3$, and $|\mathcal{RS}|$ depends on $\mathcal{T}$.

Note that if $\mathcal{T}$ represents the complete graph of interdependencies then $|\mathcal{RS}|$ is trivially small. In fact, the initial state is the one with 1 token in On_i for all

---

[5] https://github.com/numpi/kaes.

$i = 1, \ldots, n$, and when the first $TFail_j$ fires, all the tokens are removed from On_i for all i; thus, the system only has two reachable states.

Therefore, to have large $|\mathcal{RS}|$, the topology $\mathcal{T}$ of interactions is obtained as follows: first, a star topology is constructed, where, labeling the nodes from 1 to n, there exist $n - 1$ edges connecting 1 to j, for $j = 2, \ldots, n$. Then, for each node with index greater than 1, another edge connecting it to a random node is added with probability 0.2. Although artificially generated, such topologies are good representatives of topologies addressed by KAES (large number of components, loosely interconnected), and are suitable for the case study illustrated in Sect. 7.

The parameters for each M_i have been randomly selected, but aiming at obtaining a stiff CTMC. Specifically, in the performed evaluations they are:

$$\lambda_i \in [0.5, 1.5], \quad \mu_i \in [2000, 3000], \quad p_i \in [0.95, 1],$$

so that there are 4 orders of magnitude among the parameters. The tests have been repeated 100 times for each value of n, using the randomized topology described above. For large n, not all the cases could be solved using the available system memory. The percentage of cases exceeding the available memory is reported in Table 1 for each n.

The average amount of *CPU* (user and system) time (in seconds) and the average amount of the RAM memory (in GB) consumed by KAES have been quantified. The averages are computed only on those cases where KAES was successful. Computations were performed on a Intel(R) Xeon(R) CPU E5-2650 v4 @ 2.20 GHz, where each experiment had 12 CPUs and 120 GB of RAM at its disposal. As shown in Table 1, the actual memory consumption for all the values of n is much lower than the maximum available.

Note that, although not reported in the table, the standard approach was not able to complete the state space exploration for $n \geq 20$.

Table 1. Potential spaces dimensions, memory consumption, time and number of cases where the KAES approach was successful, where μ reports the average over the 100 runs and σ is the standard deviation.

n	$|\mathcal{PS}|$	Memory (Gb)		Time (s)		% Solved
		μ	σ	μ	σ	Cases
10	59049	0.90	0.08	1.17	0.81	100%
20	$3.49 \cdot 10^9$	3.07	9.68	65.83	346.24	100%
30	$2.06 \cdot 10^{14}$	8.31	19.40	193.29	619.63	91%
40	$1.22 \cdot 10^{19}$	4.42	9.97	140.89	477.67	91%
50	$7.18 \cdot 10^{23}$	7.79	17.27	299.44	840.78	84%

The method is able to solve the great majority of cases, although the rate of success decreases as the number of components increases. For n equal to 10 and 20, all cases are solved, and the lowest percentage is 84 for the most populated scenario ($n = 50$). An important observation is that time and memory consumption seem to depend on the adopted topology, and in fact they can vary significantly for the different topologies generated for a given n, as confirmed by the values of the standard deviations reported in Table 1. However, it is not straightforward to understand the phenomena leading to this result, namely whether it is strictly related to the theoretical definition of the KAES method or to its implementation (especially, how the rounding is performed since the adopted toolbox for this procedure is a general one), or to both of them. Further investigations are necessary to shed light on this aspect, so to promote refinements in the KAES methodology and/or implementation.

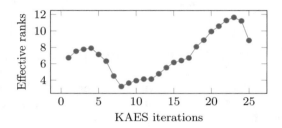

Fig. 2. Evolution of the effective ranks, representing an average of the TT-ranks of the carriages, for each iteration of KAES, with $n = 20$ and a specific topology.

To confirm the low memory consumption, in Fig. 2 the evolution of the ranks (represented using the "effective ranks", a single number that measures an average of the TT-ranks in the various modes) is reported. The ranks are considerably small, compared to $|\mathcal{PS}|$, and for all the experiments evolve in a similar way. This is a strong indicator that KAES has been well conceived as an efficient solution method.

9   Conclusions and Future Work

This paper addressed analytical modeling of large, interconnected systems by developing a new numerical evaluation approach, called KAES, to efficiently evaluate the Mean-Time-To-Absorption in CTMC with absorbing states. Resorting to powerful mathematical theories, properly combined, the symbolic representation of both the descriptor matrix and the descriptor vector is pursued to mitigate the explosion of the state space when evaluating the stochastic model. Although symbolic representation has been already applied in existing studies, such previous works focus on steady-state analysis while KAES targets limiting analysis in presence of absorbing states.

KAES has been implemented in the MATLAB evaluation environment and compared with traditional numerical solution when applied to a representative case study for the evaluation of the MTTA. Although preliminary and restricted to the studied scenario, obtained results clearly show the feasibility of KAES at increasing the size of the system under analysis, while standard numerical approaches fail due to the generated state space being too large. Moreover, the way the measure is computed guarantees a conservative assessment, which is relevant when dealing with dependability critical applications.

Of course, more experiments are needed to better understand strengths and limitations of this new technique in a variety of system scenarios, at varying both the system topology and the parameters setting. In particular, a deeper understanding of the link between TT-ranks and the topology of interactions among system components would be desirable, since the memory consumption is strictly related to TT-ranks. This extended evaluation campaign is already in progress. The obtained outcomes are expected to trigger improvements at methodology and/or implementation level.

Further and most important, the powerfulness of the adopted techniques and the conceived organization of the KAES steps make this method not restricted to the evaluation of the MTTA measure only, but adaptable to evaluate general performability related indicators. In fact, a straightforward generalization of KAES is represented by the substitution of the all-ones-vector $\mathbb{1}$ in Eq. (8) with a more general reward vector r, to promote the evaluation of other performance and dependability properties of single absorbing state CTMC, expressed as cumulative measures over the interval $[0, \infty)$. Whenever r can be expressed in terms of AND and OR conditions based on $r^{(i)}$, defined on M_i, it is possible to write r in terms of Kronecker products and sums, and apply KAES as presented in this paper.

References

1. Ajmone Marsan, M., Balbo, G., Chiola, G., Conte, G., Donatelli, S., Franceschinis, G.: An introduction to generalized stochastic Petri nets. Microelectron. Reliab. **31**(4), 699–725 (1991)
2. Babar, J., Beccuti, M., Donatelli, S., Miner, A.: GreatSPN enhanced with decision diagram data structures. In: Applications and Theory of Petri Nets, pp. 308–317 (2010)
3. Braess, D., Hackbusch, W.: Approximation of $1/x$ by exponential sums in $[1, \infty)$. IMA J. Numer. Anal. **25**(4), 685–697 (2005)
4. Brenner, L., Fernandes, P., Sales, A., Webber, T.: A framework to decompose GSPN models. In: Ciardo, G., Darondeau, P. (eds.) ICATPN 2005. LNCS, vol. 3536, pp. 128–147. Springer, Heidelberg (2005). https://doi.org/10.1007/11494744_9
5. Buchholz, P., Dayar, T., Kriege, J., Orhan, M.C.: On compact solution vectors in Kronecker-based Markovian analysis. Perform. Eval. **115**, 132–149 (2017)
6. Buchholz, P., Kemper, P.: Kronecker based matrix representations for large Markov models. In: Baier, C., Haverkort, B.R., Hermanns, H., Katoen, J.-P., Siegle, M. (eds.) Validation of Stochastic Systems. LNCS, vol. 2925, pp. 256–295. Springer, Heidelberg (2004). https://doi.org/10.1007/978-3-540-24611-4_8
7. Ciardo, G.: Data representation and efficient solution: a decision diagram approach. In: Bernardo, M., Hillston, J. (eds.) SFM 2007. LNCS, vol. 4486, pp. 371–394. Springer, Heidelberg (2007). https://doi.org/10.1007/978-3-540-72522-0_9

8. Ciardo, G., Miner, A.S.: A data structure for the efficient Kronecker solution of GSPNs. In: Proceedings 8th International Workshop on Petri Nets and Performance Models (Cat. No. PR00331), pp. 22–31 (1999)
9. Ciardo, G., Tilgner, M.: On the use of Kronecker operators for the solution of generalized stochastic Petri nets. Nasa Technical Report Server 20040110963 (1996)
10. Czekster, R.M., Fernandes, P., Vincent, J.-M., Webber, T.: Split: a flexible and efficient algorithm to vector-descriptor product. In: VALUETOOLS, pp. 83:1–83:8 (2007)
11. Dolgov, S.V., Savostyanov, D.V.: Alternating minimal energy methods for linear systems in higher dimensions. SIAM J. Sci. Comput. **36**(5), A2248–A2271 (2014)
12. Donatelli, S.: Superposed stochastic automata: a class of stochastic Petri nets with parallel solution and distributed state space. Perform. Eval. **18**(1), 21–36 (1993)
13. Donatelli, S.: Superposed generalized stochastic Petri nets: definition and efficient solution. In: Application and Theory of Petri Nets 1994, pp. 258–277 (1994)
14. Goševa-Popstojanova, K., Trivedi, K.: Stochastic modeling formalisms for dependability, performance and performability. In: Haring, G., Lindemann, C., Reiser, M. (eds.) Performance Evaluation: Origins and Directions. LNCS, vol. 1769, pp. 403–422. Springer, Heidelberg (2000). https://doi.org/10.1007/3-540-46506-5_17
15. Grasedyck, L.: Existence and computation of low Kronecker-rank approximations for large linear systems of tensor product structure. Computing **72**(3–4), 247–265 (2004)
16. Kemper, P.: Numerical analysis of superposed GSPNs. IEEE Trans. Softw. Eng. **22**(9), 615–628 (1996)
17. Kressner, D., Macedo, F.: Low-rank tensor methods for communicating Markov processes. In: Norman, G., Sanders, W. (eds.) QEST 2014. LNCS, vol. 8657, pp. 25–40. Springer, Cham (2014). https://doi.org/10.1007/978-3-319-10696-0_4
18. Kressner, D., Tobler, C.: Krylov subspace methods for linear systems with tensor product structure. SIAM J. Matrix Anal. Appl. **31**(4), 1688–1714 (2010)
19. Kutzelnigg, W.: Theory of the expansion of wave functions in a Gaussian basis. Int. J. Quant. Chem. **51**(6), 447–463 (1994)
20. Masetti, G., Robol, L.: Tensor methods for the computation of MTTF in large systems of loosely interconnected components. Technical report, ISTI-CNR Open Portal (2019). http://dcl.isti.cnr.it/tmp/tchrep-RtCv-63_CtAx_Ol19_jEN5.pdf
21. MathWorks. MATLAB R2018a. The Mathworks Inc. (2018)
22. Murata, T.: Petri nets: properties, analysis and applications. Proc. IEEE **77**(4), 541–580 (1989)
23. Oseledets, I.: DMRG approach to fast linear algebra in the TT-format. Comput. Methods Appl. Math. **11**(3), 382–393 (2011)
24. Oseledets, I.V.: Tensor-train decomposition. SIAM J. Sci. Comput. **33**(5), 2295–2317 (2011)
25. Plateau, B., Fourneau, J.-M., Lee, K.-H.: Peps: a package for solving complex Markov models of parallel systems. In: Puigjaner, R., Potier, D. (eds.) Modeling Techniques and Tools for Computer Performance Evaluation, pp. 291–305. Springer, Boston (1989). https://doi.org/10.1007/978-1-4613-0533-0_19
26. Sanders, W.H., Meyer, J.F.: A unified approach for specifying measures of performance, dependability and performability. Dependable Computing for Critical Applications. Dependable Computing and Fault-Tolerant Systems, vol. 4, pp. 215–237. Springer, Vienna (1991). https://doi.org/10.1007/978-3-7091-9123-1_10
27. Trivedi, K.S., Bobbio, A.: Reliability and Availability Engineering: Modeling, Analysis, and Applications (2017)

Tools

RenewKube: Reference Net Simulation Scaling with Renew and Kubernetes

Jan Henrik Röwekamp[✉] and Daniel Moldt

Faculty of Mathematics, Informatics and Natural Sciences,
Department of Informatics, University of Hamburg, Hamburg, Germany
roewekamp@informatik.uni-hamburg.de
http://www.informatik.uni-hamburg.de/TGI/

Abstract. When simulating reference nets, the size (places, transitions; memory and CPU consumption) of the simulation is usually not known before actual runtime. This behavior originates from the concept of net instances, which are similar to objects in object-oriented programming. The simulator Renew supports very basic distribution but the manual infrastructural setup for simulations exceeding the capabilities of one machine is left up to the modeler until now. In this work the RenewKube tool, a ready to use Kubernetes and Docker based solution, is presented, that allows to control automated scaling of simulation instances from within the net running in the Renew simulator.

Keywords: Petri net tool · High-level petri nets · Reference nets · Distributed computing · Scalability · Docker · Kubernetes

1 Introduction

In the context of simulating concurrent systems with Petri nets, a lot of different net formalisms were presented in the past decades. One of them - reference nets - allow for direct execution of a given model, as the formalism itself is expressive enough to host Java-like inscriptions in its transitions. Being a concept close to object-oriented programming, reference net simulations may vary in size during their runtime.

Excessively large simulations may at some point exceed the capabilities of the local machine they are running on. In research history there have been several approaches to run distributed versions of reference nets. The most prominent one is the Distribute plugin for the Renew simulator [16]. The setup of the infrastructure, installation and dependencies however can be tedious work. This even holds for adding a single additional instance of Renew. Our previous publications already addressed the setup problematic by wrapping Renew into virtual machines [15] as well as Docker containers [13,14].

To our best knowledge there is currently no available tool, that offers the (semi-)automated scaling of reference net simulations, especially not controlled

© Springer Nature Switzerland AG 2019
S. Donatelli and S. Haar (Eds.): PETRI NETS 2019, LNCS 11522, pp. 69–79, 2019.
https://doi.org/10.1007/978-3-030-21571-2_4

from within the simulation itself. RENEWKUBE[1] was developed to address this problem and allow for simulations, that are aware of their own scaling as well as being able to influence it. RENEWKUBE is based on Docker and Kubernetes [2].

1.1 Motivation and Scope

Due to the underlying concept of reference nets, the simulation itself is not static in size[2]. Net instances may be created freely during runtime, giving the modeler freedom to add as many transitions and places during runtime as he or she needs for modeling. Obviously, as the simulation size is not limited by the formalism like with most other net formalisms, the simulation size may at some point exceed the capabilities of a single machine.

Distributing net simulations was intensively researched in the past 20 years and results will be addressed in Sect. 3. This work, however, has a different focus and is built on top of these results.

So far the setup of the infrastructure to host a multi-instance simulation was left to the modeler. As it involves numerous tedious manual tasks, the overhead to run distributed reference net simulations (if possible at all) was considerable.

This work presents an easy to use tool, that heavily assists in running a distributed simulation. It is currently available as one-line command per node setup for Ubuntu (based) Linux distributions but works with other distributions and possibly Mac OS as well. It is also built with execution on cloud providers like Amazon AWS[3] and similar in mind. While the technology used should be natively compatible, this has not been tested and or evaluated yet, and therefore is not yet recommended to attempt.

Section 3 will briefly cover related research topics. Sections 4 to 6 present the tool itself, with Sect. 4 presenting the simulator itself, Sects. 5 and 5 addressing architectural details, and finally Sect. 6 discussing advantages and limitations. The contribution is concluded in Sect. 7.

2 Basics

The technologies RENEWKUBE is built on top of, are briefly outlined here.

Reference Nets. Reference nets are a high-level Petri net formalism introduced by Kummer [9] back in 2002, integrating Object Petri nets [18] and Object-oriented Petri Nets [10]. They follow the nets with in nets formalism [18] and are structured hierarchically. The key idea is, that tokens are treated as references to arbitrary objects or nets. Reference nets further introduce the concept of *net instances*, which is inspired by objects in object-oriented programming.

---

[1] Available at https://paose.informatik.uni-hamburg.de/paose/wiki/RenewKube.

[2] *size* refers to utilized CPU, RAM usage and also (but not primarily) hard disk usage.

[3] https://aws.amazon.com/.

Net instances may be created by transition inscriptions during simulation run-time. Therefore, for a given time a net can exist in a multitude of different variants.

Reference nets become powerful and expressive through the concept of *synchronous channels* [5,9]. A transition may be member of one or more channels. A net instance A holding the reference to another net instance B may fire a transition in A in synchronization with a transition in B, as long as they share a channel. Channels are identified by a name string.

To further illustrate the concept of reference nets consider the following example in Fig. 1. An initial net instance of NetA may fire the upper transition first, consuming an unmarked token and generating a new net instance of NetB, placing a reference to it in the upper right place. Afterwards either the transition may fire again and create a second net instance of NetB, or the transition at the bottom may fire. The transition is inscribed with a so called *downlink* `ni:ch(d)`, which means to synchronize with the net instance `ni` on the channel `ch` and unify the variable `d`. NetB has a transition, that is inscribed with an *uplink* `:ch(input)`, which is its counterpart. Note that the data transfer direction is *not* dependent on the concepts of up- or downlinks. In the example the synchronization is possible and the transitions fire simultaneously. `input` in NetB is then unified with the data of variable `d` of NetA.

Ultimately the net system blocks with a total of three net instances, with two references to net instances of NetB in the rightmost place of the single NetA instance and "data1" and "data2" tokens, one in each of the NetB instances. It cannot be determined which one ends in which instance of NetB before the actual execution and it may be different after each restart of the simulation.

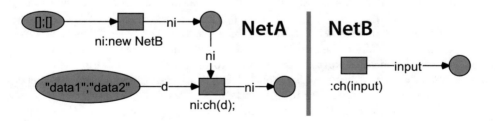

Fig. 1. A simple reference net example

Renew. The *Reference Net Workshop* (RENEW) was developed by the TGI research group (nowadays ART) at the Department of Informatics of the University of Hamburg. It features a Java based simulator implementation of the reference net formalism. Several additions and features have been and are developed for it.

The core simulator uses a concurrent but only local simulation engine. There have been however, different implementations of distributed RENEW simulations with the one used here being the Distribute RENEW plugin [16]. For further information see Sects. 3 and 4.

RENEW comes with a WYSIWYG user interface, the modeler can use to model concurrent systems. Upon completion the produced net system can directly be launched as a simulation.

Docker. Docker [2] is a Linux container technology[4] and is maintained by Docker Inc[5]. The key concepts of Docker are *images* and *containers*. It is important to understand these, as they are used in the remainders of the paper.

An *image* is a collection of binaries and is usually derived from other images. In a typical use case the image captures the minimal required software (libraries, runtimes, etc.) to run an enclosed application. Images are stored in a registry and can be pushed or pulled to target machines by a Docker daemon.

The Docker daemon may launch images as isolated processes on the system, which is then called a *container*. Containers also do not persist any data (locally) at all, though exceptions from this rule can be defined using *volumes*.

Kubernetes. Kubernetes is a cluster management system initially developed by Google, which donated it to the Cloud Native Computing Foundation[6]. Its main functionality, that is used by RENEWKUBE, is its ability to remotely launch and shut down containers. Kubernetes offers a variety of services, that are too diverse to cover here in their entirety, so only the relevant concepts for this work will be introduced:

- A *Pod* may host one or more container(s) and is the atomic unit of Kubernetes. Pods are a volatile structure and may be shut down or launched at any given time.
- A *ReplicaSet* defines a total number of pods, that the entire cluster is supposed to run. When the actual number of pods differs from the desired number, the ReplicaSet starts or stops pods to match these numbers.
- A *Deployment* usually holds a related ReplicaSet and offers more advanced and high level features like rolling updates and the like.
- A *Service* offers a kind of access layer by exposing certain parts of the cluster to either other parts, or the outside. They can target one or more Pods or Deployments. As Pods are volatile, mapping Services to Pods requires special attention and is not viable by built-in Kubernetes features.

Java RMI. Java RMI is the Java implementation of the well known remote method invocation protocol (see e.g. [11]). Currently the best available implementation of distributed reference nets (Distribute plugin [16]) is internally built on top of Java RMI. Therefore it becomes a prerequisite for RENEWKUBE in its

---

[4] In recent implementations there is Windows support as well.
[5] https://www.docker.com/.
[6] https://kubernetes.io/.

current state as well, although nowadays Java RMI is certainly not state of the art any more. While most inner workings of Java RMI are not relevant here, it is important to note, that it expects to talk to the exact same node for an established connection, as well as using non-trivial (random) port management per default, making the setup of the cluster a bit more complicated. See Sects. 5 and 6 for further information on this topic. Java RMI is transparent for the end user of RENEWKUBE.

3   Related Work

During the past years several related papers have been published, with the most relevant being briefly addressed here.

Simulation of Petri nets was researched early using place [6] or transition simulation, as well as communicating sequential processes [17]. Soon after attempts to distribute the simulation were made either as basis for distributed code fragments [7,8] or directly as net simulation [4]. A few years later reference nets were introduced by Kummer [9]. As an example for more recent publications, a net based library for parallel computing was released in 2017 [12].

The combination of cloud computing and RENEW has been addressed by Bendoukha [1]. However, remote instances were run in isolation and did not feature a full multi-machine reference net simulation. Simon [16] introduced the Distribute plugin for RENEW, that serves as the current base for RENEWKUBE. Important groundwork was done regarding the analysis of virtual machine and Docker based solutions and was published by Röwekamp [14,15]. Being early steps, these solutions did not solve the problem of (semi-)automatic scaling like RENEWKUBE does.

Architectural thoughts in the direction of REST service based distributed simulations were presented in [13] and will - once implemented - become the successor of the Distribute plugin to serve as base for RENEWKUBE.

Buchs et al. considered Docker containers for distributed model checking in 2018 [3]. The primarily addressed topic is transportability of model checkers and machine learning based tool choice but not tool-internal scaling.

4   The Simulator

The simulator interface itself is identical to RENEW. A usage guide of RENEW is available at its main website[7].

How to Obtain. RENEWKUBE can be downloaded from our website alongside with further instructions on how to install and set it up:

https://paose.informatik.uni-hamburg.de/paose/wiki/RenewKube

---

[7] http://www.renew.de.

4.1   Using the Simulator

As the major part of the simulator is RENEW, the usage is similar to the RENEW base package. The usage of the Distribute plugin is a bit different and is therefore briefly summarized here. Full instructions and workings of the plugin can be found in the respective publication [16]. Note, that these syntax and formalism changes originate from the Distribute plugin alone and are by that only indirectly related to RENEWKUBE.

Distribute Plugin. To prepare for distributed execution, a net instance needs to register itself in the RMI registry. Also, to synchronize transitions across machine borders a remote net instance needs to be known locally. Both is achieved using the transition inscriptions `DistributePlugin.registerNetInstance(...)` and `DistributePlugin.getNetInstance(...)` respectively. The exact syntax can be found in examples or [16].

Synchronization is then performed using the ! symbol instead of the usual : symbol on the downlink. Also due to limitations of the Distribute plugin it needs to be specified in which direction data transfer will happen. In contrast to the classic RENEW the syntax differs when sending data down the link or up the link[8].

To send a data token `data` from the net holding the remote net instance `r` over the channel `ch` the classic-like syntax: `r!ch(data);` and `:ch(data)` is used. To send data up the link the following syntax is used on the uplink part: `:ch() <- data` and on the downlink part: `r!ch() -> data`.

Another technical limitation arises at this point: Passed data needs to be Java serializable, otherwise it cannot be sent to another remote instance by RMI.

RENEWKUBE. RENEWKUBE itself - once set up - is almost invisible for the modeler in terms of syntax. The main interaction with RENEWKUBE is done with the `getScale();` and `setScale(<number>);` inscriptions. These can be used to get the currently available RENEW replicas in the cluster and to set a desired amount.

Another thing to bear in mind is, that due to the distributed nature of the simulation any remote net instance may cease to exist (or become unreachable) at any given time. Therefore, if tokens referring to remote workloads are used, it is advisable to only finally consume these upon completion of remote subtasks. This way it is possible to assign the workload to another remote instance, if the first instance ceases to exist. It might also be desirable to use timeouts in this context. Depending on the reliability of the underlying infrastructure a loss of remote net instances may occur more or less often.

Another necessity is to declare a net to be launched as initial remote net instance (the one the remote simulator fires up with). For now this net needs to be named *remoteStart*.

---

[8] For definition of up- and downlinks see Sect. 2.

5   Architecture

In this section a brief overview of the high-level architecture is given.

Overview. The core of RENEWKUBE consists of three types of nodes: the
RENEW with the graphical interface (following referred to as "User node"), the
cluster master and n additional worker nodes. The user node and the worker
nodes each hold an instance of the RENEW simulator, the cluster master only
addresses administration overhead. Cluster master and worker nodes run inside
a protected enclosed environment in the following referred to as the "cluster",
while the user node is positioned outside of the cluster on the users workstation.
The overview of the architecture can be found in Fig. 2.

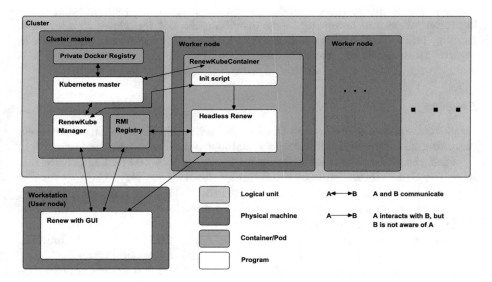

Fig. 2. Architecture of RENEWKUBE

User Node. The user node is a single running Java application, that directly
inherits the architecture of the RENEW simulator, discussed e.g. in [9]. Besides
the plugin for basic distribution it also hosts an adapter plugin responsible
for communication to the cluster master. The adapter plugin handles scaling
requests by the net system as well as reporting back the current state of the
cluster.

The additional software for the user node was implemented as modification of
the RENEW simulator, so the user is only required to launch a single executable
on their workstation, while allowing tight coupling with the running simulator
engine.

Worker Node. The worker node is the most simple component. It stays on standby until it gets recruited by the cluster master. Upon recruitment an init script will inject nets to simulate into its own copy of the RENEW simulator, requests exposure of the to be created RENEW instance from the RENEWKUBE manager (see below) and finally fires it up. The RENEW simulator then will use the distribution plugin to proceed to blend into the running simulation and extending the simulations (physical) scope.

Cluster Master. The cluster master is the most complex component. It itself consists of four services/components: The container manager (that is provided by Kubernetes) alongside with a network abstraction layer, a private Docker image registry, the Java RMI registry and finally an integration service called RENEWKUBE manager, that handles multiple things like the integration of additional physical nodes, authorization against the Kubernetes manager and relaying information and accepting tasks, providing the net templates to additional running simulations and handling individual RENEW extension containers in regards to accessibility from the user node. The RENEWKUBE manager is a stateless application based on Java Spring and the REST technology.

Process

To illustrate communication and tasks of the individual components, the process regarding installing and launching a scalable simulation is described here.

First the Kubernetes cluster manager is set up, followed by the private registry, the RENEWKUBE manager and finally the RMI registry. The private registry will be fed an image of a headless[9] RENEW alongside with an init script. After that additional worker nodes may be added to the cluster. Additional worker nodes use the RENEWKUBE manager to authorize against the master once and then are managed by Kubernetes. All of this however only happens once upon initial installation and is omitted in already set up clusters.

Now, the workstation version of RENEW is launched and upon start of a simulation, the current net system is packed and sent to the RENEWKUBE manager, which stores it in memory. Then it fires up a new *Deployment* referring to the *image* created earlier in the local Docker registry.

When more remote RENEW instances are desired, the Kubernetes master launches a new *Pod* by *ReplicaSet* reconciliation. The init script inside the *Pod* queries the RENEWKUBE manager for the current net system and also identifies itself, so the RENEWKUBE manager can set up an exposure *Service* directly for the *Pod*. This is necessary, because Java RMI requires to communicate with the same (not an equivalent) instance of a remote application every time. Therefore static routes are necessary and it is not sufficient to map an exposure *Service* to the *Deployment*. The RENEWKUBE manager uses a Kubernetes feature called *NodePorts*, that allows the *Service* (*Pod*) to be accessed on any (worker or master) node of the cluster on the same port. When the *Pod* is not hosted on the

---

[9] Headless in this context refers to running without graphical user interface.

queried node, the request is automatically and transparently proxied to the correct *node* inside the *cluster*.

The init script then proceeds to prepare the net system for the RENEW instance and launches it with the respective parameters. After that the both RENEW instances may communicate directly and unaware of the Kubernetes overhead as if they were launched on different machines by hand.

6 Evaluation

The presented simulators primary contribution is the scaling ability. The ability to scale a simulation adds great possibilities to writing simulations, like for example simulating large organizations, computing large reachability graphs, model complex agent systems with numerous agents and many more. The overhead introduced by the underlying technology (Kubernetes, Docker) might be intimidating at first but once set up it is almost entirely covered by RENEWKUBE.

A simulator with the provided functionality built right into the tool itself is to our best knowledge not yet available. Therefore RENEWKUBE lays an important foundation for large scaled Petri net simulations.

Limitations

Like every complex tool, RENEWKUBE comes with a number of limitations. Currently the simulator only allows one user node and each physical node may only run up to 100 headless RENEW instances. However, a single or a few instances usually do a good job in utilizing a local multicore system.

Another limitation arises from Java RMI, that was developed long before the cloud computing era. Because of internal limitations currently the hard limit for an entire cluster is about 65,000 instances of RENEW (after subtracting potentially used system ports). A possible solution could be virtual IPs.

7 Conclusion and Outlook

A simulator for reference nets has been presented, that features fully integrated (semi)automatic scaling capabilities. By that a base is created, that easily allows for detailed reference net simulations of a large size, that were extremely difficult or impossible to set up before. It is built on state of the art scaling technology used by enterprises and with application to cloud computing provides in mind. The evaluation of these providers however, is left to further research.

The current state of RENEWKUBE allows scaling requests send from within the simulation. However, hiding the scaling altogether is certainly an interesting topic to address in future research.

References

1. Bendoukha, S.: Multi-agent approach for managing workflows in an inter-cloud environment. Dissertation, University of Hamburg, Department of Informatics, Vogt-Kölln Str. 30, D-22527 Hamburg (2017)
2. Bernstein, D.: Containers and cloud: from LXC to Docker to Kubernetes. IEEE Cloud Comput. **1**(3), 81–84 (2014)
3. Buchs, D., Klikovits, S., Linard, A., Mencattini, R., Racordon, D.: A model checker collection for the model checking contest using docker and machine learning. In: Khomenko, V., Roux, O.H. (eds.) PETRI NETS 2018. LNCS, vol. 10877, pp. 385–395. Springer, Cham (2018). https://doi.org/10.1007/978-3-319-91268-4_21
4. Chiola, G., Ferscha, A.: Distributed simulation of Petri nets. IEEE Parallel Distrib. Technol. **1**(3), 33–50 (1993)
5. Christensen, S., Damgaard Hansen, N.: Coloured Petri Nets extended with channels for synchronous communication. In: Valette, R. (ed.) ICATPN 1994. LNCS, vol. 815, pp. 159–178. Springer, Heidelberg (1994). https://doi.org/10.1007/3-540-58152-9_10
6. Hauschildt, D.: A Petri net implementation. Fachbereichsmitteilung FBI-HH-M-145/87, University of Hamburg, Department of Computer Science, Vogt-Kölln Str. 30, D-22527 Hamburg (1987)
7. El Kaim, W., Kordon, F.: An integrated framework for rapid system prototyping and automatic code distribution. In: Proceedings of RSP, Grenoble, France, pp. 52–61. IEEE (1994)
8. Kordon, F.: Prototypage de systèmes parallèles à partir de réseaux de Petri colorés, application au langage Ada dans un environment centralisé ou réparti. Dissertation, Université P & M Curie, May 1992
9. Kummer, O.: Referenznetze. Logos Verlag, Berlin (2002)
10. Maier, C., Moldt, D.: Object coloured Petri nets - a formal technique for object oriented modelling. In: Agha, G.A., De Cindio, F., Rozenberg, G. (eds.) Concurrent Object-Oriented Programming and Petri Nets. LNCS, vol. 2001, pp. 406–427. Springer, Heidelberg (2001). https://doi.org/10.1007/3-540-45397-0_16
11. Pitt, E., McNiff, K.: Java.Rmi: The Remote Method Invocation Guide. Addison-Wesley Longman Publishing Co., Inc, Boston (2001)
12. Pommereau, F., de la Houssaye, J.: Faster simulation of (coloured) Petri nets using parallel computing. In: van der Aalst, W., Best, E. (eds.) PETRI NETS 2017. LNCS, vol. 10258, pp. 37–56. Springer, Cham (2017). https://doi.org/10.1007/978-3-319-57861-3_4
13. Röwekamp, J.H.: Investigating the Java Spring framework to simulate reference nets with Renew. Number 2018–02 in Reports/Technische Berichte der Fakultät für Angewandte Informatik der Universität Augsburg, pp. 41–46. Universität Augsburg, Fachbereich Informatik (2018)
14. Röwekamp, J.H., Moldt, D., Feldmann, M.: Investigation of containerizing distributed Petri net simulations. In: Moldt, D., Kindler, E., Rölke, H. (eds.) Petri Nets and Software Engineering. International Workshop, PNSE 2018, Bratislava, Slovakia, 25–26 June 2018. Proceedings, volume 2138 of CEUR Workshop Proceedings, pp. 133–142. CEUR-WS.org (2018)
15. Röwekamp, J.H., Moldt, D., Simon, M.: A simple prototype of distributed execution of reference nets based on virtual machines. In: Proceedings of the Algorithms and Tools for Petri Nets (AWPN) Workshop 2017, pp. 51–57, October 2017

16. Simon, M., Moldt, D.: Extending Renew's algorithms for distributed simulation. In: Cabac, L., Kristensen, L.M. Rölke, H. (eds.) Petri Nets and Software Engineering. International Workshop, PNSE 2016, Toruń, Poland, 20–21 June 2016. Proceedings, volume 1591 of CEUR Workshop Proceedings, pp. 173–192. CEUR-WS.org (2016)
17. Taubner, D.: On the implementation of Petri nets. In: Rozenberg, G. (ed.) APN 1987. LNCS, vol. 340, pp. 418–439. Springer, Heidelberg (1988). https://doi.org/10.1007/3-540-50580-6_40
18. Valk, R.: Petri nets as token objects - an introduction to elementary object. In: Desel, J., Silva, M. (eds.) ICATPN 1998. LNCS, vol. 1420, pp. 1–24. Springer, Heidelberg (1998). https://doi.org/10.1007/3-540-69108-1_1

PNemu: An Extensible Modeling Library for Adaptable Distributed Systems

Matteo Camilli$^{(\boxtimes)}$, Lorenzo Capra, and Carlo Bellettini

Department of Computer Science, Università degli Studi di Milano, Milan, Italy
{camilli,capra,bellettini}@di.unimi.it

Abstract. PNEMU is an extensible Python library primarily tailored for modeling adaptable distributed discrete-event systems by means of standard (Low- and High-level) Petri nets. The core of PNEMU is composed of a number of modules for the editing and interactive simulation of models. In particular, it supplies a number of off-the-shelf building blocks to easily formalize self-adaptation having a decentralized control through multiple interacting feedback loops. PNEMU can be used in conjunction with other software tools to efficiently compute the state space and perform formal verification activities. This paper describes the PNEMU structure, features, and usage.

Keywords: High-Level Petri nets · Self-adaptation · Decentralized control · Modeling · Simulation

1 Introduction and Objectives

Software tools play a crucial role in developing modern/complex distributed systems dealing with different operational conditions due to dynamically changing environments. In particular, guaranteeing important qualities such as correctness, safety, robustness, dependability, in such systems is a challenging task.

Self-adaptive systems [12] can dynamically and reactively evaluate their own (execution) context and adjust their behavior accordingly. A popular approach to realize adaptation is using a feedback control loop [9]. The control loop continuously reasons about the current state of the system itself and the surrounding environment to take proper actions. In a distributed (decentralized) setting, different control components may operate concurrently, possibly leading to conflicting/undesired situations. Facing such a complexity both in design and verification phases has been recognized as a major challenge in the field of self-adaptation [1]. Thus, software tools (based on formal methods) supporting both phases are highly demanded.

In this paper, we introduce an extensible Python library named PNEMU[1].

PNEMU is a modeling/simulation library for adaptable distributed discrete-event systems, having decentralized adaptation control, that leverages the theoretical foundation of High-Level Petri nets (HLPNs) [7]. Petri nets (PN)

---

[1] Available as open source software at https://github.com/SELab-unimi/pnemu.

© Springer Nature Switzerland AG 2019
S. Donatelli and S. Haar (Eds.): PETRI NETS 2019, LNCS 11522, pp. 80–90, 2019.
https://doi.org/10.1007/978-3-030-21571-2_5

are a sound and expressive (in some cases Turing complete) formal model for distributed systems. Classical PNs are, however, not adequately equipped (even in their High-level flavors) for representing self-perception and self-adaptation dynamics, typically seen in modern/complex distributed systems. Several attempts to bridge this gap have given rise to PN extensions, most with complex algebraic/functional annotations, where an enhanced modeling capability is very often not adequately supported by convenient analysis techniques and software tools (see [2,4] for a more detailed state-of-art discussion), due to an unclear/unintuitive semantics that limits the applicability of traditional PN analysis techniques and formal verification activities (such as model checking).

Of particular interest are those approaches combining higher-order tokens and the features of object-oriented languages. A representative of this class of formalisms is Reference Nets [8], supported by the RENEW tool. Even though inspired by the same principles, the PNEMU blueprint relies instead on consolidated PN formalisms. In fact, the operational semantics of models built by using PNEMU is given by a HLPN with first-order tokens (following the ISO/IEC standard [6]).

The major objectives during the development of PNEMU have been:

- *Clarity*: the modeling approach should provide a clear abstraction of dynamics associated with self-adaptive distributed systems in order to supply an easy-to-understand API, following a clear separation of concerns.
- *Extensibility*: the library should provide a baseline that can be easily extended by other developers, or exploited to build other tools.
- *Interoperability*: working with other tools should be easy, in order to exploit them to complement the currently implemented basic functionalities.

PNEMU builds on the modeling approach introduced in [2,3]. In this paper, we primarily focus on engineering aspects, such as design and implementation. concerns. We also describe the main functionalities, in particular, how the framework formalizes multiple distributed control loops by means of HLPNs The interoperability of the library with third-party software tools (in order to carry out formal verification activities) is also briefly discussed.

In Sect. 2 we sketch a few core background concepts. In Sect. 3 we describe the main elements of the PNEMU library through a simple example. In Sect. 4 we outline a number of usage scenarios. In Sect. 5 we report our conclusion and ongoing work.

2 Preliminaries

Distributed Self-adaptation. The term Self-adaptation [12] is usually employed to characterize systems that can autonomously adapt their behavior at runtime. Self-adaptive systems are able to perceive contextual changes and (re)organize their own features, services, or even structure, in response to these changes. Many models of adaptation assume a single centralized (feedback) control loop that observes the execution context and possibly changes the running system. In essence, a control loop is able to sense the state of the managed subsystem/environment by reading data from a monitor component. Gathered data are analyzed to check whether some adaptation is required. In that case, specific actions are planned, then executed by actuators.

When multiple *adaptation concerns* (or goals) [9] have to be considered (e.g, efficiency, reliability, security, etc.), a single centralized loop may not be sufficient to manage the growing complexity [1]. A way to face it is to concurrently run multiple loops over distributed components. These loops operate and communicate through a shared knowledge in a fully decentralized setting, possibly leading to conflict situations. The two main components of a distributed self-adaptive system are the *base* layer, containing the *managed subsystem* (implementing the application logic) and the surrounding *environment*, and the *managing* layer, on top of it, implementing the adaptation logic through a number of interacting feedback loops.

Petri Net Formalisms. The PNEMU framework employs two formalisms, corresponding to the aforementioned layers: the low (or base) layer is defined by a P/T system [11] (enriched with inhibitor arcs) which represents the main functionalities of a system; the high (managing) layer, in which self-adaptation aspects are gathered, is defined by interacting High-level PN (HLPN) components. The PNEMU library relies upon HLPNs as defined in the SNAKES library [10], a Python implementation of HLPNs (according to [6]). In SNAKES, tokens are Python objects, transitions are guarded by Python Boolean expressions, and arcs are annotated with Python expressions. SNAKES can be used to either perform a step-by-step simulation or exhaustively explore the state space.

3   The PNemu Library

PNEMU is primarily tailored to model distributed self-adaptive discrete-event systems by means of a clear separation of concerns. The user defines in a modular fashion an HLPN model whose architecture follows the reference two-layer model described above. Both the *base* and the *managing* layers can be specified by means of an easy-to-use API.

The rationale of our modeling approach is as follows. The base layer is formalized by a P/T system representing the common dynamics of the managed sub-system (and the surrounding environment). The P/T system is automatically encoded as the initial marking of a "special" HLPN, called *emulator*, exactly reproducing its behavior. By acting on the emulator's marking is it possible to introspect and, if needed, change the current state/structure of the emulated system. Any interaction with the emulator is safely performed through high-level subnets (using read/write library primitives) connected to the emulator by the PNEMU framework. These subnets correspond to the feedback loops of the self-adaptive system and compose the managing layer.

The library is implemented as a Python package on top of SNAKES. Figure 1 shows a static view of the PNEMU architecture. The main modules are as follows:

- **pnemu.base** contains the `Emulator` component and the structures needed to define a P/T net.
- **pnemu.manager** contains the components needed to model/execute the managing layer. In particular, the `FeedbackLoop` class allows control loops to be defined in terms of HLPNs. The class `AdaptiveNetBuilder` is used instead to connect the control loops to the emulator in order to build the overall adaptable system model.
- **pnemu.primitives** provides a number of elementary HLPNs abstracting the notions of *sensors* and *actuators* used by control loops. PNEMU comes along with a number of pre-defined `LibEntry` instances making up a basic, yet complete, set of sensors/actuators. The latter operate atomically, ensuring a consistent encoding of the base-layer.

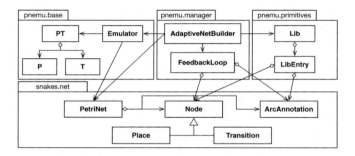

Fig. 1. UML Class diagram of the PNEMU Python package.

In the next sections, we describe the base usage of the library through a simple example of self-healing Manufacturing System (MS), whose nominal behavior is described by the P/T system in Fig. 2a. In this model, two symmetric production lines (transitions `line1`, `line2`) refine raw pieces that are then assembled to get the final product (transition `assemble`). Raw pieces are loaded from a storage into either line, two at once (transition `load`). Once all of them have been worked, the system restarts (transition `reset`). A production line is subject to failures, hence the model includes the specification of a faulty behavior (transition `fail`). A typical adaptation scenario (taking into account the *fault tolerance* adaptation concern) involves the reconfiguration of the MS upon a fault (realistically, without shutting the MS down), so that it can continue working with the available production line. The faulty line is detached from the MS, and the behavior of both the loader and the assembler is adapted accordingly. Figure 2b shows the reconfigured MS. Carrying out adaptation without blocking the system execution is far from intuitive and can lead to unexpected and/or undesired situations, such as raw piece loss or, even, deadlocks.

The complete specification of this example can be found along with the additional material provided inside the PNEMU repository.

3.1 The pnemu.base Module

This module allows the model's base layer (e.g., the P/T system in Fig. 2a) to be defined. This can be done by instantiating class PT, either by loading an existing PNML file or by editing step-by-step the base-layer net. Class PT also provides the means to carry out the interactive simulation, state-space exploration, and visualization of the base layer.

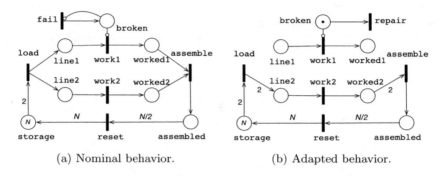

(a) Nominal behavior. (b) Adapted behavior.

Fig. 2. A fault-tolerant manufacturing system.

Once the base layer has been defined, we need to initialize the *emulator* component, that reifies the base layer into the high layer. An instance of class Emulator implements the HLPN shown in Fig. 3, whose marking encodes the base layer. Each place of the emulator, in fact, matches an element of the definition of a P/T system. As an example, place mark holds a multiset of type-P tokens which represents the P/T system's current marking. The guard of transition move and the annotations on incident arcs encode the enabling and firing rule for P/T systems, respectively [2]: there is an exact correspondence between enabled firing modes of move and base layer's transitions enabled in the P/T marking encoded by place mark.

Figure 4 (line 3) shows the usage of PNEMU to perform these preliminary steps, i.e., base-layer definition and subsequent emulator initialization.

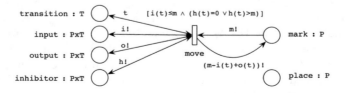

Fig. 3. The emulator model. The operator "!" denotes the *flush* function.

3.2 The `pnemu.manager` Module

The components of this module are used to specify (arbitrarily complex) feedback loops in terms of HLPNs. Each loop interacts with the base layer by means of *read/write* primitives acting on the emulator's marking. This way, the managing layer can monitor-analyze the execution of the managed subsystem, then plan-execute changes accordingly, in order to achieve specific adaptation goals. A loop is incrementally built by defining the structure and the annotations of a HLPN, with intuitive method-calls (Fig. 4, lines 6–15). For instance the net `loop` (line 6) contains the places `init` and `sample`, connected by a transition whose label "`lib.getTokens(p)`→`n`" denotes a (call to a) *read* library primitive having the corresponding signature. The firing of this transition causes the assignment of the current marking of `p` (the base layer place passed as an argument) to the output variable `n` (line 12). This primitive is used to sample the status of the faulty line, upon which (i.e., when place `broken` is marked) the managed subsystem reconfigures itself.

Transitions having labels `rmOArc` and `addOArc` (lines 14–15) are instead two examples of *write* primitives (i.e., *actuators*). The former is used to temporarily detach the faulty line (by removing the arc from `load` to `line1`). The latter makes the pair of loaded pieces enter the available line (by increasing the weight of the input arc from `load` to `line2`). Figure 5a shows a portion of the HLPN that formalizes the feedback loop. It is worth noting that a change on the base layer is triggered just after verifying the presence of a token in `broken`. In such a case, loading of pieces is temporarily suspended (by means of `addHArc` primitive), pending pieces on faulty line are moved to the other line, and so on, until the final reconfiguration, shown in Fig. 2b, is eventually reached. As already mentioned, all the changes operated by means of write primitives take place concurrently with base layer emulation.

In our example, the specification of the managing subsystem contains another loop, taking into account the *load balance* concern: this loop allows the MS nominal behavior to be restored after the faulty line has been repaired, so that the raw pieces can be evenly distributed among the available production lines.

Once the feedback loops have been created, the overall adaptive system model is made up using the `AdaptiveNetBuilder` (lines 17–20). The builder component plugs each feedback loop to the emulator net by connecting the emulator's transition `move` to the initial places of any loop net. This way, whenever the base layer enters a new state (i.e., transition `move` fires), the loops are triggered. The framework transparently handles (at net level) the interaction between base layer and feedback loops according to the *Balking pattern* (i.e., new triggers are ignored while a loop is being executed). It is also possible to associate a loop with *observable* events (i.e., specific firing transitions) of the base layer.

The `build` method call (line 20) returns a SNAKES HLPN (i.e., a `PetriNet` object) that can be in turn inspected, visualized, executed, or analyzed, as discussed in Sect. 4.

```
1    from snakes.nets import *
2    from pnemu import PT, Emulator, AdaptiveNetBuilder
3    emulator = Emulator(PT(name='ms', pnml='my/path/ms.pnml'))
4    emulator.get_marking().get('mark')
5    # the previous line returns: MultiSet(['storage'] * 10)
6    loop = FeedbackLoop(name='fault—tolerance')
7    loop.add_place(name='init')
8    loop.add_place(name='sample')
9    read_primitive = 'lib.getTokens("broken")—>n'
10   loop.add_transition(name=read_primitive)
11   loop.add_input_arc(src='init', dst=read_primitive, Variable('b'))
12   loop.add_output_arc(src=read_primitive, dst='sample', Variable('n'))
13   loop.add_transition(name='lib.addHArc("broken","load",1)')
14   loop.add_transition(name='lib.rmOArc("line1","load",1)')
15   loop.add_transition(name='lib.addOArc("line2","load",1)')
16   ...
17   net = AdaptiveNetBuilder(emulator)
18     .add_loop(control_loop=loop, initial_places=['init'])
19     .add_loop(control_loop=loop2, initial_places=['init2'])
20     .build()
21   mode = net.transition('move').modes()[1]
22   # mode = Substitution(m=MultiSet(['storage'] * 2), t='fail', ...)
23   net.transition('move').fire(mode)
24   assert net.get_marking()('init') == MultiSet(['fail'] * 1)
```

Fig. 4. A Python code snippet showing an example of the PNEMU API usage.

3.3   The pnemu.primitives Module

This module contains a basic, yet complete, set of primitives that constitute an easy-to-use API through which to do base layer introspection/manipulation. These primitives implement read, add, removal, operations on any structural or state elements of the emulated P/T system. The primitives are implemented as high-level transitions wrapped into instances of LibEntry class. As an example, Fig. 5b shows the *definition* and *instantiation* of getTokens. The definition contains the signature and the net elements of the primitive. In the example, the getTokens reads the content of the base layer's place mark into m, and returns the multiplicity (i.e., the marking) m(p) of the P/T place denoted by the argument p (a free variable). An instantiation instead takes place at any primitive call by AdaptiveNetBuilder, during the construction of the HLPN representing the whole self-adaptive system. Each transition representing a primitive call in a loop-net is *superposed* with the corresponding definition, by carrying out a simple *term matching* in their signatures. For example, when superposing the occurrence of getTokens in the loop-net with its definition, the free variable p in the definition is matched by the constant "broken", whereas the free variable n of the loop's transition is matched by expression m(p). The arguments of a primitive call can be either supplied "in place", as shown in Fig. 5a, or indirectly, by means of variables annotating input arcs of the calling transition.

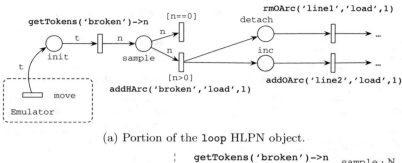

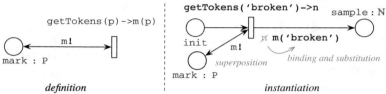

(a) Portion of the `loop` HLPN object.

(b) The `getTokens` read primitive *definition* and *instantiation*.

Fig. 5. Managing subsystem example.

Table 1 lists all the pre-defined read/write primitives provided by PNEMU. This set can be easily extended by adding new user-defined `LibEntry` objects. The class `LibEntry` allows new, primitive or even complex, library functions to be created by defining their structure and inscribing it with arbitrary (standard or user-defined) Python functions.

4   Usage Scenarios

We are releasing PNEMU as open-source software. Here are some typical scenarios where PNEMU could be helpful.

Modeling and Validation. Software tools equipped to formally represent and analyse self-adaptive/dynamic systems are highly demanded [1]. The PNEMU library provides a general and consolidated two-layer modeling framework based on the use of multiple, distributed feedback loops. It directly supports inspection (of both the state and the structure) of both layers, interactive simulation and visualization of model components. In particular, at each step of the token game it is possible to check for enabled transitions, firing modes, make transitions fire, and perform on-the-fly changes to the model's state/structure. A simple example of interactive simulation is shown in Fig. 4 (lines 21–24): line 21 shows the firing modes for the transition `move`, among which there is the one representing base level's transition `fail`; line 23 makes `move` fire with this mode; line 24 checks that after the firing of the base layer's transition `fail`, the *fault tolerance* control loop is correctly triggered.

Table 1. Pre-defined read/write primitives of PNemu.

Name	Arguments	Outcome
Read primitives		
getTokens	p : P	Returns the number of tokens held by p
getMarking	-	returns the multiset of P-elements representing the P/T marking
getPlaces/getTransitions	-	Returns the set of P-/T-elements
exists	e : P or T	Returns True if the given element exists
pre/post	e : P or T	Returns the set of P-/T-elements representing the pre-/post-set of e
iMult/oMult/hMult	p : P, t : T	Returns the multiplicity of the input/output/inhibitor arc between p and t
Write primitives		
addPlace/addTransition	p : P/t : T	Add the given place/transition p/t
rmPlace/rmTransition	p : P/t : T	Remove the given place/transition p/t
setTokens	p : P, n :$\mathbb{N}$	Set the number of tokens of p to n
addIArc/addOArc/addHArc	p : P, t; T, n :$\mathbb{N}$	Add the given input/output/inhibitor arc between p and t with multiplicity n
rmIArc/rmOArc/rmHArc	p : P, t; T, n :$\mathbb{N}$	Remove n occurrences of the given input/output/inhibitor arc between p and t

Formal Verification. The modeling approach supported by PNemu is based on well established formalisms (i.e., P/T systems and HLPNs following the ISO/IEC standard). This makes it possible the usage of consolidated analysis techniques and/or existing software tools. As an example, since the module `pnemu.manager` produces SNAKES HLPNs (`snakes.nets.PetriNet` objects) we can exploit the Neco [5] model checker (designed to work along with SNAKES) to verify the correctness of models with respect to requirements expressed in Linear Temporal Logic (LTL). For instance, it is possible to check reachability, safety, and liveness properties; local (i.e., relating the base layer only) and/or global (considering both layers) invariants, and *adaptation integrity* constraints [9].

As a simple example, we verified the LTL safety property $\Box(\neg \texttt{deadlock})$ (deadlock-freedom) on the self-healing MS. Another property we checked is:

$$\Box(\texttt{mark} > [\texttt{storage}, \texttt{storage}, \texttt{broken}] \rightarrow \Diamond(\texttt{mark} > [\texttt{assembled}]))$$

i.e., the ability to assemble raw pieces even when a production line is down.

5 Conclusion

We introduced PNEMU, a Python library based on standard HLPNs and supporting modeling/validation of adaptable distributed systems through a two-layers approach allowing for a *clear* separation of concerns between the managed and the managing subsystems. The library has been designed to be easily *extensible* with additional user-defined features, such as read (sensors) and write (actuators) primitives, and user-defined Python functions to inscribe net elements. Finally, PNEMU can *interoperate* with other third-party tools. The full support to the PNML language to define the base layer, makes it possible to leverage existing visual editors. In addition to modeling and validation (currently supported by PNEMU), formal verification can be carried out by means of existing state space builders/model checkers.

The library has been released as open-source software and it is currently considered as "beta software". We plan to enhance PNEMU with the ability to verify structural properties by means of library primitives that compute structural relations (e.g., conflict, causal connection, mutual exclusion) on P/T elements. Moreover, we want to integrate the NECO model checker and enhance it with the ability to discover conflicting control loops (i.e., loops that apply conflicting changes to achieve different adaptation goals).

Acknowledgments. The authors would like to thank Giuliana Franceschinis for her helpful advices and comments on early drafts of this paper.

References

1. Arcaini, P., Riccobene, E., Scandurra, P.: Formal design and verification of self-adaptive systems with decentralized control. ACM Trans. Auton. Adapt. Syst. **11**(4), 25:1–25:35 (2017). https://doi.org/10.1145/3019598
2. Camilli, M., Bellettini, C., Capra, L.: A high-level petri net-based formal model of distributed self-adaptive systems. In: Proceedings of the 12th European Conference on Software Architecture: Companion Proceedings. ECSA 2018, pp. 40:1–40:7. ACM, New York (2018). https://doi.org/10.1145/3241403.3241445
3. Capra, L.: A pure spec-inscribed pn model for reconfigurable systems. In: 2016 13th International Workshop on Discrete Event Systems (WODES), pp. 459–465. IEEE Computer Society, May 2016. https://doi.org/10.1109/WODES.2016.7497888
4. Ding, Z., Zhou, Y., Zhou, M.: Modeling self-adaptive software systems with learning petri nets. IEEE Trans. Syst. Man Cybern. Syst. **46**(4), 483–498 (2016). https://doi.org/10.1109/TSMC.2015.2433892
5. Fronc, Ł., Pommereau, F.: Building Petri nets tools around Neco compiler. In: Proceedings of PNSE 2013, pp. 1–7, 06 2013
6. Systems and software engineering - High-level Petri nets - Part 1: Concepts, definitions and graphical notation. Standard, International Organization for Standardization, Geneva, CH (Dec 2004), ISO/IEC 15909-1:2004
7. Jensen, K., Rozenberg, G. (eds.): High-level Petri Nets: Theory and Application. Springer-Verlag, London (1991). https://doi.org/10.1007/978-3-319-19488-2_13

8. Kummer, O., et al.: An extensible editor and simulation engine for petri nets: renew. In: Cortadella, J., Reisig, W. (eds.) Applications and Theory of Petri Nets 2004, pp. 484–493. Springer, Heidelberg (2004)

9. de Lemos, R., Garlan, D., Ghezzi, C., Giese, H. (eds.): Software Engineering for Self-Adaptive Systems III. Assurances. LNCS, vol. 9640. Springer, Cham (2017). https://doi.org/10.1007/978-3-319-74183-3

10. Pommereau, F.: Snakes: a flexible high-level petri nets library (tool paper). In: Devillers, R., Valmari, A. (eds.) Application and Theory of Petri Nets and Concurrency, pp. 254–265. Springer, Cham (2015)

11. Reisig, W.: Petri Nets: An Introduction. Springer, New York (1985)

12. Salehie, M., Tahvildari, L.: Self-adaptive software: landscape and research challenges. ACM Trans. Auton. Adapt. Syst. 4(2), 14:1–14:42 (2009). https://doi.org/10.1145/1516533.1516538

CoRA: An Online Intelligent Tutoring System to Practice Coverability Graph Construction

Jan Martijn E. M. van der Werf$^{(\boxtimes)}$ and Lucas Steehouwer

Department of Information and Computing Science,
Utrecht University, Utrecht, The Netherlands
`j.m.e.m.vanderwerf@uu.nl`, `lucas@architecturemining.org`

Abstract. While teaching Petri nets, many students face difficulties in constructing coverability graphs from Petri nets. Providing students with individual feedback becomes infeasible in large classes.

In this paper, we present CoRA: the Coverability and Reachability graph Assistant. It is an online intelligent tutoring system designed to support users in constructing a coverability graph for a Petri net. Its main goal is to provide additional tutorial support to students, so they can practice on their own and ask questions to staff when required. CoRA is capable of giving personalized feedback; whenever a user submits a solution CoRA provides targeted feedback stating what is correct in what is not. CoRA's feedback is designed to be both guiding and informational; a user should be able to understand what went wrong and how they can improve their graph.

Keywords: Petri nets · Coverability graph · Education · Intelligent tutoring

1 Introduction

Since several years, the course Information Systems is taught as a mandatory course to first year Bachelor students Information Sciences. The course serves as an introduction in process modeling and analysis. For the theoretical aspects, the book of Van der Aalst & Stahl (2011) is used [1]. Many properties of Petri nets are explored during the course, and students are required to make statements about the nets they produce for assignments. One of the skills the students have to learn is to construct a coverability graph from a given Petri net [11]. Over the years, we observed that many students struggle with this topic and require additional guidance. As each year the course has over 180 participants, providing sufficient individual guidance and feedback becomes unfeasible. Therefore, we searched for a more creative solution, allowing students to practice converting Petri nets to coverability graphs in their own time, at their own pace, while still giving adequate feedback to support the learning process of the student.

© Springer Nature Switzerland AG 2019
S. Donatelli and S. Haar (Eds.): PETRI NETS 2019, LNCS 11522, pp. 91–100, 2019.
https://doi.org/10.1007/978-3-030-21571-2_6

Software-based solutions for these kinds of problems come in the form of Intelligent Tutoring Systems (ITS) [12]. These are systems which provide feedback, guidance, and other supporting factors one would find in a typical educational environment focusing on a specific topic [12]. Over the years many ITSs have been built. For example, [5] reports on the tool LISPITS, an ITS which was used to teach students the contents of a LISP course at Carnegie Melon University. In their experiment, students using LISPITS and the students receiving human tutoring worked through the material much faster than the students learning on their own. LISPITS students were slower than the students tutored by lecturers, but not by much: 11.4 hours versus 15 hours, whereas the third group took 26.5 h on average to go through the material [2]. Hence, an implemented ITS can provide sufficient tutorial support to students.

In this paper, we present CoRA: the Coverability and Reachability graph Assistant. CoRA is an ITS that focuses on providing tutorial support on how to convert a Petri net to a coverability graph. Software to automatically infer coverability graphs for a given Petri net already exist for many years. For example, the Low Level Analyzer (LoLA) [15] supports coverability analysis, including various reduction techniques. Similarly, the open-source framework ProM [16] has plugins that create a coverability graph inference. However, these tools focus on automatic model checking, rather than on providing feedback for learning coverability graph construction.

The remainder of this paper is structured as follows. In the next section, we introduce the tool by presenting a typical use case showing most of the steps a student would go through when practicing coverability graphs. In Sect. 3, we present the principles behind CoRA: coverability-graph validation and feedback generation. The tool has been tested and evaluated by a selected group of students, on which we report in Sect. 4. Last, Sect. 5 concludes the paper discussing future work.

2   CoRA to Assist Students

Over the years that the course Information Systems has been taught, we observed that many students had problems with constructing coverability graphs. For example, students often find it difficult to traverse over the state space of the Petri net in a structured manner. As a consequence, students overlook many states. Another example is that many students simply do not know when to introduce the symbol ω. CoRA supports students by providing them feedback during the construction process. It is a web-based tool and runs in most browsers[1].

Upon entering the website, the user registers with a new user name. Next, the user needs to upload a Petri net in LoLA format [15]. We deliberately chose to not include a Petri net editor, but instead rely on existing tools, such as Yasper [9], and WoPeD [8]. After a short introduction to the main interactions with the tool, the coverability construction phase starts.

---

[1] We tested the tool in Google Chrome and Mozilla Firefox.

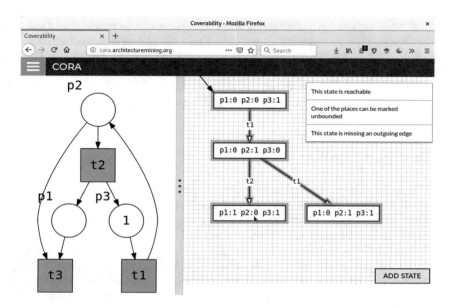

Fig. 1. Screenshot of CoRA, showing a partial coverability graph with feedback.

After uploading the file, a workspace as presented in Fig. 1 is shown. The uploaded Petri net is shown on the left. The grid canvas on the right is where the student has to construct the coverability graph. Initially, the canvas is empty, and the student can start adding states. By double clicking on a state, the student can update the state, by setting the token values for the different places. Similarly, arcs can be labeled with an action by double clicking on it, and selecting a transition from a drop-down list.

Constructing the coverability graph can be done in any order, contrary to feedback generation, which requires an initial state. Therefore, as long as no state has been marked as the initial, the only feedback the tool gives is that it does not know the initial state. Once the student sets the initial state, the tool starts providing feedback.

Colors are used to indicate whether a state or arc is correct (green), partially correct (orange), or incorrect (red). Upon hovering on an element, the tool displays the generated feedback as a list. For example, in Fig. 1, the first two states are green, indicating that these are correct, i.e., that they are reachable, have the correct token values, and the correct outgoing transitions. The last two states are orange, showing that the state is only partially correct. In this example, The student has highlighted state $[p1 : 1, p2 : 0, p3 : 1]$. As this state covers the initial marking, the tool hints that one of the places can be marked as unbounded. Additionally, outgoing transitions are missing. Although missing edges gives an incorrect result, we color the state as partially correct, to provide better feedback. Transition $t1$ from this state is colored red, hinting that the transition is not enabled in this state.

After each action by the student, the tool validates the model and generates new feedback. This process continues until the student has a correct coverability graph, i.e., that all states and transitions are colored green. By default, the tool provides immediate feedback. This is typically preferred for the target audience: novice learners [12]. Students can also choose to set feedback to manual. They then can decide themselves when to submit the graph, and when to receive feedback.

3   Design of CoRA

CoRA is a web-based tool, consisting of a client for modeling, and a server for validation and feedback generation. CoRA-client provides a coverability-graph modeling environment which uses the API exposed by CoRA-server. CoRA-client runs in the browser as a combination of HTML, JavaScript and CSS. Its JavaScript library is written in Typescript, a programming language with static-typing, which can be compiled to JavaScript. If immediate feedback is switched on, CoRA-client sends the complete coverability graph to CoRA-server after each user action. The server then validates it, and returns the generated feedback.

CoRA-Server is built on the Slim-framework[2], a bare-bones PHP framework for creating web applications, and implements a REST-service [7]. CoRA-Server runs on any server with support for PHP7. It uses a MySQL database for storage.

The tool is publicly available at http://CoRA.architecturemining.org/, its source code can be found at GitHub[3].

3.1   Analyzing Coverability Graphs

In theory, composing a coverability graph for a given Petri net is nothing more than following Algorithm 1. To find all paths from the initial marking to the newly added marking we use a graph traversal algorithm like Breadth First Search [6]. From a tutorial perspective, we have an initial state (an empty graph) and a goal state (a correct coverability graph). The student "only" needs to learn the correct strategy to arrive at a goal state from the initial state [12]. However, a coverability graph is not unique for a Petri net. Consequently, the traversal and discovery of nodes in the graph may happen in any order. This needs to be taken into account when generating automatic feedback.

To show that the order of traversal determines the resulting coverability graph, consider the example depicted in Fig. 2, taken from [1]. Starting in initial marking $[p_1]$, two transitions are enabled, t_1 and t_2. Firing t_1 results in $[p_2]$, transition t_1 in $[p_3]$. From $[p_3]$, firing transition t_4 brings us back to $[p_2]$. Now, we fire transition t_3, resulting in marking $[p_2, p_3]$. This marking clearly covers the earlier visited marking $[p_2]$, showing that place p_3 is unbounded. Similarly, this marking covers the already visited marking $[p_3]$ as well. This results in the left

---

[2] https://www.slimframework.com.
[3] https://github.com/ArchitectureMining/CoRA.

Algorithm 1. Generating a coverability graph

1: **procedure** GENERATECOVERABILITYGRAPH($(P, T, F), m_0$)
2:     $s_0 \leftarrow m_0$; $V \leftarrow \emptyset$; $E \leftarrow \emptyset$
3:     $O \leftarrow \emptyset$ ▷ Frontier: Set of markings which still have to be expanded
4:     $Q \leftarrow$ Queue of markings   ▷ Frontier provides $O(1)$ lookup of markings
5:     Q.Enqueue(s_0)
6:     **while** Q not empty **do**
7:         $s \leftarrow Q$.Dequeue() ; $O \leftarrow O \setminus \{ s \}$; $V \leftarrow V \cup \{ s \}$
8:         $R \leftarrow$ All paths from s_0 to s
9:         **for all** $r \in R$ **do**
10:            **for all** $m \in r$ **do**
11:                **if** $s \geq m$ **then**
12:                    **for all** $p \in P$ **do**
13:                        **if** $s(p) > m(p)$ **then** $s(p) = \omega$
14:            **for all** $t \in T$ **do**
15:                **if** t is enabled for s **then**
16:                    $s' \leftarrow \forall p \in P : s(p) - F(p, t) + F(t, p)$
17:                    **if** $s' \notin V \wedge s' \notin O$ **then** ▷ Add the state to the Queue and Frontier
18:                        $O \leftarrow O \cup \{ s' \}$; Q.Enqueue(s')
19:                    $E \leftarrow E \cup \{ (s, s', t) \}$ ▷ Add the discovered edge
20:     **return** $G = (V, E, s_0)$

---

coverability graph depicted in Fig. 3 would we have followed a different strategy in generating a coverability graph, e.g. by first continuing with transition t_1, we would not yet have discovered the unboundedness of place p_2, resulting in the middle coverability graph of Fig. 3.

Another challenge for feedback generation is that marking a place as unbounded does not have to happen immediately. Students may discover only later in the process that a place could already be marked unbounded, and unfold the graph unnecessarily deep, as shown in the third example of Fig. 3. Note that the third example in Fig. 3 cannot be produced by Algorithm 1. Still, the delivered coverability graph is correct. When a student omits to mark a place as unbounded while it is possible to do so, then the feedback generation algorithm needs to adapt to this. It needs some way of "remembering" that this place can be marked as unbounded and that all markings in the postset of the current marking can also mark this place as such.

3.2   Providing Feedback

The main goal of CoRA is providing useful feedback on coverability graphs. Designing good feedback however, is not an easy task. Feedback can inform the user about his or her progress and can also guide the user to the correct solution, for example by giving pointers [10]. These two types of feedback, informational and guiding feedback, can overlap. The goal of CoRA is providing both these forms of feedback; it should provide information to the users about which elements of the coverability graph are correct and incorrect, but CoRA should also

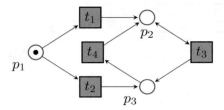

Fig. 2. Petri net example taken from [1]

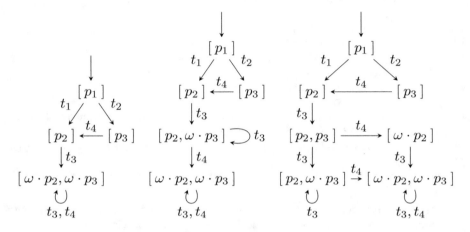

Fig. 3. Three correct coverability graphs of the Petri net of Fig. 2, the first and second followed a different strategy in traversing the states. In the third, unbounded places were discovered at a later stage.

provide messages which suggest a corrective measure if required. For example, if a state misses an outgoing edge, a message should be given stating that an edge is missing. This message shows that something is not right (an edge is missing), but also provides a hint on how this problem should be solved (adding an edge).

There are three kinds of feedback messages that we can give with CoRA. For each element of the graph we can decide whether it is a correct, partially correct, or incorrect element. For example, the user can model two reachable states, but the label of the edge between those states is not correct. CoRA gives feedback that this edge represents the wrong transition.

CoRA can also provide warnings. Warnings are intended for elements that are not incorrect, but do provide a risk to make mistakes later on. For example, students can make coverability graphs with duplicate states. Having duplicate states in a coverability graph does not make the graph incorrect, but it does make it unnecessarily difficult for the student to not make mistakes.

Another example is the introduction on unboundedness of places. Instead of following Algorithm 1, CoRA checks for each state created by the user if this state covers some state on the path from the initial state to the state under analysis. If it is indeed a cover, but no ω has been introduced, the user receives feedback that it is possible to introduce an ω in this state.

Next to the kinds of feedback, there is a difference on when feedback should be provided: for beginners it is preferred that feedback is given instantly, whereas for people with some experience it is better to provide delayed feedback [12]. CoRA supports both forms. At first, most of the students will not be familiar with composing coverability graphs and will therefore need immediate guidance, whereas at later stages it can be more beneficial for students to only get feedback when they feel like they need it or when they are finished.

With this knowledge in mind, we designed a set of messages to be shown to the user. Messages can be assigned to particular elements of the graph, as well as to the graph as a whole. Table 1 shows the feedback messages that we designed for CoRA. All feedback messages have been constructed to be both informational as well as guiding. However, further research is required to validate these messages.

4 Initial Evaluation

CoRA was presented and pitched during the Information Systems course. This provided an opportunity to test CoRA immediately with its target audience: students. CoRA was introduced during one of the lectures and a small demo was given on how to use it. After this demo a set of Petri nets was given to the students to practice with. Students were also asked to fill in a questionnaire after practicing a few times. The questionnaire was presented in an on-line form, and was available at all times. This questionnaire contained the System Usability Scale (SUS) [4]. The SUS provides a simple way to measure the usability of a system. The Net Promoter Score (NPS) [13] was included as it is correlated to the SUS ($r = 0.61$, [14]), and thus provides an extra validation. The recommendations section of the questionnaire had questions on how to improve CoRA, as well as a question whether they would recommend CoRA to fellow students.

Only ten responses to the survey were recorded, probably since the tool was presented only in one of the last lectures before the exam, Eight participants were following the Information Systems course while taking the survey. These eight participants did state that they found CoRA to be at least somewhat supportive in their efforts to understand the subject of coverability graphs. With a score of 3.25 out of 5 for this question the response was quite neutral, with a slight favor towards being helpful. For the System Usability Scale scores ranged between 62.5 and 95, with an average score of 70.25, and a median of 76.25. According to [3] this is often an acceptable score, but there certainly is room for improvement. The scores for the NPS seem to indicate this as well. NPS scores range between -100% and 100%. With a score of 20% we have more promoters than detractors, but clearly the score could be much better.

Eight of the participants provided suggestions on improving CoRA. The main complaint was that it was not possible to upload a second Petri net, to keep practicing. Instead, they needed to restart CoRA-client each time they wanted to start a new session, which also means going through the hoops of registering

Table 1. Types of feedback and hints the tool provides

Type of feedback	Hint provided to user
Initial state	
No initial state	No state has been declared as the initial state
Incorrect initial state	The initial state of the graph is not equal to the initial marking of the Petri net
Correct initial state	The initial state of the graph and the initial marking of the Petri net are identical
Unbounded place	At least one of the places in the initial state is marked as unbounded. This is impossible!
States	
Reachable from preset	This state is reachable from its preset
Duplicate state	This state occurs multiple times in the graph
Omitted ω	It is possible to introduce an ω in this state, but this has not been done
Not reachable from preset	This state is not reachable from at least one of the states in its preset
Edge missing	This state is missing an outgoing edge
Not reachable from initial state	This state is not reachable from the initial state
Omitted ω from preset	At least one of the states in the preset has at least one place marked as unbounded, while this state does not mark this place as such
Edges	
Enabled and correct post	The edge's label corresponds to an enabled transition and points to a reachable state
Duplicate edge	There are multiple edges with the same label originating from the same state
Enabled, correct post state, wrong label	The transition corresponding to the label is enabled and the target of the label is reachable, but firing this transition does not lead to this state
Enabled, incorrect post state	The transition corresponding to the label is enabled, but the target state is not reachable from the source state
Disabled	The transition corresponding to the label is disabled
Disabled, correct post state	The transition corresponding to the label is disabled, but the target state is reachable by firing a different transition

a new account again. Participants also indicated that there were a few bugs
with the modeler, especially regarding opening menu's for editing elements of
the graph.

5   Conclusions

In this paper, we presented CoRA: the Coverability and Reachability graph
Assistant. It is an online intelligent tutoring system to practice constructing
coverability graphs, and provides feedback based on the users' input.

Initial evaluation shows that the tool has the desired potential, but requires
some improvements. In the near future, we plan to extend the tool with a framework to support multiple exercises per session, and to add PNML support.

Future work lies in assessing the quality of the feedback messages the tool
provides. The tool stores all intermediate stages of the coverability graph, allowing to study which feedback is given, and how this feedback is perceived by
students. Another direction of research lies in studying the possibilities in the
realm of gamification and applied games. Games provide a continuous loop of
feedback messages [10]. CoRA only provides feedback when the client requests
it. With a game, feedback could be provided constantly, possibly providing better support to the user. Currently, CoRA just assumes the user is a novice. A
possible avenue of further research, perhaps in combination with the idea of
gamification, would be to automatically infer the experience level of a user and
adapt the provided feedback based on this level.

CoRA gives feedback on the conversion of Petri nets to coverability graphs.
There are many other analysis techniques students need to learn. Future work
therefore also includes the development of tutoring systems for other techniques,
such as invariants. These solutions could then be integrated into a electronic
learning environment, providing the students an environment where they can
practice all sorts of techniques regarding Petri nets, including coverability analysis with CoRA.

References

1. van der Aalst, W.M.P., Stahl, C.: Modeling Business Processes–A Petri Net-
   Oriented Approach. Cooperative Information Systems Series. MIT Press (2011)
2. Anderson, J.R., Boyle, C.F., Reiser, B.J.: Intelligent tutoring systems. Science
   228(4698), 456–462 (1985)
3. Bangor, A., Kortum, P.T., Miller, J.T.: An empirical evaluation of the system
   usability scale. Int. J. Hum.-Comput. Interact. **24**(6), 574–594 (2008)
4. Brooke, J., et al.: SUS-a quick and dirty usability scale. In: Usability Evaluation
   in Industry, vol. 189, no. 194, pp. 4–7 (1996)
5. Corbett, A.T., Anderson, J.R.: Lisp intelligent tutoring system: research in skill
   acquisition. In: Computer-Assisted Instruction and Intelligent Tutoring Systems:
   Shared Goals and Complementary Approaches, pp. 73–109 (1992)
6. Cormen, T.H., Leiserson, C.E., Rivest, R.L., Stein, C.: Introduction to Algorithms.
   MIT Press, Cambridge (2009)

7. Fielding, R.T., Taylor, R.N.: Architectural styles and the design of network-based software architectures, vol. 7. University of California, Irvine Doctoral dissertation (2000)
8. Freytag, T., Sänger, M.: WoPeD - an educational tool for workflow nets. In: Proceedings of the BPM Demo Sessions, CEUR Workshop Proceedings, vol. 1295, pp. 31–35 (2014). CEUR-WS.org
9. van Hee, K.M., Oanea, O., Post, R.D.J., Somers, L.J., van der Werf, J.M.E.M.: Yasper: a tool for workflow modeling and analysis. In: ACSD, pp. 279–282 (2006)
10. Kapp, K.M.: The Gamification of Learning and Instruction: Game-based Methods and Strategies for Training and Education. Wiley (2012)
11. Karp, R.M., Miller, R.E.: Parallel program schemata. J. Comput. Syst. Sci. **3**(2), 147–195 (1969)
12. Passier, H.J.M.: Aspects of Feedback in Intelligent Tutoring Systems for Modeling Education. PhD thesis, Open University, The Netherlands (2013)
13. Reichheld, F.F.: The one number you need to grow. Harvard Bus. Rev. **81**(12), 46–55 (2003)
14. Sauro, J.: Does better usability increase customer loyalty? (2010). https://measuringu.com/usability-loyalty/. Accessed 05 July 2018
15. Schmidt, K.: LoLA a low level analyser. In: Nielsen, M., Simpson, D. (eds.) ICATPN 2000. LNCS, vol. 1825, pp. 465–474. Springer, Heidelberg (2000). https://doi.org/10.1007/3-540-44988-4_27
16. Verbeek, H.M.W., Buijs, J.C.A.M., van Dongen, B.F., van der Aalst, W.M.P.: XES, XESame, and ProM 6. In: Soffer, P., Proper, E. (eds.) CAiSE Forum 2010. LNBIP, vol. 72, pp. 60–75. Springer, Heidelberg (2011). https://doi.org/10.1007/978-3-642-17722-4_5

Tools for Curry-Coloured Petri Nets

Michael Simon[✉], Daniel Moldt, Dennis Schmitz, and Michael Haustermann

Faculty of Mathematics, Informatics and Natural Sciences,
Department of Informatics, University of Hamburg, Hamburg, Germany
michael.simon@informatik.uni-hamburg.de
http://www.informatik.uni-hamburg.de/TGI/

Abstract. The Curry-coloured Petri net (CCPN) simulation combines Petri nets with the purely functional logic programming language Curry. The most notable aspects of the *CCPN* simulator are the absence of side effects, the use of logic program evaluation for the transition binding search and a concurrent simulation. Furthermore, the inscribed programs can be non-deterministic.

In this contribution we present the tools that were developed so far: simulator, editor, reachability graph and library.

Keywords: High-level Petri nets · Coloured Petri nets · Curry-coloured Petri nets · Tools · Simulator · Editor · Library

1 Introduction

Contemporary *Coloured Petri net*[1] (CPN) [7] tools and simulators have problems regarding side effects and outdated implementations.

This contribution presents a concurrent CPN simulator written in Haskell and an editor to visually model CPNs as well as to observe and control their simulation.[2] The simulator is also provided as a library to facilitate the integration into other Haskell programs. Additionally, a tool for generating and visually inspecting reachability graphs is presented.

CPNs generalize the Petri net formalism to tokens of multiple *colors*. In the context of programming languages a token's color can be interpreted as a value the token carries. A place marking is a multiset of token colors. Transitions can also be fired in different *bindings* (also called *modes*). A binding defines the colors of the tokens that are taken in and put out. The binding itself is defined by inscriptions to the transition and its environment.

The inscriptions might only define bindings for a subset of all possible combinations of ingoing tokens. Thus, multiple combinations of ingoing tokens might have to be tried in a *binding search*. Optimizations of this binding search are

---

[1] We spell *Coloured Petri net* in its original form but otherwise use American English.

[2] A guide and the resources for the installation of the *CCPN* tool set can be found at https://git.informatik.uni-hamburg.de/tgipublic/ccpn/ccpn/wikis/pn19.

© Springer Nature Switzerland AG 2019
S. Donatelli and S. Haar (Eds.): PETRI NETS 2019, LNCS 11522, pp. 101–110, 2019.
https://doi.org/10.1007/978-3-030-21571-2_7

fundamental to reduce the simulation's runtime complexity. They are a major focus of CPN simulator implementations.

An important aspect of CPN simulation is the control of side effects in the binding search. The evaluation of the inscriptions in the search may not include side effects as these cannot be reverted in general. Despite this challenge, most popular CPN simulators employ inscription languages that fail to prevent side effects in the search. The modeler is left with the obligation to ensure side effect freedom of the inscriptions.

In order to address the challenges of side effects and an efficient binding search concerning the transition inscriptions in CPNs, this contribution introduces the *Curry-coloured Petri nets* (CCPN) formalism. For a short introduction to the formalism and its simulation see [13] and for an in-depth description see [15]. The formalism allows transition inscriptions in *Curry* [5]. Curry is a general purpose purely functional logic programming language that combines tightly controlled side effects and non-deterministic evaluation. It is the product of active research and one of the best developed functional logic programming languages. One primary focus of this research is the efficient and robust evaluation of non-deterministic programs. This is utilized for the *CCPN* transition binding search and is therefore an important feature missing from other implementations with purely functional programming languages like Haskell.

The presented tools that support the modeling, simulation, control and observation of CCPNs are – through the integration into the RENEW GUI – easy to use. The binding search is quite efficient and the side effect freedom is naturally inherent in the *CCPN* formalism. After an explanation of the objectives in Sect. 2 other CPN simulators are discussed in Sect. 3 and the required functionality is described in Sect. 4. Section 5 gives an overview of the architecture and Sect. 6 illustrates a simple example of a simulation. Finally, we conclude in Sect. 7.

2 Objectives

One objective is to provide a formalism that is expressive enough for general purpose programming. Furthermore, a concurrent simulation algorithm should realize the potential of the *CCPN* formalism to be used to implement concurrent programs. The operational semantics of the inscription language should be formally defined. This allows the existing extensive research on CPNs to be used for further research on verification and analysis of the *CCPN* formalism. Another important objective is to solve the problems that other CPN simulators have (see Sect. 3), in particular the prevention of side effects.

One of the most distinguishing features of existing CPN simulators are the inscription languages they employ. In this paper we present a CPN simulator that employs Curry as an inscription language. As a purely functional language Curry isolates the occurrence of side effects to the execution of special data types. Functions are pure, their evaluation does not include side effects.

As a logic programming language the Curry evaluation system can be used to implement an efficient binding search with unification. This only requires the binding search to be formulated as a non-deterministic Curry program. It also allows the inscribed program itself to introduce non-determinism by established Curry concepts.

The choice of a general purpose language is a good basis for the development of a CPN formalism and corresponding tool set that can themselves be used for general purpose programming. The inclusion of a concurrent simulator into the tool set is important to realize the potential of the *CCPN* formalism to be used as a general purpose concurrent programming language. To take full advantage of the intuitive graphical Petri net representation a GUI for editing and simulation is important. A highly modular simulation library with multiple interfaces to the simulation allows it to be flexibly used and extended by client Haskell programs as well as other programs via XML. There is a formal definition of Curry's operational semantics [1] and preexisting research on formal verification [2] (see [15, p. 23]). Also, Curry's expressive static type system with type inference catches many potential errors in the compilation. Furthermore, it provides powerful concepts for abstract and generic programming that help to create smaller nets.

3   Comparison

This section presents other CPN simulators. In particular, it compares them to the *CCPN* simulator regarding the objectives defined in Sect. 2.

The *Java reference net* formalism in RENEW, the *Scheme reference net* formalism in RENEW as well as the *CPN Tools* provide a full-fledged simulation environment but fail to prevent side effects. This inability is forced upon them by their inscription language selection. All three tools base their inscription language on a preexisting language that does not offer a type system that tracks and controls side effects. This is a clear disadvantage compared to pure functional languages such as Haskell or Curry.

The *HCPN* simulator successfully prevents side effects by the selection of Haskell as an inscription language but does not provide a full-fetched simulation environment.

RENEW
RENEW [10] offers multiple formalisms, most prominently the *Java reference net* formalism [11]. It is a CPN formalism with a Java-like inscription language, Java objects as tokens and the possibility to contain marked nets as tokens. It offers a *synchronous channel* concept to combine multiple transitions into a single binding search and fire them synchronously in one atomic action. Channel parameters allow the bidirectional exchange of information via unification. Channels are the only concept in the inscription language that can be used to express non-determinism. Otherwise, the inscribed programs are completely deterministic.

Because arbitrary Java methods can be called by the inscription language, there is no way to prevent side effects in general. The modeler must ensure that normal inscriptions only call methods without side effects. Methods with side effects should only be called from within *action inscriptions*. These are executed after the binding search is finished and under the assumption that the binding is guaranteed to be fired.[3] Because the return values of methods called from action inscriptions are only known after the binding search, the non-deterministic part of the inscribed program cannot depend on them. This part includes the selection of ingoing tokens and the binding of synchronous channels. It limits the expressiveness of the inscription language.

The semantics of the inscription language is based on the Java semantics but implemented by a depth-first search with unification much like logic programming language evaluation [11,14]. Thus, there are subtle differences between the inscription language and Java's semantics. This can lead to unintended behavior that is confusing and unintuitive for programmers familiar with Java (see [14]).

In the binding search, objects get unified that are equal according to their `equals` method. Furthermore, the `hashcode` method's return value is used to store objects as tokens in the place marking. To be handled correctly by RENEW, the tokens' methods must be pure: they must always return the same value when called with the same parameters throughout the object's lifetime. This is not the case for Java objects in general. In Java programming it is common to implement mutable objects that change their state and with it the equality relation and the hashcode. Thus, modeling in RENEW requires a pure programming style that is unusual in the context of Java development and the wrapping of code that does not adhere to this style.

Scheme Reference Nets in RENEW

In [4] Delgado Friedrichs describes the *Scheme reference net* formalism. It is built upon the Java reference net formalism by the integration of a Scheme interpreter and logic library into RENEW's simulation algorithms. Because Scheme is a functional language, a pure programming style is not as foreign as in Java development. However, Scheme does not enforce purity: it allows side effects to occur anywhere in the execution. The Scheme reference net formalism plugin prevents some side effects by excluding a list of known Scheme procedures with side effects but cannot prevent side effects in general.

Since 2007 no changes to the integrated Scheme interpreter and logic library have been released. Therefore, future research and practical use of the formalism would require considerable effort to port it to new underlying libraries.

CPN Tools

The *CPN Tools* [8,9] offer a CPN formalism with an inscription language based on *Standard ML*, which is a functional language that shares a common ancestry with Haskell but its functions are impure. Thus, similar to RENEW, the CPN Tools leave it to the modeler to refrain from using functions with side effects.

Standard ML is not under active development anymore. Although, similar to Curry, its operational semantics are formally specified, the last specification was released in 1997.

---

[3] The occurrence of Java exceptions breaks this assumption which is a problem.

Haskell-Coloured Petri Nets

In [12] Reinke presents a simulator for CPNs with Haskell inscriptions. It is implemented as a RENEW plugin that generates a Haskell program for a given CPN. When executed, the program prints out a simulation log containing the names of fired transitions and text representations of reached markings. The implementation is minimalistic and the binding search is based on the standard functionality of the Haskell list type. It generates and tests all possible input token combinations for all transitions without any optimization.

Because Haskell is a pure language side effects are prevented.

4 Functionality

The CURRYCPN RENEW plugin offers two main interfaces to the *CCPN* simulator: the *Curry Net* formalism and the *Curry Reachability Graph* command. The *Curry Net* formalism integrates the *CCPN* simulator to be used similar to other simulators that are directly built into RENEW's Java implementation. In the background, it starts a separate *CCPN* simulator process. Proxy Java classes pass on standard simulation commands such as *start*, *step*, *stop* or *terminate* via an XML interface. The *load* command is sent first and includes a full PNML [6] representation of the CCPN to simulate.

The *Curry Reachability Graph* command instructs the *CCPN* simulator process to generate a complete reachability graph and send it back via XML. The graph layout is then automatically generated and the graph is displayed. In the graph representation each marking represented by the whole net instance with its marking can be inspected by double-clicking.

The *CCPN* simulator can also be used as a library by client Haskell programs. It exposes the same functionality as offered by the XML interface to the CURRYCPN RENEW plugin. Additionally, it offers monadic[4] interfaces for finer grained control of the simulation. The *CCPN* simulator is implemented as a highly modular library with the aim to be easily extendable. Most of the development efforts were spent on the simulator as the basis of all tools. The modular architecture is described in more detail in Sect. 5.

---

[4] A *monad* is a versatile Haskell type class that can be implemented by specific data types. It is used by the *CCPN* simulator to offer simulation actions that operate on a simulation state and are easy to combine and mix with user-provided pure code. The properties of monads guarantee that this can be achieved without directly exposing the simulation state and its implementation to the library user. The *CCPN* simulator provides multiple monad data types that encapsulate different aspects of the simulation state. The high-level monad type in the *SimulationMonad* module additionally encapsulates the IO side effects in the compilation of CCPN inscriptions. See [16] for a detailed introduction to monads.

5   Architecture

Figure 1 is an overview of the *CCPN* simulator's software components. Components depend on those below them. The names of the components that were developed specifically for the *CCPN* simulator are highlighted in bold font in the figure: the *ccpn* and *ccpn-runtime* Haskell packages as well as the CURRYCPN RENEW plugin. The architecture is split into three parts with thick outlines that are run as separate processes. The CURRYCPN RENEW plugin is executed inside a RENEW process. It and all RENEW plugins it depends on, are colored green. The main *CCPN* simulator, highlighted red, is started by the CURRYCPN plugin as another process. An XML interface is used for the interaction between the processes. The main *CCPN* simulator is composed of the *ccpn* and *ccpn-runtime* Haskell packages as well as some packages provided by the *KiCS2* Curry compiler [3] that are colored blue. It also uses the *KiCS2* compiler directly as a *kics2c* process.

Figure 2 is a dependency diagram of the modules of the *CCPN* simulator highlighted red in Fig. 1. For the sake of clarity only the transitive reduction of the dependencies is represented. The modules in the *ccpn* Haskell package are white, the *ccpn-runtime* package modules light red and modules from the *KiCS2* compiler blue. Encircled in orange is the simulation core; in blue (because of the *KiCS2* dependency) are the modules concerned with the compilation of CCPN inscriptions. The remaining modules are high-level and concerned with both. Four modules are in both circles because they are shared dependencies of both, the simulation core and the CCPN compilation. Only the CCPN compilation depends on packages of the *KiCS2* compiler. This dependency is contained in the transition binding searches. The simulation core is agnostic about the implementation of the transition binding searches and thus has no *KiCS2* dependency.

The inscriptions are compiled by the *Compiler* and *UncompiledNet* modules. First, the functionality of the *UncompiledNet* module parses the inscriptions to internal representation types of the *curry-base* Curry frontend package. Then, it generates a Curry module file with a function for each transition and one for each initial place marking.

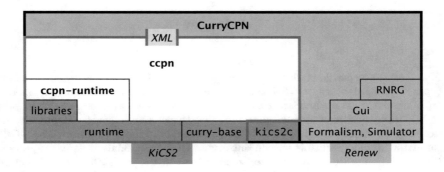

Fig. 1. Overview of the architecture. (Color figure online)

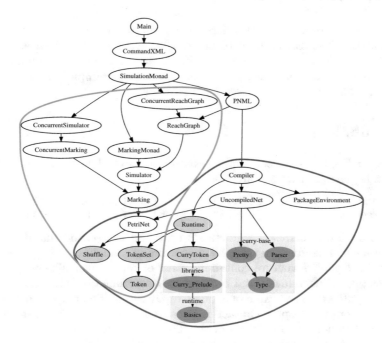

Fig. 2. The *CCPN* simulator architecture.

In the *Compiler* module, the module file is then compiled to Haskell code executing the *KiCS2* Curry compiler as a separate process. In the next step, using the library of the *GHC* Haskell compiler, the Haskell code is dynamically compiled and loaded into the running Haskell process. The functions generated for the transitions are wrapped to represent the full binding search as a Curry program. Finally, the *KiCS2* evaluation system is used to evaluate this program to a list of all possible deterministic transition bindings. The *Runtime* module defines the wrapping and evaluation and needs to be in the compilation context when these are dynamically compiled and loaded. The modules in red in Fig. 2 are in the separate *ccpn-runtime* package to allow the *Runtime* module to be dynamically imported at runtime.

The result of the whole compilation is a CCPN defined by the data structures in the *PetriNet* module. All *KiCS2*-dependencies are contained in the compilation. The simulation core is completely independent from *KiCS2* and the implementation of the binding search.

The *SimulationMonad* module offers a high-level monad interface to the simulation. It implements everything also accessible by the XML interface. This includes the initialization of a simulation by compiling and loading in the CCPN. The low-level interfaces and simulation implementations only work on simulations that are already set up.

The *MarkingMonad* module offers a low-level monad interface to the sequential simulation.[5] It can be used to fire specific transitions, as well as take and insert tokens from/to the marking. The *ConcurrentSimulator* module offers similar functionality for concurrent simulations. For examples using the *CCPN* simulator as a library, see [15, Section 8.2].

6   Example

The net in Fig. 3a is a very simple example CCPN. The following transition function is generated for the arc inscriptions:

```
CCPN__trans__word  :: [[Char]]           -> ([[Char]], ())
CCPN__trans__word     [a ++ " " ++ b] = ([a, b]   , ())
```

The first line declares the function's type and the second line declares the function itself. The function takes a list containing a single string as a parameter and returns a tuple structure containing a list of two strings. The output tokens are these two string values a and b. They are determined by matching the a ++ " " ++ b pattern to the value of a single input token. The pattern matches any string that contains a space character. Because the pattern is non-deterministic, the whole function is non-deterministic. This can be seen in the first transition firing: a can either be "these", or "these are" and b can be correspondingly "are words", or "words". You can find the two resulting markings in the reachability graph depicted in Fig. 3b.

A feature of Curry that distinguishes it from Haskell is the ability to match against patterns that include function applications. In this case, the pattern includes two applications of the list concatenation operator ++ that is built into the Curry's standard library (the *Prelude*):

```
(++)  :: [a] -> [a] -> [a]
[]       ++ ys        = ys
(x:xs)  ++ ys        = x :  xs++ys
```

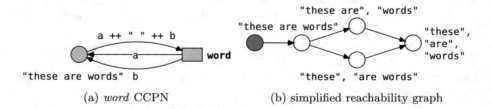

(a) *word* CCPN   (b) simplified reachability graph

Fig. 3. The *word* example.

---

[5] It offers a monad transformer that forms a monad based on any **MonadPlus** instance type given by the library user. This can be used to explore multiple markings for the same simulation steps instead of just a single simulation trace at a time.

The first line is the type: the ++ operator takes two lists and returns a third. All lists contain elements of the same generic type. In the a ++ " " ++ b pattern the lists are strings and the element type is Char. The last two lines are the rules that define the function. To match the pattern, the Curry evaluation environment applies them backwards in a search: the return value is known and the parameters are sought. This is an important feature of Curry as a logic programming language. It allows the CCPN's inscription to be very concise.

As you can see in Fig. 3b, the *word* CCPN has a deterministic final marking, but the intermediate markings are non-deterministic. The red circle does not represent a marking but points to all initial markings. There might be multiple, because place inscriptions may be non-deterministic. In this case there is only one. The reachability graph was generated by the CURRYCPN RENEW plugin. In the RENEW GUI a marking can be explored by double-clicking. This displays the marked CCPN. In Fig. 3b, additional text representations were added to the markings for demonstrative purposes.

For more usage examples including the usage of the Haskell library, see [15].

7 Conclusion

This contribution presents a tool set for the newly developed *Curry-coloured Petri nets* (CCPN) formalism. It introduces the functional logic programming language Curry with its powerful paradigm as the inscription language for CPNs for the first time. On the one hand, the *CCPN* formalism and tool set extend the Curry programming language by CPN concepts to realize a graphical concurrent modeling language. On the other hand, they extend the CPN formalism by Curry as an inscription language to realize side effect freedom, as well as a clean and versatile implementation of the binding search. The main components are the editor, the simulator, the reachability graph tool and the library.

By embedding the *CCPN* simulator in the RENEW environment a full-fledged graphical editor can be used to model CCPNs and to visualize their reachability graph. Due to the highly modular implementation in Haskell, the simulator can also be integrated into Haskell programs as a library and be extended by them.

The side effect freedom of the inscriptions is guaranteed by Curry's strong type system. The strict purely functional programming style allows inscriptions to be very abstract and concise. While e.g. for RENEW's Java reference nets the binding search implementation is highly complex, the *CCPN* simulator integrates the flexible *KiCS2* evaluation environment. Further development on the evaluation environment will directly benefit the *CCPN* binding search. Coupling of non-determinism of Curry and Petri nets as well as the concurrency concepts and their implementations raises relevant research questions.

Currently, hierarchical and nets-within-nets concepts are missing from the formalism. However, the architecture and design of the system are well prepared to incorporate these concepts. While the formal analysis of Java reference nets is restricted by Java's lack of control over side effects, the *CCPN* formalism's potential for formal analysis is very promising. Furthermore, case studies on the development of practical applications and empirical comparisons with other CPN simulators would be very helpful in identifying requirements for further research and development.

The *CCPN* tool set extends the multi-formalism capabilities of RENEW. It integrates a formalism that greatly complements the preexisting ecosystem. Further work on this integration will enable a case-by-case decision on the most suitable formalism for each part of a bigger system model.

References

1. Albert, E., Hanus, M., Huch, F., Oliver, J., Vidal, G.: Operational semantics for declarative multi-paradigm languages. J. Symb. Comput. **40**(1), 795–829 (2005)
2. Antoy, S., Hanus, M., Libby, S.: Proving non-deterministic computations in agda. In: Proceedings of WFLP 2016. Electronic Proceedings in Theoretical Computer Science, vol. 234, pp. 180–195. Open Publishing Association (2017)
3. Braßel, B., Hanus, M., Peemöller, B., Reck, F.: KiCS2: a new compiler from curry to Haskell. In: Kuchen, H. (ed.) WFLP 2011. LNCS, vol. 6816, pp. 1–18. Springer, Heidelberg (2011). https://doi.org/10.1007/978-3-642-22531-4_1
4. Delgado Friedrichs, F.: Referenznetze mit Anschriften in Scheme. Diploma thesis, University of Hamburg, Department of Informatics, September 2007
5. Hanus, M. (ed.): Curry: an integrated functional logic language (vers. 0.9.0) (2016). http://www.curry-language.org
6. Hillah, L., Kindler, E., Kordon, F., Petrucci, L., Trèves, N.: A primer on the Petri Net Markup Language and ISO/IEC 15909–2. Petri Net Newsl. **76**, 9–28 (2009)
7. Jensen, K.: Coloured Petri Nets: a high level language for system design and analysis. In: Rozenberg, G. (ed.) ICATPN 1989. LNCS, vol. 483, pp. 342–416. Springer, Heidelberg (1991). https://doi.org/10.1007/3-540-53863-1_31
8. Jensen, K., Kristensen, L.M., Wells, L.: Coloured Petri Nets and CPN Tools for modelling and validation of concurrent systems. IJSTTT **9**(3), 213–254 (2007)
9. Kristensen, L.M., Christensen, S.: Implementing Coloured Petri Nets using a functional programming language. High. Order Symb. Comput. **17**(3), 207–243 (2004)
10. Kummer, O., Wienberg, F., Duvigneau, M., Cabac, L., Haustermann, M., Mosteller, D.: Renew - The Reference Net Workshop, June 2016. http://www.renew.de/, release 2.5
11. Kummer, O.: Referenznetze. Logos Verlag, Berlin (2002)
12. Reinke, C.: Haskell-Coloured Petri Nets. In: Koopman, P., Clack, C. (eds.) IFL 1999. LNCS, vol. 1868, pp. 165–180. Springer, Heidelberg (2000). https://doi.org/10.1007/10722298_10
13. Simon, M., Moldt, D.: About the development of a Curry-Coloured Petri net simulator. In: Lorenz, R., Bergenthum, R. (eds.) AWPN 2018, pp. 53–54 (2018). https://opus.bibliothek.uni-augsburg.de/opus4/41861
14. Simon, M.: Concept and implementation of distributed simulations in Renew. Bachelor thesis, University of Hamburg, Department of Informatics, March 2014
15. Simon, M.: Curry-Coloured Petri Nets: a concurrent simulator for Petri Nets with purely functional logic program inscriptions. Master thesis, University of Hamburg, Department of Informatics, April 2018
16. Wadler, P.: Comprehending monads. Math. Struct. Comput. Sci. **2**(4), 461–493 (1992)

Synthesis

Articulation of Transition Systems and Its Application to Petri Net Synthesis

Raymond Devillers[✉]

Université Libre de Bruxelles, Boulevard du Triomphe C.P. 212,
1050 Bruxelles, Belgium
rdevil@ulb.ac.be

Abstract. In order to speed up the synthesis of Petri nets from labelled transition systems, a divide and conquer strategy consists in defining LTS decomposition techniques and corresponding PN composition operators to recombine the solutions of the various components. The paper explores how an articulation decomposition, possibly combined with a product and addition technique developed in previous papers, may be used in this respect and generalises sequence operators, as well as looping ones.

Keywords: Labelled transition systems · Composition · Decomposition · Petri net synthesis

1 Introduction

Instead of analysing a given system to check if it satisfies a set of desired properties, the synthesis approach tries to build a system "correct by construction" directly from those properties. In particular, more or less efficient algorithms have been developed to build a bounded Petri net (possibly of some subclass) the reachability graph of which is isomorphic to (or close to) a given finite labelled transition system [2,7,10,11,15].

The synthesis problem is usually polynomial in terms of the size of the LTS, with a degree between 2 and 5 depending on the subclass of Petri nets one searches for [2,3,7,10], but can also be NP-complete [4]. Hence the interest to apply a "divide and conquer" synthesis strategy when possible. The general idea is to decompose the given LTS into components, to synthesise each component separately and then to recombine the results in such a way to obtain a solution to the global problem. This has been applied successfully to disjoint products of LTS, which correspond to disjoint sums of Petri nets [12,13]. But it has also been observed that such products may be hidden inside other kinds of components, for instance in sequences of LTS, as illustrated in Fig. 1 (borrowed from [12]; the initial states are slightly fatter than the other ones), and developed in the algebra of Petri nets[1] [8,9].

---

[1] Note that this theory uses labelled Petri nets, where several transitions may have the same label, or multiset of labels, while here we shall only consider unlabelled Petri nets.

© Springer Nature Switzerland AG 2019
S. Donatelli and S. Haar (Eds.): PETRI NETS 2019, LNCS 11522, pp. 113–126, 2019.
https://doi.org/10.1007/978-3-030-21571-2_8

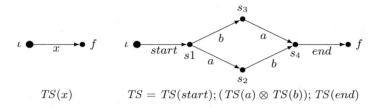

$$TS(x) \qquad\qquad TS = TS(start); (TS(a) \otimes TS(b)); TS(end)$$

Fig. 1. Combination of sequence operators with a product.

We shall here develop this last point, but we shall also generalise the idea of sequences into an *articulation* operator. The difference is that it will sometimes be possible to come back to the first component after having executed the second one, and repeat the alternation between these two components. A side consequence will be that the articulation operator will not always be anti-commutative, like the sequence is, and that it includes a choice as well as a looping feature.

The structure of the paper is as follows. After recalling the bases of labelled transition systems, a new articulation operator is introduced, and its basic properties are analysed. Then, the bases of the Petri net synthesis problem are recalled, and Sect. 4 shows how synthesis applies to the components of an articulation. Next, it is shown how articulations may be used to simplify a synthesis problem, by composing a solution of the given system from the solutions of the articulated components. A procedure is then detailed to show how to decompose a given LTS into articulated components, when possible. As usual, the last section concludes.

2   Labelled Transition Systems and Articulations

A classic way for representing the possible (sequential) evolutions of a dynamic system is through its labelled transition system [1].

Definition 1. LABELLED TRANSITION SYSTEMS
A *labelled transition system* (LTS for short) with initial state is a tuple $TS = (S, \rightarrow, T, \iota)$ with node (or state) set S, edge label set T, edges $\rightarrow \subseteq (S \times T \times S)$, and an initial state $\iota \in S$. We shall denote $s[t\rangle$ for $t \in T$ if there is an arc labelled t from s, $[t\rangle s$ if there is an arc lalbelled t going into s, and $s[\alpha\rangle s'$ if there is a path labelled $\alpha \in T^*$ from s to s'. Such a path will also be called an evolution of the LTS (from s to s').

Two LTSs $TS_1 = (S_1, \rightarrow_1, T, \iota_1)$ and $TS_2 = (S_2, \rightarrow_2, T, \iota_2)$ with the same label set T are (state-)isomorphic, denoted $TS_1 \equiv_T TS_2$ (or simply $TS_1 \equiv TS_2$ if T is clear from the context), if there is a bijection $\zeta \colon S_1 \rightarrow S_2$ with $\zeta(\iota_1) = \iota_2$ and $(s, t, s') \in \rightarrow_1 \Leftrightarrow (\zeta(s), t, \zeta(s')) \in \rightarrow_2$, for all $s, s' \in S_1$ and $t \in T$. We shall usually consider LTSs up to isomorphism.

We shall also assume each LTS is totally reachable (i.e., $\forall s \in S \exists \alpha \in T^*$: $\iota[\alpha\rangle s$), that it is weakly live (i.e., each label $t \in T$ occurs at least once in $\rightarrow$), and $T \neq \emptyset$.

Let $T_1 \subseteq T$. We shall denote by $adj(T_1) = \{s \in S | \exists t \in T_1 : s[t\rangle \text{ or } [t\rangle s\}$ the *adjacency set* of T_1, i.e., the set of states connected to T_1. Let $\emptyset \subset T_1 \subset T$, $T_2 = T \setminus T_1$ and $s \in S$. We shall say that TS is *articulated*[2] by T_1 and T_2 around s if $adj(T_1) \cap adj(T_2) = \{s\}$, $\forall s_1 \in adj(T_1) \exists \alpha_1 \in T_1^* : \iota[\alpha_1\rangle s_1$ and $\forall s_2 \in adj(T_2) \exists \alpha_2 \in T_2^* : s[\alpha_2\rangle s_2$.

Let $TS_1 = (S_1, \rightarrow_1, T_1, \iota_1)$ and $TS_2 = (S_2, \rightarrow_2, T_2, \iota_2)$ two (totally reachable) LTSs with $T_1 \cap T_2 = \emptyset$ and $s \in S_1$. Thanks to isomorphisms we may assume that $S_1 \cap S_2 = \{s\}$ and $\iota_2 = s$. We shall then denote by $TS_1 \lhd s \rhd TS_2 = (S_1 \cup S_2, T_1 \cup T_2, \rightarrow_1 \cup \rightarrow_2, \iota_1)$ the *articulation* of TS_1 and TS_2 around s.   □

Several easy but interesting properties may be derived for this articulation operator.

Note first that this operator is only defined up to isomorphism since we may need to rename the state sets (usually the right one, but we may also rename the left one, or both). The only constraint is that, after the relabellings, s is the unique common state of TS_1 and TS_2, and is the state where the two systems are to be articulated. Figure 2 illustrates this operator. It also shows that the articulation highly relates on the state around which the articulation takes part. It may also be observed that, if $TS_0 = (\{\iota\}, \emptyset, \emptyset, \iota)$ is the trivial empty LTS, we have that, for any state s of TS, $TS \lhd s \rhd TS_0 \equiv TS$, i.e., we have a kind of right neutral trivial articulation. Similarly, $TS_0 \lhd \iota \rhd TS \equiv TS$, i.e., we have a kind of left neutral trivial articulation. However, these neutrals will play no role in the following of this paper, so that we shall exclude them from our considerations (that is why we assumed the edge label sets to be non-empty).

Corollary 1. BOTH FORMS OF ARTICULATION ARE EQUIVALENT
If $TS = (S, \rightarrow, T, \iota)$ is articulated by T_1 and T_2 around s, then the structures $TS_1 = (adj(T_1), \rightarrow_1, T_1, \iota)$ and $TS_2 = (adj(T_2), \rightarrow_2, T_2, s)$, where $\rightarrow_1$ is the restriction of $\rightarrow$ to T_1 (i.e., $\rightarrow_1 = \rightarrow \cap adj(T_1) \times T_1 \times adj(T_1)$), and similarly for $\rightarrow_2$, are (totally reachable) LTSs, $TS \equiv_{T_1 \uplus T_2} TS_1 \lhd s \rhd TS_2$ (in that case we do not need to apply a relabelling to TS_1 and TS_2).

Conversely, $TS_1 \lhd s \rhd TS_2$ is articulated by the label sets of TS_1 and TS_2 around s, if these LTSs are totally reachable.   □

Corollary 2. EVOLUTIONS OF AN ARTICULATION
If $TS \equiv TS_1 \lhd s \rhd TS_2$, $\iota[\alpha\rangle s'$ is an evolution of TS iff it is an alternation of evolutions of TS_1 and TS_2 separated by occurrences of s, i.e., either $\alpha \in T_1^*$ or $\alpha = \alpha_1 \alpha_2 \ldots \alpha_n$ such that $\alpha_i \in T_1^*$ if i is odd, $\alpha_i \in T_2^*$ if i is even, $\iota[\alpha_1\rangle s$ and $\forall i \in \{1, 2, \ldots, n-1\} : [\alpha_i\rangle s[\alpha_{i+1}\rangle.$ □

---

[2] This notion has some similarity with the cut vertices (or articulation points) introduced for connected unlabelled undirected graphs, whose removal disconnects the graph. They have been used for instance to decompose such graphs into biconnected components [14, 16].

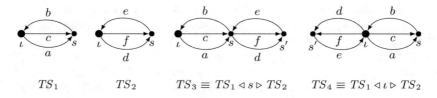

TS_1 TS_2 $TS_3 \equiv TS_1 \triangleleft s \triangleright TS_2$ $TS_4 \equiv TS_1 \triangleleft \iota \triangleright TS_2$

Fig. 2. Some articulations.

For instance, in TS_3 from Fig. 2, a possible evolution is $\iota[abc\rangle s[fede\rangle s[b\rangle\iota$, but also equivalently $\iota[a\rangle s[\varepsilon\rangle s[bc\rangle s[fe\rangle s[\varepsilon\rangle[de\rangle s[b\rangle\iota$ (where ε is the empty sequence).

Corollary 3. ASSOCIATIVITY OF ARTICULATIONS
Let us assume that TS_1, TS_2 and TS_3 are three LTSs with label sets T_1, T_2 and T_3 respectively, pairwise disjoint. Let s_1 be a state of TS_1 and s_2 be a state of TS_2. Then, $TS_1 \triangleleft s_1 \triangleright (TS_2 \triangleleft s_2 \triangleright TS_3) \equiv_{T_1 \cup T_2 \cup T_3} (TS_1 \triangleleft s_1 \triangleright TS_2) \triangleleft s_2' \triangleright TS_3$, where s_2' corresponds in $TS_1 \triangleleft s_1 \triangleright TS_2$ to s_2 in TS_2 (let us recall that the articulation operator may rename the states of the second operand).   □

This is illustrated by Fig. 3.

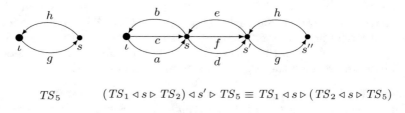

TS_5 $(TS_1 \triangleleft s \triangleright TS_2) \triangleleft s' \triangleright TS_5 \equiv TS_1 \triangleleft s \triangleright (TS_2 \triangleleft s \triangleright TS_5)$

Fig. 3. Associativity of articulations.

Corollary 4. COMMUTATIVE ARTICULATIONS
If $TS_1 = (S_1, \to_1, T, \iota_1)$ and $TS_2 = (S_2, \to_2, T, \iota_2)$ with disjoint label sets (i.e., $T_1 \cap T_2 = \emptyset$), then $TS_1 \triangleleft \iota_1 \triangleright TS_2 \equiv_{T_1 \cup T_2} TS_2 \triangleleft \iota_2 \triangleright TS_1$.   □

For instance, in Fig. 2, $TS_4 \equiv TS_1 \triangleleft \iota \triangleright TS_2 \equiv TS_2 \triangleleft \iota \triangleright TS_1$.

Corollary 5. COMMUTATIVE ASSOCIATIVITY OF ARTICULATIONS
Let us assume that TS_1, TS_2 and TS_3 are three LTSs with label sets T_1, T_2 and T_3 respectively, pairwise disjoint. Let s_2 and s_3 be two states of TS_1 ($s_2 = s_3$ is allowed). Then, $(TS_1 \triangleleft s_2 \triangleright TS_2) \triangleleft s_3 \triangleright TS_3 \equiv_{T_1 \cup T_2 \cup T_3} (TS_1 \triangleleft s_3 \triangleright TS_3) \triangleleft s_2 \triangleright TS_2$ (Fig. 4).   □

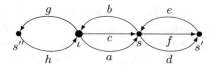

$$(TS_1 \lhd s \rhd TS_2) \lhd \iota \rhd TS_5 \equiv (TS_1 \lhd \iota \rhd TS_5) \lhd s \rhd TS_2$$

Fig. 4. Commutative associativity of articulations.

Corollary 6. SEQUENCE ARTICULATIONS

If $TS_1 = (S_1, \to_1, T, \iota_1)$ and $TS_2 = (S_2, \to_2, T, \iota_2)$ with disjoint label sets (i.e., $T_1 \cap T_2 = \emptyset$), if $\forall s_1 \in S_1 \exists \alpha_1 \in T_1^* : s_1[\alpha_1 \rangle s$ (s is a home state in TS_1) and $\nexists t_1 \in T_1 : s[t_1 \rangle$ (s is a dead end in TS_1), then $TS_1 \lhd s \rhd TS_2$ behaves like a sequence, i.e., once TS_2 has started, it is no longer possible to execute T_1.

The same occurs when ι_2 does not occur in a non-trivial cycle, i.e., $\iota_2[\alpha_2\rangle \iota_2 \wedge \alpha_2 \in T_2^* \Rightarrow \alpha_2 = \varepsilon$: once TS_2 has started, it is no longer possible to execute T_1.

□

This is illustrated in Fig. 5. It may be observed that sequences in [9] correspond to the intersection of both cases.

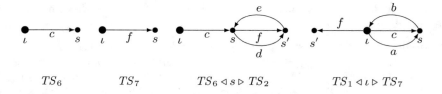

Fig. 5. Sequential articulations.

3   Petri Nets and Synthesis

Definition 2. PETRI NETS

An *initially marked Petri net* (PN for short) is denoted as $N = (P, T, F, M_0)$ where P is a set of places, T is a disjoint set of transitions $(P \cap T = \emptyset)$, F is the flow function $F : ((P \times T) \cup (T \times P)) \to \mathbb{N}$ specifying the arc weights, and M_0 is the initial marking (where a marking is a mapping $M : P \to \mathbb{N}$, indicating the number of tokens in each place).

Two Petri nets $N_1 = (P_1, T, F_1, M_0^1)$ and $N_2 = (P_2, T, F_2, M_0^2)$ with the same transition set T are isomorphic, denoted $N_1 \equiv_T N_2$ (or simply $N_1 \equiv N_2$ if T is clear from the context), if there is a bijection $\zeta : P_1 \to P_2$ such that, $\forall p_1 \in P_1, t \in T : M_0^1(p_1) = M_0^2(\zeta(p_1))$, $F_1(p_1, t) = F_2(\zeta(p_1), t)$ and $F_1(t, p_1) = F_2(t, \zeta(p_1))$. We shall usually consider Petri nets up to isomorphism.

A transition $t \in T$ is enabled at a marking M, denoted by $M[t\rangle$, if $\forall p \in P\colon M(p) \geq F(p, t)$. The firing of t leads from M to M', denoted by $M[t\rangle M'$, if $M[t\rangle$ and $M'(p) = M(p) - F(p, t) + F(t, p)$. This can be extended, as usual, to $M[\sigma\rangle M'$ for sequences $\sigma \in T^*$, and $[M\rangle$ denotes the set of markings reachable from M. The net is bounded if there is $k \in \mathbb{N}$ such that $\forall M \in [M_0\rangle, p \in P\colon M(p) \leq k$.

The reachability graph $RG(N)$ of N is the labelled transition system with the set of vertices $[M_0\rangle$, initial state M_0, label set T, and set of edges $\{(M, t, M') \mid M, M' \in [M_0\rangle \wedge M[t\rangle M'\}$. If an LTS TS is isomorphic to the reachability graph of a Petri net N, we say[3] that TS is *solvable* and that N *solves* TS. A synthesis problem consists in finding a PN solution for a given LTS, when possible.

Let M_1 and M_2 be two reachable markings of some Petri net N. We shall say that M_1 is dominated by M_2 if $M_1 \lneq M_2$, i.e., M_1 is distinct from M_2 and componentwise not greater. $\qquad\square$

Corollary 7. INDEPENDENCE FROM ISOMORPHISMS
Let N_1 and N_2 be two Petri nets. If $N_1 \equiv_T N_2$, then $RG(N_1) \equiv_T RG(N_2)$, so that if N_1 solves some LTS TS, N_2 also solves TS.
Let TS_1 and TS_2 be two LTS. If $TS_1 \equiv_T TS_2$ and some Petri net N solves TS_1, then N also solves TS_2. $\qquad\square$

4   Petri Net Synthesis and Articulation

We shall first see that if an articulation is solvable, then each component is individually solvable too.

Proposition 1. SYNTHESIS OF COMPONENTS OF AN ARTICULATION
If $TS = (S, \rightarrow, T_1 \uplus T_2, \iota)$ is articulated by T_1 and T_2 around s, so that $T \equiv TS_1 \lhd s \rhd TS_2$ with $TS_1 = (adj(T_1), \rightarrow_1, T_1, \iota)$ and $TS_2 = (adj(T_2), \rightarrow_2, s)$ (see Corollary 1), and is PN-solvable, so is each component TS_1 and TS_2. Moreover, in the corresponding solution for TS_1, the marking corresponding to s is not dominated by any other reachable marking. The same happens for the marking corresponding to ι_2 in TS_2 if the latter is finite.

Proof: Let $N = (P, T, F, M_0)$ be a solution for TS. It is immediate that $N_1 = (P, T_1, F_1, M_0)$, where F_1 is the restriction of F to T_1, is a solution for TS_1 (but there may be many other ones).

Similarly, if M is the marking of N (and N_1) corresponding to s, it may be seen that $N_2 = (P, T_2, F_2, M)$, where F_2 is the restriction of F to T_2, is a solution for TS_2 (but there may be many other ones).

Moreover, if $s[t_2\rangle$ for some label $t_2 \in T_2$ and M' is a marking of N_1 corresponding to some state s' in TS_1 with $M' \gneq M$, then $s \neq s'$, $s'[t_2\rangle$ and s is not the unique articulation between T_1 and T_2.

If M' is a reachable marking of N_2 with $M' \gneq M$, then, it is well known that PN_2 is unbounded, hence TS_2 may not be finite. $\qquad\square$

---

[3] Note that an LTS may be unsolvable, but if it is solvable there are many solutions, sometimes with very different structures.

Note that there may also be solutions to TS_1 (other than N_1) such that the marking M corresponding to s is dominated, but not if the LTS is reversible, i.e., if $\forall s_1 \in S_1 \exists \alpha_1 \in T_1^* : s_1[\alpha_1\rangle\iota_1$, due to the same infiniteness argument as above. This is illustrated in Fig. 6.

5   Recomposition

The other way round, let us now assume that $TS = TS_1 \triangleleft s \triangleright TS_2$ is an articulated LTS and that it is possible to solve TS_1 and TS_2. Is it possible from that to build a solution of TS?

To do that, we shall add the constraint already observed in Proposition 1 that, in the solution of TS_1 as well as in the one of TS_2, the marking corresponding to s is not dominated by another reachable marking. If this is satisfied we shall say that the solution is *adequate* (with respect to s). Hence, in the treatment of the system in Fig. 6, we want to avoid considering the solution N_1' of TS_1; on the contrary, N_1 or N_1'' will be acceptable.

If TS_2 is finite, as already mentioned, it is immediate that the initial marking M_0^2 (corresponding to s) in the solution of TS_2 is not dominated by any reachable marking, otherwise there is a path $M_0^2[\alpha\rangle M$ in that solution such that $M_0^2 \lneqq M$ and an infinite path $M_0^2[\alpha^\infty\rangle$, hence also an infinite path $\iota_2[\alpha^\infty\rangle$ in TS_2, contradicting the finiteness assumption.

If TS_1 is finite and reversible, from a similar argument, no marking reachable in the solution of TS_1 is dominated by another one, so that the constraint on s is satisfied. Otherwise, it is possible to force such a solution (if there is one) in the following way:

Proposition 2. Forcing an adequate solution for TS_1
Let us add to TS_1 an arc $s[u\rangle s$ where u is a new fresh label. Let TS_1' be the LTS so obtained. If TS_1' is not solvable, there is no adequate solution. Otherwise, solve TS_1' and erase u from the solution. Let N_1 be the net obtained with the procedure just described: it is a solution of TS_1 with the adequate property that the marking corresponding to s is not dominated by another one.

Proof: If there is an adequate solution N_1 of TS_1, with a marking M corresponding to s, let us add a new transition u to it with, for each place p of N_1, $W(p, u) = M(p) = W(u, p)$: the reachability graph of this new net is (isomorphic to) TS_1' since u is enabled by marking M (or any larger one, but there is none) and does not modify the marking. Hence, if there is no adequate solution of TS_1, there is no solution of TS_1'.

Let us now assume there is a solution N_1' of TS_1'. The marking M corresponding to s is not dominated otherwise there would be a loop $M'[s\rangle M'$ elsewhere in the reachability graph of N_1', hence also in TS_1'. Hence, dropping u in N_1' will lead to an adequate solution of TS_1.   □

For instance, when applied to TS_1 in Fig. 6, this will lead to N_1'', and not N_1' (N_1 could also be produced, but it is likely that a 'normal' synthesis procedure will not construct the additional isolated place).

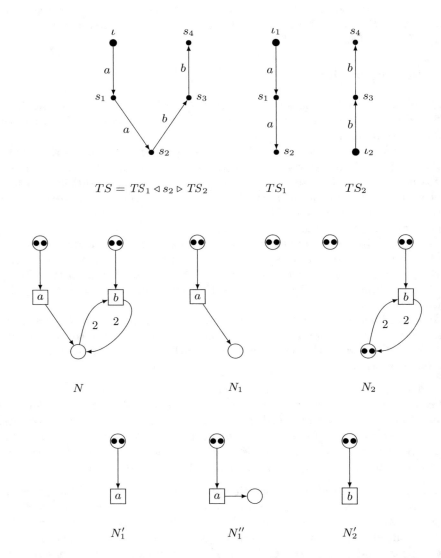

Fig. 6. The lts TS is articulated around s_2, with $T_1 = \{a\}$ and $T_2 = \{b\}$, hence leading to TS_1 and TS_2. It is solved by N, and the corresponding solutions for TS_1 and TS_2 are N_1 and N_2, respectively. TS_1 also has the solution N_1' but the marking corresponding to s_2 is then empty, hence it is dominated by the initial marking (as well as by the intermediate one). This is not the case for the other solution N_1'' (obtained from N_1 by erasing the useless isolated place: we never claimed that N_1 is a minimal solution). TS_2 also has the solution N_2'.

Now, to understand how one may generate a solution for TS from the ones obtained for TS_1 and TS_2, let us first consider the example illustrated in Fig. 7. This leads to the following construction.

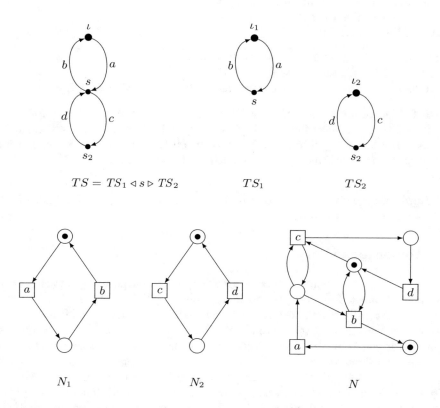

Fig. 7. The lts TS is articulated around s, with $T_1 = \{a, b\}$ and $T_2 = \{c, d\}$, hence leading to TS_1 and TS_2. It is solved by N, and the corresponding solutions for TS_1 and TS_2 are N_1 and N_2, respectively. In N, we may recognise N_1 and N_2, connected by two kinds of side conditions: the first one connects the label b out of s in TS_1 to the initial marking of N_2, the other one connects the label c out of ι_2 in TS_2 to the marking of N_1 corresponding to s.

Construction

Let $TS = TS_1 \triangleleft s \triangleright TS_2$ be an articulation of the LTS TS around s for the partition $T = T_1 \uplus T_2$.

Let N_1 be a Petri net solution of TS_1, with a non-dominated marking M_1 corresponding to s, and N_2 be a Petri net solution of TS_2, with an initial marking M_2 that we know to be non-dominated.

Let us assume that the places of N_1 and N_2 are disjoint, which is possible since we consider nets up to isomorphism, and let us put them side by side.

For each transition t_1 out of s in TS_1, and each place p_2 such that $M_2(p_2) > 0$, let us create a side condition $F(t_1, p_2) = F(p_2, t_1) = M_2(p_2)$.
For each transition t_2 out of ι_2 in TS_2, and each place p_1 such that $M_1(p_1) > 0$, let us create a side condition $F(t_2, p_1) = F(p_1, t_2) = M_1(p_1)$.
The result is a Petri net N.
End of Construction

Proposition 3. SYNTHESIS OF ARTICULATION
If TS_1 or TS_2 are not solvable, so is TS.
    Otherwise, the net N constructed as above is a solution of TS.

Proof: The property arises from the observation that N_1 with the additional side conditions behaves like the original N_1 provided that, when we reach M_1, N_2 does not leave M_2. Similarly, N_2 with the added side conditions behaves like the original N_2 provided N_1 reached M_1 and stays there, until N_2 returns to M_2.   □

Note that we do not claim this is the only solution, but the goal is to find a solution when there is one.

6   Decomposition

It remains to show when and how an LTS may be decomposed by an articulation (or several ones). Let us thus consider some LTS $TS = (S, \rightarrow, T, \iota)$. We may assume it is totally reachable (the states which may not be reached from ι play no role in the evolutions of the system) and that the label set T is finite (otherwise, it may happen that the finest decomposition is infinite. Usually we shall also assume that the state set S is also finite, otherwise there may be a problem to implement the procedure we are about to describe in a true algorithm. We may also assume it is deterministic, i.e., $(s[t\rangle \wedge s[t'\rangle) \Rightarrow t = t'$ and $([t\rangle s \wedge [t'\rangle s) \Rightarrow t = t'$ for any state $s \in S$ and labels $t, t' \in T$, otherwise there may be no unlabelled Petri net solution.

First, we may observe that, for any two distinct labels $t, t' \in T$, if $|adj(\{t\}) \cap adj(\{t'\})| > 1$, t and t' must belong to a same subset for defining an articulation (if any). Let us extend the function adj to non-empty subsets of labels by stating $adj(T') = \cup_{t \in T'} adj(t)$ when $\emptyset \subset T' \subset T$. We then have that, if $\emptyset \subset T_1, T_2 \subset T$ and we know that all the labels in T_1 must belong to a same subset for defining an articulation, and similarly for T_2, $|adj(T_1) \cap adj(T_2)| > 1$ implies that $T_1 \cup T_2$ must belong to a same subset of labels defining an articulation (if any). If we get the full set T, that means that there is no possible articulation (but trivial ones, that we excluded from this study).

Hence, starting from any partition $\mathcal{T}$ of T (initially, if $T = \{t_1, t_2, \ldots, t_n\}$, we shall start from the finest partition $\mathcal{T} = \{\{t_1\}, \{t_2\}, \ldots, \{t_n\}\}$), we shall construct the finest partition compatible with the previous rule:

while there is $T_1, T_2 \in \mathcal{T}$ such that $T_1 \neq T_2$ and $|adj(T_1) \cap adj(T_2)| > 1$, replace T_1 and T_2 in $\mathcal{T}$ by $T_1 \cup T_2$.

At the end, if $\mathcal{T} = \{T\}$, we may stop with the result: *there is no non-trivial articulation*.

Otherwise, we may define a finite bipartite undirected graph whose nodes are the members of the partition $\mathcal{T}$ and some states of S, such that if $T_i, T_j \in \mathcal{T}, T_i \neq T_j$ and $adj(T_i) \cap adj(T_j) = \{s\}$, there is a node s in the graph, connected to T_i and T_j (and this is the only reason to have a state as a node of the graph). Since TS is weakly live and totally reachable, this graph is connected, and each state occurring in it has at least two neighbours (on the contrary, a subset of labels may be connected to a single state). Indeed, since TS is weakly live, $\cup_{T' \in \mathcal{T}} adj(T') = S$. Each state s occurring as a node in the graph is connected to at least two members of the $\mathcal{T}$, by the definition of the introduction of s in the graph. Let T_1 be the member of $\mathcal{T}$ such that $\iota \in adj(T_1)$, let T_i be any other member of $\mathcal{T}$, and let us consider a path $\iota[\alpha\rangle$ ending with some $t \in T_i$ (we may restrict our attention to a short such path, but this is not necessary): each time there is a sequence $t't''$ in α such that t' and t'' belong to two different members T' and T'' of $\mathcal{T}$, we have $[t'\rangle s[t''\rangle$, where s is the only state-node connected to T' and T'', hence in the graph we have $T' \to s \to T''$. This will yield a path in the constructed graph going from T_1 to T_i, hence the connectivity.

If there is a cycle in this graph, that means that there is no way to group the members of $\mathcal{T}$ in this cycle in two subsets such that the corresponding adjacency sets only have a single common state. Hence we need to fuse all these members, for each such cycle, leading to a new partition, and we also need to go back to the refinement of the partition in order to be compatible with the intersection rule, and to the construction of the graph.

Finally, we shall get an acyclic graph G, with at least three nodes (otherwise we stopped the articulation algorithm).

We shall now define a procedure $articul(SG)$ that builds an LTS expression based on articulations from a subgraph SG of G with a chosen state-node root. We shall then apply it recursively to G, leading finally to an articulation-based (possibly complex) expression equivalent to the original LTS TS.

The basic case will be that, if SG is a graph composed of a state s connected to a subset node T_i, $articul(SG)$ will be the LTS $TS_i = (adj(T_i), T_i, \to_i, s)$ (as usual $\to_i$ is the projection of $\to$ on T_i; by construction, it will always be the case that $s \in adj(T_i)$).

First, if ι is a state-node of the graph, G then has the form of a star with root ι and a set of satellite subgraphs $G_1, G_2, \ldots, G_n$ (n is at least 2). Let us denote by SG_i the subgraph with root ι connected to G_i: the result will then be the (commutative, see Corollary 4) articulation around ι of all the LTSs $articul(SG_i)$.

Otherwise, let T_1 be the (unique) label subset in the graph such that $\iota \in adj(T_1)$. G may then be considered as a star with T_1 at the center, surrounded by subgraphs $SG_1, SG_2, \ldots, SG_n$ (here n may be 1), each one with a root s_i connected to T_1 (we have here that $s_i \in adj(T_1)$, and we allow $s_i = s_j$): the result is then $((\ldots((adj(T_1), T_1, \to_1, \iota) \triangleleft s_1 \triangleright articul(SG_1)) \triangleleft s_2 \triangleright articul(SG_2)) \ldots) \triangleleft s_n \triangleright articul(SG_n))$. Note that, if $n > 1$, the order in which we consider the subgraphs is irrelevant from Corollary 5.

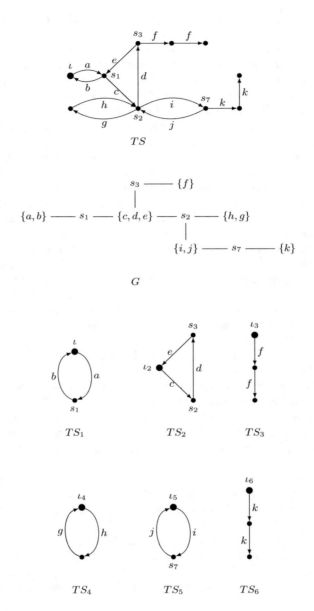

Fig. 8. The lts TS leads to the graph G. The corresponding components are TS_1 to TS_6, which may easily be synthesised; note that, from the total reachability of TS, they are all totally reachable themselves. This leads to the articulated expression below.

Finally, if a subgraph starts from a state s', followed by a subset T', itself followed by subgraphs $SG_1, SG_2, \ldots, SG_n$ ($n \geq 1$; if it is 0 we have the base case), each one with a root s_i connected to T' (we have here that $s' \in adj(T')$, and we allow $s_i = s_j$): the result is then $((\ldots((adj(T'), T', \rightarrow', s') \lhd s_1 \rhd articul(SG_1)) \lhd s_2 \rhd articul(SG_2)) \ldots) \lhd s_n \rhd articul(SG_n))$. Again, if $n > 1$, the order in which we consider the subgraphs is irrelevant from Corollary 5.

This procedure is illustrated in Fig. 8.

7 Concluding Remarks

We have developed a theory around a new operator acting on labelled transition systems, that we called articulation. Its main algebraic properties have been exhibited, and it was shown how this may be used to construct syntheses from the solutions of the various components. Since the latter are simpler than the original LTS (when articulation is possible), it is also much simpler to synthesise them (when possible), hence speeding up the global synthesis, or the detection that this is not possible (while pointing at the culprit components). A procedure has also been devised to decompose a given LTS into articulated components, when possible.

It remains to build effectively the corresponding procedures, and to incorporate them in some existing tool, like APT [11].

Other possible issues are to examine how this may be specialised for some subclasses of Petri nets, like Choice-Free ones, where each place has at most one outgoing transition: this is exactly the class of Petri nets allowing fully distributed implementations [6], and they present interesting behavioural properties [5,10] which could be exploited.

Another possible extension would be to look how these articulations behave in the context of approximate solutions devised when an exact synthesis is not possible, in the spirit of the notions and procedures developed in [15].

Finally, other kinds of LTS operators could be searched for, having interesting decomposition procedures, and corresponding to compositions of component solutions allowing to speed up synthesis problems.

Acknowledgements. The author thanks Eike Best as well as the anonymous referees for their useful remarks and suggestions.

References

1. Arnold, A.: Finite Transition Systems - Semantics of Communicating Systems. Prentice Hall International Series in Computer Science. Prentice Hall, Hertfordshire (1994)
2. Badouel, E., Bernardinello, L., Darondeau, P.: Petri Net Synthesis. TTCSAES. Springer, Heidelberg (2015). https://doi.org/10.1007/978-3-662-47967-4

3. Badouel, E., Bernardinello, L., Darondeau, P.: Polynomial algorithms for the synthesis of bounded nets. In: Mosses, P.D., Nielsen, M., Schwartzbach, M.I. (eds.) CAAP 1995. LNCS, vol. 915, pp. 364–378. Springer, Heidelberg (1995). https://doi.org/10.1007/3-540-59293-8_207

4. Badouel, E., Bernardinello, L., Darondeau, P.: The synthesis problem for elementary net systems is NP-complete. Theor. Comput. Sci. **186**(1–2), 107–134 (1997). https://doi.org/10.1016/S0304-3975(96)00219-8

5. Best, E., Devillers, R., Schlachter, U., Wimmel, H.: Simultaneous Petri net synthesis. Sci. Ann. Comput. Sci. **28**(2), 199–236 (2018)

6. Best, E., Darondeau, P.: Petri net distributability. In: Clarke, E., Virbitskaite, I., Voronkov, A. (eds.) PSI 2011. LNCS, vol. 7162, pp. 1–18. Springer, Heidelberg (2012). https://doi.org/10.1007/978-3-642-29709-0_1

7. Best, E., Devillers, R.: Characterisation of the state spaces of live and bounded marked graph Petri nets. In: Dediu, A.-H., Martín-Vide, C., Sierra-Rodríguez, J.-L., Truthe, B. (eds.) LATA 2014. LNCS, vol. 8370, pp. 161–172. Springer, Cham (2014). https://doi.org/10.1007/978-3-319-04921-2_13

8. Best, E., Devillers, R., Koutny, M.: Petri Net Algebra. Monographs in Theoretical Computer Science. An EATCS Series. (2001). https://doi.org/10.1007/978-3-662-04457-5

9. Best, E., Devillers, R., Koutny, M.: The box algebra = Petri nets + process expressions. Inf. Comput. **178**(1), 44–100 (2002). https://doi.org/10.1006/inco.2002.3117

10. Best, E., Devillers, R., Schlachter, U.: Bounded choice-free Petri net synthesis: algorithmic issues. Acta Informatica **55**(7), 575–611 (2018)

11. Best, E., Schlachter, U.: Analysis of Petri nets and transition systems. In: Proceedings 8th Interaction and Concurrency Experience, ICE 2015, Grenoble, France, 4–5th June 2015, pp. 53–67 (2015). https://doi.org/10.4204/EPTCS.189.6

12. Devillers, R.: Factorisation of transition systems. Acta Informatica **55**(4), 339–362 (2018)

13. Devillers, R., Schlachter, U.: Factorisation of Petri net solvable transition systems. In: Application and Theory of Petri Nets and Concurrency - 39th International Conference, PETRI NETS 2018, Bratislava, Slovakia, pp. 82–98 (2018)

14. Hopcroft, J.E., Tarjan, R.E.: Efficient algorithms for graph manipulation [H] (algorithm 447). Commun. ACM **16**(6), 372–378 (1973)

15. Schlachter, U.: Over-approximative Petri net synthesis for restricted subclasses of nets. In: Klein, S.T., Martín-Vide, C., Shapira, D. (eds.) LATA 2018. LNCS, vol. 10792, pp. 296–307. Springer, Cham (2018). https://doi.org/10.1007/978-3-319-77313-1_23

16. Westbrook, J., Tarjan, R.E.: Maintaining bridge-connected and biconnected components on-line. Algorithmica **7**(5&6), 433–464 (1992). https://doi.org/10.1007/BF01758773

Hardness Results for the Synthesis of b-bounded Petri Nets

Ronny Tredup[(✉)]

Universität Rostock, Institut für Informatik, Theoretische Informatik,
Albert-Einstein-Straße 22, 18059 Rostock, Germany
ronny.tredup@uni-rostock.de

Abstract. Synthesis for a type τ of Petri nets is the following search problem: For a transition system A, find a Petri net N of type τ whose state graph is isomorphic to A, if there is one. To determine the computational complexity of synthesis for types of bounded Petri nets we investigate their corresponding decision version, called feasibility. We show that feasibility is NP-complete for (pure) b-bounded P/T-nets if $b \in \mathbb{N}^+$. We extend (pure) b-bounded P/T-nets by the additive group $\mathbb{Z}_{b+1}$ of integers modulo $(b+1)$ and show feasibility to be NP-complete for the resulting type. To decide if A has the *event state separation property* is shown to be NP-complete for (pure) b-bounded and group extended (pure) b-bounded P/T-nets. Deciding if A has the *state separation property* is proven to be NP-complete for (pure) b-bounded P/T-nets.

1 Introduction

Synthesis for a Petri net type τ is the task to find, for a given transition system (TS, for short) A, a Petri net N of this type such that its state graph is isomorphic to A if such a net exists. The decision version of synthesis is called τ-*feasibility*. It asks whether for a given TS A a Petri net N of type τ exists whose state graph is isomorphic to A.

Synthesis for Petri nets has been investigated and applied for many years and in numerous fields: It is used to extract concurrency and distributability data from sequential specifications like transition systems or languages [5]. Synthesis has applications in the field of process discovery to reconstruct a model from its execution traces [1]. In [9], it is employed in supervisory control for discrete event systems and in [6] it is used for the synthesis of speed-independent circuits. This paper deals with the computational complexity of synthesis for types of *bounded* Petri nets, that is, Petri nets for which there is a positive integer b restricting the number of tokens on every place in any reachable marking.

In [2,4], synthesis has been shown to be solvable in polynomial time for bounded and pure bounded P/T-nets. The approach provided in [2,4] guarantees a (pure) bounded P/T-net to be output if such a net exists. Unfortunately, it does not work for preselected bounds. In fact, in [3] it has been shown that feasibility is NP-complete for 1-bounded P/T-nets, that is, if the bound $b = 1$ is chosen a

© Springer Nature Switzerland AG 2019
S. Donatelli and S. Haar (Eds.): PETRI NETS 2019, LNCS 11522, pp. 127–147, 2019.
https://doi.org/10.1007/978-3-030-21571-2_9

priori. In [15,17], it was proven that this remains true even for strongly restricted input TSs. In contrast, [12] shows that it suffices to extend pure 1-bounded P/T-nets by the additive group $\mathbb{Z}_2$ of integers modulo 2 to bring the complexity of synthesis down to polynomial time. The work of [16] confirms also for other types of 1-bounded Petri nets that the presence or absence of interactions between places and transitions tip the scales of synthesis complexity. However, some questions in the area of synthesis for Petri nets are still open. Recently, in [11] the complexity status of synthesis for (pure) b-bounded P/T-nets, $2 \leq b$, has been reported as unknown. Furthermore, it has not yet been analyzed whether extending (pure) b-bounded P/T-nets by the group $\mathbb{Z}_{b+1}$ provides also a tractable superclass if $b \geq 2$.

In this paper, we show that feasibility for (pure) b-bounded P/T-nets, $b \in \mathbb{N}^+$, is NP-complete. This makes their synthesis NP-hard. Moreover, we introduce (pure) $\mathbb{Z}_{b+1}$-extended b-bounded P/T-nets, $b \geq 2$. This type origins from (pure) b-bounded P/T-nets by adding interactions between places and transitions simulating addition of integers modulo $b+1$. This extension is a natural generalization of Schmitt's approach [12], which does this for $b = 1$. In contrast to the result of [12], this paper shows that feasibility for (pure) $\mathbb{Z}_{b+1}$-extended b-bounded P/T-nets remains NP-complete if $b \geq 2$.

To prove the NP-completeness of feasibility we use its well known close connection to the so-called *event state separation property* (ESSP) and *state separation property* (SSP). In fact, a TS A is feasible with respect to a Petri net type if and only if it has the type related ESSP *and* SSP [4]. The question of whether a TS A has the ESSP or the SSP also defines decision problems. The possibility to decide efficiently if A has at least one of both properties serves as quick-fail pre-processing mechanisms for feasibility. Moreover, if A has the ESSP then synthesizing Petri nets up to language equivalence is possible [4]. This makes the decision problems ESSP and SSP worth to study. In [8], both problems have been shown to be NP-complete for pure 1-bounded P/T-nets. This has been confirmed for almost trivial inputs in [15,17].

This paper shows feasibility, ESSP and SSP to be NP-complete for b-bounded P/T-nets, $b \in \mathbb{N}^+$. Moreover, feasibility and ESSP are shown to remain NP-complete for (pure) $\mathbb{Z}_{b+1}$-extended b-bounded P/T-nets if $b \geq 2$. Interestingly, [13] shows that SSP is decidable in polynomial time for (pure) $\mathbb{Z}_{b+1}$-extended b-bounded P/T-nets, $b \in \mathbb{N}^+$. So far, this is the first net family where the provable computational complexity of SSP is different to feasibility and ESSP.

All presented NP-completeness proofs base on a reduction from the monotone one-in-three 3-SAT problem which is known to be NP-complete [10]. Every reduction starts from a given boolean input expression φ and results in a TS A_φ. The expression φ belongs to monotone one-in-three 3-SAT if and only if A_φ has the (target) property ESSP, SSP or feasibility, respectively.

This paper is organized as follows: Sect. 2 introduces the formal definitions and notions. Section 3 introduces the concept of unions applied in by our proofs. Section 4 provides the reductions and proves their functionality. A short conclusion completes the paper. This paper is an extended abstract of the technical report [14]. The proofs that had to be removed due to space limitation are given in [14].

2   Preliminaries

See Figs. 1 and 2 for an example of the notions defined in this section. A *transition system* (TS for short) $A = (S, E, \delta)$ consists of finite disjoint sets S of states and E of events and a partial *transition function* $\delta : S \times E \to S$. Usually, we think of A as an edge-labeled directed graph with node set S where every triple $\delta(s, e) = s'$ is interpreted as an e-labeled edge $s \xrightarrow{e} s'$, called *transition*. We say that an event e *occurs* at state s if $\delta(s, e) = s'$ for some state s' and abbreviate this with $s \xrightarrow{e}$. This notation is extended to words $w' = wa$, $w \in E^*, a \in E$ by inductively defining $s \xrightarrow{\varepsilon} s$ for all $s \in S$ and $s \xrightarrow{w'} s''$ if and only if $s \xrightarrow{w} s'$ and $s' \xrightarrow{a} s''$. If $w \in E^*$ then $s \xrightarrow{w}$ denotes that there is a state $s' \in S$ such that $s \xrightarrow{w} s'$. An *initialized* TS $A = (S, E, \delta, s_0)$ is a TS with an initial state $s_0 \in S$ where every state is *reachable*: $\forall s \in S, \exists w \in E^* : s_0 \xrightarrow{w} s$. The language of A is the set $L(A) = \{w \in E^* \mid s_0 \xrightarrow{w}\}$. In the remainder of this paper, if not explicitly stated otherwise, we assume all TSs to be initialized and we refer to the components of an (initialized) TS A consistently by $A = (S_A, E_A, \delta_A, s_{0,A})$.

The following notion of *types of nets* has been developed in [4]. It allows us to uniformly capture several Petri net types in one general scheme. Every introduced Petri net type can be seen as an instantiation of this general scheme. A type of nets τ is a TS $\tau = (S_\tau, E_\tau, \delta_\tau)$ and a Petri net $N = (P, T, f, M_0)$ of type τ, τ-net for short, is given by finite and disjoint sets P of places and T of transitions, an initial marking $M_0 : P \longrightarrow S_\tau$, and a flow function $f : P \times T \to E_\tau$. The meaning of a τ-net is to realize a certain behavior by cascades of firing transitions. In particular, a transition $t \in T$ can fire in a marking $M : P \longrightarrow S_\tau$ and thereby produces the marking $M' : P \longrightarrow S_\tau$ if for all $p \in P$ the transition $M(p) \xrightarrow{f(p, t)} M'(p)$ exists in τ. This is denoted by $M \xrightarrow{t} M'$. Again, this notation extends to sequences $\sigma \in T^*$. Accordingly, $RS(N) = \{M \mid \exists \sigma \in T^* : M_0 \xrightarrow{\sigma} M\}$ is the set of all reachable markings of N. Given a τ-net $N = (P, T, f, M_0)$, its behavior is captured by the TS $A_N = (RS(N), T, \delta, M_0)$, called the state graph of N, where for every reachable marking M of N and transition $t \in T$ with $M \xrightarrow{t} M'$ the transition function δ of A_N is defined by $\delta(M, t) = M'$.

The following notion of τ-regions allows us to define the type related ESSP and SSP. If τ is a type of nets then a τ-region of a TS A is a pair of mappings (sup, sig), where $sup : S_A \longrightarrow S_\tau$ and $sig : E_A \longrightarrow E_\tau$, such that, for each transition $s \xrightarrow{e} s'$ of A, we have that $sup(s) \xrightarrow{sig(e)} sup(s')$ is a transition of τ. Two distinct states $s, s' \in S_A$ define an *SSP atom* (s, s'), which is said to be τ-solvable if there is a τ-region (sup, sig) of A such that $sup(s) \neq sup(s')$. An event $e \in E_A$ and a state $s \in S_A$ at which e does not occur, that is $\neg s \xrightarrow{e}$, define an *ESSP atom* (e, s). The atom is said to be τ-solvable if there is a τ-region (sup, sig) of A such that $\neg sup(s) \xrightarrow{sig(e)}$. A τ-region solving an ESSP or a SSP atom (x, y) is a *witness* for the τ-solvability of (x, y). A TS A has the τ-ESSP (τ-SSP) if all its ESSP (SSP) atoms are τ-solvable. Naturally, A is said to be τ-feasible if it has the τ-ESSP and the τ-SSP. The following fact is well known from [4, p.161]: A set $\mathcal{R}$ of τ-regions of A contains a witness for all ESSP and SSP atoms if and only if the *synthesized τ-net* $N_A^{\mathcal{R}} = (\mathcal{R}, E_A, f, M_0)$ has a state graph that is isomorphic to A. The flow function of $N_A^{\mathcal{R}}$ is defined by $f((sup, sig), e) = sig(e)$ and its initial marking is $M_0((sup, sig)) = sup(s_{0,A})$ for all $(sup, sig) \in \mathcal{R}, e \in E_A$. The regions of $\mathcal{R}$ become places and the events of E_A become transitions of $N_A^{\mathcal{R}}$. Hence, for a τ-feasible TS A where $\mathcal{R}$ is known, we can synthesize a net N with state graph isomorphic to A by constructing $N_A^{\mathcal{R}}$.

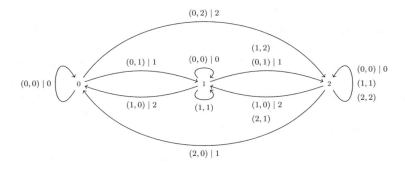

Fig. 1. The types $\tau_0^2, \tau_1^2, \tau_2^2$ and τ_3^2. τ_0^2 is sketched by the (m, n)-labeled transitions where edges with different labels represent different transitions. Discard from τ_0^2 the $(1, 1), (1, 2), (2, 1)$ and $(2, 2)$ labeled transitions to get τ_1^2 and add for $i \in \{0, 1, 2\}$ the i-labeled transitions and remove $(0, 0)$ to have τ_2^2. Discarding $(1, 1), (1, 2), (2, 1), (2, 2)$ leads from τ_2^2 to τ_3^2.

In this paper, we deal with the following b-bounded types of Petri nets:

1. The type of *b-bounded P/T-nets* is defined by $\tau_0^b = (\{0, \ldots, b\}, \{0, \ldots, b\}^2, \delta_{\tau_0^b})$ where for $s \in S_{\tau_0^b}$ and $(m, n) \in E_{\tau_0^b}$ the transition function is defined by $\delta_{\tau_0^b}(s, (m, n)) = s - m + n$ if $s \geq m$ and $s - m + n \leq b$, and undefined otherwise.

2. The type τ_1^b of *pure b-bounded P/T-nets* is a restriction of τ_0^b-nets that discards all events (m, n) from $E_{\tau_0^b}$ where both, m and n, are positive. To be exact, $\tau_1^b = (\{0, \ldots, b\}, E_{\tau_0^b} \setminus \{(m, n) \mid 1 \leq m, n \leq b\}, \delta_{\tau_1^b})$, and for $s \in S_{\tau_1^b}$ and $e \in E_{\tau_0^b}$ we have $\delta_{\tau_1^b}(s, e) = \delta_{\tau_0^b}(s, e)$.

3. The type τ_2^b of $\mathbb{Z}_{b+1}$-*extended b-bounded P/T-nets* origins from τ_0^b by extending the event set $E_{\tau_0^b}$ with the elements $0, \ldots, b$. The transition function additionally simulates the addition modulo $(b+1)$. More exactly, this type is defined by $\tau_2^b = (\{0, \ldots, b\}, (E_{\tau_0^b} \setminus \{(0, 0)\}) \cup \{0, \ldots, b\}, \delta_{\tau_2^b})$ where for $s \in S_{\tau_2^b}$ and $e \in E_{\tau_2^b}$ we have that $\delta_{\tau_2^b}(s, e) = \delta_{\tau_0^b}(s, e)$ if $e \in E_{\tau_0^b}$ and, otherwise, $\delta_{\tau_2^b}(s, e) = (s + e) \mod (b + 1)$.

In order to prove the functionality of the constituents and to convey the corresponding intuition without becoming too technical, we proceed as follows. On the one hand, we precisely define the constituents of the unions for arbitrary bound b and input instance $\varphi = \{C_0, \ldots, C_{m-1}\}$, $C_i = \{X_{i,0}, X_{i,1}, X_{i,2}\}$, $i \in \{0, \ldots, m-1\}$, $V(\varphi) = \{X_0, \ldots, X_{m-1}\}$, and prove their functionality. On the other hand, we provide for comprehensibility full examples for the types $\tau \in \{\tau_0^b, \tau_1^b\}$ and the unions U_τ and W. The illustrations also provide a τ-region solving the corresponding key atom. For a running example, the input instance is $\varphi_0 = \{C_0, \ldots, C_5\}$ with clauses $C_0 = \{X_0, X_1, X_2\}$, $C_1 = \{X_2, X_0, X_3\}$, $C_2 = \{X_1, X_3, X_0\}$, $C_3 = \{X_2, X_4, X_5\}$, $C_4 = \{X_1, X_5, X_4\}$, $C_5 = \{X_4, X_3, X_5\}$ that allows the one-in-three model $\{X_0, X_4\}$. A full example for $\tau \in \{\tau_2^b, \tau_3^b\}$ is given in [13]. For further simplification, we reuse gadgets for several unions as far as possible. This is not always possible as small differences between two types of nets imply huge differences in the possibilities to build corresponding regions: The more complex (the transition function of) the considered types, the more difficult the task to connect the solvability of the key atom with the signature of the interface events, respectively to connect the signature of the interface events with an implied model. Moreover, the more difficult these tasks, the more complex the corresponding gadgets. Hence, less complex gadgets are useless for more complex types. Reversely, the more complex the gadgets the more possibilities to solve all ESSP atoms and all SSP atoms are needed. Hence, more complex gadgets are not useful for less complex types. At the end, some constituents may differ only slightly at first glance but their differences have a crucial and necessary impact.

Note, that some techniques of the proof of Theorem 1 are very general advancements of our previous work [15, 17]. For example, like in [15, 17] the proof of Theorem 1 bases on reducing cubic monotone one-in-three 3-SAT. Moreover, we apply unions as part of *component design* [7]. However, the reductions in [15, 17] fit only for the basic type τ_1^1 and they are already useless for τ_0^1. They fit even less for τ_0^b and τ_1^b if $b \geq 2$ and certainly not for their group extensions.

We proceed as follows. Sections 4.1 and 4.2 introduce the keys $K_{\tau_0^b}, K_{\tau_1^b}, K$ and translators $T_{\tau_0^b}, T_{\tau_1^b}, T$ and prove their functionality. Sections 4.3 and 4.4 present $K_{\tau_2^b}, K_{\tau_3^b}$ and $T_{\tau_2^b}, T_{\tau_3^b}$ and carry out how they work. Section 4.5 proves that the keys and translators collaborate properly.

4.1   The Unions $K_{\tau_0^b}$ and $K_{\tau_1^b}$ and K

Let $\tau \in \{\tau_0^b, \tau_1^b\}$. The aim of K_τ and K is summarized by the next lemma:

Lemma 3. *The keys K_τ and K implement the interface events $k_0, \ldots, k_{6m-1}$ and provide a key atom a_τ and α, respectively, such that the following is true:*

1. *(Completeness) If (sup_K, sig_K) is τ-region of K_τ, respectively of K, that solves α_τ, respectively α, then $sig_K(k_0) = \cdots = sig_K(k_{6m-1}) = (0, b)$ or $sig_K(k_0) = \cdots = sig_K(k_{6m-1}) = (b, 0)$.*

2. (Existence) *There is a τ-region (sup_K, sig_K) of K_τ, respectively of K, that solves a_τ, respectively α, such that $sig_K(k_0) = \cdots = sig_K(k_{6m-1}) = (0, b)$.*

Firstly, we introduce the keys $K_{\tau_0^b}, K_{\tau_1^b}$ and K and show that they satisfy Lemma 3.1. Secondly, we present corresponding τ-regions which prove Lemma 3.2.

The union $K_{\tau_0^b}$ contains the following TS H_0 which provides the ESSP atom $(k, h_{0,4b+1})$:

$$H_0 = h_{0,0} \xrightarrow{k} \cdots \xrightarrow{k} h_{0,b} \xrightarrow{z} \cdots \xrightarrow{z} h_{0,2b} \xrightarrow{o_0} h_{0,2b+1} \xrightarrow{k} \cdots \xrightarrow{k} h_{0,3b+1} \Big\downarrow z$$

$$h_{0,6b+1} \xleftarrow{k} \cdots \xleftarrow{k} h_{0,5b+1} \xleftarrow{o_1} \cdots \xleftarrow{o_1} h_{0,4b+1} \xleftarrow{z} \cdots$$

$K_{\tau_0^b}$ also installs for $j \in \{0, \ldots, 6m-1\}$ the TS $D_{j,0}$ providing interface event k_j:

$$D_{0,j} = d_{j,0,0} \xrightarrow{o_0} d_{j,0,1} \xrightarrow{k_j} d_{j,0,2} \xrightarrow{o_1} d_{j,0,3} \cdots \xrightarrow{o_1} d_{j,0,b+3}$$

Overall, $K_{\tau_0^b} = (H_0, D_{0,0}, \ldots, D_{6m-1,0})$.

Proof of Lemma 3.1 for τ_0^b. For $j \in \{0, \ldots, 6m-1\}$ the TSs H_0 and $D_{j,0}$ interact as follows: If (sup_K, sig_K) is a region of $K_{\tau_0^b}$ solving $(k, h_{0,4b+1})$ then either $sig_K(o_0) = (0, b)$ *and* $sig_K(o_1) = (0, 1)$ or $sig_K(o_0) = (b, 0)$ *and* $sig_K(o_1) = (1, 0)$. By $\xrightarrow{o_0} d_{j,0,1}$, $d_{j,0,2} \xrightarrow{o_1}$ and Lemma 1, if $sig_K(o_0) = (0, b), sig_K(o_1) = (0, 1)$ then $sup_K(d_{j,0,1}) = b$ and $sup_K(d_{j,0,2}) = 0$. This implies $sig_K(k_j) = (b, 0)$. Similarly, $sig_K(o_0) = (b, 0), sig_K(o_1) = (1, 0)$ implies $sup_K(d_{j,0,1}) = 0$ and $sup_K(d_{j,0,2}) = b$ yielding $sig_K(k_j) = (0, b)$. Hence, it is $sig_K(k_0) = \cdots = sig_K(k_{6m-1}) = (b, 0)$ or $sig_K(k_0) = \cdots = sig_K(k_{6m-1}) = (0, b)$.

To prove Lemma 3.1 for $K_{\tau_0^b}$ it remains to argue that a τ_0^b-region (sig, sup) of $K_{\tau_0^b}$ solving $(k, h_{0,4b+1})$ satisfies $sig_K(o_0) = (0, b), sig_K(o_1) = (0, 1)$ or $sig_K(o_0) = (b, 0), sig_K(o_1) = (1, 0)$. Let $E_0 = \{(m, m) \mid 0 \le m \le b\}$.

By definition, if $sig(k) = (m, m) \in E_0$ then $sup(h_{0,3b+1}), sup(h_{0,5b+1}) \ge m$. Event (m, m) occurs at every state s of τ_0^b satisfying $s \ge m$. Hence, by $\neg h_{0,4b+1} \xrightarrow{(m,m)}$, we get $sup(h_{0,4b+1}) < m$. Observe, that z occurs always b times in a row. Therefore, by $sup(h_{0,3b+1}) \ge m$, $sup(h_{0,4b+1}) < m$ and Lemma 1, we have $sup(z) = (1, 0)$, $sig(o_1) = (0, 1)$ and immediately obtain $sup(h_{0,2b}) = 0$ and $sup(h_{0,3b+1}) = b$. Moreover, by $sig(k) = (m, m)$ and $sup(h_{0,3b+1}) = b$ we get $sup(h_{0,2b+1}) = b$ implying with $sup(h_{0,2b}) = 0$ that $sig(o_0) = (0, b)$. Thus, we have $sig(o_0) = (0, b)$ and $sig(o_1) = (0, 1)$.

Otherwise, if $sig(k) \notin E_0$, then Lemma 1 ensures $sig(k) \in \{(1, 0), (0, 1)\}$. If $sig(k) = (0, 1)$ then, by $s \xrightarrow{(0,1)}$ for every state $s \in \{0, \ldots, b-1\}$ of τ_0^b, we have $sup(h_{0,4b+1}) = b$. Moreover, again by $sig(k) = (0, 1)$ we have $sup(h_{0,b}) = sup(h_{0,3b+1}) = b$ and $sup(h_{0,2b+1}) = sup(h_{0,5b+1}) = 0$. By $sup(h_{0,3b+1}) = sup(h_{0,4b+1}) = b$ we have $sig(z) \in E_0$ which together with $sup(h_{0,b}) = b$ implies $sup(h_{0,2b}) = b$. Thus, by $sup(h_{0,2b}) = b$ and $sup(h_{0,2b+1}) = 0$, it is $sig(o_0) = (b, 0)$. Moreover, by $sup(h_{0,4b+1}) = b$ and $sup(h_{0,5b+1}) = 0$, we conclude $sig(o_1) = (1,0)$. Hence, we have $sig(o_0) = (b,0)$ and $sig(o_1) = (1,0)$. Similar arguments show that $sig_K(k) = (1,0)$ implies $sig(o_0) = (0,b)$ and $sig(o_1) = (0,1)$. Overall, this proves the announced signatures of o_0 and o_1. Hence, $K_{\tau_0^b}$ satisfies Lemma 3.1.   □

The union $K_{\tau_1^b}$ uses the next TS H_1 to provide the key atom $(k, h_{1,2b+4})$:

$$H_1 = h_{1,0} \xrightarrow{k} \cdots \xrightarrow{k} h_{1,b} \xrightarrow{z_0} h_{1,b+1} \xrightarrow{o_0} h_{1,b+2} \xrightarrow{k} \cdots \xrightarrow{k} h_{1,2b+2} \xrightarrow{z_1} h_{1,2b+3} \xrightarrow{z_0} h_{1,2b+4}$$

$$h_{1,3b+5} \xleftarrow{k} \cdots \xleftarrow{k} h_{1,2b+5} \xrightarrow{o_2} h_{1,2b+4}$$

Furthermore, $K_{\tau_1^b}$ contains for $j \in \{0, \ldots, 6m-1\}$ the TS $D_{j,1}$ which provides the interface event k_j:

$$D_{j,1} = d_{j,1,0} \xrightarrow{o_0} d_{j,1,1} \xrightarrow{k_j} d_{j,1,2} \xrightarrow{o_2} d_{j,1,3}$$

Altogether, $K_{\tau_1^b} = U(H_1, D_{0,1}, \ldots, D_{6m-1,1})$.

Proof of Lemma 3.1 for τ_1^b. For $j \in \{0, \ldots, 6m-1\}$ the TSs H_1 and $D_{j,1}$ interact as follows: If (sup_K, sig_K) is a τ_1^b-region of $K_{\tau_1^b}$ solving $(k, h_{1,2b+4})$ then either $sig_K(o_0) = sig_K(o_2) = (b,0)$ or $sig_K(o_0) = sig_K(o_2) = (0,b)$. Clearly, $sig_K(o_0) = sig_K(o_2) = (b,0)$, respectively $sig_K(o_0) = sig_K(o_2) = (0,b)$, implies $sig_K(k_0) = \cdots = sig_K(k_{6m-1}) = (0,b)$, respectively $sig_K(k_0) = \cdots = sig_K(k_{6m-1}) = (b,0)$.

We argue that the τ_1^b-solvability of $(k, h_{1,2b+4})$ implies the announced signatures of o_0, o_2. If (sup_K, sig_K) is a τ_1^b-region that solves $(k, h_{1,2b+4})$ then, by definition of τ_1^b and Lemma 1, we get $sig_K(k) \in \{(1,0), (0,1)\}$. Let $sig_K(k) = (0,1)$. The event $(0,1)$ occurs at every $s \in \{0, \ldots, b-1\}$ of τ_1^b. Hence, $\neg sup_K(h_{1,2b+4}) \xrightarrow{(0,1)}$ implies $sup_K(h_{1,2b+4}) = b$. Moreover, k occurs b times in a row. Thus, by $sig_K(k) = (0,1)$ and Lemma 1, we obtain $sup_K(h_{1,b}) = b$ and $sup_K(h_{1,b+2}) = sup_K(h_{1,2b+5}) = 0$. This implies, by $h_{1,2b+4} \xrightarrow{o_2} h_{1,2b+5}$, $sup(h_{1,2b+4}) = b$ and $sup(h_{1,2b+5}) = 0$, that $sig(o_2) = (b,0)$. Hence, by $sup_K(h_{1,b}) = sup_K(h_{1,2b+4}) = b$, $h_{1,b} \xrightarrow{z_0}$ and $\xrightarrow{z_0} h_{1,2b+4}$, we get $sig(z_0) = (0,0)$. Finally, by $sup(h_{1,b}) = b$, $h_{1,b} \xrightarrow{z_0}$ and $sig(z_0) = (0,0)$ we deduce $sup(h_{1,b+1}) = b$. Hence, $h_{1,b+1} \xrightarrow{o_0} h_{1,b+2}$, $sup(h_{1,b+1}) = b$ and $sup(h_{1,b+1}) = 0$ yield $sig(o_0) = (b,0)$. Altogether, we have that $sig(o_0) = sig(o_2) = (b,0)$. Similarly, one verifies that $sig_K(k) = (1,0)$ results in $sig(o_0) = sig(o_2) = (0,b)$. This proves Lemma 3.1 for $K_{\tau_1^b}$.   □

The union K uses the following TS H_2 to provide the key atom $(h_{2,0}, h_{2,b})$:

$$H_2 = h_{2,0} \xrightarrow{k} \cdots \xrightarrow{k} h_{2,b} \xrightarrow{o_0} h_{2,b+1} \xrightarrow{k} \cdots \xrightarrow{k} h_{2,2b+1} \xrightarrow{o_2} h_{2,2b+2} \xrightarrow{k} \cdots \xrightarrow{k} h_{2,3b+2}$$

K also contains the TSs $D_{0,1}, \ldots, D_{6m-1,1}$, thus $K = (H_2, D_{0,1}, \ldots, D_{6m-1,1})$.

Proof of Lemma 3.1 for τ_2^b. K works as follows: The event k occurs b times in a row at $h_{2,0}$. Thus, by Lemma 1, a region (sup_K, sig_K) solving $(h_{2,0}, h_{2,b})$ satisfies

$sig_K(k) \in \{(1,0),(0,1)\}$. If $sig_K(k) = (1,0)$ then $sup_K(h_{2,b}) = sup_K(h_{2,2b+1}) = b$ and $sup_K(h_{2,b+1}) = sup_K(h_{2,2b+2}) = 0$ implying $sig_k(o_0) = sig_k(o_2) = (b,0)$. Otherwise, if $sig_K(k) = (0,1)$ then $sup_K(h_{2,b}) = sup_K(h_{2,2b+1}) = 0$ and $sup_K(h_{2,b+1}) = sup_K(h_{2,2b+2}) = b$ which implies $sig_k(o_0) = sig_k(o_2) = (0,b)$. As already discussed for $K_{\tau_1^b}$, we have that $sig_k(o_0) = sig_k(o_2) = (b,0)$ $(sig_k(o_0) = sig_k(o_2) = (0,b))$ implies $sig_K(k_j) = (0,b)$ $(sig_K(k_j) = (b,0))$ for $j \in \{0,\ldots,6m-1\}$. Hence, Lemma 3.1 is true for K.   □

It remains to show that $K_{\tau_0^b}, K_{\tau_1^b}$ and K satisfy the objective of *existence*:

Proof of Lemma 3.2. We present corresponding regions. Let S and E be the set of all states and of all events of $K, K_{\tau_0^b}$ and $K_{\tau_1^b}$, respectively. We define mappings $sig : E \longrightarrow E_{\tau_1^b}$ and $sup : S \longrightarrow S_{\tau_1^b}$ by:

$$sig(e) = \begin{cases} (0,b), & \text{if } e \in \{k_0,\ldots,k_{6m-1}\} \\ (0,1), & \text{if } e = k \\ (0,0), & \text{if } e \in \{z,z_0,z_1\} \\ (1,0), & \text{if } e = o_1 \\ (b,0), & \text{if } e \in \{o_0,o_2\} \end{cases} \qquad sup(s) = \begin{cases} 0, & \text{if } s \in \{h_{0,0}, h_{1,0}, h_{2,0}\} \\ b, & \text{if } s \in \{d_{j,0,0}, d_{j,1,0}\} \\ & \text{and } 0 \le j \le 6m-1 \end{cases}$$

By $sig_K, sig_{K_{\tau_0^b}}$ and $sig_{K_{\tau_1^b}}$ $(sup_K, sup_{K_{\tau_0^b}}$ and $sup_{K_{\tau_1^b}})$ we denote the restriction of sig (sup) to the events (states) of K, $K_{\tau_0^b}$ and $K_{\tau_1^b}$, respectively. As sup defines the support of every corresponding initial state, by Lemma 1, we obtain fitting regions (sup_K, sig_K), $(sup_{K_{\tau_0^b}}, sig_{K_{\tau_0^b}})$ and $(sup_{K_{\tau_1^b}}, sig_{K_{\tau_1^b}})$ that solve the corresponding key atom. Figure 3 sketches this region for $K_{\tau_1^2}$ and K.   □

4.2   The Translators $T_{\tau_0^b}$ and $T_{\tau_1^b}$ and T

In this subsection, we present translator T, which we also use as $T_{\tau_0^b}$ and $T_{\tau_1^b}$, that is, $T_{\tau_0^b} = T_{\tau_1^b} = T$.

For every $i \in \{0,\ldots,m-1\}$ the clause $C_i = \{X_{i,0}, X_{i,1}, X_{i,2}\}$ is translated into the following three TSs which use the variables of C_i as events:

$$T_{i,0} = t_{i,0,0} \xrightarrow{k_{6i}} t_{i,0,1} \xrightarrow{X_{i,0}} \cdots \xrightarrow{X_{i,0}} t_{i,0,b+1} \xrightarrow{x_i} t_{i,0,b+2} \xrightarrow{X_{i,2}} \cdots \xrightarrow{X_{i,2}} t_{i,0,2b+2} \xrightarrow{k_{6i+1}} t_{i,0,2b+3}$$

$$T_{i,1} = t_{i,1,0} \xrightarrow{k_{6i+2}} t_{i,1,1} \xrightarrow{X_{i,1}} \cdots \xrightarrow{X_{i,1}} t_{i,1,b+1} \xrightarrow{p_i} t_{i,1,b+2} \xrightarrow{k_{6i+3}} t_{i,1,b+3}$$

$$T_{i,2} = t_{i,2,0} \xrightarrow{k_{6i+4}} t_{i,2,1} \xrightarrow{x_i} t_{i,2,2} \xrightarrow{p_i} t_{i,2,3} \xrightarrow{k_{6i+5}} t_{i,2,4}$$

Altogether, $T = U(T_{0,0}, T_{0,1}, T_{0,2}, \ldots, T_{m-1,0}, T_{m-1,1}, T_{m-1,2})$. Figure 3 provides an example for T where $b = 2$ and $\varphi = \varphi_0$. In accordance to our general approach and Lemma 3 the following lemma states the aim of T:

Lemma 4. *Let* $\tau \in \{\tau_0^b, \tau_1^b\}$.

1. (Completeness) *If* (sup_T, sig_T) *is a* τ-*region of* T *such that* $sig_T(k_0) = \cdots = sig_T(k_{6m-1}) = (0,b)$ *or* $sig_T(k_0) = \cdots = sig_T(k_{6m-1}) = (b,0)$ *then* φ *has a one-in-three model.*

$$t_{0,0,0} \xrightarrow{k_0} t_{0,0,1} \xrightarrow{X_0} t_{0,0,2} \xrightarrow{X_0} t_{0,0,3} \xrightarrow{x_0} t_{0,0,4} \xrightarrow{X_2} t_{0,0,5} \xrightarrow{X_2} t_{0,0,6} \xrightarrow{k_1} t_{0,0,7}$$
[0] (0,2) [2] (1,0) [1] (1,0) [0] (0,0) [0] (0,0) [0] (0,0) [0] (0,2) [2]

$$t_{0,1,0} \xrightarrow{k_2} t_{0,1,1} \xrightarrow{X_1} t_{0,1,2} \xrightarrow{X_1} t_{0,1,3} \xrightarrow{p_0} t_{0,1,4} \xrightarrow{k_3} t_{0,1,5} \qquad t_{0,2,0} \xrightarrow{k_4} t_{0,2,1} \xrightarrow{x_0} t_{0,2,2} \xrightarrow{p_0} t_{0,2,3} \xrightarrow{k_5} t_{0,2,4}$$
[0] (0,2) [2] (0,0) [2] (0,0) [2] (2,0) [0] (0,2) [2]   [0] (0,2) [2] (0,0) [2] (2,0) [0] (0,2) [2]

$$t_{1,0,0} \xrightarrow{k_6} t_{1,0,1} \xrightarrow{X_2} t_{1,0,2} \xrightarrow{X_2} t_{1,0,3} \xrightarrow{x_1} t_{1,0,4} \xrightarrow{X_3} t_{1,0,5} \xrightarrow{X_3} t_{1,0,6} \xrightarrow{k_7} t_{1,0,7}$$
[0] (0,2) [2] (0,0) [2] (0,0) [2] (2,0) [0] (0,0) [0] (0,0) [0] (0,2) [2]

$$t_{1,1,0} \xrightarrow{k_8} t_{1,1,1} \xrightarrow{X_0} t_{1,1,2} \xrightarrow{X_0} t_{1,1,3} \xrightarrow{p_1} t_{1,1,4} \xrightarrow{k_9} t_{1,1,5} \qquad t_{1,2,0} \xrightarrow{k_{10}} t_{1,2,1} \xrightarrow{x_1} t_{1,2,2} \xrightarrow{p_1} t_{1,2,3} \xrightarrow{k_{11}} t_{1,2,4}$$
[0] (0,2) [2] (1,0) [1] (1,0) [0] (0,0) [0] (0,2) [2]   [0] (0,2) [2] (2,0) [0] (0,0) [0] (0,2) [2]

$$t_{2,0,0} \xrightarrow{k_{12}} t_{2,0,1} \xrightarrow{X_1} t_{2,0,2} \xrightarrow{X_1} t_{2,0,3} \xrightarrow{x_2} t_{2,0,4} \xrightarrow{X_0} t_{2,0,5} \xrightarrow{X_0} t_{2,0,6} \xrightarrow{k_{13}} t_{2,0,7}$$
[0] (0,2) [2] (0,0) [2] (0,0) [2] (0,0) [2] (1,0) [1] (1,0) [0] (0,2) [2]

$$t_{2,1,0} \xrightarrow{k_{14}} t_{2,1,1} \xrightarrow{X_3} t_{2,1,2} \xrightarrow{X_3} t_{2,1,3} \xrightarrow{p_2} t_{2,1,4} \xrightarrow{k_{15}} t_{2,1,5} \qquad t_{2,2,0} \xrightarrow{k_{16}} t_{2,2,1} \xrightarrow{x_2} t_{2,2,2} \xrightarrow{p_2} t_{2,2,3} \xrightarrow{k_{17}} t_{2,2,4}$$
[0] (0,2) [2] (0,0) [2] (0,0) [2] (2,0) [0] (0,2) [2]   [0] (0,2) [2] (0,0) [2] (2,0) [0] (0,2) [2]

$$t_{3,0,0} \xrightarrow{k_{18}} t_{3,0,1} \xrightarrow{X_2} t_{3,0,2} \xrightarrow{X_2} t_{3,0,3} \xrightarrow{x_3} t_{3,0,4} \xrightarrow{X_5} t_{3,0,5} \xrightarrow{X_5} t_{3,0,6} \xrightarrow{k_{19}} t_{3,0,7}$$
[0] (0,2) [2] (0,0) [2] (0,0) [2] (2,0) [0] (0,0) [0] (0,0) [0] (0,2) [2]

$$t_{3,1,0} \xrightarrow{k_{20}} t_{3,1,1} \xrightarrow{X_4} t_{3,1,2} \xrightarrow{X_4} t_{3,1,3} \xrightarrow{p_3} t_{3,1,4} \xrightarrow{k_{21}} t_{3,1,5} \qquad t_{3,2,0} \xrightarrow{k_{22}} t_{3,2,1} \xrightarrow{x_3} t_{3,2,2} \xrightarrow{p_3} t_{3,2,3} \xrightarrow{k_{23}} t_{3,2,4}$$
[0] (0,2) [2] (1,0) [1] (1,0) [0] (0,0) [0] (0,2) [2]   [0] (0,2) [2] (2,0) [0] (0,0) [0] (0,2) [2]

$$t_{4,0,0} \xrightarrow{k_{24}} t_{4,0,1} \xrightarrow{X_1} t_{4,0,2} \xrightarrow{X_1} t_{4,0,3} \xrightarrow{x_4} t_{4,0,4} \xrightarrow{X_4} t_{4,0,5} \xrightarrow{X_4} t_{4,0,6} \xrightarrow{k_{25}} t_{4,0,7}$$
[0] (0,2) [2] (0,0) [2] (0,0) [2] (0,0) [2] (1,0) [1] (1,0) [0] (0,2) [2]

$$t_{4,1,0} \xrightarrow{k_{26}} t_{4,1,1} \xrightarrow{X_5} t_{4,1,2} \xrightarrow{X_5} t_{4,1,3} \xrightarrow{p_4} t_{4,1,4} \xrightarrow{k_{27}} t_{4,1,5} \qquad t_{4,2,0} \xrightarrow{k_{28}} t_{4,2,1} \xrightarrow{x_4} t_{4,2,2} \xrightarrow{p_4} t_{4,2,3} \xrightarrow{k_{29}} t_{4,2,4}$$
[0] (0,2) [2] (0,0) [2] (0,0) [2] (2,0) [0] (0,2) [2]   [0] (0,2) [2] (0,0) [2] (2,0) [0] (0,2) [2]

$$t_{5,0,0} \xrightarrow{k_{30}} t_{5,0,1} \xrightarrow{X_4} t_{5,0,2} \xrightarrow{X_4} t_{5,0,3} \xrightarrow{x_5} t_{5,0,4} \xrightarrow{X_5} t_{5,0,5} \xrightarrow{X_5} t_{5,0,6} \xrightarrow{k_{31}} t_{5,0,7}$$
[0] (0,2) [2] (1,0) [1] (1,0) [0] (0,0) [0] (0,0) [0] (0,0) [0] (0,2) [2]

$$t_{5,1,0} \xrightarrow{k_{32}} t_{5,1,1} \xrightarrow{X_3} t_{5,1,2} \xrightarrow{X_3} t_{5,1,3} \xrightarrow{p_5} t_{5,1,4} \xrightarrow{k_{33}} t_{5,1,5} \qquad t_{5,2,0} \xrightarrow{k_{34}} t_{5,2,1} \xrightarrow{x_5} t_{5,2,2} \xrightarrow{p_5} t_{5,2,3} \xrightarrow{k_{35}} t_{5,2,4}$$
[0] (0,2) [2] (0,0) [2] (0,0) [2] (2,0) [0] (0,2) [2]   [0] (0,2) [2] (0,0) [2] (2,0) [0] (0,2) [2]

T { (braces grouping the above twelve rows)

$K_{\tau_0^2}$ {

$$h_{2,0} \xrightarrow{k} h_{2,1} \xrightarrow{k} h_{2,2} \xrightarrow{z} h_{2,3} \xrightarrow{z} h_{2,4} \xrightarrow{o_0} h_{2,5} \xrightarrow{k} h_{2,6} \xrightarrow{k} h_{2,7} \xrightarrow{z} h_{2,8} \xrightarrow{z} h_{2,9} \xrightarrow{o_1} h_{2,10} \xrightarrow{o_1} h_{2,11}$$
[0] (0,1) [1] (0,1) [2] (0,0) [2] (0,0) [2] (2,0) [0] (0,1) [1] (0,1) [2] (0,0) [2] (0,0) [2] (1,0) [1] (1,0)   [0]

$$\downarrow k$$

$$d_{0,0,0} \xrightarrow{o_0} d_{0,0,1} \xrightarrow{k_0} d_{0,0,2} \xrightarrow{o_2} d_{0,0,3} \xrightarrow{o_2} d_{0,0,4} \cdots d_{35,0,0} \xrightarrow{o_0} d_{35,0,1} \xrightarrow{k_{35}} d_{35,0,2} \xrightarrow{o_2} d_{35,0,3} \xrightarrow{o_2} d_{35,0,4} \qquad h_{2,13} \xleftarrow{k} h_{2,12}$$
[2] (2,0) [0] (0,2) [2] (1,0) [1] (1,0) [0]   [2] (2,0) [0] (0,2) [2] (1,0) [1] (1,0)   [2] (0,1) [1]

$K_{\tau_1^2}$ {

$$h_{1,0} \xrightarrow{k} h_{1,1} \xrightarrow{k} h_{1,2} \xrightarrow{z_0} h_{1,3} \xrightarrow{o_0} h_{1,4} \xrightarrow{k} h_{1,5} \xrightarrow{k} h_{1,6} \xrightarrow{z_1} h_{1,7} \xrightarrow{z_0} h_{1,8} \xrightarrow{o_1} h_{1,9} \xrightarrow{k} h_{1,10} \xrightarrow{k} h_{1,11}$$
[0] (0,1) [1] (0,1) [2] (0,0) [2] (2,0) [0] (0,1) [1] (0,1) [2] (0,0) [2] (0,0) [2] (2,0) [0] (0,1) [1] (0,1) [2]

$$d_{0,1,0} \xrightarrow{o_0} d_{0,1,1} \xrightarrow{k_0} d_{0,1,2} \xrightarrow{o_2} d_{0,1,3} \cdots d_{35,1,0} \xrightarrow{o_0} d_{35,1,1} \xrightarrow{k_{35}} d_{35,1,2} \xrightarrow{o_2} d_{35,1,3}$$
[2] (2,0) [0] (0,2) [2] (2,0) [0]   [2] (2,0) [0] (0,2) [2] (2,0) [0]

K {

$$h_{2,0} \xrightarrow{k} h_{2,1} \xrightarrow{k} h_{2,2} \xrightarrow{o_0} h_{2,3} \xrightarrow{k} h_{2,4} \xrightarrow{k} h_{2,5} \xrightarrow{o_2} h_{2,6} \xrightarrow{k} h_{2,7} \xrightarrow{k} h_{2,8}$$
[0] (0,1) [1] (0,1) [2] (2,0) [0] (0,1) [1] (0,1) [2] (2,0) [0] (0,1) [1] (0,1) [2]

Fig. 3. Constituents K_τ, K, T for $\tau \in \{\tau_0^2, \tau_1^2\}$ and φ_0. TSs are defined by bold drawn states, edges and events. Labels with reduced opacity correspond to region (sup, sig) defined in Sects. 4.1, 4.2: $sup(s)$ is presented in square brackets below state s and $sig(e)$ is depicted below every e-labeled transition. The model of φ_0 is $\{X_0, X_4\}$.

2. (Existence) *If φ has a one-in-three model then there is a τ-region (sup_T, sig_T) of T such that $sig_T(k_0) = \cdots = sig_T(k_{6m-1}) = (0, b)$.*

Proof. To fulfill its destiny, T works as follows. By definition, if (sup_T, sig_T) is a region of T then $\pi_{i,0}, \pi_{i,1}, \pi_{i,2}$, defined by

$$\pi_{i,0} = sup_T(t_{i,0,1}) \xrightarrow{sig_T(X_{i,0})} \cdots \xrightarrow{sig_T(X_{i,0})} sup_T(t_{i,0,b+1}) \xrightarrow{sig_T(x_i)} sup_T(t_{i,0,b+2}) \xrightarrow{sig_T(X_{i,2})} \cdots \xrightarrow{sig_T(X_{i,2})} sup_T(t_{i,0,2b+2})$$

$$\pi_{i,1} = sup_T(t_{i,1,1}) \xrightarrow{sig_T(X_{i,1})} \cdots \xrightarrow{sig_T(X_{i,1})} sup_T(t_{i,1,b+1}) \xrightarrow{sig_T(p_i)} sup_T(t_{i,1,b+2})$$

$$\pi_{i,2} = sup_T(t_{i,2,1}) \xrightarrow{sig_T(x_i)} sup_T(t_{i,2,2}) \xrightarrow{sig_T(p_i)} sup_T(t_{i,2,3})$$

are directed labeled paths of τ. For every $i \in \{0, \ldots, m-1\}$, the events $k_{6i}, \ldots, k_{6i+5}$ belong to the interface. By Lemma 3.1, K_τ and K ensure the following: If (sup_K, sig_K) is a region of K_τ, respectively K, that solves the key atom a_τ, respectively α, then either $sig_K(k_0) = \cdots = sig_K(k_{6m-1}) = (0, b)$ or $sig_K(k_0) = \cdots = sig_K(k_{6m-1}) = (b, 0)$. For every transition $s \xrightarrow{k_j} s'$, the first case implies $sup(s) = 0$ and $sup(s') = b$ while the second case implies $sup(s) = b$ and $sup(s') = 0$, where $j \in \{0, \ldots, 6m-1\}$. Hence, a τ-region (sup_T, sig_T) of T being compatible with (sup_K, sig_K) satisfies exactly one of the next conditions:

(1) $sig_T(k_0) = \cdots = sig_T(k_{6m-1}) = (0, b)$ and for every $i \in \{0, \ldots, m-1\}$ the paths $\pi_{i,0}, \pi_{i,1}, \pi_{i,2}$ start at b and terminate at 0.

(2) $sig_T(k_0) = \cdots = sig_T(k_{6m-1}) = (b, 0)$ and for every $i \in \{0, \ldots, m-1\}$ the paths $\pi_{i,0}, \pi_{i,1}, \pi_{i,2}$ start at 0 and terminate at b.

The construction of T ensures that if (1), respectively if (2), is satisfied then there is for every $i \in \{0, \ldots, m-1\}$ exactly one variable event $X \in \{X_{i,0}, X_{i,1}, X_{i,2}\}$ such that $sig(X) = (1, 0)$, respectively $sig(X) = (0, 1)$. Each triple $T_{i,0}, T_{i,1}, T_{i,2}$ corresponds exactly to the clause C_i. Hence, $M = \{X \in V(\varphi) | sig_T(X) = (1, 0)\}$ or $M = \{X \in V(\varphi) | sig_T(X) = (0, 1)\}$, is a one-in-three model of φ, respectively. Having sketched the plan to satisfy Lemma 4.1, it remains to argue that the deduced conditions (1), (2) have the announced impact on the variable events.

For a start, let (2) be satisfied and $i \in \{0, \ldots, m-1\}$. By $sig_T(k_{6i}) = \cdots = sig_T(k_{6i+5}) = (0, b)$ we have that $sup_T(t_{i,0,1}) = sup_T(t_{i,1,1}) = sup_T(t_{i,1,1}) = b$ and $sup_T(t_{i,0,2b+2}) = sup_T(t_{i,1,b+2}) = sup_T(t_{i,1,3}) = 0$. Notice, for every event $e \in \{X_{i,0}, X_{i,1}, X_{i,2}, x_i, p_i\}$ there is a state s such that $s \xrightarrow{e}$ and $sup_T(s) = b$ or such that $\xrightarrow{e} s$ and $sup_T(s) = 0$. Consequently, if $(m, n) \in E_\tau$ and $m < n$ then $sig(e) \neq (m, n)$. This implies the following condition:

(3) If $e \in \{X_{i,0}, X_{i,1}, X_{i,2}, x_i, p_i\}$ and $s \xrightarrow{e} s'$ then $sup_T(s) \geq sup_T(s')$.

Moreover, every variable event $X_{i,0}, X_{i,1}, X_{i,2}$ occurs b times consecutively in a row. Hence, by Lemma 1, we have:

(4) If $X \in \{X_{i,0}, X_{i,1}, X_{i,2}\}$, $sig_T(X) = (m, n)$ and $m \neq n$ then $(m, n) = (1, 0)$.

The paths $\pi_{i,0}, \pi_{i,1}, \pi_{i,2}$ of τ start at b and terminate at 0. Hence, by definition of τ, for every $\pi \in \{\pi_{i,0}, \pi_{i,1}, \pi_{i,2}\}$ there has to be an event e_π, which occurs at π, such that $sig_T(e_\pi) = (m, n)$ with $m > n$.

If for $\pi \in \{\pi_{i,0}, \pi_{i,1}\}$ it is true that $e_\pi \notin \{X_{i,0}, X_{i,1}, X_{i,2}\}$ then for $X \in \{X_{i,0}, X_{i,1}, X_{i,2}\}$ we have $sig_T(X) = (m, m)$ for some $m \in \{0, \ldots, b\}$.

This yields $sup(t_{i,0,b+1}) = sup(t_{i,1,b+1}) = b$ and $sup(t_{i,0,b+2}) = 0$ which with $sup(t_{i,1,b+2}) = 0$ implies $sig_T(x_i) = sig_T(p_i) = (b,0)$. By $sig_T(x_i) = (b,0)$, we obtain $sup(t_{i,2,2}) = 0$ and, by $sig_T(p_i) = (b,0)$, we obtain $sup(t_{i,2,2}) = b$, a contradiction. Consequently, by Condition 4, there has to be an event $X \in \{X_{i,0}, X_{i,1}, X_{i,2}\}$ such that $sig_T(X) = (1,0)$. We discuss all possible cases to show that X is unambiguous.

If $sig_T(X_{i,0}) = (1,0)$ then, by Lemma 1, we have that $sup_T(t_{i,0,b+1}) = 0$. By (3), this implies that $sup_T(t_{i,0,b+2}) = \cdots = sup_T(t_{i,0,2b+1}) = 0$ and $sig_T(x_i) = sig_T(X_{i,2}) = (0,0)$. Moreover, $sig_T(x_i) = (0,0)$ and $sup(t_{i,2,1}) = b$ imply $sup(t_{i,2,2}) = b$ which with $sup(t_{i,2,3}) = 0$ implies $sig_T(p_i) = (b,0)$. By $sig_T(p_i) = (b,0)$ we obtain $sup(t_{i,1,b+1}) = b$ which, by Lemma 1 and contraposition shows that $sig_T(X_{i,1}) \neq (1,0)$. Hence, we have $sig_T(X_{i,1}) \neq (1,0)$.

If $sig_T(X_{i,2}) = (1,0)$ then, by Lemma 1, we have that $sup_T(t_{i,0,b+2}) = b$. Again by (3), this implies that $sup_T(t_{i,0,1}) = \cdots = sup_T(t_{i,0,b+2}) = b$ and $sig_T(x_i) = (m,m)$, $sig_T(X_{i,0}) = (m',m')$ for some $m,m' \in \{0,\ldots,b\}$. Especially, we have that $sig_T(X_{i,0}) \neq (1,0)$. Moreover, by $sig_T(x_i) = (m,m)$, we obtain $sup_T(t_{i,2,2}) = b$ implying with $sup_T(t_{i,2,3}) = 0$ that $sig_T(p_i) = (b,0)$. As in the previous case this yields $sig_T(X_{i,1}) \neq (1,0)$.

Finally, if $sig_T(X_{i,1}) = (1,0)$ then, by Lemma 1, we get $sup_T(t_{i,1,b+1}) = 0$. By $sup_T(t_{i,1,b+1}) = sup_T(t_{i,1,b+2}) = 0$ we conclude $sig_T(p_i) = (0,0)$ which with $sup_T(t_{i,2,3}) = 0$ implies $sup_T(t_{i,2,2}) = 0$. Using $sup_T(t_{i,2,1}) = b$ and $sup_T(t_{i,2,2}) = 0$ we obtain $sig_T(x_i) = (b,0)$ implying that $sup_T(t_{i,0,b+1}) = b$ and $sup_T(t_{i,0,b+2}) = 0$. By (3), this yields $sup_T(t_{i,0,1}) = \cdots = sup_T(t_{i,0,b+1}) = b$ and $sup_T(t_{i,0,b+2}) = \cdots = sup_T(t_{i,0,2b+2}) = b$ which, by Lemma 1, implies $sig_T(X_{i,0}) \neq (1,0)$ and $sig_T(X_{i,2}) \neq (1,0)$.

So far, we have proven that if (1) is satisfied then for every $i \in \{0,\ldots,m-1\}$ there is exactly one variable event $X \in \{X_{i,0}, X_{i,1}, X_{i,2}\}$ such that $sig_T(X) = (1,0)$. Consequently, the set $M = \{X \in V(\varphi) | sig_T(X) = (1,0)\}$ is a one-in-three model of φ. One verifies, by analogous arguments, that (2) implies for every $i \in \{0,\ldots,m-1\}$ that there is exactly one variable event $X \in \{X_{i,0}, X_{i,1}, X_{i,2}\}$ with $sig_T(X) = (0,1)$, which makes $M = \{X \in V(\varphi) | sig_T(X) = (0,1)\}$ a one-in-three model of φ. Hence, a τ-region of T_τ that satisfies (1) or (2) implies a one-in-three model of φ.

Reversely, if M is a one-in-three model of φ then there is a τ-region (sup_T, sig_T) satisfying (1) which, by Lemma 1, is completely defined by $sup_T(t_{i,0,0}) = sup_T(t_{i,1,0}) = sup_T(t_{i,1,0}) = 0$ for $i \in \{0,\ldots,m-1\}$ and

$$sig_T(e) = \begin{cases} (0,b), & \text{if } e \in \{k_0,\ldots,k_{6m-1}\} \\ (0,0), & \text{if } e \in V(\varphi) \setminus M \\ (0,0), & \text{if } (e = p_i, X_{i,1} \in M) \text{ or } (e = x_i, X_{i,1} \notin M), 0 \leq i \leq m-1 \\ (1,0), & \text{if } e \in M \\ (b,0), & \text{if } (e = x_i, X_{i,1} \in M) \text{ or } (e = p_i, X_{i,1} \notin M), 0 \leq i \leq m-1 \end{cases}$$

See Fig. 3, for a sketch of this region for $\tau \in \{\tau_0^2, \tau_1^2\}$, φ_0 and $M = \{X_0, X_4\}$. This proves Lemma 4. $\qquad\square$

4.3 The Key Unions $K_{\tau_2^b}$ and $K_{\tau_3^b}$

The unions $U_{\tau_2^b}, U_{\tau_3^b}$ install the same key. More exactly, if $\tau \in \{\tau_2^b, \tau_3^b\}$ then K_τ uses only the TS H_3 to provide key atom $(k, h_{3,1,b-1})$ and the interface k and z:

$$H_3 = h_{3,0,0} \xrightarrow{k} \cdots \xrightarrow{k} h_{3,0,b-1} \xrightarrow{k} h_{3,0,b}$$
$$u \downarrow \qquad \qquad \qquad \qquad \nearrow z$$
$$h_{3,1,0} \xrightarrow{k} \cdots \xrightarrow{k} h_{3,1,b-1}$$

The next lemma summarizes the intention behind K_τ:

Lemma 5. *Let* $\tau \in \{\tau_2^b, \tau_3^b\}$ *and* $E_0 = \{(m,m) | 1 \le m \le b\} \cup \{0\}$.

1. *(Completeness) If* (sup_K, sig_K) *is a τ-region that solves* $(k, h_{3,1,b-1})$ *in* K_τ *then* $sig(k) \in \{(1,0), (0,1)\}$ *and* $sig_K(z) \in E_0$.
2. *(Existence) There is a τ-region* (sup_K, sig_K) *of* K_τ *solving* $(k, h_{3,1,b-1})$ *such that* $sig(k) = (0,1)$ *and* $sig_K(z) = 0$.

Proof. For the first statement, we let (sup_K, sig_K) be a region solving α_τ. By $\xrightarrow{k} h_{3,1,b-1}$ and $\neg sup_K(h_{3,1,b-1}) \xrightarrow{sig_K(k)}$ we immediately have $sig(K) \notin E_0$. Moreover, for every group event $e \in \{0, \ldots, b\}$ and every state s of τ we have that $s \xrightarrow{e}$. Hence, by $\neg sup_K(h_{3,1,b-1}) \xrightarrow{sig_K(k)}$ we have $sig_K(k) \notin \{0, \ldots, b\}$. The event k occurs b times in a row. Therefore, by Lemma 1, we have that $sig_K(k) \in \{(1,0), (0,1)\}$ and if $sig_K(k) = (1,0)$ then $sup_K(h_{3,0,b}) = 0$ and if $sig_K(k) = (0,1)$ then $sup_K(h_{3,0,b}) = b$. If $s \in \{0, \ldots, b-1\}$ then $s \xrightarrow{(0,1)}$ is true. Furthermore, every state $s \in \{1, \ldots, b\}$ satisfies $s \xrightarrow{(1,0)}$. Consequently, by $\neg sup_K(h_{3,1,b-1}) \xrightarrow{sig_K(k)}$, if $sig_K(k) = (0,1)$ then $sup_K(h_{3,1,b-1}) = b$ and if $sig_K(k) = (1,0)$ then $sup_K(h_{3,1,b-1}) = 0$. This implies for (sup_K, sig_K) that $sig_K(z) \in E_0$ and proves Lemma 5.1. For Lemma 5.2 we easily verify that (sup_K, sig_K) with $sig_K(k) = (0,1)$, $sig_K(u) = 1$, $sig_K(z) = 0$ and $sup_K(h_{3,0,0}) = 0$ properly defines a solving τ-region. $\qquad\square$

4.4 The Translators $T_{\tau_2^b}$ and $T_{\tau_3^b}$

In this section we introduce $T_{\tau_2^b}$ which is used for $U_{\tau_2^b}$ and $U_{\tau_3^b}$, that is, $T_{\tau_3^b} = T_{\tau_2^b}$. Let $\tau \in \{\tau_2^b, \tau_3^b\}$. Firstly, the translator T_τ contains for every variable X_j of φ, $j \in \{0, \ldots, m-1\}$, the TSs F_j, G_j below, that apply X_j as event:

$$F_j = f_{j,0,0} \xrightarrow{k} \cdots \xrightarrow{k} f_{j,0,b} \qquad G_j = g_{j,0} \xrightarrow{k} \cdots \xrightarrow{k} g_{j,b} \xrightarrow{X_j} g_{j,b+1}$$
$$v_j \downarrow \qquad \qquad \uparrow X_j$$
$$f_{j,1,0} \xrightarrow{k} \cdots \xrightarrow{k} f_{j,1,b-1}$$

Secondly, translator T_τ implements for every clause $C_i = \{X_{i,0}, X_{i,1}, X_{i,2}\}$ of φ, $i \in \{0, \ldots, m-1\}$, the following TS T_i that applies the variables of C_i as events :

$$T_i = \quad t_{i,0} \xrightarrow{\; k \;} \cdots \xrightarrow{\; k \;} t_{i,b} \xrightarrow{\; X_{i,0} \;} t_{i,b+1} \xrightarrow{\; X_{i,1} \;} t_{i,b+2} \xrightarrow{\; X_{i,2} \;} t_{i,b+3} \xrightarrow{\; z \;} t_{i,b+4} \xrightarrow{\; k \;} \cdots \xrightarrow{\; k \;} t_{i,2b+4}$$

Altogether, we have $T_\tau = (F_0, G_0, \ldots, F_{m-1}, G_{m-1}, T_0, \ldots, T_{m-1})$. The next lemma summarizes the functionality of T_τ:

Lemma 6. *If $\tau \in \{\tau_2^b, \tau_3^b\}$ then the following conditions are true:*

1. *(Completeness) If (sup_T, sig_T) is a τ-region of T_τ such that $sig_T(z) \in E_0$ and $sig_T(k) = (0,1)$, respectively $sig_T(k) = (1,0)$, then φ is one-in-three satisfiable.*

2. *(Existence) If φ is one-in-three satisfiable then there is a τ-region (sup_T, sig_T) of T_τ such that $sig_T(z) = 0$ and $sig_T(k) = (0,1)$.*

Proof. Firstly, we argue for Lemma 6.1. Let (sup_T, sig_T) be a region of T_τ which satisfies $sig_T(z) \in E_0, sig_T(k) \in \{(1,0), (0,1)\}$. By definition, π_i defined by

$$\pi_i = \quad sup_T(t_{i,b}) \xrightarrow{\; sig_T(X_{i,0}) \;} sup_T(t_{i,b+1}) \xrightarrow{\; sig_T(X_{i,1}) \;} sup_T(t_{i,b+2}) \xrightarrow{\; sig_T(X_{i,2}) \;} sup_T(t_{i,b+3})$$

is a directed labeled path in τ. By $sig_T(z) \in E_0$ and $t_{i,b+3} \xrightarrow{z} t_{i,b+4}$ we obtain that $sup_T(t_{i,b+3}) = sup_T(t_{i,b+4})$. Moreover, k occurs b times in a row at $t_{i,0}$ and $t_{i,b+4}$. By Lemma 1, this implies if $sig_T(k) = (1,0)$ then $sup_T(t_{i,b}) = b$ and $sup_T(t_{i,b+4}) = 0$ and if $sig_T(k) = (0,1)$ then $sup_T(t_{i,b}) = 0$ and $sup_T(t_{i,b+4}) = b$. Altogether, we obtain that the following conditions are true: If $sig_T(z) \in E_0, sig_T(k) = (1,0)$ then path p_i starts a 0 and terminates at b and if $sig_T(z) \in E_0, sig_T(k) = (0,1)$ then the path p_i starts a b and terminates at 0.

By definition of τ, both conditions imply that there has to be at least one event $X \in \{X_{i,0}, X_{i,1}, X_{i,2}\}$ whose signature satisfies $sig_T(X) \notin E_0$. Again, our intention is to ensure that for exactly one such variable event the condition $sig_T(X) \notin E_0$ is true. Here, the TSs $F_0, G_0, \ldots, F_{m-1}, G_{m-1}$ come into play. The aim of $F_0, G_0, \ldots, F_{m-1}, G_{m-1}$ is to restrict the possible signatures for the variable events as follows: If $sig_T(k) = (1,0)$ then $X \in V(\varphi)$ implies $sig_T(X) \in E_0 \cup \{b\}$ and if $sig_T(k) = (0,1)$ then $X \in V(\varphi)$ implies $sig_T(X) \in E_0 \cup \{1\}$.

We now argue, that the introduced conditions ensure that there is exactly one variable event $X \in \{X_{i,0}, X_{i,1}, X_{i,2}\}$ with $sig_T(X) \notin E_0$. Remember that, by definition, if $sig_T(X) \in E_0$ then $sig_T^-(X) + sig_T^+(X) = |sig_T(X)| = 0$.

For a start, let $sig_T(z) \in E_0, sig_T(k) = (1,0)$, implying that p_i starts at b and terminates at 0, and assume $sig_T(X) \in E_0 \cup \{b\}$. By Lemma 1, we obtain:

$$(|sig_T(X_{i,0})| + |sig_T(X_{i,1})| + |sig_T(X_{i,2})|) \equiv b \bmod (b+1) \tag{1}$$

Clearly, if $sig_T(X_{i,0}), sig_T(X_{i,1}), sig_T(X_{i,2}) \in E_0$, then we obtain a contradiction to (1) by $|sig_T(X_{i,0})| = |sig_T(X_{i,1})| = |sig_T(X_{i,2})| = 0$. Hence, there has to be at least one variable event $X \in \{X_{i,0}, X_{i,1}, X_{i,2}\}$ with $sig_T(X) = b$.

If there are two different variable events $X, Y \in \{X_{i,0}, X_{i,1}, X_{i,2}\}$ such that $sig_T(X) = sig_T(Y) = b$ and $sig_T(Z) \in E_0$ for $Z \in \{X_{i,0}, X_{i,1}, X_{i,2}\} \setminus \{X, Y\}$ then, by symmetry and transitivity, we obtain:

$$b \equiv (|sig_T(X_{i,0})| + |sig_T(X_{i,1})| + |sig_T(X_{i,2})|) \bmod (b+1) \quad |(1) \tag{2}$$
$$(|sig_T(X_{i,0})| + |sig_T(X_{i,1})| + |sig_T(X_{i,2})|) \equiv 2b \bmod (b+1) \quad |\text{assumpt.} \tag{3}$$
$$b \equiv 2b \bmod (b+1) \quad |(2), (3) \tag{4}$$
$$2b \equiv (b-1) \bmod (b+1) \quad |\text{def.} \equiv \tag{5}$$
$$b \equiv (b-1) \bmod (b+1) \quad |(4), (5) \tag{6}$$
$$\exists m \in \mathbb{Z} : m(b+1) = 1 \quad |(6) \tag{7}$$

By (7) we obtain $b = 0$, a contradiction. Similarly, if we assume that $|sig_T(X_{i,0})| = |sig_T(X_{i,1})| = |sig_T(X_{i,2})| = b$ then we obtain

$$(|sig_T(X_{i,0})| + |sig_T(X_{i,1})| + |sig_T(X_{i,2})|) \equiv 3b \bmod (b+1) \quad |\text{assumpt.} \tag{8}$$
$$b \equiv 3b \bmod (b+1) \quad |(2), (8) \tag{9}$$
$$3b \equiv (b-2) \bmod (b+1) \quad |\text{def.} \equiv \tag{10}$$
$$b \equiv (b-2) \bmod (b+1) \quad |(9), (10) \tag{11}$$
$$\exists m \in \mathbb{Z} : m(b+1) = 2 \quad |(11) \tag{12}$$

By (12), we have $b \in \{0, 1\}$ which contradicts $b \geq 2$. Consequently, if $sig_T(z) \in E_0$ and $sig_T(k) = (1, 0)$ and $sig_T(X) \in E_0 \cup \{b\}$ then there is exactly one variable event $X \in \{X_{i,0}, X_{i,1}, X_{i,2}\}$ with $sig_T(X) \notin E_0$.

If we continue with $sig_T(z) \in E_0$, $sig_T(k) = (0, 1)$ and $sig_T(X) \in E_0 \cup \{1\}$ then we find the following equation to be true:

$$(|sig_T(X_{i,0})| + |sig_T(X_{i,1})| + |sig_T(X_{i,2})|) \equiv 0 \bmod (b+1) \tag{13}$$

Analogously to the former case one argues that the assumption that not exactly one variable event $X \in \{X_{i,0}, X_{i,1}, X_{i,2}\}$ is equipped with the signature 1, that is, $sig_T(X) \notin E_0$, leads to the contradiction $b \in \{0, 1\}$. Altogether, we have shown that if (sup_T, sig_T) is a region such that $sig_T(k) \in \{(0, 1), (1, 0)\}$ and $sig_T(z) \in E_0$ and if the TSs $F_0, G_0, \ldots, F_{m-1}, G_{m-1}$ do as announced then there is exactly one variable event $X \in \{X_{i,0}, X_{i,1}, X_{i,2}\}$ for every $i \in \{0, \ldots, m-1\}$ such that $sig_T(X) \notin E_0$. By other words, in that case we have that the set $M = \{X \in V(\varphi) | sig_T(X) \notin E_0\}$ defines a one-in-three model of φ.

Hence, to complete the arguments for Lemma 6.1, it remains to argue for the announced functionality of $F_0, G_0, \ldots, F_{m-1}, G_{m-1}$. Let $j \in \{0, \ldots, m-1\}$. We argue for X_j that if $sup_T(k) = (1, 0)$ then $sup_T(X_j) \in E_0 \cup \{b\}$ and if $sup_T(k) = (0, 1)$ then $sup_T(X_j) \in E_0 \cup \{1\}$, respectively.

To begin with, let $sig_T(k) = (1, 0)$. The event k occurs b times in a row at $f_{j,0,0}$ and $g_{j,0}$ and $b - 1$ times in a row at $f_{j,1,0}$. By Lemma 1 this implies $sup_T(f_{j,0,b}) = sup_T(g_{j,b}) = 0$ and $sup_T(f_{j,1,b-1}) \in \{0, 1\}$. Clearly, if $sup_T(f_{j,0,b}) = sup_T(f_{j,1,b-1}) = 0$ then $sig_T(X_j) \in E_0$. We argue, $sup_T(f_{j,1,b-1}) = 1$ implies $sig_T(X_j) = b$.

Assume, for a contradiction, that $sig_T(X_j) \neq b$. If $sig_T(X_j) = (m, m)$ for some $m \in \{1, \ldots, b\}$ then $-sig_T^-(X_j) + sig_T^+(X_j) = |sig_T(X_j)| = 0$. By Lemma 1 this contradicts $sup_T(f_{j,0,b}) \neq sup_T(f_{j,1,b-1})$. If $sig_T(X_j) = (m, n)$ with $m \neq n$ then the $|sig_T(X_j)| = 0$. By Lemma 1, we have $sup_T(f_{j,0,b}) = sup_T(f_{j,1,b-1}) - sig_T^-(X_j) + sig_T^+(X_j)$ implying $sig_T(X_j) = (1, 0)$. But, by $sup_T(g_{j,b}) = 0$ and $\neg 0 \xrightarrow{(1,0)}$ in τ, this contradicts $sup_T(g_{j,b}) \xrightarrow{sig_T(X_j)}$. Finally, if $sig_T(X_j) = e \in \{0, \ldots, b-1\}$ then we have $1 + e \not\equiv 0 \bmod (b+1)$. Again, this is a contradiction to $sup_T(f_{j,1,b-1}) \xrightarrow{sig_T(X_j)} sup_T(f_{j,0,b})$. Hence, we have $sig_T(X_j) = b$. Overall, it is proven that if $sup_T(k) = (1, 0)$ then $sup_T(X_j) \in E_0 \cup \{b\}$.

To continue, let $sig_T(k) = (0, 1)$. Similar to the former case, by Lemma 1, we obtain that $sup_T(f_{j,0,b}) = sup_T(g_{j,b}) = b$ and $sup_T(f_{j,1,b-1}) \in \{b-1, b\}$. If $sup_T(f_{j,1,b-1}) = b$ then $sig_T(X_j) \in E_0$. We show that $sup_T(f_{j,1,b-1}) = b-1$ implies $sig_T(X_j) = 1$: Assume $sig_T(X_j) = (m, n) \in E_\tau$. If $m = n$ or if $m > n$ then, by $sup_T(f_{j,0,b}) = sup_T(f_{j,1,b-1}) - sig_T^-(X_j) + sig_T^+(X_j)$, we have $sup_T(f_{j,0,b}) < b$, a contradiction. If $m < n$ then, by $sup_T(g_{j,b+1}) = sup_T(g_{j,b}) - sig_T^-(X_j) + sig_T^+(X_j)$, we get the contradiction $sup_T(g_{j,b+1}) > b$. Hence, $sig_T(X_j) \in \{0, \ldots, b\}$. Again, $sig_T(X_j) = e \in \{0, 2 \ldots, b\}$ implies $(b-1+e) \not\equiv b \bmod (b+1)$ which contradicts $sup_T(f_{j,0,b}) = sup_T(f_{j,1,b-1}) + |sig_T(X_j)|$. Consequently, we obtain $sig_T(X_j) = 1$ which shows that $sup_T(k) = (0, 1)$ implies $sup_T(X_j) \in E_0 \cup \{1\}$. Altogether, this proves Lemma 6.1.

To complete the proof Lemma 6, we show its second condition to be true. To do so, we start from a one-in-three model $M \subseteq V(\varphi)$ of φ and define the following τ-region (sup_T, sig_T) of T_τ that satisfies Lemma 6.2: For $e \in E_{T_\tau}$ we define $sig_T(e) =$

$$\begin{cases} (0,1), & \text{if } e = k \\ 0, & \text{if } e \in \{z\} \cup (V(\varphi) \setminus M) \text{ or } e = v_j \text{ and } X_j \in M, 0 \leq j \leq m-1 \\ 1, & \text{if } e \in M \cup \{u\} \text{ or } e = v_j \text{ and } X_j \notin M, 0 \leq j \leq m-1 \end{cases}$$

By Lemma 1, having sig_T, it is sufficient to define the values of the initial states of the constituent of T_τ. To do so, we define $sup_T(f_{j,0,0}) = sup_T(g_{j,0}) = t_{j,0} = 0$ for $j \in \{0, \ldots, m-1\}$. One easily verifies that (sup_T, sig_T) is a well defined region of T_τ. Finally, that proves Lemma 6.   □

4.5 The Liaison of Key and Translator

The following lemma completes our reduction and finally proves Theorem 1:

Lemma 7 (Suffiency)

1. Let $\tau \in \{\tau_0^b, \tau_1^b, \tau_2^b, \tau_3^b\}$. U_τ is τ-feasible, respectively has the τ-ESSP, if and only if there is a τ-region of U_τ solving its key atom α_τ if and only if φ has a one-in-three model.
2. Let $\tau' \in \{\tau_0^b, \tau_1^b\}$. W has the τ'-SSP if and only if there is a τ'-region of W solving its key atom α if and only if φ has a one-in-three model.

Proof. By Lemma 3, Lemma 4, respectively Lemma 5, Lemma 6, the respective key atoms are solvable if and only if φ is one-in-three satisfiable. Clearly, if all corresponding atoms are solvable the key atom is, too. Hence, it remains to prove that the τ-solvability (τ'-solvability) of the key atom α_τ (α) implies the τ-ESSP and τ-SSP for U_τ (τ'-SSP for W). The corresponding proofs are in [14].   □

5   Conclusions

In this paper, we show that τ-feasibility and τ-ESSP, $\tau \in \{\tau_0^b, \ldots, \tau_3^b\}$, are NP-complete. This makes τ-synthesis NP-hard. Moreover, we argue that the τ-SSP, $\tau' \in \{\tau_0^b, \tau_1^b\}$, is also NP-complete. It is future work to investigate if there are superclasses of (group extended) (pure) b-bounded P/T-nets where synthesis is tractable. Moreover, one may search for parameters of the net-types or the input TSs for which the decision problems are *fixed parameter tractable*.

Acknowledgements. I would like to thank Christian Rosenke and Uli Schlachter for their precious remarks. Also, I'm thankful to the anonymous reviewers for their helpful comments.

References

1. Aalst, W.M.P.: Process Mining Discovery - Conformance and Enhancement of Business Processes. Springer, Berlin (2011). https://doi.org/10.1007/978-3-642-19345-3
2. Badouel, E., Bernardinello, L., Darondeau, P.: Polynomial algorithms for the synthesis of bounded nets. In: Mosses, P.D., Nielsen, M., Schwartzbach, M.I. (eds.) CAAP 1995. LNCS, vol. 915, pp. 364–378. Springer, Heidelberg (1995). https://doi.org/10.1007/3-540-59293-8_207
3. Badouel, E., Bernardinello, L., Darondeau, P.: The synthesis problem for elementary net systems is NP-complete. Theor. Comput. Sci. **186**(1–2), 107–134 (1997). https://doi.org/10.1016/S0304-3975(96)00219-8
4. Badouel, E., Bernardinello, L., Darondeau, P.: Petri Net Synthesis. TTCSAES. Springer, Heidelberg (2015). https://doi.org/10.1007/978-3-662-47967-4
5. Badouel, E., Caillaud, B., Darondeau, P.: Distributing finite automata through petri net synthesis. Formal Asp. Comput. **13**(6), 447–470 (2002). https://doi.org/10.1007/s001650200022
6. Cortadella, J., Kishinevsky, M., Kondratyev, A., Lavagno, L., Yakovlev, A.: A region-based theory for state assignment in speed-independent circuits. IEEE Trans. CAD Integr. Circ. Syst. **16**(8), 793–812 (1997). https://doi.org/10.1109/43.644602
7. Garey, M.R., Johnson, D.S.: Computers and Intractability: A Guide to the Theory of NP-Completeness. W. H. Freeman (1979)
8. Hiraishi, K.: Some complexity results on transition systems and elementary net systems. Theor. Comput. Sci. **135**(2), 361–376 (1994). https://doi.org/10.1016/0304-3975(94)90112-0
9. Holloway, L.E., Krogh, B.H., Giua, A.: A survey of petri net methods for controlled discrete event systems. Discrete Event Dyn. Syst. **7**(2), 151–190 (1997). https://doi.org/10.1023/A:1008271916548

10. Moore, C., Robson, J.M.: Hard tiling problems with simple tiles. Discrete Comput. Geom. **26**(4), 573–590 (2001). https://doi.org/10.1007/s00454-001-0047-6
11. Schlachter, U., Wimmel, H.: k-bounded petri net synthesis from modal transition systems. In: CONCUR. LIPIcs, vol. 85, pp. 6:1–6:15. Schloss Dagstuhl - Leibniz-Zentrum fuer Informatik (2017). https://doi.org/10.4230/LIPIcs.CONCUR.2017.6
12. Schmitt, V.: Flip-flop nets. In: Puech, C., Reischuk, R. (eds.) STACS 1996. LNCS, vol. 1046, pp. 515–528. Springer, Heidelberg (1996). https://doi.org/10.1007/3-540-60922-9_42
13. Tredup, R.: Fixed parameter tractability and polynomial time results for the synthesis of b-bounded petri nets. In: Donatelli, S., Haar, S. (eds.) PETRI NETS 2019. LNCS, vol. 11522, pp. 148–168. Springer, Cham (2019)
14. Tredup, R.: Hardness results for the synthesis of b-bounded petri nets (technical report). CoRR abs/1904.01094 (2019)
15. Tredup, R., Rosenke, C.: Narrowing down the hardness barrier of synthesizing elementary net systems. In: CONCUR. LIPIcs, vol. 118, pp. 16:1–16:15. Schloss Dagstuhl - Leibniz-Zentrum fuer Informatik (2018). https://doi.org/10.4230/LIPIcs.CONCUR.2018.16
16. Tredup, R., Rosenke, C.: The Complexity of synthesis for 43 boolean petri net types. In: Gopal, T.V., Watada, J. (eds.) TAMC 2019. LNCS, vol. 11436, pp. 615–634. Springer, Cham (2019). https://doi.org/10.1007/978-3-030-14812-6_38
17. Tredup, R., Rosenke, C., Wolf, K.: Elementary net synthesis remains NP-complete even for extremely simple inputs. In: Khomenko, V., Roux, O.H. (eds.) PETRI NETS 2018. LNCS, vol. 10877, pp. 40–59. Springer, Cham (2018). https://doi.org/10.1007/978-3-319-91268-4_3

Fixed Parameter Tractability
and Polynomial Time Results
for the Synthesis of *b*-bounded Petri Nets

Ronny Tredup[(✉)]

Institut für Informatik, Theoretische Informatik, Universität Rostock,
Albert-Einstein-Straße 22, 18059 Rostock, Germany
ronny.tredup@uni-rostock.de

Abstract. Synthesis for a type τ of Petri nets is the problem of finding, for a given transition system (TS, for short) A, a Petri net N of this type whose state graph is isomorphic to A if such a net exists. The decision version of this search problem, called τ-feasibility, asks if, for a given TS A, there exists a Petri net N of type τ with a state graph isomorphic to A. In this case, A is called τ-feasible. A's feasibility is equivalent to fulfilling two so-called *separation properties*. In fact, a transition system A is τ-feasible if and only if it satisfies the type related *state separation property* (SSP) and *event state separation property* (ESSP). Both properties, SSP and ESSP, define decision problems. In this paper, we introduce for $b \in \mathbb{N}$ the type of restricted $\mathbb{Z}_{b+1}$-extended b-bounded P/T-nets and show that synthesis and deciding ESSP and SSP for this type is doable in polynomial time. Moreover, we demonstrate that, given a TS A, deciding if A has the SSP can be done in polynomial time for the types of (pure) $\mathbb{Z}_{b+1}$-extended b-bounded P/T-nets. Finally, we exhibit that deciding if a TS A is feasible or has the ESSP for the types of (pure) $\mathbb{Z}_{b+1}$-extended b-bounded P/T-nets is *fixed parameter tractable* if the number of occurrences of events is considered as parameter.

1 Introduction

Synthesis for a Petri net type τ (τ-synthesis, for short) is the task to find for a transition system A a Petri net N of type τ (τ-net, for short) with a state graph isomorphic to A. The associated decision version, which we call feasibility for τ (τ-feasibility, for short), asks whether there is a corresponding τ-net for the input A. If such a net exists then we call A τ-feasible. A's feasibility is equivalent to fulfilling two so-called separation properties. More exactly, A is τ-feasible if and only if it has the *state separation property* and the *event state separation property* for τ (τ-SSP and τ-ESSP, for short) [5]. Both, τ-SSP and τ-ESSP, define decision problems asking whether the input A has the τ-SSP or the τ-ESSP, respectively.

Petri net synthesis has been investigated for many years and is applied in numerous fields. It yields implementations which are correct by design and allows

© Springer Nature Switzerland AG 2019
S. Donatelli and S. Haar (Eds.): PETRI NETS 2019, LNCS 11522, pp. 148–168, 2019.
https://doi.org/10.1007/978-3-030-21571-2_10

extracting concurrency and distributability information from sequential specifications as transition systems and languages [6,7]. Further application areas of Petri net synthesis currently cover, among others, the reconstruction of a model from its execution traces (process discovery), supervisory control for discrete event systems and the synthesis of speed-independent circuits [1,9,13].

A type of Petri nets is called *bounded* if there is a positive integer b which is not exceeded by the number of tokens on any place in every reachable marking. This paper deals with the computational complexity of synthesis, feasibility, SSP and ESSP for b-*bounded* Petri net types, that is, bounded Petri nets where b is predetermined.

In [3,5], Badouel et al. showed that Synthesis, feasibility, SSP and ESSP for the type of bounded and pure bounded place/transition nets (P/T-nets, for short) are solvable in polynomial time if no bound b is preselected. On the contrary, SSP, ESSP and feasibility are NP-complete for pure 1-bounded P/T-nets [4,12]. This remains true even for strongly restricted input transition systems [19,21]. In [18], we showed that feasibility, SSP and ESSP are NP-complete for (pure) b-bounded P/T-nets for arbitrary $b \in \mathbb{N}^+$.

In [16], Schmitt advanced the pure 1-bounded P/T-net type by an interaction between places and transitions simulating addition of integers modulo 2. This brings the complexity of synthesis, feasibility, ESSP and SSP for the resulting pure $\mathbb{Z}_2$-extended 1-bounded P/T-nets down to polynomial time. On the contrary, we proved in [18] that extending (pure) b-bounded P/T-nets by $\mathbb{Z}_{b+1}$ yields no tractable type if $b \geq 2$. In particular, feasibility and ESSP for (pure) $\mathbb{Z}_{b+1}$-extended b-bounded P/T-nets, $b \geq 2$, are NP-complete. We continued research on the impact of interactions on the computational complexity of synthesizing 1-bounded Petri nets in [20]. Here, we investigated 43 1-bounded types purely defined by interactions which they have or not. While for 37 of them synthesis is tractable, feasibility and ESSP for the remaining 7 are NP-complete.

Results of [2,8] show that putting restrictions on the sought nets's (syntactical) structure can have a positive impact on the complexity of synthesis. In particular, in [2], Agostini et al. proposed a polynomial time synthesis algorithm for Free-Choice Acyclic pure 1-bounded P/T-nets having applications in workflow models. Moreover, in [8], Best et al. showed that it suffices to check certain structural properties of the input A if the sought net is a pure b-bounded live marked graph. Whether A has these properties or not is decidable in polynomial time [14].

In this paper, we examine whether there are also types of b-bounded P/T-nets for which synthesis is tractable if $b \geq 2$. We affirm this question and propose the restricted $\mathbb{Z}_{b+1}$-extended b-bounded P/T-nets, $b \in \mathbb{N}$. This paper shows, that synthesis, feasibility, ESSP and SSP are solvable in polynomial time for this type. Moreover, our results prove that deciding whether a transition system (TS, for short) A has the SSP for the types of (pure) $\mathbb{Z}_{b+1}$-extended b-bounded P/T-nets, $b \in \mathbb{N}$, is also doable in polynomial time. Notice, that this discovers the first Petri net type where the provable computational complexity of SSP is different to ESSP and feasibility.

To decide whether a TS A is τ-feasible or has the τ-ESSP, where τ corresponds to (pure) $\mathbb{Z}_{b+1}$-extended b-bounded P/T-nets, $b \geq 2$, is NP-complete [18]. Hence, this problem is considered inherently hard to solve algorithmically. Consequently, one expects that every corresponding decision algorithm has an exponential running time if complexity is measured in terms of the input size of A only. In this paper, we analyze the computational complexity of feasibility and ESSP for these types in finer detail. To do so, we apply parameterization, a typical approach of modern complexity theory to tackle hard problems. The running time of parameterized algorithms is not only expressed in the input's size, but it also takes the parameters into account. The number k of *occurrences of events*, the maximum number of different transitions at which an event occur, is one of the most obvious parts of a TS which can be considered as a parameter. We show that feasibility and ESSP related to the types of (pure) $\mathbb{Z}_{b+1}$-extended b-bounded P/T-nets are only exponential in the size of k while polynomial in the size of the input. Hence, both problems are *fixed parameter tractable* if k is considered as parameter. This result could not be foreseen with certainty. In fact, in [19], we showed that feasibility, ESSP and SSP remain NP-complete for pure 1-bounded P/T-nets even if every event occurs at most *twice*. Hence, related to pure 1-bounded P/T-nets, these problems parameterized by k are not fixed parameter tractable as long as P $\neq$ NP.

2   Preliminaries

A *transition system* (TS for short) $A = (S, E, \delta)$ consists of finite disjoint sets S of states and E of events and a partial *transition function* $\delta : S \times E \rightarrow S$. Usually, we think of A as an edge-labeled directed graph with node set S where every triple $\delta(s, e) = s'$ is interpreted as an e-labeled edge $s \xrightarrow{e} s'$, called *transition*. We say that an event e *occurs* at state s if $\delta(s, e) = s'$ for some state s' and abbreviate this with $s \xrightarrow{e}$. This notation is extended to words $w' = wa$, $w \in E^*, a \in E$ by inductively defining $s \xrightarrow{\varepsilon} s$ for all $s \in S$ and $s \xrightarrow{w'} s''$ if and only if $s \xrightarrow{w} s'$ and $s' \xrightarrow{a} s''$. If $w \in E^*$ then $s \xrightarrow{w}$ denotes that there is a state $s' \in S$ such that $s \xrightarrow{w} s'$. An *initialized* TS $A = (S, E, \delta, s_0)$ is a TS with an initial state $s_0 \in S$ where every state is *reachable*: $\forall s \in S, \exists w \in E^* : s_0 \xrightarrow{w} s$. The language of A is the set $L(A) = \{w \in E^* \mid s_0 \xrightarrow{w}\}$. In the remainder of this paper, if not explicitly stated otherwise, we assume all TSs to be initialized and we refer to the components of an (initialized) TS A consistently by $A = (S_A, E_A, \delta_A, s_{0,A})$.

The following notion of *types of nets* has been developed in [5]. It allows us to uniformly capture several Petri-net types in one general scheme. Every introduced Petri-net type can be seen as an instantiation of this general scheme. A type of nets τ is a TS $\tau = (S_\tau, E_\tau, \delta_\tau)$ and a Petri net $N = (P, T, f, M_0)$ of type τ, τ-net for short, is given by finite and disjoint sets P of places and T of transitions, an initial marking $M_0 : P \longrightarrow S_\tau$, and a flow function $f : P \times T \rightarrow E_\tau$. The meaning of a τ-net is to realize a certain behavior by cascades of firing transitions. In particular, a transition $t \in T$ can fire in a marking $M : P \longrightarrow S_\tau$

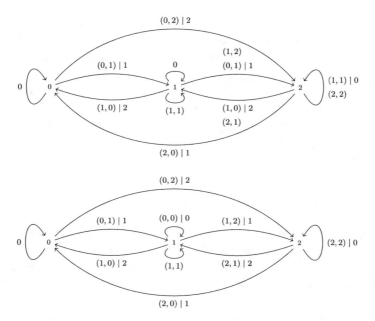

Fig. 1. Top: The type τ_2^2. Edges with several labels represent different transitions. Discarding the $(1,1)$, $(1,2)$, $(2,1)$ and $(2,2)$ labeled transitions yields τ_3^2. Bottom: The type τ_4^2.

and thereby produces the marking $M' : P \longrightarrow S_\tau$ if for all $p \in P$ the transition $M(p) \xrightarrow{f(p,t)} M'(p)$ exists in τ. This is denoted by $M \xrightarrow{t} M'$. Again, this notation extends to sequences $\sigma \in T^*$. Accordingly, $RS(N) = \{M \mid \exists \sigma \in T^* : M_0 \xrightarrow{\sigma} M\}$ is the set of all reachable markings of N. Given a τ-net $N = (P, T, f, M_0)$, its behavior is captured by the TS $A_N = (RS(N), T, \delta, M_0)$, called the state graph of N, where for every reachable marking M of N and transition $t \in T$ with $M \xrightarrow{t} M'$ the transition function δ of A_N is defined by $\delta(M, t) = M'$. The following types of b-bounded P/T-nets and pure b-bounded P/T-nets build the basis of the announced $\mathbb{Z}_{b+1}$-extensions:

0. The type of *b-bounded P/T-nets* is defined by $\tau_0^b = (\{0, \ldots, b\}, \{0, \ldots, b\}^2, \delta_{\tau_0^b})$ where for $s \in S_{\tau_0^b}$ and $(m, n) \in E_{\tau_0^b}$ the transition function is defined by $\delta_{\tau_0^b}(s, (m, n)) = s - m + n$ if $s \geq m$ and $s - m + n \leq b$, and undefined otherwise.

1. The type of *pure b-bounded P/T-nets* is a restriction of τ_0^b-nets that discards all events (m, n) from $E_{\tau_0^b}$ where both, m and n, are positive. To be exact, $\tau_1^b = (\{0, \ldots, b\}, E_{\tau_0^b} \setminus \{(m, n) \mid 1 \leq m, n \leq b\}, \delta_{\tau_1^b})$, and for $s \in S_{\tau_1^b}$ and $e \in E_{\tau_1^b}$ we have $\delta_{\tau_1^b}(s, e) = \delta_{\tau_0^b}(s, e)$.

Having τ_0^b and τ_1^b, their following $\mathbb{Z}_{b+1}$-extension allows them to simulate the addition of integers modulo $b + 1$.

2. The type of $\mathbb{Z}_{b+1}$-*extended b-bounded P/T-nets* τ_2^b arises from τ_0^b such that, firstly, $E_{\tau_2^b}$ extends the event set $E_{\tau_0^b}$ by the elements $0, \ldots, b$ and, secondly, the transition function $\delta_{\tau_2^b}$ extends $\delta_{\tau_0^b}$ by the addition of integers modulo $b+1$. More exactly, $\tau_2^b = (\{0, \ldots, b\}, (E_{\tau_0^b} \setminus \{(0,0)\}) \cup \{0, \ldots, b\}, \delta_{\tau_2^b})$ where for $s \in S_{\tau_2^b}$ and $e \in E_{\tau_2^b}$ we have that $\delta_{\tau_2^b}(s,e) = \delta_{\tau_0^b}(s,e)$ if $e \in E_{\tau_0^b}$ and $\delta_{\tau_2^b}(s,e) = (s+e) \bmod (b+1)$ if $e \in \{0, \ldots, b\}$.

3. The type of $\mathbb{Z}_{b+1}$-*extended pure b-bounded P/T-nets* is defined by $\tau_3^b = (\{0, \ldots, b\}, (E_{\tau_1^b} \setminus \{(0,0)\}) \cup \{0, \ldots, b\}, \delta_{\tau_3^b})$ where for $s \in S_{\tau_3^b}$ and $e \in E_{\tau_3^b}$ the transition function is given by $\delta_{\tau_3^b}(s,e) = \delta_{\tau_1^b}(s,e)$ if $e \in E_{\tau_1^b}$ and $\delta_{\tau_3^b}(s,e) = (s+e) \bmod (b+1)$ if $e \in \{0, \ldots, b\}$.

The new Petri net τ_4^b type arises as a restriction of τ_2^b:

4. The type of *restricted* $\mathbb{Z}_{b+1}$-*extended b-bounded P/T-nets* $\tau_4^b = (S_{\tau_2^b}, E_{\tau_2^b}, \delta_{\tau_4^b})$, $b \in \mathbb{N}^+$, origins from τ_2^b and has the same state set, $S_{\tau_4^b} = S_{\tau_2^b}$, and the same event set, $E_{\tau_4^b} = E_{\tau_2^b}$, but a restricted transition function $\delta_{\tau_4^b}$. In particular, the transition function $\delta_{\tau_4^b}$ restricts $\delta_{\tau_2^b}$ in way that for $s \in S_{\tau_4^b}$ and $(m,n) \in E_{\tau_4^b}$ we have that $\delta_{\tau_4^b}(s,(m,n)) = \delta_{\tau_2^b}(s,(m,n))$ if $s = m$ and, otherwise, if $s \neq m$ then $\delta_{\tau_4^b}(s,(m,n))$ remains undefined. Hence, every $(m,n) \in E_{\tau_4^b}$ occurs exactly once in τ_4^b. Furthermore, if $(s,e) \in \{0, \ldots, b\}^2$ then $\delta_{\tau_4^b}(s,e) = \delta_{\tau_2^b}(s,e)$.

In the remainder of this paper we assume $\tau \in \{\tau_2^b, \tau_3^b, \tau_4^b\}$ and $b \in \mathbb{N}^+$, unless stated otherwise. Notice, that τ_4^1 coincides with Schmitt's type [16]. Figure 1 gives a graphical representation of τ_2^2, τ_3^2 and τ_4^2. The following notion of τ-regions allows us, on the one hand, to define the type related ESSP and SSP and, on the other hand, to reveal in which way we are able to obtain a τ-net N for a given TS A if it exists. Figure 2 shows examples of all subsequently introduced terms.

If τ is a type of nets then a τ-*region* of a TS A is a pair of mappings (sup, sig), where $sup : S_A \longrightarrow S_\tau$ and $sig : E_A \longrightarrow E_\tau$, such that, for each transition $s \xrightarrow{e} s'$ of A, we have that $sup(s) \xrightarrow{sig(e)} sup(s')$ is a transition of τ. If (sup, sig) is a τ-region of A then for $e \in E_A$ we define $sig^-(e) = m$, $sig^+(e) = n$ and $|sig(e)| = 0$ if $sig(e) = (m,n) \in E_\tau$ and, otherwise, $sig^-(e) = sig^+(e) = 0$ and $|sig(e)| = sig(e)$ if $sig(e) \in \{0, \ldots, b\}$. Hence, by definition of τ, (sup, sig) is a τ-region if and only if $s \xrightarrow{e} s'$ entails $sup(s') = (sup(s) - sig^-(e) + sig^+(e) + |sig(e)|) \bmod (b+1)$.

Two distinct states $s, s' \in S_A$ define an *SSP atom* (s,s'), which is said to be τ-*solvable* if there is a τ-region (sup, sig) of A such that $sup(s) \neq sup(s')$. An event $e \in E_A$ and a state $s \in S_A$ at which e does not occur, that is $\neg s \xrightarrow{e}$, define an *ESSP atom* (e,s). The atom is said to be τ-solvable if there is a τ-region (sup, sig) of A such that $\neg sup(s) \xrightarrow{sig(e)}$. A τ-region solving an ESSP or a SSP atom (x,y) is a *witness* for the τ-solvability of (x,y). A TS A has the

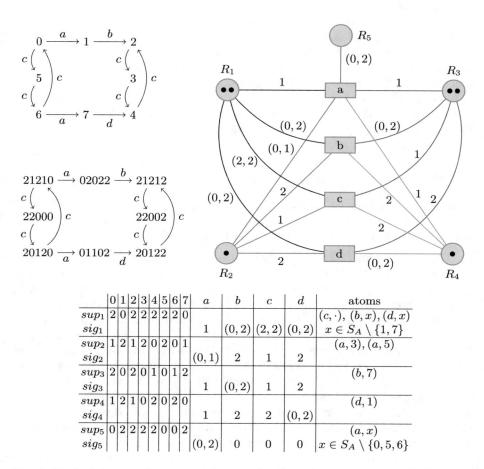

	0	1	2	3	4	5	6	7	a	b	c	d	atoms
sup_1	2	0	2	2	2	2	2	0					$(c, \cdot), (b, x), (d, x)$
sig_1									1	$(0,2)$	$(2,2)$	$(0,2)$	$x \in S_A \setminus \{1,7\}$
sup_2	1	2	1	2	0	2	0	1					$(a,3), (a,5)$
sig_2									$(0,1)$	2	1	2	
sup_3	2	0	2	0	1	0	1	2					$(b,7)$
sig_3									1	$(0,2)$	1	2	
sup_4	1	2	1	0	2	0	2	0					$(d,1)$
sig_4									1	2	2	$(0,2)$	
sup_5	0	2	2	2	2	0	0	2					(a,x)
sig_5									$(0,2)$	0	0	0	$x \in S_A \setminus \{0,5,6\}$

Fig. 2. Upper left, top: Input TS A. Bottom: The table depicts the set of τ_4^2-regions $\mathcal{R} = \{R_1, R_2, R_3, R_4, R_5\}$ where $R_i = (sup_i, sig_i)$ for $i \in \{1, \ldots, 5\}$. All regions of $\mathcal{R}$ are also τ_2^2-regions and all of $\mathcal{R} \setminus \{R_1\}$ are also τ_3^2-regions. One verifies that $\mathcal{R}$ contains a witness for every ESSP atom and every SSP atom of A. Hence, if $\tau \in \{\tau_2^2, \tau_4^2\}$ then A and the synthesized τ-net $N_A^\mathcal{R}$ has a state graph that is isomorphic to A. However, the set $\{R_2, \ldots, R_5\}$ contains no witness for the solvability of $(c, 1)$ and R_1 is no τ_3^2-region. The ESSP atom $(c, 1)$ is not τ_3^2-solvable at all, hence, A is not τ_3^2-feasible. Upper right: The graphical representation of the synthesized τ-net $N_A^\mathcal{R} = (\mathcal{R}, \{a, b, c\}, f, 21210)$, where $f(R_i, x) = sig_i(x)$ for every $x \in \{a, b, c\}$ and $M_0(R_1) \cdots M_0(R_5) = sup_1(0) \ldots sup_5(0)$. For readability, 0-labeled flow arcs for the representation of $f(R_5, x) = 0$ for $x \in \{b, c, d\}$ are neglected and flow arcs to the same place are drawn in the same color. Upper left, bottom: The state graph $A_{N_A^\mathcal{R}}$ of $N_A^\mathcal{R}$ where the reachable markings (states) are represented by 5-tuples $M(R_1) \cdots M(R_5)$. Obviously, $A_{N_A^\mathcal{R}}$ is isomorphic to A.

τ-ESSP (τ-SSP) if all its ESSP (SSP) atoms are τ-solvable. Naturally, A is said to be τ-feasible if it has the τ-ESSP and the τ-SSP.

The following fact is well known from [5, p. 161]: A set $\mathcal{R}$ of τ-regions of A contains a witness for all ESSP and SSP atoms if and only if the *synthesized* τ-net $N_A^{\mathcal{R}} = (\mathcal{R}, E_A, f, M_0)$ has a state graph that is isomorphic to A. The flow function of $N_A^{\mathcal{R}}$ is defined by $f((sup, sig), e) = sig(e)$ and its initial marking is $M_0((sup, sig)) = sup(s_{0,A})$ for all $(sup, sig) \in \mathcal{R}, e \in E_A$. The regions of $\mathcal{R}$ become places and the events of E_A become transitions of $N_A^{\mathcal{R}}$. Hence, for a τ-feasible TS A where $\mathcal{R}$ is known, we can synthesize a net N with state graph isomorphic to A by constructing $N_A^{\mathcal{R}}$.

3   Polynomial Time Results

Theorem 1. *Solving τ_4^b-synthesis for a TS A or deciding if A has the τ_4^b-ESSP is doable in polynomial time. Moreover, for $\tau \in \{\tau_2^b, \tau_3^b, \tau_4^b\}$ one can decide in polynomial time whether a given TS A has the τ-SSP.*

The contribution of Theorem 1 is threefold. Firstly, in [18] it has been shown that deciding the τ-ESSP and τ-feasibility is a NP-complete problem for $\tau \in \{\tau_2^b, \tau_3^b\}$. Hence, by showing that deciding the τ-SSP for $\tau \in \{\tau_2^b, \tau_3^b\}$ is doable in polynomial time, Theorem 1 discovers the first Petri net types where the provable computational complexity of SSP is different to ESSP and feasibility.

In [16], Schmitt advanced pure 1-bounded P/T-nets by the additive group of integers modulo 2 and discovered a tractable superclass. In [18], we showed that lifting this approach to (pure) b-bounded P/T-nets where $b \geq 2$ do not lead to superclasses with a tractable synthesis problem. Thus, Theorem 1 proposes the first tractable type of b-bounded Petri nets, $b \geq 2$, so far. Finally, Theorem 1 gives us insight into which of the *τ-net properties*, $\tau \in \{\tau_0^b, \tau_1^b\}$, cause the synthesis' hardness. In particular, flow arc relations (events in τ) between places and transitions in a τ-net define conditions when a transition is able to fire. For example, if N is a τ-net with transition t and place p such that $f(p, t) = (1, 0)$ then the firing of t in a marking M requires $M(p) \geq 1$. By Theorem 1, the hardness of finding a τ-net N for A origins from the potential possibility of τ-nets to satisfy such conditions by multiple markings $M(p) \in \{1, \ldots, b\}$. In fact, the definition of τ_4^b implies that $f(p, t) = (m, n)$ requires $M(p) = m$ for the firing of t and prohibits the possibility of multiple choices. By Theorem 1, this makes τ_4^b-synthesis tractable. It should be noted that the results of [4,12] show that the restriction to "unambiguous markings" of p satisfying conditions defined by $f(p, t)$ does *not guarantee* tractability.

While the question of whether there are superclasses of $\tau_0^b, \tau_1^b, b \geq 2$, for which synthesis is doable in polynomial time remains unanswered, the following lemma shows that the type τ_4^b yields at least a tractable superclasses of Schmitt's type τ_4^1 [16]. In particular, if $b < b'$ then the class of τ_4^b-nets is strictly more comprehensive than the class of $\tau_4^{b'}$-nets.

Lemma 1. *If $b < b' \in \mathbb{N}^+$ and if $\mathcal{T}$ is the set of τ_4^b-feasible TSs and $\mathcal{T}'$ the set of $\tau_4^{b'}$-feasible TSs then $\mathcal{T} \subset \mathcal{T}'$.*

Proof. To proof the lemma, we consider a TS A which is $\tau_4^{b'}$-feasible but not τ_4^b-feasible. Let A defined by $A = (\{s_0, \ldots, s_{b'}\}, \{a\}, \delta, s_0)$ the TS with transition function $\delta(s_i, a) = s_{i+1}$ for $i \in \{0, \ldots, b'-1\}$ and $\delta(s_{b'}, a) = s_0$. By other words, A is a directed labeled cycle $s_0 \xrightarrow{a} \ldots \xrightarrow{a} s_{b'} \xrightarrow{s} s_0$ where every transition is labeled by a. Notice, that A has no ESSP atom and, hence, the τ-ESSP for every type of nets. Consequently, A is τ-feasible if and only if it has the τ-SSP.

Assume, for a contradiction, that A is τ_4^b-feasible. By $b < b'$, A provides the SSP atom (s_0, s_{b+1}) and A's τ_4^b-feasibility implies that there is a τ_4^b-region (sup, sig) solving it. If $sig(a) = (m, n)$ then $sup(s_1) = sup(s_0) - m + n \neq sup(s_0)$ and, by definition of τ_4^b, $\neg sup(s_1) \xrightarrow{(m,n)}$. This is a contradiction to $s_1 \xrightarrow{a}$. Hence, $sig(a) \in \{1, \ldots, b\}$. By induction, $sup(s_{b+1}) = sup(s_0) + (b+1) \cdot sig(a) = sup(s_0) \bmod (b+1)$ implying $sup(s_{b+1}) = sup(s_0)$. Thus, (sup, sig) does not solve (s_0, s_{b+1}), which proves that A not to be τ_4^b-feasible.

On the contrary, it is easy to see that the $\tau_4^{b'}$-region (sup, sig), which is defined by $sup(s_0) = 0$, $sig(e) = 1$ and $sup(s_{i+1}) = sup(s_i) + sig(a)$ for $i \in \{0, \ldots, b'-1\}$, solves every SSP atom of A. Hence, A is $\tau_4^{b'}$-feasible.   □

3.1   Abstract Regions and Fundamental Cycles

Unless otherwise stated, in the remainder of this paper we assume that A is a (non-trivial) TS with at least two states, $|S_A| \geq 2$ and event set $E_A = \{e_1, \ldots, e_n\}$. Recall that $\tau \in \{\tau_2^b, \tau_3^b, \tau_4^b\}$ and $b \in \mathbb{N}^+$.

The proof of Theorem 1 bases on a generalization of the approach used in [16] that reduces ESSP and SSP to systems of linear equations modulo $b + 1$. It exploits that the solvability of such systems is decidable in polynomial time which is the statement of the following lemma borrowed from [11]:

Lemma 2 ([11]). *If $A \in \mathbb{Z}_{b+1}^{k \times n}$ and $c \in \mathbb{Z}_{b+1}^k$ then deciding if there is an element $x \in \mathbb{Z}_{b+1}^n$ such that $Ax = c$ is doable in time $\mathcal{O}(nk \cdot max\{n, k\})$.*

Essentially, our generalization composes for every ESSP atom and every SSP atom $\alpha = (x, y)$ of A, respectively, a system of equations modulo $b + 1$ which is solvable if and only if α is τ-solvable. Moreover, a solution of the corresponding system shall provide a τ-region of A that solves α. On the one hand, this approach ensures that having a solution for every system defined by single ESSP atoms and SSP atoms implies the τ-ESSP and τ-SSP for A, respectively. On the other hand, it provides a τ-solving region for every atom in question and, hence, a set $\mathcal{R}$ of τ-regions that witnesses the τ-ESSP and τ-SSP of A. Thus, $\mathcal{R}$ allows us to construct the synthesized net $N_A^{\mathcal{R}}$ with a state graph isomorphic to A. In the following, we establish the notions of *abstract regions* and *fundamental cycles* which make such a translation possible.

We proceed by deducing the notion of abstract regions. Our starting point is the goal to obtain regions (sup, sig) of A as solutions of linear equation systems modulo $b + 1$. By definition, (sup, sig) is a τ-region of A if and only if for every transition $s \xrightarrow{e} s'$ it is true that

$$sup(s') = (sup(s) - sig^-(e) + sig^+(e) + |sig(e)|) \bmod (b+1) \qquad (1)$$

Hence, installing for every transition $s \xrightarrow{e} s'$ the corresponding Eq. 1 yields a linear system of equations whose solutions are regions of A. If (sup, sig) is a solution of this system such that $sig(e) = (m, n) \in E_\tau \setminus \{0, \ldots, b\}$ for $e \in E_A$ then, by definition, for every transition $s \xrightarrow{e} s'$ it has to be true that $m \leq sup(s)$ and $sup(s') - m + n \leq b$. Unfortunately, the conditions $m \leq sup(s)$ and $sup(s') - m + n \leq b$ can not be tested in the group $\mathbb{Z}_{b+1}$. To cope with this obstacle, we abstract from elements $(m, n) \in E_\tau$ by restricting to regions (solutions) that identify (m, n) with the unique element $x \in \{0, \ldots, b\}$ such that $x = (n - m) \bmod (b + 1)$. This leads to the notion of *abstract τ-regions*. A τ-region (sup, sig) of A is called abstract if sig's codomain restricts to $\{0, \ldots, b\}$, that is, $sig : E_A \longrightarrow \{0, \ldots, b\}$. If (sup, sig) is an abstract region, then we call sig an *abstract* signature. For the sake of clarity, we denote abstract signatures by abs instead of sig and abstract regions by (sup, abs) instead of (sup, sig).

By definition, two mappings $sup, abs : \{0, \ldots, b\} \longrightarrow \{0, \ldots, b\}$ define an abstract τ-region if and only if for every transition $s \xrightarrow{e} s'$ of A it is true that

$$sup(s') = (sup(s) + abs(e)) \bmod (b + 1) \tag{2}$$

Obviously, for abstract regions Eq. 1 reduces to Eq. 2. Installing for every transition $s \xrightarrow{e} s'$ of A its corresponding Eq. 2 yields a system modulo $b + 1$ whose solutions are abstract regions. Uncomfortably, such systems require to deal with sup and abs simultaneously. It is better to first obtain abs independently of sup and then define sup with the help of abs. The following observations show how to realize this idea.

By induction and Eq. 2, one immediately obtains that (sup, abs) is an abstract region if and only if for every directed labeled path $p = s_{0,A} \xrightarrow{e'_1} \ldots \xrightarrow{e'_m} s_m$ of A from the initial state $s_{0,A}$ to the state s_m the *path equation* holds:

$$sup(s_m) = (sup(s_{0,A}) + abs(e'_1) + \cdots + abs(e'_m)) \bmod (b + 1) \tag{3}$$

To exploit Eq. 3 we, firstly, identify every abstract signature abs with the unique element $abs = (abs(e_1), \ldots, abs(e_n)) \in \mathbb{Z}_{b+1}^n$. Secondly, we say that $\psi_{b+1}^p = (\#_{e_1}^p, \ldots, \#_{e_n}^p) \in \mathbb{Z}_{b+1}^n$ is the Parikh-vector of p that counts the number $\#_{e_i}^p$ of occurrences of every event $e_i \in E_A$ on the path p modulo $(b + 1)$. Thirdly, for two elements $v, w \in \mathbb{Z}_{b+1}^n$ we define $v \cdot w = v_1 w_1 + \cdots + v_n w_n$. As a result, considering p and abs as elements of $\mathbb{Z}_{b+1}^n$ allows us to reformulate the path equation by $sup(s_m) = (sup(s_{0,A}) + \psi_{b+1}^p \cdot abs) \bmod (b + 1)$. Especially, if p, p' are two different paths from $s_{0,A}$ to s_m then $\psi_{b+1}^p \cdot abs = \psi_{b+1}^{p'} \cdot abs$. Thus, the support sup is fully determined by $sup(s_{0,A})$ and abs. By the validity of the path equation, every abstract signature abs implies $b + 1$ different abstract τ-regions of A, one for every $sup(s_{0,A}) \in \{0, \ldots, b\}$. Altogether, we have argued that the challenge of finding abstract regions of A reduces to the task of finding A's abstract signatures. In the following, we deduce the notion of fundamental cycles defined by chords of a spanning tree of A which enables us to find abstract

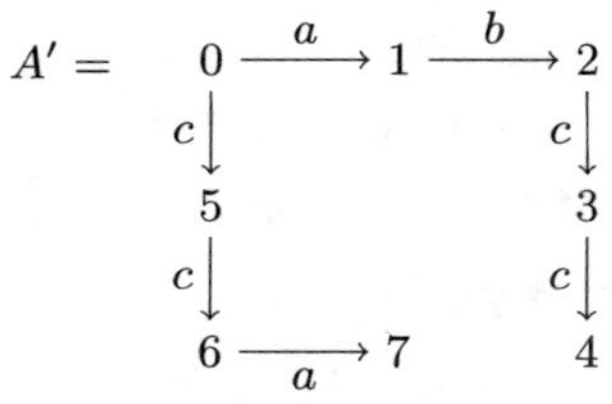

$$A' = \quad 0 \xrightarrow{\ a\ } 1 \xrightarrow{\ b\ } 2$$

$$c\downarrow \qquad\qquad\qquad c\downarrow$$

$$5 \qquad\qquad\qquad 3$$

$$c\downarrow \qquad\qquad\qquad c\downarrow$$

$$6 \xrightarrow[\ a\ ]{} 7 \qquad\qquad 4$$

Fig. 3. A spanning tree A' of running example TS A introduced in Fig. 2. The unique Parikh vectors $\psi_0, \ldots \psi_7$ of A' (written as rows) are given by $\psi_0 = (0,0,0,0), \psi_1 = (1,0,0,0), \psi_2 = (1,1,0,0), \psi_3 = (1,1,1,0), \psi_4 = (1,1,2,0), \psi_5 = (0,0,1,0), \psi_6 = (0,0,2,0)$ and $\psi_7 = (1,0,2,0)$. The transitions $\delta_A(7,d) = 4$, $\delta_A(4,c) = 2$ and $\delta_A(6,c) = 0$ of A define the chords of A'. The corresponding fundamental cycles are given by $\psi_t = \psi_7 + (0,0,0,1) - \psi_4 = (0,2,0,1)$ and $\psi_{t'} = \psi_4 + (0,0,1,0) - \psi_2 = (0,0,0,0)$ and $\psi_{t''} = \psi_6 + (0,0,1,0) - \psi_0 = (0,0,0,0)$. Hence, if $abs = (x_a, x_b, x_c, x_d)$ then $\psi_t \cdot abs = 0 \cdot x_a + 2 \cdot x_b + 0 \cdot x_c + x_d = 2 \cdot x_b + x_d$. By $\psi_{t'} \cdot abs = \psi_{t''} \cdot abs = 0$ for every map abs, only the equation $2 \cdot x_b + x_d = 0$ contributes to the basic part of every upcoming system.

signatures. For readability, we often write $x = y_1 + \cdots + y_\ell \mod (b+1)$ instead of $x = (y_1 + \cdots + y_\ell) \mod (b+1)$

A *spanning tree* A' of A is a sub-transition system $A' = (S_A, E_A, \delta_{A'}, s_{0,A})$ of A with a restricted transition function $\delta_{A'}$ such that, firstly, $\delta_{A'}(s,e) = s'$ entails $\delta_A(s,e) = s'$ and, secondly, for every $s \in S_{A'}$ there is *exactly* one path $p = s_{0,A} \xrightarrow{e_1} \ldots \xrightarrow{e_m} s$ in A'. By other words, the underlying undirected graph of A' is a tree in the common graph-theoretical sense. Every transition $s \xrightarrow{e} s'$ of A which is not in A' is called a *chord* (of A'). The chords of A' are exactly the edges that induce a cycle in A''s underlying undirected graph. This gives rise to the following notion of fundamental cycles. For $e_i \in \{e_1, \ldots, e_n\}$ we define $1_i = (x_1, \ldots, x_n)^t \in \mathbb{Z}_{b+1}^n$, where $x_j = 1$ if $j = i$ and, else $x_j = 0$. If $t = s \xrightarrow{e_i} s'$ is a chord of A' then there are *unique* paths p from $s_{0,A}$ to s and p' from $s_{0,A}$ to s' in A' such that t corresponds to the unique element $\psi_t = \psi_{b+1}^p + 1_i - \psi_{b+1}^{p'} \in \mathbb{Z}_{b+1}^n$, called the *fundamental cycle* of t.

The following lemma teaches us how to use fundamental cycles to generate abstract signatures of A:

Lemma 3. *If A' is a spanning tree of a TS A with chords $t_1, \ldots, t_k$ then $abs \in \mathbb{Z}_{b+1}^n$ is an abstract signature of A if and only if $\psi_{t_i} \cdot abs = 0$ for all $i \in \{1, \ldots, k\}$. Two different spanning trees A' and A'' provide equivalent systems of equations.*

Proof. We start with proving the first statement. *If*: Let $abs \in \mathbb{Z}_{b+1}^n$ such that $\psi_{t_i} \cdot abs = 0$ for all $i \in \{1, \ldots, k\}$ and $sup(s_{0,A}) \in \{0, \ldots, b\}$. If $s \in S_{A'}$ then there is a unique path $p = s_{0,A} \xrightarrow{e'_1} \ldots \xrightarrow{e'_m} s_m = s$ in A' from $s_{0,A}$ to s. By defining $sup(s) = sup(s_{0,A}) + \psi_{b+1}^p \cdot abs$ we obtain inductively that every transition $s \xrightarrow{e} s'$ of A' satisfies $sup(s') = sup(s) + abs(e)$. It remains to prove that this definition is consistent with the remaining transitions of A, the chords of A'. Let $t = s \xrightarrow{e} s'$

be a chord of A' and let $p = s_{0,A} \xrightarrow{e'_1} \ldots \xrightarrow{e'_m} = s$ and $p' = s_{0,A} \xrightarrow{e''_1} \ldots \xrightarrow{e''_\ell} = s'$ be the unique paths from $s_{0,A}$ to s and s' in A', respectively. By $sup(s) = sup(s_{0,A}) + \psi^p_{b+1} \cdot abs$ and $sup(s') = sup(s_{0,A}) + \psi^{p'}_{b+1} \cdot abs$ we have that

$$0 = \psi_t \cdot abs \qquad\qquad\qquad \Longleftrightarrow$$

$$0 = (-\psi^{p'}_{b+1} + 1_i + \psi^p_{b+1}) \cdot abs \qquad\qquad \Longleftrightarrow$$

$$0 = -\psi^{p'}_{b+1} \cdot abs + abs(e) + \psi^p_{b+1} \cdot abs \qquad\qquad \Longleftrightarrow$$

$$\psi^{p'}_{b+1} \cdot abs = abs(e) + \psi^p_{b+1} \cdot abs \qquad\qquad \Longleftrightarrow$$

$$sup(s_{0,A}) + \psi^{p'}_{b+1} \cdot abs = sup(s_{0,A}) + \psi^p_{b+1} \cdot abs + abs(e) \qquad \Longleftrightarrow$$

$$sup(s') = sup(s) + abs(e)$$

where $0 = \psi_t \cdot abs$ is true by assumption. Hence, abs is an abstract signature of A and the proof shows how to get a corresponding abstract region (sup, abs) of A.

Only-if: If abs is an abstract region of A then we have $sup(s') = sup(s) + abs(e)$ for every transition in A. Hence, if $t = s \xrightarrow{e} s'$ a chord of a spanning tree A' of A then working backwards the equivalent equalities above proves $\psi_t \cdot abs = 0$.

The second statement is implied by the first: If A', A'' are two spanning trees of A with fundamental cycles $\psi^{A'}_{t_1}, \ldots, \psi^{A'}_{t_k}$ and $\psi^{A''}_{t'_1}, \ldots, \psi^{A''}_{t'_k}$, respectively, then we have for $abs \in \mathbb{Z}^n_{b+1}$ that $\psi^{A'}_{t_i} \cdot abs = 0, i \in \{1, \ldots, k\}$ if and only if abs is an abstract signature of A if and only if $\psi^{A''}_{t'_i} \cdot abs = 0, i \in \{1, \ldots, k\}$. $\qquad\square$

In the following, justified by Lemma 3, we assume A' to be a fixed spanning tree of A with chords $t_1, \ldots, t_k$. By $M_{A'}$ we denote the system of equations $\psi_{t_i} \cdot abs = 0, i \in \{1, \ldots, k\}$. Moreover, by ψ_s we abridge for $s \in S_A$ the Parikh-vector ψ^p_{b+1} of the unique path $s_{0,A} \xrightarrow{e'_1} \ldots \xrightarrow{e'_m} s$ in A'. A spanning tree of A is computable in polynomial time: As δ_A is a function, A has at most $|E||S_A|^2$ transitions and A' contains $|S_A| - 1$ transitions. Thus, by $2 \leq |S_A|$, A' has at most $|E||S_A|^2 - 1$ chords. Consequently, a spanning tree A' of A is computable in time $\mathcal{O}(|E||S_A|^3)$ [17].

To get polynomial time solvable systems of equations, we have restricted ourselves to equations like Eq. 2. This restriction results in the challenge to compute abstract signatures of A. By Lemma 3, abstract signatures of A are solutions of $M_{A'}$. An (abstract) τ-region (sup, abs) of A arises from abs by defining $sup(s_{0,A})$ and $sup(s) = sup(s_{0,A}) + \psi_s \cdot abs$, $s \in S(A)$. However, if (s, s') is a SSP atom of A then $sup(s) \neq sup(s')$ is not implied. Moreover, by definition, to τ-solve ESSP atoms (e, s) we need (concrete) τ-regions (sup, sig) such that $sig : E_A \longrightarrow E_\tau$. The next section shows how to extend $M_{A'}$ to get τ-solving regions.

3.2   The Proof of Theorem 1

This section shows how to extend $M_{A'}$ for a given (E)SSP atom α to get a system M_α, whose solution yields a region solving α if there is one.

If α is an SSP atom (s, s') then we only need to assure that the (abstract) region (sup, abs) built on a solution of $M_{A'}$ satisfies $sup(s) \neq sup(s')$. By $sup(s) = sup(s_{A,0}) + \psi_s \cdot abs$ and $sup(s') = sup(s_{A,0}) + \psi_{s'} \cdot abs$, it is sufficient to extend $M_{A'}$ in a way that ensures $\psi_s \cdot abs \neq \psi_{s'} \cdot abs$. The next lemma proves this claim.

Lemma 4. *If $\tau \in \{\tau_2^b, \tau_3^b, \tau_4^b\}$ then a τ-SSP atom (s, s') of A is τ-solvable if and only if there is an abstract signature abs of A with $\psi_s \cdot abs \neq \psi_{s'} \cdot abs$.*

Proof. If: If abs is an abstract signature with $\psi_s \cdot abs \neq \psi_{s'} \cdot abs$ then the τ-region (sup, abs) with $sup(s_{0,A}) = 0$ and $sup(s) = \psi_s \cdot abs$ satisfies $sup(s) \neq sup(s')$. *Only-if:* If (sup, sig) is a τ-region then we obtain a corresponding abstract τ-region (sup, abs) as defined in Lemma 6. Clearly, abs is an abstract signature and satisfies the path equations. Consequently, by $sup(s_0) + \psi_s \cdot abs = sup(s) \neq sup(s') = sup(s_0) + \psi_{s'} \cdot abs$, we have that $\psi_s \cdot abs \neq \psi_{s'} \cdot abs$. $\square$

The next lemma applies Lemma 4 to get a polynomial time algorithm which decides the τ-SSP if $\tau \in \{\tau_2^b, \tau_3^b, \tau_4^b\}$.

Lemma 5. *If $\tau \in \{\tau_2^b, \tau_3^b, \tau_4^b\}$ then to decide whether a TS A has the τ-SSP is doable in time $\mathcal{O}(|E_A|^3 \cdot |S_A|^6 \cdot)$.*

Proof. If $\alpha = (s, s')$ is a SSP atom of A then the (basic) part $M_{A'}$ of M_α consists of at most $|E| \cdot |S_A|^2 - 1$ equations for the fundamental cycles. To satisfy $\psi_s \cdot abs \neq \psi_{s'} \cdot abs$, we add the equation $(\psi_s - \psi_{s'}) \cdot abs = q$, where initially $q = 1$, and get (the first possible) M_α. A solution of M_α provides an abstract region satisfying $\psi_s \neq \psi_{s'}$. By Lemma 4, this proves the solvability of α. If M_α is not solvable then we modify M_α to M_α' simply by incrementing q and try to solve M_α'. Either we get a solution or we modify M_α' to M_α'' by incrementing q again. By Lemma 4, if (s, s') is solvable then there is a $q \in \{1, \ldots, b\}$ such that the corresponding (modified) system has a solution. Hence, after at most b iterations we can decide whether (s, s') is solvable or not. Consequently, we have to solve at most b linear systems with at most $|E_A| \cdot |S_A|^2$ equations for (s, s'). The value b is not part of the input. Thus, by Lemma 2, this is doable in $\mathcal{O}(|E_A|^3 \cdot |S_A|^4)$ time. We have at most $|S_A|^2$ different SSP atoms to solve. Hence, we can decide the τ-SSP in time $\mathcal{O}(|E_A|^3 \cdot |S_A|^6)$. $\square$

As a next step, we let $\tau = \tau_4^b$ and prove the polynomial time decidability of τ-ESSP. But before that we need the following lemma that tells us how to obtain abstract regions from (concrete) regions:

Lemma 6. *If (sup, sig) is a τ-region of a TS A then we obtain a corresponding abstract τ-region (sup, abs) by defining abs for $e \in E_A$ as follows: If $sig(e) = (m, n)$ then $abs(e) = -m + n \mod (b + 1)$ and, otherwise, if $sig(e) \in \{0, \ldots, b\}$ then $abs(e) = sig(e)$.*

Proof. We have to show that $s \xrightarrow{e} s'$ in A entails $sup(s) \xrightarrow{abs(e)} sup(s')$ in τ. If $abs(e) = sig(e) \in \{0, \ldots, b\}$ this is true as (sup, sig) is a τ-region.

If $sig(e) = (m,n)$ then, by definition, we have $sup(s') = sup(s) - m + n \bmod (b+1)$ implying $sup(s') - sup(s) = -m + n \bmod (b+1)$. By $abs(e) = -m + n \bmod (b+1)$ and symmetry, we get $-m + n = abs(e) \bmod (b+1)$ and, by transitivity, we obtain $sup(s') - sup(s) = abs(e) \bmod (b+1)$ which implies $sup(s') = sup(s) + abs(e) \bmod (b+1)$. Thus $sup(s) \xrightarrow{abs(e)} sup(s')$.   □

Let α be an ESSP atom (e,s) and let $s_1, \ldots, s_k$ be the sources of e in A. By definition, a τ-region (sup, sig) solves α if and only if $sig(e) = (m,n)$ and $\neg sup(s) \xrightarrow{sig(e)}$ for a $(m,n) \in E_\tau$. By definition of τ, every element $(m,n) \in E_\tau$ occurs at exactly one state in τ and this state is m. Hence, $sup(s_1) = \cdots = sup(s_k) = m$ and $sup(s) \neq m$. We base the following lemma on this simple observation. It provides necessary and sufficient conditions that an *abstract* region must fulfill to imply a *solving* (concrete) region.

Lemma 7. *Let* $\tau = \tau_4^b$ *and* A *be a TS and let* $s_1 \xrightarrow{e} s_1', \ldots, s_k \xrightarrow{e} s_k'$ *be the e-labeled transitions in* A, *that is, if* $s' \in S_A \setminus \{s_1, \ldots, s_k\}$ *then* $\neg s' \xrightarrow{e}$. *The atom* (e,s) *is* τ-*solvable if and only if there is an event* $(m,n) \in E_\tau$ *and an abstract region* (sup, abs) *of* A *such that the following conditions are satisfied:*

1. $abs(e) = -m + n \bmod (b+1)$,
2. $\psi_{s_1} \cdot abs = m - sup(s_{A,0}) \bmod (b+1)$,
3. $(\psi_{s_1} - \psi_{s_i}) \cdot abs = 0 \bmod (b+1)$ *for* $i \in \{2, \ldots, k\}$,
4. $(\psi_{s_1} - \psi_s) \cdot abs \neq 0 \bmod (b+1)$.

Proof. If: Let (sup, abs) be an abstract region that satisfies the conditions 1–4. We obtain a τ-solving region (sup, sig) with (the same support and) the signature sig defined by $sig(e') = abs(e')$ if $e' \neq e$ and $sig(e') = (m,n)$ if $e' = e$. To argue that (sup, sig) is a τ-region we have to argue that $q \xrightarrow{e'} q'$ in A implies $sup(q) \xrightarrow{sig(e')} sup(q')$. As (sup, abs) is an abstract region this is already clear for transitions $q \xrightarrow{e'} q'$ where $e' \neq e$. Moreover, (sup, abs) satisfies $\psi_{s_1} \cdot abs = m - sup(s_{A,0}) \bmod (b+1)$ and the path equation holds, that is, $sup(s_1) = sup(s_{A,0}) + \psi_{s_1} \cdot abs \bmod (b+1)$ which implies $sup(s_1) = m$. Consequently, by definition of τ, we have $sup(s_1) \xrightarrow{(m,n)} n$ in τ. Furthermore, by $abs(e) = -m + n \bmod (b+1)$ we have $m + abs(e) = n \bmod (b+1)$. Hence, by $sup(s_1) \xrightarrow{abs(e)} sup(s_1')$, we conclude $sup(s_1') = n$ and, thus, $sup(s_1) \xrightarrow{(m,n)} sup(s_1')$. By $(\psi_{s_1} - \psi_{s_i}) \cdot abs = 0 \bmod (b+1)$ for $i \in \{2, \ldots, k\}$, we obtain that $sup(s_1) = \cdots = sup(s_k) = m$. Therefore, similar to the discussion for $s_1 \xrightarrow{e} s_1'$, we obtain by $sup(s_i) \xrightarrow{abs(e)} sup(s_i')$ that the transitions $sup(s_i) \xrightarrow{(m,n)} sup(s_i')$ are present in τ for $i \in \{2, \ldots, k\}$. Consequently, (sup, sig) is a τ-region.

Finally, by $(\psi_{s_1} - \psi_s) \cdot abs \neq 0 \bmod (b+1)$, have that $sup(s_1) \neq sup(s)$ and, thus, $\neg sup(s) \xrightarrow{sig(e)}$. This proves (e,s) to be τ-solvable by (sup, sig).

Only-if: Let (sup, sig) be a τ-region that solves (e,s) implying, by definition, $\neg sup(s) \xrightarrow{sig(e)}$. We use (sup, sig) to define a corresponding abstract τ-region

(sup, abs) in accordance to Lemma 6. If $sig(e) \in \{0, \ldots, b\}$ then $sup(s) \xrightarrow{sig(e)}$, a contradiction. Hence, it is $sig(e) = (m, n) \in E_\tau$ such that $sup(s_i) \xrightarrow{(m, n)}$ for $i \in \{1, \ldots, k\}$ and $\neg sup(s) \xrightarrow{(m, n)}$. This immediately implies $sup(s) \neq sup(s_1)$ and, hence, $(\psi_{s_1} - \psi_s) \cdot abs \neq 0 \mod (b + 1)$. By $sup(s_i) \xrightarrow{(m, n)} sup(s_i')$ and definition of τ, we have that $sup(s_i) = m$ and $sup(s_i') = n$ for $i \in \{1, \ldots, k\}$ implying $(\psi_{s_1} - \psi_{s_i}) \cdot abs = 0 \mod (b + 1)$ for $i \in \{2, \ldots, k\}$. Moreover, by $sup(s_1) \xrightarrow{abs(e)} sup(s_1')$ we have $abs(e) = sup(s_1') - sup(s_1) \mod (b + 1)$. Hence, it is $abs(e) = -m + n \mod (b + 1)$. Finally, by the path equation, we have $sup(s_1) = sup(s_{A,0}) + \psi_{s_1} \cdot abs \mod (b + 1)$ which with $sup(s_1) = m$ implies $\psi_{s_1} \cdot abs = m - sup(s_{A,0}) \mod (b + 1)$. This proves the lemma. $\qquad \square$

The next lemma's proof exhibits a polynomial time decision algorithm for the τ_4^b-ESSP: Given a TS A and a corresponding ESSP atom α, the system $M_{A'}$ is extended to a system M_α. If M_α has a solution abs then it implies a region (sup, abs) satisfying the conditions of Lemma 9. By Lemma 9, this implies α's solvability. Reversely, by Lemma 9, if α is solvable then there is an abstract region (sup, abs) which satisfies the conditions (1–4). The abstract signature abs is the solution of a corresponding equation system M_α. Hence, we get a solvable M_α if and only if α is solvable. We argue that the number of possible systems is bounded polynomially in the size of A. The solvability of every system is also decidable in polynomial time. Consequently, by the at most $|E_A| \cdot |S_A|$ ESSP atoms to solve, this yields the announced decision procedure.

Lemma 8. *If a TS A has the τ_4^b-ESSP is decidable in time $\mathcal{O}(|E_A|^4 \cdot |S_A|^5)$.*

Proof. To estimate the computational complexity of deciding the τ_4^b-ESSP for A observe that A has at most $|S_A| \cdot |E_A|$ ESSP atoms to solve. Hence, the maximum costs of deciding the τ_4^b-ESSP for A equals $|S_A| \cdot |E_A|$-times the maximum effort for a single atom.

To decide the τ-solvability of a single ESSP atom (e, s), we compose systems in accordance to Lemma 7. The maximum costs can be estimated as follows: The (basic) part $M_{A'}$ of M_α has at most $|E_A| \cdot |S_A|^2$ equations. Moreover, e occurs at most at $|S_A| - 1$ states. This makes at most $|S_A|$ equations to ensure that e's sources will have the same support, the third condition of Lemma 7. According to the first and the second condition, we choose an event $(m, n) \in E_\tau$, a value $sup(s_{A,0}) \in \{0, \ldots, b\}$, define $abs(e) = -m + n \mod (b + 1)$ and add the corresponding equation $\psi_{s_1} \cdot abs = m - sup(s_{A,0})$. For the fourth condition we choose a fixed value $q \in \{1, \ldots, b\}$ and add the equation $(\psi_{s_1} - \psi_s) \cdot abs = q$. Hence, the system has at most $2 \cdot |E_A| \cdot |S_A|^2$ equations.

By Lemma 2, one checks in time $\mathcal{O}(|E_A|^3 \cdot |S_A|^4)$ if such a system has a solution. Notice, we use that $2 \cdot |E_A| \cdot |S_A|^2 = max\{|E_A|, 2 \cdot |E_A| \cdot |S_A|^2\}$. There are at most $(b + 1)^2$ possibilities to choose a corresponding $(m, n) \in E_\tau$ and only $b + 1$ possible values for x and for q, respectively. Hence, for a fixed atom (e, s), we have to solve at most $(b + 1)^4$ such systems and b is not part of the input. Consequently, we can decide in time $\mathcal{O}(|E_A|^3 \cdot |S_A|^4)$ if (e, s) is solvable.

A provides at most $|S_A| \cdot |E_A|$ ESSP atoms. Hence, the τ_4^b-ESSP for A is decidable in time $\mathcal{O}(|E_A|^4 \cdot |S_A|^5)$.   □

The following lemma completes the proof of Theorem 1 and shows that τ_4^b-synthesis is doable in polynomial time.

Corollary 1. *To construct for a TS A a τ_4^b-net N with a state graph A_N isomorphic to A if it exists is doable in time $\mathcal{O}(|E_A|^3 \cdot |S_A|^5 \cdot max\{|E_A|, |S_A|\})$.*

Proof. By [5], if $\mathcal{R}$ is a set of regions of A containing for each ESSP and SSP atom of A a solving region, respectively, then the τ-net $N_A^{\mathcal{R}} = (\mathcal{R}, E_A, f, M_0)$, where $f((sup, sig), e) = sig(e)$ and $M_0((sup, sig)) = sup(s_{0,A})$ for $(sup, sig) \in \mathcal{R}, e \in E_A$, has a state graph isomorphic to A. Hence, the corollary follows from Lemmas 5 and 8.   □

3.3 Examples

We pick up our running example TS A introduced in Fig. 2 and its spanning tree A' presented in Fig. 3. We present two steps of the method given by Lemma 8 for the type τ_4^2 and check τ_4^2-solvability of the ESSP atom $(c, 1)$.

For a start, we choose $(m, n) = (0, 1)$ and $sup(0) = 0$ and determine $abs(c) = -0 + 1 = 1$ which yields $abs = (x_a, x_b, 1, x_d)$. We have to add $\psi_0 \cdot abs = m - sup(0) = 0$ which, by $\psi_0 = (0, 0, 0, 0)$, is always true and do not contribute to the system. Moreover, for $i \in \{0, 2, 3, 4, 5, 6\}$, we add the equation $(\psi_0 - \psi_i) \cdot abs = 0$. We have $\psi_0 - \psi_6 = (0, 0, -2, 0)$ and $(0, 0, -2, 0) \cdot abs = 0 \cdot x_a - 0 \cdot x_b - 2 - 0 \cdot x_d = 0$ yields a contradiction. Hence, $(c, 1)$ is not solvable by a region (sup, sig) where $sup(0) = 0$ and $sig(c) = (0, 1)$. Similarly, we obtain that the system corresponding to $sup(0) \in \{1, 2\}$ and $sig(c) = (0, 1)$ is also not solvable.

For another try, we choose $(m, n) = (2, 2)$ and $sup(0) = 2$. In accordance to the first and the second condition of Lemma 7 this determines $abs = (x_a, x_b, 0, x_d)$ and yields the equation $\psi_0 \cdot abs = m - sup(0) = 2 - 2 = 0$ which is always true. For the fourth condition, we pick $q = 2$ and add the equation $(\psi_0 - \psi_1) \cdot abs = 2 \cdot x_a = 2$. Finally, for the third condition, we add for $i \in \{0, 2, 3, 4, 5, 6\}$ the equation $(\psi_0 - \psi_i) \cdot abs = 0$ and obtain the following system of equations modulo $(b + 1)$:

$$
\begin{aligned}
\psi_t \cdot abs = && 2 \cdot x_b & + x_d & = 0 \\
(\psi_0 - \psi_1) \cdot abs = 2 \cdot x_a & & & & = 2 \\
(\psi_0 - \psi_2) \cdot abs = 2 \cdot x_a & + 2 \cdot x_b & & & = 0 \\
(\psi_0 - \psi_3) \cdot abs = 2 \cdot x_a & + 2 \cdot x_b & +2 \cdot 0 & & = 0 \\
(\psi_0 - \psi_4) \cdot abs = 2 \cdot x_a & + 2 \cdot x_b & +1 \cdot 0 & & = 0 \\
(\psi_0 - \psi_5) \cdot abs = && 2 \cdot 0 & & = 0 \\
(\psi_0 - \psi_6) \cdot abs = && 1 \cdot 0 & & = 0
\end{aligned}
$$

This system is solvable by $abs = (1, 2, 0, 2)$. We construct a region in accordance to the proof of Lemma 7: By $sup(0) = 2$ we obtain $sup(1) = 2 + \psi_1 \cdot abs = 2 + (1, 0, 0, 0) \cdot (1, 2, 0, 2) = 0$. Similarly, by $sup(i) = 2 + \psi_i \cdot abs$ for $i \in \{2, \ldots, 7\}$ we obtain $sup(2) = sup(3) = sup(4) = sup(5) = sup(6) = 2$ and $sup(7) = 0$. Hence, by defining $sig(c) = (2, 2)$, $sig(a) = 1$, $sig(b) = 2$ and $sig(d) = 2$ we obtain a fitting τ_4^b-region (sup, sig) that solves $(c, 1)$.

A closer look shows, that this support equals sup_1 which is presented in Fig. 2 and allows the signature sig_1, hence, $(sup, sig_1) = (sup_1, sig_1)$. The τ_4^b-region (sup, sig_1) solves a lot of further ESSP and SSP atoms. This observation reveals a first possible improvement of the method introduced by Lemma 8 and suggest, given a solution abs, to map as many events of A to a signature different from $0, \ldots, b$ as possible.

4 Fixed Parameter Tractability Results

Classical complexity theory measures the computational complexity of decision problems only in the size of the input. In [18], we showed that deciding if a TS A is τ-feasible or has the τ-ESSP, respectively, is NP-complete for $\tau \in \{\tau_2^b, \tau_3^b\}$. Thus, both problems are intractable from the perspective of classical complexity. Unfortunately, measuring the complexity purely in the size of A tells us nothing about the "source" of this negative result. On the contrary, parameterized complexity, developed by Downey and Fellows [10], allows us to study in which way different parameters of a TS A influence the complexity. This makes a finer analysis possible. Moreover, if we find a parameter, typically small on input instances of real-world applications, then algorithms, exponential in the size of the parameter but polynomial in the size of A, may work well in practice.

Formally, we say that a (decision) problem P is *fixed paramter tractable* with respect to parameter k if there exists an algorithm that solves P in time $\mathcal{O}(f(k)n^c)$, where f is some computable function, n is the size of the input and c is a constant independent from parameter k.

Let A be a TS and let for $e \in E_A$ the set $S_e = \{s \in S_A \mid s \overset{e}{\longrightarrow}\}$ containing the states of A at which e occur. The (maximum) *number of occurrences of events* is defined by $k = max\{|S_e| \mid e \in E_A\}$. In [19] it has been shown that deciding τ_1^1-feasibility and τ_1^1-ESSP is NP-complete even if $k = 2$. If there is a $\mathcal{O}(f(k)|A|^c)$-time algorithm for these problems then, for $k = 2$, it runs in polynomial time in A's size. This is because $f(2)$ is a constant. Thus, τ_1^1-feasibility and τ_1^1-ESSP, parameterized by k, are not fixed parameter tractable as long as P$\neq$NP.

On the contrary, the main result of this paper discovers that τ-ESSP and τ-feasibility parameterized by k are fixed parameter tractable. This reveals, that the number of occurrences of events is a *structural property of the input A* that makes τ-ESSP and τ-feasibility problems inherently hard to solve.

Theorem 2. *Let $\tau \in \{\tau_2^b, \tau_3^b\}$ and let A be a TS system with number of occurrences of events k. The τ-ESSP and the τ-feasibility are fixed parameter tractable with respect to parameter k.*

Given an ESSP atom α of a TS A, the following lemma provides conditions which an abstract τ-region of A satisfies if and only if α is τ-solvable. Moreover, it teaches us how to gain a corresponding τ-solving region from an abstract region satisfying the conditions.

Lemma 9. *Let $\tau \in \{\tau_2^b, \tau_3^b\}$, let (e, s) be an ESSP atom of A and let $s_1, \ldots, s_k$ be the sources of e in A, that is, $s_i \xrightarrow{e}$ for $i \in \{1, \ldots, k\}$ and if $s' \in S_A \backslash \{s_1, \ldots, s_k\}$ then $\neg s' \xrightarrow{\cdot e}$.*

The ESSP atom (e, s) is τ-solvable if and only if there is an event $(m, n) \in E_\tau$ and an abstract region (sup, abs) of A that satisfies the following conditions:

1. *$abs(e) = -m + n \bmod (b + 1)$,*
2. *$\psi_{s_i} \cdot abs = sup(s_i) - sup(s_{A,0}) \bmod (b + 1)$ and $m \leq sup(s_i) \leq b + m - n$ for $i \in \{1, \ldots, k\}$,*
3. *$\psi_s \cdot abs = sup(s) - sup(s_{A,0}) \bmod (b + 1)$ and $0 \leq sup(s) \leq m - 1$ or $b + m - n + 1 \leq sup(s) \leq b$*

Proof. If: Let (sup, abs) be an abstract τ-region of A satisfying (1)–(3). We get a τ-solving region (sup, sig) as follows: For $e' \in E_A$ we define $sig(e') = abs(e')$ if $e' \neq e$ and, otherwise, we set $sig(e') = (m, n)$ if $e' = e$. Firstly, we show that (sup, sig) is a region and, secondly, we argue that it τ-solves (e, s).

We have to show, that $q \xrightarrow{e'} q'$ in A implies $sup(q) \xrightarrow{sig(e')} sup(q')$ in τ. If $e' \neq e$, then this is true by (sup, abs) being a τ-region. It remains to show that $s_i \xrightarrow{e} s_i'$ implies $sup(s_i) \xrightarrow{(m, n)} sup(s_i')$ for $i \in \{0, \ldots, k\}$. By $m \leq sup(s_i) \leq b + m - n$ and the definition of τ, there is an $s_\tau \in \{0, \ldots, b\}$ with $sup(s_i) \xrightarrow{(m, n)} s_\tau$. This implies $s_\tau = sup(s_i) - m + n \bmod (b + 1)$. The assumption $abs(e) = -m + n \bmod (b + 1)$ yields $s_\tau = sup(s_i) + abs(e) \bmod (b + 1)$. Hence, we have that $sup(s_i) + abs(e) = sup(s_i) - m + n \bmod (b + 1)$. By $sup(s_i) \xrightarrow{abs(e)} sup(s_i')$ we get $sup(s_i') = sup(s_i) + abs(e) \bmod (b + 1)$ such that $sup(s_i') = sup(s_i) - m + n \bmod (b + 1)$. Consequently, $s_\tau = sup(s_i')$ implying $sup(s_i) \xrightarrow{(m, n)} sup(s_i')$ making (sup, sig) a τ-region.

Moreover, by $0 \leq sup(s) \leq m - 1$ or $b + m - n + 1 \leq sup(s) \leq b$ we have that $\neg sup(s) \xrightarrow{(m, n)}$ such that (sup, sig) τ-solves (e, s).

Only-If: Let (sup, sig) be a τ-region that solves (e, s). In accordance to Lemma 6, we define the τ-abstract region (sup, abs) originating from (sup, sig). We argue that (sup, abs) satisfies the conditions (1)–(3).

As (sup, sig) τ-solves (e, s) there is an event $(m, n) \in E_\tau$ such that $sup(s_i) \xrightarrow{(m, n)}$, $i \in \{1, \ldots, k\}$, and $\neg sup(s) \xrightarrow{(m, n)}$. By abs's definition, $abs = -m + n \bmod (b + 1)$ implying the first condition. Moreover, (sup, abs) satisfies the path equation. Hence, we have $sup(s_i) = sup(s_{A,0}) + \psi_{s_i} \cdot abs \bmod (b + 1)$ implying $\psi_{s_i} \cdot abs = sup(s_i) - sup(s_{A,0}) \bmod (b + 1)$ for $i \in \{1, \ldots, k\}$. Furthermore, by $sup(s_i) \xrightarrow{(m, n)}$ and τ's definition, we have $m \leq sup(s_i) \leq b + m - n$. Thus, the second condition is satisfied. Similarly, the path equation implies $\psi_s \cdot abs = sup(s) - sup(s_{A,0}) \bmod (b + 1)$ and, by $\neg sup(s) \xrightarrow{(m, n)}$, we obtain

$0 \leq sup(s) \leq m - 1$ or $b + m - n + 1 \leq sup(s) \leq b$. Hence, the third condition is also true. □

The following lemma shows that deciding the τ-ESSP for A is only exponential in parameter k but polynomial in the size of the input.

Lemma 10. *If $\tau \in \{\tau_2^b, \tau_3^b\}$, then to decide for a k-fold TS A whether it has the τ-ESSP is possible in time $\mathcal{O}((b+1)^{k+4} \cdot |E_A|^4 \cdot |S_A|^5)$.*

Proof. An ESSP atom (e, s) of A is a τ-solvable if and only if there is an abstract region (sup, abs) of A that satisfying the conditions of Lemma 9. Using Lemma 9, to decide the solvability of (e, s) we iteratively construct systems of linear equations M_α. There is an abstract region (sup, abs), fulfilling the conditions, if and only if at least one M_α is solvable by abs. A single system to be computed modulo $b + 1$ is obtained as follows:

Firstly, it implements the basic part $M_{A'}$ requiring at most $|E||S_A|^2 - 1$ equations.

Secondly, we choose an event $(m, n) \in E_A$ and a value $sup(s_{A,0}) \in \{0, \dots, b\}$ and, in accordance to Lemma 9.1, set $abs(e) = -m + n \bmod (b + 1)$. Thus, the number of unknown becomes $|E_A| - 1$.

Thirdly, in accordance to Lemma 9.2, we choose for every source s' of e in A $(s' \xrightarrow{e})$ a value $sup(s')$ satisfying $m \leq sup(s') \leq b + m - n$. After that we add the equation $\psi_{s'} \cdot abs = sup(s') - sup(s_{A,0})$. By definition of k, there are at most k sources of e. This yields at most k additional equations.

Finally, we choose $sup(s)$ such that $0 \leq sup(s) \leq m - 1$ or $b + m - n + 1 \leq sup(s) \leq b$, respectively. Then we add the equation $\psi_s \cdot abs = sup(s) - sup(s_{A,0})$. Now, a solution satisfies the condition of Lemma 9.3.

Altogether, by Lemma 9, this defines a fitting system whose solvability proves the τ-solvability of (e, s). Moreover, the system has at most $|E_A| \cdot |S_A|^2 + k \leq 2 \cdot |E_A| \cdot |S_A|^2$ equations.

We estimate how many such systems must be maximally resolved for a single atom: By definition of τ, we have at most $(b + 1)^2$ possible choices for $(m, n) \in \{0, \dots, b\}^2$, and at most $b+1$ different values for $sup(s_A, 0) \in \{0, \dots, b\}$, respectively. Furthermore, having (m, n) and $sup(s_{A,0})$ already chosen, there are at most $b + 1$ possible choices for $sup(s)$ with $0 \leq sup(s) \leq m - 1$ or $b + m - n + 1 \leq sup(s) \leq b$. Similarly, for every source s' of e we have at most $b + 1$ choices for $sup(s')$ with $m \leq sup(s') \leq b + m - n$. By definition of k this makes at most $(b+1)^k$ different possible choices for the sources of k. Altogether, we have at most $(b + 1)^{k+4}$ possibilities to define a system of linear equations whose solvability implies the τ-solvability of (e, s). Moreover, each system has at most size $2 \cdot |E_A| \cdot |S_A|^2$.

Hence, by Lemmas 2 and 9 and $|E_A| \leq |E_A| \cdot |S_A|^2$, we can decide in time $\mathcal{O}((b+1)^{k+4} \cdot |E_A|^3 \cdot |S_A|^4)$ if the atom (e, s) is τ-solvable. Consequently, by the at most $|E_A| \cdot |S_A|$ different ESSP atoms of A, we can decide whether A has the τ-ESSP in $\mathcal{O}((b+1)^{k+4} \cdot |E_A|^4 \cdot |S_A|^5)$ time. □

If $\tau \in \{\tau_2^b, \tau_3^b\}$ then, by Lemmas 5 and 9, deciding if a TS A has the τ-SSP and the τ-ESSP is doable in time $\mathcal{O}(|E_A|^3 \cdot |S_A|^6)$ and $\mathcal{O}((b+1)^{k+4} \cdot |E_A|^4 \cdot |S_A|^5)$, respectively. Thus, the following corollary is justified and completes the proof of Theorem 2.

Corollary 2. *If $\tau \in \{\tau_2^b, \tau_3^b\}$ then to decide if a TS A has the τ-feasibility is doable in time $\mathcal{O}((b+1)^{k+4} \cdot |E_A|^3 \cdot |S_A|^5 \cdot max\{|E_A|, |S_A|\})$.*

5   Conclusion

In this paper, we investigate the computational complexity of synthesis, feasibility, ESSP and SSP for several types of b-bounded P/T-nets, $b \in \mathbb{Z}_{b+1}$. We introduce the new Petri net type of restricted $\mathbb{Z}_{b+1}$-extended b-bounded P/T-nets and show that for this type synthesis and all corresponding decision problems are solvable in polynomial time. Moreover, we show that SSP is decidable in polynomial time for the types of (pure) $\mathbb{Z}_{b+1}$-extended b-bounded P/T-nets. Finally, we prove that feasibility and ESSP for (pure) $\mathbb{Z}_{b+1}$-extended b-bounded P/T-nets are fixed parameter tractable if the (maximum) number of occurrences of events is considered as parameter.

It remains for future work to search for other parameters that makes feasibility for Petri net types fixed parameter tractable. Moreover, the question whether there are tractable superclasses of (pure) b-bounded P/T-nets is still open. One might also investigate the computational complexity for other Petri nets related synthesis problems: The exact complexity status of synthesis up to *language equiavalence* is unknown. In [3], Badouel et al. proposed an algorithm that requires exponential space. Another open question has been stated in [15]: Schlachter et al. suggested to characterize the complexity of synthesis for b-bounded P/T-nets from *modal* transitions systems. Here, the task is to find, for a given modal TS M, a Petri net N that implements M. So far, we are at least aware of some (new) lower and upper bounds:

Conjecture 1. Let $b \geq 2$. Deciding, for a given TS A, if there is a (pure) b-bounded P/T-net N such that its state graph has the same language as A is NP-hard. Moreover, the problem is in PSPACE. To decide for a given modal TS M if there exists a (pure) b-bounded P/T-net N that implements M is NP-hard.

Acknowledgements. I would like to thank Uli Schlachter for his helpful remarks and for simplifying the proof of Lemma 1. Also, I'm thankful to the anonymous reviewers for their valuable comments.

References

1. van der Aalst, W.M.P.: Process Mining - Discovery, Conformance and Enhancement of Business Processes. Springer, Heidelberg (2011). https://doi.org/10.1007/978-3-642-19345-3
2. Agostini, A., De Michelis, G.: Improving flexibility of workflow management systems. In: van der Aalst, W., Desel, J., Oberweis, A. (eds.) Business Process Management. LNCS, vol. 1806, pp. 218–234. Springer, Heidelberg (2000). https://doi.org/10.1007/3-540-45594-9_14
3. Badouel, E., Bernardinello, L., Darondeau, P.: Polynomial algorithms for the synthesis of bounded nets. In: Mosses, P.D., Nielsen, M., Schwartzbach, M.I. (eds.) CAAP 1995. LNCS, vol. 915, pp. 364–378. Springer, Heidelberg (1995). https://doi.org/10.1007/3-540-59293-8_207
4. Badouel, E., Bernardinello, L., Darondeau, P.: The synthesis problem for elementary net systems is NP-complete. Theor. Comput. Sci. **186**(1–2), 107–134 (1997). https://doi.org/10.1016/S0304-3975(96)00219-8
5. Badouel, E., Bernardinello, L., Darondeau, P.: Petri Net Synthesis. Texts in Theoretical Computer Science. An EATCS Series. Springer, Heidelberg (2015). https://doi.org/10.1007/978-3-662-47967-4
6. Badouel, E., Caillaud, B., Darondeau, P.: Distributing finite automata through Petri net synthesis. Formal Asp. Comput. **13**(6), 447–470 (2002). https://doi.org/10.1007/s001650200022
7. Best, E., Darondeau, P.: Petri net distributability. In: Clarke, E., Virbitskaite, I., Voronkov, A. (eds.) PSI 2011. LNCS, vol. 7162, pp. 1–18. Springer, Heidelberg (2012). https://doi.org/10.1007/978-3-642-29709-0_1
8. Best, E., Devillers, R.: Characterisation of the state spaces of live and bounded marked graph Petri nets. In: Dediu, A.-H., Martín-Vide, C., Sierra-Rodríguez, J.-L., Truthe, B. (eds.) LATA 2014. LNCS, vol. 8370, pp. 161–172. Springer, Cham (2014). https://doi.org/10.1007/978-3-319-04921-2_13
9. Cortadella, J., Kishinevsky, M., Kondratyev, A., Lavagno, L., Yakovlev, A.: A region-based theory for state assignment in speed-independent circuits. IEEE Trans. CAD Integr. Circ. Syst. **16**(8), 793–812 (1997). https://doi.org/10.1109/43.644602
10. Downey, R.G., Fellows, M.R.: Fundamentals of Parameterized Complexity. Texts in Computer Science. Springer, London (2013). https://doi.org/10.1007/978-1-4471-5559-1
11. Goldmann, M., Russell, A.: The complexity of solving equations over finite groups. Inf. Comput. **178**(1), 253–262 (2002). https://doi.org/10.1006/inco.2002.3173
12. Hiraishi, K.: Some complexity results on transition systems and elementary net systems. Theor. Comput. Sci. **135**(2), 361–376 (1994). https://doi.org/10.1016/0304-3975(94)90112-0
13. Holloway, L.E., Krogh, B.H., Giua, A.: A survey of Petri net methods for controlled discrete event systems. Discret. Event Dyn. Syst. **7**(2), 151–190 (1997). https://doi.org/10.1023/A:1008271916548
14. Schlachter, U.: (2019, private correspondance)
15. Schlachter, U., Wimmel, H.: k-bounded Petri net synthesis from modal transition systems. In: CONCUR. LIPIcs, vol. 85, pp. 6:1–6:15. Schloss Dagstuhl - Leibniz-Zentrum fuer Informatik (2017). https://doi.org/10.4230/LIPIcs.CONCUR.2017.6

16. Schmitt, V.: Flip-flop nets. In: Puech, C., Reischuk, R. (eds.) STACS 1996. LNCS, vol. 1046, pp. 515–528. Springer, Heidelberg (1996). https://doi.org/10.1007/3-540-60922-9_42

17. Tarjan, R.E.: Finding optimum branchings. Networks **7**(1), 25–35 (1977). https://doi.org/10.1002/net.3230070103

18. Tredup, R.: Hardness results for the synthesis of b-bounded Petri nets. In: Donatelli, S., Haar, S. (eds.) PETRI NETS 2019. LNCS, vol. 11522, pp. 127–147. Springer, Cham (2019)

19. Tredup, R., Rosenke, C.: Narrowing down the hardness barrier of synthesizing elementary net systems. In: CONCUR. LIPIcs, vol. 118, pp. 16:1–16:15. Schloss Dagstuhl - Leibniz-Zentrum fuer Informatik (2018). https://doi.org/10.4230/LIPIcs.CONCUR.2018.16

20. Tredup, R., Rosenke, C.: The complexity of synthesis for 43 Boolean Petri net types. In: Gopal, T.V., Watada, J. (eds.) TAMC 2019. LNCS, vol. 11436, pp. 615–634. Springer, Cham (2019). https://doi.org/10.1007/978-3-030-14812-6_38

21. Tredup, R., Rosenke, C., Wolf, K.: Elementary net synthesis remains NP-complete even for extremely simple inputs. In: Khomenko, V., Roux, O.H. (eds.) PETRI NETS 2018. LNCS, vol. 10877, pp. 40–59. Springer, Cham (2018). https://doi.org/10.1007/978-3-319-91268-4_3

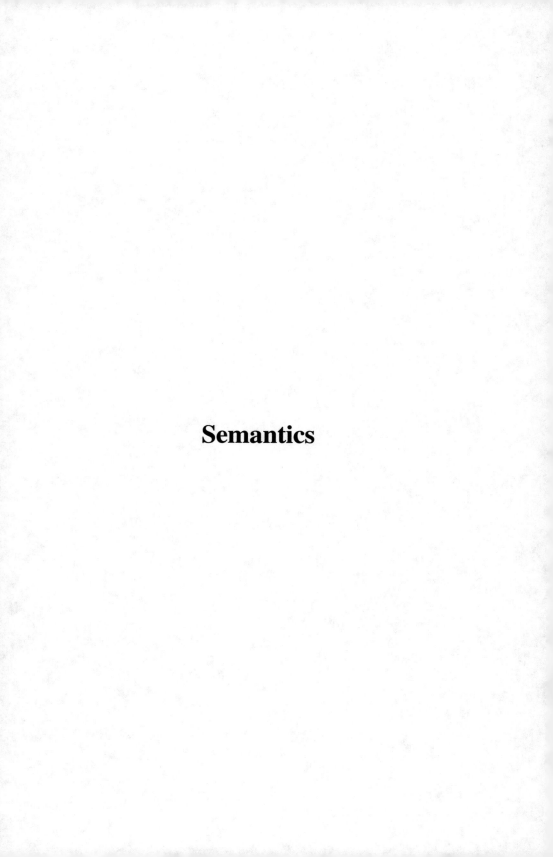

Semantics

Reversing Steps in Petri Nets

David de Frutos Escrig[1], Maciej Koutny[2], and Łukasz Mikulski[3(✉)]

[1] Dpto. Sistemas Informáticos y Computación, Facultad de Ciencias Matemáticas,
Universidad Complutense de Madrid, Madrid, Spain
defrutos@sip.ucm.es
[2] School of Computing, Newcastle University, Newcastle upon Tyne NE4 5TG, UK
maciej.koutny@ncl.ac.uk
[3] Faculty of Mathematics and Computer Science,
Nicolaus Copernicus University, Chopina 12/18, Toruń, Poland
lukasz.mikulski@mat.umk.pl

Abstract. In reversible computations one is interested in the development of mechanisms allowing to undo the effects of executed actions. The past research has been concerned mainly with reversing single actions. In this paper, we consider the problem of reversing the effect of the execution of groups of actions (steps).

Using Petri nets as a system model, we introduce concepts related to this new scenario, generalising notions used in the single action case. We then present a number of properties which arise in the context of reversing of steps of executed transitions in place/transition nets. We obtain both positive and negative results, showing that dealing with steps makes reversibility more involved than in the sequential case. In particular, we demonstrate that there is a crucial difference between reversing steps which are sets and those which are true multisets.

Keywords: Petri net · Reversible computation · Step semantics

1 Introduction

Reversibility of (partial) computations has been extensively studied during the past years, looking for mechanisms that allow to (partially) undo some actions executed during a process, that for some reason we need to cancel. As a result, the execution can then continue from a consistent state as if that suppressed action had not been executed at all. In particular, these mechanisms allow for the correct implementation of transactions [7,8], that are partial computations which either are totally executed or they are not executed at all. This includes the modification of information in data bases, so that we never include an 'incomplete' set of related updates that would produce an inconsistent state. In such a state one could infer some pieces of information that do not match, due to the fact that the modification procedure has not been satisfactorily completed. Another typical example would be the transactions between financial institutions, for instance, when transferring money, or nowadays any e-commerce platform, where the payments received should match the distributed goods [6].

© Springer Nature Switzerland AG 2019
S. Donatelli and S. Haar (Eds.): PETRI NETS 2019, LNCS 11522, pp. 171–191, 2019.
https://doi.org/10.1007/978-3-030-21571-2_11

Within the domain of Formal Methods, reversibility has been studied, for instance, in the framework of process calculi [15,17], event structures [18], DNA-computing [5], category theory [9], as well as within the field of quantum computing [20]. In the latter case, reversibility plays a central role due to the inherent reversibility of the mechanisms on which quantum computing is based. On the other hand, in Petri nets reversibility is usually understood as a global property. Historically it was considered in a sense closer to its meaning in process calculi [13], but such a local reversibility within the framework of *Petri Nets* has not been yet extensively studied. This is quite surprising as the formalization of transitions by means of pairs of *precondition* and *postcondition* places gives one an immediate way of defining the *reversal* of a transition simply by interchanging those two sets. There are, however, some more recent approaches that either focus on the *structural* study of Petri Nets [14], or on their *algebraic* study by means of *invariants* [16].

The approach presented in this paper is more *operational*, and extends the study of reversing (sequential) transitions initiated in [4], where it was shown that the apparent simplicity of this approach is far from trivial, mainly due to the difficulty of avoiding a situation that the added reversing transitions are fired in an *inconsistent* way; for instance, before the transition to be reversed was fired at all. [3] continued the study considering the particular case of *bounded* Petri nets, and distinguishing between *strict* reverses and *effect* reverses. The latter produce the effect of reversing the original transitions, but possibly with increasing or reducing the conditions checked for the reversed firing. It was shown that some transition systems which can be *solved* by a bounded net allow the reversal of their transitions by means of single reversing transitions, while in some other cases the reversal is only possible if we allow the *splitting* of reverses. This means that one can have a collection of reverses for the same transition, and each of them will be only fired at some of the markings, where the reversal of the original transition must be possible.

In [3] only the sequential (*interleaving*) semantics of nets was considered and, in fact, several of the presented examples were just (finite) *trace* systems, taking advantage of the results presented in [2,12], where binary words representable by Petri net were characterised. The latter problem and its consequences for reversibility has been recently further investigated in [11].

In this paper, we initiate the study of *step reversing* assuming the *step* semantics of Petri nets. We assume that the transition systems to be synthesized include the information about the multisets of enabled transitions that should be fireable in parallel. The reversal of the transitions should preserve this step information so that the simultaneous firing of several reverse transitions should exactly correspond to the original steps at the system represented by a Petri net.

Using Petri nets as a system model, we introduce concepts related to this new scenario, generalising notions used in the single action case. Since our aim now is to reverse steps, the simple definition which worked in the sequential case is no longer sufficient. When looking for the adequate generalization defining *step reversing*, we have found that two (non equivalent) definitions look 'natural'.

The former only allows steps which comprise either the original actions, or the reverse actions (*direct reversibility*). The latter allows also mixing of these two kinds of actions (*mixed reversibility*). It turns out that these two ways of interpreting reversibility of steps cause very big differences. Crucially, it appears that the direct reversibility cannot be implemented for steps which are true multisets, and so in such a case one has to aim at mixed reversibility. In this way we have found a striking difference between reversing steps which are sets and those which are true multisets (when autoconcurrency of actions in system executions is allowed). However, we still have a general positive result which shows that whenever sequential reversing is possible, once the steps of the system have been satisfactorily represented, we obtain also a sound reversal of those steps.

The paper is organised as follows. In the next section we recall a number of notions and notations used throughout the paper. We also introduce the direct and mixed step reversibility. In Sect. 4, we show that the direct reversibility cannot be achieved in the presence of autoconcurrency. The following section presents our positive results about lifting of sequential reversibility to step reversibility, by taking into account autoconcurrency. In Sect. 6, we develop results which show that in many cases the reversibility problem can be reduced to the net synthesis problem. The paper ends with some concluding remarks.

2   Preliminaries

Multisets. A *multiset* over a finite set X is a mapping $\alpha : X \to \mathbb{N}$, where $\mathbb{N}$ is the set of non-negative integers. The set of all multisets over X is denoted by $mult(X)$. $\alpha + \beta$ and $\alpha - \beta$ denote the multiset sum and difference, i.e., $(\alpha + \beta)(x) = \alpha(x) + \beta(x)$ and $(\alpha - \beta)(x) = \alpha(x) - \beta(x)$, for every $x \in X$. Note that $\alpha - \beta$ is defined provided that $\beta \leq \alpha$ which means that $\beta(x) \leq \alpha(x)$, for every $x \in X$. The size of α is defined as $|\alpha| = \sum_{x \in X} \alpha(x)$, and the support as the set $supp(\alpha) = \{x \in X \mid \alpha(x) \geq 1\}$. We also denote $x \in \alpha$ if $\alpha(x) \geq 1$. For $Y \subseteq X$, $\alpha \cap Y$ denotes multiset β (still over X) such that $\beta(y) = \alpha(y)$, for $y \in Y$, and $\beta(x) = 0$, for $x \in X \setminus Y$. Subsets of X can be identified with multisets which return values in $\{0, 1\}$, and its elements with singleton sets (i.e., multisets of size one). The empty (multi)set is denoted by $\varnothing$; a multiset α such that $\alpha(a) = 2$, $\alpha(b) = 1$, and $\alpha(X \setminus \{a, b\}) = \{0\}$, can be denoted by (aab); and a^k denotes multiset α such that $\alpha(a) = k$ and $\alpha(X \setminus \{a\}) = \{0\}$.

Step Transition Systems. A *step transition system* is defined as a tuple $STS = (S, T, \to, s_0)$ such that S is a nonempty set of *states*, T is a finite set of *actions*, $\to \ \subseteq S \times mult(T) \times S$ is the set of *arcs* (also called transitions), and $s_0 \in S$ is the *initial state*. The labels in $mult(T)$ represent simultaneous executions of groups of actions, called *steps*. Rather than $(s, \alpha, r) \in \to$, we can denote $s \xrightarrow{\alpha}_{STS} r$ or $s \xrightarrow{\alpha} r$ or $r \xleftarrow{\alpha} s$. Moreover, $s \xrightarrow{\alpha}_{STS}$ or $s \xrightarrow{\alpha}$ means that there is some r such that $s \xrightarrow{\alpha}_{STS} r$.

A state r is *reachable* from state s if there are steps $\alpha_1, \ldots, \alpha_k$ ($k \geq 0$) and states $s_1, \ldots, s_{k+1}$ such that $s = s_1 \xrightarrow{\alpha_1} s_2 \ldots s_k \xrightarrow{\alpha_k} s_{k+1} = r$. We denote this by $s \xrightarrow{\alpha_1 \ldots \alpha_k}_{STS} r$ or $s \xrightarrow{\alpha_1 \ldots \alpha_k} r$ or $r \xleftarrow{\alpha_k \ldots \alpha_1} s$. The set of all states from which a state s is reachable is denoted by $pred(s)$, and s is a *home state* if $pred(s) = S$. Moreover, a set of states $S' \subseteq S$ is a *home cover* of STS if $S = \bigcup_{s \in S'} pred(s)$.

STS is *state-finite* if S is finite, *step-finite* if $\{\alpha \mid s \xrightarrow{\alpha} s'\}$ is finite, and *finite* if it is both state- and step-finite (and so $\to$ is also finite).

Step transition systems are intended here to capture (step) reachability graphs of Petri nets. Of course, not every step transition system can be such a graph. To reflect this, we formulate first the following properties of STS:

FD *forward deterministism*
  $s' \xleftarrow{\alpha} s \xrightarrow{\alpha} s''$ implies $s' = s''$. Then we can define $s \oplus \alpha = s'$.
BD *backward deterministism*
  $s' \xrightarrow{\alpha} s \xleftarrow{\alpha} s''$ implies $s' = s''$. Then we can define $s \ominus \alpha = s'$.
REA *reachability*
  $s_0 \in pred(s)$, for every $s \in S$.
SEQ *sequentialisability*
  $s \xrightarrow{\alpha}$ implies $s \xrightarrow{\alpha_1 \ldots \alpha_k}$, whenever $\alpha = \sum_{i=1}^{k} \alpha_i$.
EL *empty loops*
  $s \xrightarrow{\varnothing} s$, for every $s \in S$.[1]
FC *forward confluence*
  $s' \xleftarrow{\alpha_k \ldots \alpha_1} s \xrightarrow{\beta_1 \ldots \beta_m} s''$ implies $s' = s''$, whenever $\sum_{i=1}^{k} \alpha_i = \sum_{j=1}^{m} \beta_j$.
BC *backward confluence*
  $s' \xrightarrow{\alpha_1 \ldots \alpha_k} s \xleftarrow{\beta_m \ldots \beta_1} s''$ implies $s' = s''$, whenever $\sum_{i=1}^{k} \alpha_i = \sum_{j=1}^{m} \beta_j$.

Any STS satisfying the above properties will be called, in this paper, a *well-formed step transition system* (or WFST-system). Note that FC and BC respectively generalise FD and BD, and EL & FD means that $s \xrightarrow{\varnothing} s' \iff s = s'$.

Proposition 1. *Let $STS = (S, T, \to, s_0)$ be a WFST-system and $s \in S$. If $s \oplus \alpha$ is defined and $\beta + \gamma \leq \alpha$, then $s \oplus \beta, s \oplus (\beta + \gamma)$ and $(s \oplus \beta) \oplus \gamma$ are also defined, and $(s \oplus \beta) \oplus \gamma = s \oplus (\beta + \gamma)$.*

Proof. By $s \xrightarrow{\alpha}$ and *SEQ* & *FD*, we have $s \xrightarrow{\beta} s \oplus \beta \xrightarrow{\gamma} (s \oplus \beta) \oplus \gamma$ as well as $s \xrightarrow{\beta + \gamma} s \oplus (\beta + \gamma)$. Hence, by *FC*, $(s \oplus \beta) \oplus \gamma = s \oplus (\beta + \gamma)$. $\square$

Being well-formed does not still characterise step transition systems defined by PT-nets. A complete characterisation can be obtained using, e.g., theory of regions [1,10]. However, we will not need here such a characterisation, since we are only interested in obtaining sufficient conditions for the representability of step transition systems by PT-nets, starting from the existing results about the representability of ordinary (sequential) transition systems.

Let $STS = (S, T, \to, s_0)$ and $STS' = (S', T', \to', s_0')$ be step transition systems. Then STS is:

---

[1] Arcs labelled with the empty multiset will not be usually depicted.

- a *sub-system* of STS' if $S \subseteq S'$, $T \subseteq T'$, $\rightarrow \subseteq \rightarrow'$, and $s_0 = s_0'$. We denote this by $STS \blacktriangleleft STS'$.
- *included* in STS', if $T \subseteq T'$, and there is a bijection ψ with the domain containing S such that $\{(\psi(s), \alpha, \psi(s')) \mid s \xrightarrow{\alpha} s'\} \subseteq \rightarrow'$, $\psi(S) = S'$, and $\psi(s_0) = s_0'$. We denote this by $STS \lhd_\psi STS'$ or $STS \lhd STS'$.
- *isomorphic* with STS' if $STS \lhd_\psi STS'$ and $STS' \lhd_{\psi^{-1}} STS$, for some ψ.[2] We denote this by $STS \simeq_\psi STS'$ or $STS \simeq STS'$.

We also define three ways of removing transitions from a step transition system:

$$
\begin{aligned}
STS^{seq} &= (S, T, \{(s, \alpha, r) \mid s \xrightarrow{\alpha} r \wedge |\alpha| \leq 1\}, & s_0) \\
STS^{set} &= (S, T, \{(s, \alpha, r) \mid s \xrightarrow{\alpha} r \wedge supp(\alpha) = \alpha\}, & s_0) \\
STS^{spike} &= (S, T, \{(s, \alpha, r) \mid s \xrightarrow{\alpha} r \wedge |supp(\alpha)| \leq 1\}, s_0) \,.
\end{aligned}
$$

That is, STS^{seq} is obtained by only retaining singleton steps and $\varnothing$, STS^{set} by only retaining steps which are sets, and STS^{spike} by removing all steps which use more than one action. Then STS is a *sequential / set / spiking* step transition system if respectively $STS = STS^{seq}$ / $STS = STS^{set}$ / $STS = STS^{spike}$.[3]

For a step transition system $S = (S, T, \rightarrow, s_0)$ and $T' \subseteq T$, the subsystem of S induced by T' is $STS|_{T'} = (S, T', \{(s, \alpha, s') \mid s \xrightarrow{\alpha} s' \wedge \alpha \in mult(T')\}, s_0)$.[4]

Place/Transition-Nets. A *Place/Transition net* (or PT-net) [19] is a tuple $N = (P, T, F, M_0)$, where P is a finite set of *places*, T is a disjoint finite set of *transitions* (or actions), F is the *flow function* $F \colon ((P \times T) \cup (T \times P)) \rightarrow \mathbb{N}$ specifying the arc weights, and M_0 is the *initial marking* (where a *marking* — a global state — is a multiset over P). Moreover, (P, T, F) is an *unmarked* PT-net.

Multisets over T—called again *steps*—represent executions of groups of transitions. The *effect* of a step α is a multiset of places $eff_N(\alpha) = post_N(\alpha) - pre_N(\alpha)$, where, for every $p \in P$:

$$
pre_N(\alpha)(p) = \sum_{t \in T} \alpha(t) \cdot F(p, t) \text{ and } post_N(\alpha)(p) = \sum_{t \in T} \alpha(t) \cdot F(t, p) \,.
$$

A step α is *enabled* at a marking M if $M \geq pre_N(\alpha)$. We denote this by $M[\alpha\rangle_N$. The *firing* of such a step M leads to marking $M' = M + eff_N(\alpha)$. We denote this by $M[\alpha\rangle_N M'$. Note that $M[\alpha\rangle_N$ implies $M[\beta\rangle_N$, for every $\beta \leq \alpha$. Moreover, $M[\alpha + \beta\rangle_N$ implies: $M [\alpha\rangle_N M + eff_N(\alpha) [\beta\rangle_N M + eff_N(\alpha + \beta)$. The set $reach_N$ of *reachable* markings is the smallest set of markings such that $M_0 \in reach_N$ and if $M \in reach_N$ and $M[\alpha\rangle_N M'$, for some α, then $M' \in reach_N$. The overall behaviour of N can be captured by its *concurrent reachability graph* defined as $CRG_N = (reach_N, T, \{(M, \alpha, M') \mid M \in reach_N \wedge M[\alpha\rangle_N M'\}, M_0)$. CRG_N is a WFST-system, and $M \xrightarrow{\alpha}_N M'$ will denote that $M \xrightarrow{\alpha}_{CRG_N} M'$.

---

[2] If STS and STS' are well-formed, then ψ is unique due to *FD & REA*.

[3] If STS is well-formed, then STS^{seq}, STS^{set}, and STS^{spike} satisfy *REA* due to *REA & SEQ*.

[4] Note that $STS|_{T'}$ may be not *REA* even for STS that is *REA*.

A step transition system STS is *solvable* if there is a PT-net N such that $STS \simeq CRG_N$. Moreover, step transition systems $STS_r = (S_r, T, \rightarrow_r, s_r)$ (for $r \in R$) are *simultaneously solvable* if there are PT-nets $N_r = (P, T, F, M_r)$ (for $r \in R$) and a bijection $\psi : \bigcup_{r \in R} S_r \rightarrow \bigcup_{r \in R} reach_{N_r}$ such that $STS_r \simeq_\psi CRG_{N_r}$, for every $r \in R$. (Note that the S_r's need not be disjoint.)

For a PT-net $N = (P, T, F, M_0)$ and $T' \subseteq T$, the PT-(sub)net of N induced by T' is $N|_{T'} = (P, T', F_{(P \times T') \cup (T' \times P)}, M_0)$.

3   Reversing Steps

A reverse of an action or net transition x will be denoted by $\overline{x}$, and for a multiset $X = (x_1 \ldots x_k)$ with $k \geq 0$, we denote $\overline{X} = (\overline{x}_1 \ldots \overline{x}_k)$.

Reversing in Transition Systems. We introduce three ways in which one can modify a step transition system in order to capture the effect of reversing actions.

The *direct/set/mixed reverse* of a step transition system $STS = (S, T, \rightarrow, s_0)$ satisfying *SEQ & FD* is respectively given by:

$$STS^{rev} = (S, T \cup \overline{T}, \rightarrow \cup \rightarrow_{rev}, s_0)$$
$$STS^{srev} = (S, T \cup \overline{T}, \rightarrow \cup \rightarrow_{srev}, s_0)$$
$$STS^{mrev} = (S, T \cup \overline{T}, \rightarrow_{mrev}, \quad s_0) , \quad \text{where:}$$

$$\rightarrow_{rev} = \{(s \oplus \alpha, \overline{\alpha}, s) \mid s \xrightarrow{\alpha}\}$$
$$\rightarrow_{srev} = \{(s \oplus \alpha, \overline{\alpha}, s) \mid s \xrightarrow{\alpha} \wedge supp(\alpha) = \alpha\}$$
$$\rightarrow_{mrev} = \{(s \oplus \alpha, \overline{\alpha} + \beta, s \oplus \beta) \mid s \xrightarrow{\alpha+\beta}\} .$$

Therefore, $\rightarrow_{rev}$ reverses *all* the (original) *steps*: $\rightarrow_{srev}$ *only* reverses the steps that are *sets*; and finally $\rightarrow_{mrev}$ introduces *partial* reverses, which means *mixed* steps, including both original and reversed actions.

Figure 1 illustrates the idea of mixed reversing. Note that $s \oplus \alpha$ and $s \oplus \beta$ above are well-defined states in STS due to *SEQ & FD*.

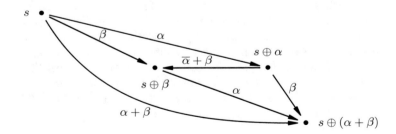

Fig. 1. A mixed reverse transition $s \oplus \alpha \xrightarrow{\overline{\alpha}+\beta}_{mrev} s \oplus \beta$ derived from $s \xrightarrow{\alpha+\beta}$.

Proposition 2. *Let STS be a* WFST*-system, and* α, β *be steps of its actions.*

1. $STS \blacktriangleleft STS^{srev} \blacktriangleleft STS^{rev} \blacktriangleleft STS^{mrev}.$
2. $s \xrightarrow{\beta}_{STS^{mrev}} s'$ *iff* $s \xrightarrow{\beta}_{STS} s'.$
3. $s \xrightarrow{\overline{\alpha}}_{STS^{mrev}} s'$ *iff* $s \xrightarrow{\overline{\alpha}}_{STS^{rev}} s'.$
4. $s \xrightarrow{\alpha+\beta}_{STS}$ *implies* $s \oplus \alpha \xrightarrow{\overline{\delta}+\gamma}_{STS^{mrev}} s \oplus (\gamma + \alpha - \delta),$
   for all $\delta \leq \alpha$ *and* $\gamma \leq \beta.$

Proof. (1) Clearly, $STS \blacktriangleleft STS^{srev} \blacktriangleleft STS^{rev}$. Finally, $STS^{rev} \blacktriangleleft STS^{mrev}$, because we can take $\alpha = \varnothing$ and then $s \oplus \alpha = s$ using that STS satisfies $FD \& EL$; and also $\rightarrow_{rev} \subseteq \rightarrow_{mrev}$ can be obtained in a similar way, taking $\beta = \varnothing$.

(2) By part (1), we only need to show that $s \xrightarrow{\beta}_{STS^{mrev}} s'$ implies $s \xrightarrow{\beta}_{STS} s'$. Indeed, given that $\beta = \overline{\varnothing} + \beta$ and STS satisfies FD, the former implies that there is $r \in S$ such that $r \xrightarrow{\varnothing+\beta}_{STS} r \oplus \beta$, $s = r \oplus \varnothing$, and $s' = r \oplus \beta$. Hence, by $EL \& BD$ for STS, $s = r$. Thus $s \xrightarrow{\beta}_{STS} s'$.

(3) By part (1), we only need to show that $s \xrightarrow{\overline{\alpha}}_{STS^{mrev}} s'$ implies $s \xrightarrow{\overline{\alpha}}_{STS^{rev}} s'$. Indeed, given that $\overline{\alpha} = \overline{\alpha} + \varnothing$ and STS satisfies FD, the former implies that there is $r \in S$ such that $r \xrightarrow{\alpha+\varnothing}_{STS} r \oplus \alpha$, $s = r \oplus \alpha$, and $s' = r \oplus \varnothing$. Hence, by $EL \& BD$ for STS, $s' = r$. Thus $s \xrightarrow{\alpha}_{STS} s'$ so $s' \xrightarrow{\overline{\alpha}}_{STS^{rev}} s$.

(4) By $s \xrightarrow{\alpha+\beta}_{STS}$ and $SEQ \& FD$ for STS, $s \xrightarrow{\alpha-\delta}_{STS} s \oplus (\alpha - \delta) \xrightarrow{\delta+\gamma}_{STS}$. Hence, by the definition of STS^{mrev},

$$(s \oplus (\alpha - \delta)) \oplus \delta \xrightarrow{\overline{\delta}+\gamma}_{STS^{mrev}} (s \oplus (\alpha - \delta)) \oplus \gamma .$$

Moreover, by $s \xrightarrow{\alpha+\beta}_{STS}$ and Proposition 1,

$$(s \oplus (\alpha - \delta)) \oplus \delta = s \oplus ((\alpha - \delta) + \delta) = s \oplus \alpha$$
$$(s \oplus (\alpha - \delta)) \oplus \gamma = s \oplus ((\alpha - \delta) + \gamma) = s \oplus (\gamma + \alpha - \delta) . \qquad \square$$

In general, STS^{rev}, STS^{srev}, and STS^{mrev} need not be well-formed even though STS was. However, the only properties which may fail to carry over from STS are the two versions of confluence.

Example 1. The step transition system in Fig. 2(a) is well-formed. However, adding reversals destroys forward confluence, as demonstrated in Fig. 2(b). Reversing the arcs results in a symmetric counterexample for the preservation of backward confluence. $\diamondsuit$

Proposition 3. *If STS satisfies* $FD \& BD \& REA \& SEQ \& EL$, *then the step transition systems* STS^{mrev}, STS^{srev}, *and* STS^{rev} *also satisfy them.*

Fig. 2. Reversing does not preserve confluence.

Proof. By Proposition 2(1), the result follows immediately for *EL* and *REA*. For the remaining three properties, by Proposition 2(2,3), it suffices to show it for STS^{mrev}. To this end, suppose that:

$$s \xrightarrow{\alpha+\beta}_{STS} \qquad s \oplus \alpha \xrightarrow{\overline{\alpha}+\beta}_{STS^{mrev}} s \oplus \beta$$
$$s' \xrightarrow{\alpha+\beta}_{STS} \qquad s' \oplus \alpha \xrightarrow{\overline{\alpha}+\beta}_{STS^{mrev}} s' \oplus \beta \ .$$

Then, by *SEQ & FD* for *STS*, we have:

$$s \xrightarrow{\alpha}_{STS} s \oplus \alpha \qquad s \xrightarrow{\beta}_{STS} s \oplus \beta$$
$$s' \xrightarrow{\alpha}_{STS} s' \oplus \alpha \qquad s' \xrightarrow{\beta}_{STS} s' \oplus \beta \ .$$

Suppose now that $s \oplus \alpha = s' \oplus \alpha$. Then, by *BD* for *STS*, $s = s'$. Hence, by *FD* for *STS*, $s \oplus \beta = s' \oplus \beta$. As a result, *FD* holds for STS^{mrev}. The proof of *BD* is symmetric.

To prove *SEQ* for STS^{mrev}, it suffices to consider $k = 2$. Suppose that:

$$s \xrightarrow{\alpha_1+\alpha_2+\beta_1+\beta_2}_{STS} \text{ and } s \oplus (\alpha_1 + \alpha_2) \xrightarrow{\overline{\alpha}_1+\overline{\alpha}_2+\beta_1+\beta_2}_{STS^{mrev}} s \oplus (\beta_1 + \beta_2) \ .$$

Then, by *SEQ* for *STS*, we have $s \oplus \alpha_2 \xrightarrow{\alpha_1+\beta_1}_{STS}$ and $s \oplus \beta_1 \xrightarrow{\alpha_2+\beta_2}_{STS}$. Hence, by the definition of STS^{mrev},

$$(s \oplus \alpha_2) \oplus \alpha_1 \xrightarrow{\overline{\alpha}_1+\beta_1}_{STS^{mrev}} (s \oplus \alpha_2) \oplus \beta_1$$
$$(s \oplus \beta_1) \oplus \alpha_2 \xrightarrow{\overline{\alpha}_2+\beta_2}_{STS^{mrev}} (s \oplus \beta_1) \oplus \beta_2 \ .$$

Moreover, by Proposition 1, we have:

$$s \oplus (\alpha_2 + \alpha_1) = (s \oplus \alpha_2) \oplus \alpha_1 \qquad\qquad (s \oplus \beta_1) \oplus \beta_2 = s \oplus (\beta_1 + \beta_2)$$
$$(s \oplus \alpha_2) \oplus \beta_1 = s \oplus (\alpha_2 + \beta_1) = (s \oplus \beta_1) \oplus \alpha_2 \ .$$

Hence, we obtain:

$$s \oplus (\alpha_1 + \alpha_2) \xrightarrow{\overline{\alpha}_1+\beta_1}_{STS^{mrev}} s \oplus (\alpha_2 + \beta_1) \xrightarrow{\overline{\alpha}_2+\beta_2}_{STS^{mrev}} s \oplus (\beta_1 + \beta_2) \ ,$$

which means that *SEQ* holds for STS^{mrev}. □

Proposition 4. $s \xrightarrow{\overline{\alpha}+\beta}_{STS^{mrev}} s'$ *iff* $s' \xrightarrow{\alpha+\overline{\beta}}_{STS^{mrev}} s$, *for every* WFST-*system STS*.

Proof. Since both implications really state the same, it suffices to show any of them. Suppose that $s \xrightarrow{\overline{\alpha}+\beta}_{STS^{mrev}} s'$. Then there is r such that $r \xrightarrow{\alpha+\beta}_{STS}$, $s = r \oplus \alpha$, and $s' = r \oplus \beta$ and then we only need to swap the roles of α and β to conclude $r \oplus \beta \xrightarrow{\alpha+\overline{\beta}}_{STS^{mrev}} r \oplus \alpha$ □

Reversing in Nets. Due to the natural decomposability character of steps made up of net transitions, adding reverses to PT-nets is done at the level of transitions rather than steps:

- A PT-net N with *reverses* is such that, for each original transition t, there is a reverse transition $\overline{t}$ with the opposite *effect*, i.e., $\mathit{eff}_N(t) = -\mathit{eff}_N(\overline{t})$.
- A PT-net N with *strict reverses* is such that, for each original transition t, there is a reverse transition $\overline{t}$ with the opposite *connectivity*, i.e., $\mathit{pre}_N(\overline{t}) = \mathit{post}_N(t)$ and $\mathit{post}_N(\overline{t}) = \mathit{pre}_N(t)$.

Proposition 5. *If STS is a solvable step transition system, then STS^{rev} and STS^{mrev} are* WFST*-systems.*

Proof. Since $STS = (S, T, \to, s_0)$ is solvable, there is a PT-net $N = (P, T, F, M_0)$ and a bijection $\psi : S \to \mathit{reach}_N$ such that $STS \simeq_\psi CRG_N$. Hence, since CRG_N is well-formed, STS is also well-formed. Below $M_s = \psi(s)$, for every $s \in S$.

It suffices to show that STS^{mrev} is well-formed, and, by Proposition 3, we only need to check that FC and BC hold for STS^{mrev}. Suppose that: $s \xrightarrow{\alpha+\beta}_{STS}$ and $s \oplus \alpha \xrightarrow{\overline{\alpha}+\beta}_{STS^{mrev}} s \oplus \beta$. Then, by *FD & BD* for CRG_N, we have:

$$M_{s \oplus \beta} = M_s + \mathit{eff}_N(\beta) = M_{s \oplus \alpha} - \mathit{eff}_N(\alpha) + \mathit{eff}_N(\beta) .$$

Therefore, if $s \xrightarrow{(\overline{\alpha}_1+\beta_1)...(\overline{\alpha}_k+\beta_k)}_{STS^{mrev}} s'$, then:

$$M_{s'} = M_s - \sum_{i=1}^{k} \mathit{eff}_N(\alpha_i) + \sum_{i=1}^{k} \mathit{eff}_N(\beta_i) .$$

Hence both FC and BC hold for STS^{mrev}. □

As an immediate consequence, we obtain the following a characterisation.

Corollary 1. *If STS is a* WFST*-system, but STS^{mrev} is not, then STS is not solvable.*

4 Multisets and Mixed Reversibility

Our investigation of step reversibility starts with a straightforward but pivotal result stating that, in the domain of PT-nets, direct reversibility cannot handle steps which are true multisets.

Proposition 6. *Let STS be a* WFST-*system which is not a set transition system. Then STS^{rev} is not solvable.*

Proof (See Fig. 3(a)). Let $STS = (S, T, \rightarrow, s_0)$ and $N = (P, T \cup \overline{T}, F, M_0)$ be a PT-net such that $STS^{rev} \simeq_\psi CRG_N$.

Suppose that $v \xrightarrow{\alpha}$ and $(aa) \leq \alpha$. Then, since STS satisfies SEQ, there are $w, q \in S$ such that $v \xrightarrow{(aa)} w$ and $v \xrightarrow{a} q$.

Let $M_x = \psi(x)$, for $x \in \{v, w, q\}$. By $STS^{rev} \simeq_\psi CRG_N$, the step $(a\overline{a})$ is not enabled at M_q. Hence, there must be $p \in P$ such that

$$M_q(p) < F(p, \overline{a}) + F(p, a) . \tag{1}$$

On the other hand, (aa) is enabled at M_v, and $(\overline{a}\overline{a})$ is enabled at M_w. Hence $M_v(p) \geq 2 \cdot F(p, a)$ and $M_w(p) \geq 2 \cdot F(p, \overline{a})$. We also have:

$$M_w(p) = M_v(p) + 2 \cdot F(a, p) - 2 \cdot F(p, a)$$
$$M_q(p) = M_v(p) + \quad F(a, p) - \quad F(p, a) .$$

Thus we obtain:

$$2 \cdot F(p, \overline{a}) + 2 \cdot F(p, a) \leq M_v(p) + M_w(p)$$
$$= 2 \cdot M_v(p) + 2 \cdot F(a, p) - 2 \cdot F(p, a) ,$$

and so $F(p, \overline{a}) + F(p, a) \leq M_v(p) + F(a, p) - F(p, a) = M_q(p)$, yielding a contradiction with (1). $\square$

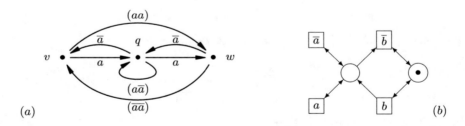

Fig. 3. An illustration of the proof of Proposition 6 (a), and PT-net generating concurrent reachability graph which is not step-finite (b).

A result similar to Proposition 6 does not hold for STS^{mrev} since, in this case it may contain, in particular, the mixed step $(a\overline{a})$ that was needed in the proof of the last result (a suitable counterexample can be provided by a WFST-system which 'executes' the diamond of (aa)). Hence, in the case of step (but not set) transition systems, it makes sense to investigate mixed reversibility, rather than direct reversibility, which we have proved to be impossible.

Proposition 7. *Let STS be a step-finite* WFST-*system. If STS^{mrev} is solvable, then STS^{srev} is also solvable.*

Proof. Since STS is step-finite, there is $k \geq 1$ such that $|\alpha| \leq k$, whenever $s \xrightarrow{\alpha}_{STS}$. Moreover, since STS^{mrev} is solvable, there exists a PT-net $N = (P, T \cup \overline{T}, F, M_0)$ such that $STS^{mrev} \simeq_\psi CRG_N$. We then modify N, getting a new net N', by adding to P a set of fresh places $P' = \{p_{tu} \mid t \in T \wedge u \in \overline{T}\}$. Each p_{tu} is such that $M_0(p_{tu}) = k$ and has four non-zero connections:

$$F(t, p_{tu}) = F(p_{tu}, t) = 1 \text{ and } F(u, p_{tu}) = F(p_{tu}, u) = k .$$

For the obtained PT-net N', $STS^{srev} \simeq_{\psi'} CRG_{N'}$ where, for every state s of STS, $\psi'(s) = \psi(s) + \sum_{p \in P'} p^k$. □

Example 2. The last result no longer holds if we drop the assumption that STS is step-finite. Consider, for example, $STS = (\{s_0, s_1, \dots\}, \{a, b\}, \rightarrow, s_0)$, where:

$$\rightarrow = \{(s_i, a^j, s_i) \mid i \geq 0 \wedge j \leq i\} \cup \{(s_i, b + a^j, s_{i+1}) \mid i \geq 0 \wedge j \leq i\} .$$

Then STS^{mrev} is solvable by the PT-net in Fig. 3(b), but STS^{srev} is not solvable by any PT-net $N = (P, \{a, b, \overline{a}, \overline{b}\}, F, M_0)$. Indeed, if N existed, then it would have distinct reachable markings $M_0, M_1, \dots$ such that, for all $i \geq 0$:

$$(i)\ M_i \xrightarrow{b}_N M_{i+1}, \quad (ii)\ M_i \xrightarrow{a^i}_N M_i, \quad (iii)\ M_i \xrightarrow{\overline{a}}_N M_i \text{ (for } i > 0),$$

but we would *not* have $(iv)\ M_i \xrightarrow{(a\overline{a})}_N M_i$.

We now observe that (i) means that $M_0 \leq M_1 \leq \dots$. Hence, (iv) together with the finiteness of P, implies that there is $p \in P$ such that $F(p, a) + F(p, \overline{a}) > M_0(p) = M_1(p) = \dots$. But (iii) implies $F(p, \overline{a}) \leq M_0(p) = M_1(p) = \dots$, and from (ii) we obtain $F(p, a) = 0$, getting $F(p, a) + F(p, \overline{a}) \leq M_0(p)$, which contradicts our first inequation.

Corollary 2. *Let STS be a well-formed set transition system. If STS^{mrev} is solvable, then STS^{rev} is also solvable.*

Proof. As a set transition system, STS is step-finite and $STS^{rev} = STS^{srev}$. Hence the result follows from Proposition 7. □

5   Reversibility and Plain Solvability

The feasibility of reversing steps in WFST-systems can in some cases be replaced by checking the solvability of the original transition system, and the solvability of its pure reversed version(s). The latter are formalised in the following way.

Let $STS = (S, T, \rightarrow, s_0)$ be a WFST-system and $r \in S$.

Then we define the step transition system $\overline{STS}_r = (pred(r), \overline{T}, \rightarrow_r, r)$, where:

$$\rightarrow_r = \{(s', \overline{\alpha}, s) \mid s' \in pred(r) \wedge s \xrightarrow{\alpha} s'\} .$$

It is easy to check that $\overline{STS}_r$ is also well-formed. Moreover, since STS satisfies REA, $s_0 \in pred(r)$ and it is reachable in $\overline{STS}_r$ from every state of the latter.

Theorem 1. *Let R be a home cover of a WFST-system STS. Then STS^{mrev} is solvable iff STS is solvable and $\overline{STS}_r$ (for all $r \in R$) are simultaneously solvable.*

Proof. Note that $S = \bigcup_{r \in R} S_r$, as R is a home cover. In the proof below, we will use the following notation, where $r \in R$:

$$STS = (S, T, \rightarrow, s_0) \qquad STS^{rev} = (S, T \cup \overline{T}, \rightarrow_{rev}, s_0)$$
$$STS^{mrev} = (S, T \cup \overline{T}, \rightarrow_{mrev}, s_0) \qquad \overline{STS}_r = (S_r, \overline{T}, \rightarrow_r, r) .$$

($\Longrightarrow$) Suppose that $N = (P, T, F, M_0)$ is such that $STS^{mrev} \simeq_\psi CRG_N$.

To show that STS is solvable, let $N' = N|_T$. Then $STS \simeq_\psi CRG_{N'}$. Indeed, we first note that $\psi(s_0) = M_0$. Suppose now that $s \in S$ and $\psi(s) \in reach_{N'}$. Let us see that the execution of transitions is preserved in both directions by ψ:

(i) $s \xrightarrow{\alpha} s'$. Then, by Proposition 2(2), we have $s \xrightarrow{\alpha}_{mrev} s'$. Hence, by $STS^{mrev} \simeq_\psi CRG_N$, we have $\psi(s) \xrightarrow{\alpha}_N \psi(s')$. Moreover, the enabling and firing of steps over T are exactly the same in N and N'. Hence $\psi(s) \xrightarrow{\alpha}_{N'} \psi(s')$.

(ii) $\psi(s) \xrightarrow{\alpha}_{N'} M$. Then, as the enabling and firing of steps over T are exactly the same in N and N', $\psi(s) \xrightarrow{\alpha}_N M$. Hence, by $STS^{mrev} \simeq_\psi CRG_N$, $M \in \psi(S)$ and $s \xrightarrow{\alpha}_{mrev} \psi^{-1}(M)$. Thus, by Proposition 2(2), $s \xrightarrow{\alpha} \psi^{-1}(M)$.

To show that the $\overline{STS}_r$'s are simultaneously solvable, let us take N_r as the net $N|_{\overline{T}}$ with the initial marking set to $\psi(r)$, for every $r \in R$. Then $\overline{STS}_r \simeq_\psi CRG_{N_r}$. Indeed, we first note that the initial states of $\overline{STS}_r$ and CRG_{N_r} are related by ψ. Suppose now that s is a state in $\overline{STS}_r$ such that $\psi(s) \in reach_{N_r}$. Again we have:

(i) $s \xrightarrow{\overline{\alpha}}_r s'$. Then $s \xrightarrow{\overline{\alpha}}_{rev} s'$ and so, by Proposition 2(3), we have $s \xrightarrow{\overline{\alpha}}_{mrev} s'$. Hence we have, by $STS^{mrev} \simeq_\psi CRG_N$, $\psi(s) \xrightarrow{\overline{\alpha}}_N \psi(s')$. Moreover, the enabling and firing of steps over $\overline{T}$ are exactly the same in N and N_r. Hence $\psi(s) \xrightarrow{\overline{\alpha}}_{N_r} \psi(s')$.

(ii) $\psi(s) \xrightarrow{\overline{\alpha}}_{N'} M$. Then, as the enabling and firing of steps over $\overline{T}$ are exactly the same in N and N_r, we have $\psi(s) \xrightarrow{\overline{\alpha}}_N M$. Hence, by $STS^{mrev} \simeq_\psi CRG_N$, we have $M \in \psi(S)$ and $s \xrightarrow{\overline{\alpha}}_{STS^{mrev}} \psi^{-1}(M)$. Thus, by Proposition 2(3), $s \xrightarrow{\overline{\alpha}}_r \psi^{-1}(M)$.

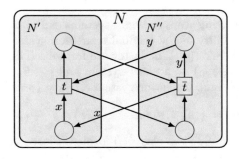

Fig. 4. An illustration of the proof of Theorem 1.

($\Longleftarrow$) Since STS is solvable, there is a PT-net $N' = (P', T, F', M'_0)$ such that $STS \simeq_{\psi'} CRG_{N'}$. And, since the $\overline{STS}_r$'s are simultaneously solvable, there are PT-nets $N_r = (P'', \overline{T}, F'', M_r)$ (for all $r \in R$) and $\psi'' : S \to \bigcup_{r \in R} reach_{N_r}$ such that $\overline{STS}_r \simeq_{\psi''} CRG_{N_r}$, for every $r \in R$. Note that $\psi'(s_0) = M'_0$ and $\psi''(r) = M_r$, for every $r \in R$. Clearly, we may assume that $P' \cap P'' = \varnothing$.

Let $N = (P' \cup P'', T \cup \overline{T}, F, M_0)$ the PT-net, where $M_0 = M'_0 + \psi''(s_0) = \psi'(s_0) + \psi''(s_0)$. Now taking $N'' = (P'', \overline{T}, F'', \varnothing)$, for every $t \in T$ we have:

$$pre_N(t) = pre_{N'}(t) + post_{N''}(\overline{t}) \quad post_N(t) = post_{N'}(t) + pre_{N''}(\overline{t})$$
$$pre_N(\overline{t}) = pre_{N''}(\overline{t}) + post_{N'}(t) \quad post_N(\overline{t}) = post_{N''}(\overline{t}) + pre_{N'}(t) . \tag{2}$$

Note that N is a PT-net with strict reverses (see Fig. 4). Moreover, for all $t \in T$ and $r \in R$:

$$pre_{N''}(\overline{t}) = pre_{N_r}(\overline{t}) \quad post_{N''}(\overline{t}) = post_{N_r}(\overline{t}) . \tag{3}$$

Let ψ be a mapping with the domain S which, for every $s \in S$, returns $\psi'(s) + \psi''(s)$. Note that ψ is well-defined since R is a home cover of STS, and that $\psi(s_0) = M_0$.

We first show that $STS^{rev} \simeq_{\psi} STS'$, where STS' is just CRG_N but removing from it all the arcs labelled by the mixed steps (i.e., steps of the form $\alpha + \overline{\beta}$, for $\alpha, \beta \neq \varnothing$) deleted. (Note that this does not produce unreachable states since CRG_N satisfies SEQ.) Indeed, we first note that the initial states of STS^{rev} and STS' are related by ψ. Suppose now that $s \in S$ and $\psi(s) \in reach_N$. Once again we see that the execution of transitions is preserved in both directions by ψ:

(i) $s \xrightarrow{\alpha}_{rev} s'$. Then, by $STS \simeq_{\psi'} CRG_{N'}$, we have $\psi'(s) \xrightarrow{\alpha}_{N'} \psi'(s')$. Moreover, $s' \xrightarrow{\overline{\alpha}}_{rev} s$. Hence, since R is a home cover, there is $r \in R$ such that $s' \xrightarrow{\overline{\alpha}}_r s$. Thus, by $\overline{STS}_r \simeq_{\psi''} CRG_{N_r}$, we have $\psi''(s') \xrightarrow{\overline{\alpha}}_{N_r} \psi''(s)$, and so $\psi''(s) \geq post_{N_r}(\overline{\alpha})$. Hence, by (2) and (3), we have:

$$\psi(s) = \psi'(s) + \psi''(s) \geq pre_{N'}(\alpha) + post_{N_r}(\overline{\alpha}) = pre_N(\alpha) .$$

As a result, $\psi(s) \xrightarrow{\alpha} \psi(s) + eff_N(\alpha)$. Moreover, by (2) and (3), we have:

$\psi(s) + eff_N(\alpha)$
$\quad = \psi'(s) + \psi''(s) + post_N(\alpha) - pre_N(\alpha)$
$\quad = \psi'(s) + \psi''(s) + (post_{N'}(\alpha) + pre_{N''}(\overline{\alpha})) - (pre_{N'}(\alpha) + post_{N''}(\overline{\alpha}))$
$\quad = (\psi'(s) + post_{N'}(\alpha) - pre_{N'}(\alpha)) + (\psi''(s) - post_{N''}(\overline{\alpha}) + pre_{N''}(\overline{\alpha}))$
$\quad = \psi'(s') + \psi''(s') .$

Hence $\psi(s) \xrightarrow{\alpha} \psi(s')$.

(ii) $s \xrightarrow{\overline{\alpha}}_{rev} s'$. Then $s' \xrightarrow{\alpha}_{rev} s$ and so, by Case 1, $\psi(s') \xrightarrow{\alpha}_N \psi(s)$. Hence, since N is PT-net with strict reverses, $\psi(s) \xrightarrow{\overline{\alpha}}_N \psi(s')$.

(iii) $\psi(s) \xrightarrow{\alpha}_N M$. Then

$$\psi'(s) + \psi''(s) = \psi(s) \geq pre_N(\alpha) = pre_{N'}(\alpha) + post_{N''}(\overline{\alpha}) .$$

$$M = \psi'(s) + \psi''(s) + post_{N'}(\alpha) + pre_{N''}(\overline{\alpha}) - (pre_{N'}(\alpha) + post_{N''}(\overline{\alpha})) \, .$$

Hence, since $P' \cap P'' = \varnothing$, $\psi'(s) \geq pre_{N'}(\alpha)$ and $\psi''(s) \geq post_{N''}(\overline{\alpha})$. Moreover, we have:

$$M \cap P' = \psi'(s) + post_{N'}(\alpha) - pre_{N'}(\alpha)$$
$$M \cap P'' = \psi''(s) + pre_{N''}(\overline{\alpha}) - post_{N''}(\overline{\alpha}) \, .$$

Thus $\psi'(s) \xrightarrow{\alpha}_{N'} M \cap P'$. Hence, by $STS \simeq_{\psi'} CRG_{N'}$, we obtain $M \cap P' \in \psi'(S)$ and $s \xrightarrow{\alpha}_{rev} s'$, where $\psi'(s') = M \cap P'$. We need to show that $\psi(s) = M$, and this would follow from $\psi''(s') = M \cap P''$. Indeed, we have $s' \xrightarrow{\overline{\alpha}}_{rev} s$, and so there is $r \in R$ such that $s' \in S_r$. Now, by $\overline{STS}_r \simeq_{\psi''} CRG_{N_r}$, $\psi''(s') \xrightarrow{\overline{\alpha}}_{N_r} \psi''(s)$. But this means that $\psi''(s) = \psi''(s') + post_{N''}(\overline{\alpha}) - pre_{N''}(\overline{\alpha})$. Thus

$$\psi''(s') = \psi''(s) - post_{N''}(\overline{\alpha}) + pre_{N''}(\overline{\alpha}) = M \cap P'' \, .$$

(iv) $\psi(s) \xrightarrow{\overline{\alpha}}_N M$. Then we have:

$$\psi'(s) + \psi''(s) = \psi(s) \geq pre_N(\overline{\alpha}) = pre_{N''}(\overline{\alpha}) + post_{N'}(\alpha)$$
$$M = \psi'(s) + \psi''(s) + post_{N''}(\overline{\alpha}) + pre_{N'}(\alpha) - (pre_{N''}(\overline{\alpha}) + post_{N'}(\alpha)) \, .$$

Hence, since $P' \cap P'' = \varnothing$, $\psi'(s) \geq post_{N'}(\alpha)$ and $\psi''(s) \geq pre_{N''}(\overline{\alpha})$. Moreover, we have:

$$M \cap P' = \psi'(s) + pre_{N'}(\alpha) - post_{N'}(\alpha)$$
$$M \cap P'' = \psi''(s) + post_{N''}(\overline{\alpha}) - pre_{N''}(\overline{\alpha}) \, .$$

Thus $\psi''(s) \xrightarrow{\overline{\alpha}}_{N''} M \cap P''$. Hence, since R is a home cover, there is $r \in R$ such that $s \in S_r$. Thus, by $\overline{STS}_r \simeq_{\psi''} CRG_{N_r}$, $M \cap P'' \in \psi''(S)$ and $s \xrightarrow{\overline{\alpha}}_{rev} s'$, where $\psi''(s') = M \cap P''$. We need to show that $\psi(s) = M$, and this would follow from $\psi'(s') = M \cap P'$. Indeed, we have $s' \xrightarrow{\overline{\alpha}}_{rev} s$. Hence, by $STS \simeq_{\psi'} CRG_{N'}$, we obtain $\psi'(s') \xrightarrow{\overline{\alpha}}_{N'} \psi'(s)$. But this means that

$$\psi'(s) = \psi'(s') + post_{N'}(\alpha) - pre_{N'}(\alpha) \, ,$$

and so we obtain: $\psi'(s') = \psi'(s) - post_{N'}(\alpha) + pre_{N'}(\alpha) = M \cap P'$.

Now in order to conclude $STS^{mrev} \simeq_{\psi} CRG_N$ we only need to consider the case of mixed transitions:

(i) $s \xrightarrow{\alpha+\beta}_{rev}$ and $s \oplus \alpha \xrightarrow{\overline{\alpha}+\beta}_{mrev} s \oplus \beta$. Then $s \xrightarrow{\alpha}_{rev} s \oplus \alpha$ and $s \xrightarrow{\beta}_{rev} s \oplus \beta$. Thus, by $STS^{rev} \simeq_{\psi} STS'$,

$$\psi(s) \xrightarrow{\alpha+\beta}_N \quad \psi(s) \xrightarrow{\alpha}_{rev} \psi(s \oplus \alpha) \quad \psi(s) \xrightarrow{\beta}_{rev} \psi(s \oplus \beta) \, .$$

Hence, we have $\psi(s) \geq pre_N(\alpha + \beta) = pre_N(\alpha) + pre_N(\beta)$. Thus

$$\psi(s) \xrightarrow{\alpha}_N \psi(s) + eff_N(\alpha) = \psi(s) + post_N(\alpha) - pre_N(\alpha)$$
$$= \psi(s) + pre_N(\overline{\alpha}) - pre_N(\alpha) \geq pre_N(\overline{\alpha} + \beta) \, .$$

Moreover, by FD, $\psi(s \oplus \alpha) = \psi(s) + \text{eff}_N(\alpha) \geq \text{pre}_N(\overline{\alpha} + \beta)$. And, finally, $\psi(s) + \text{eff}_N(\alpha) + \text{eff}_N(\overline{\alpha} + \beta) = \psi(s) + \text{eff}_N(\beta)$.

(ii) $\psi(s) \xrightarrow{\overline{\alpha}+\beta}_N M$. Then we have $\psi(s) \xrightarrow{\overline{\alpha}}_N \psi(s) \oplus \overline{\alpha}$ and, using that $post(\overline{\alpha}) = pre_N(\alpha)$, $\psi(s) \oplus \overline{\alpha} \xrightarrow{\alpha+\beta}_N$. Thus, from $STS^{rev} \simeq_\psi STS'$ it follows that $\psi^{-1}(\psi(s) \oplus \overline{\alpha}) \xrightarrow{\overline{\alpha}+\beta}_{rev}$. Hence

$$\psi^{-1}(\psi(s) \oplus \overline{\alpha}) \oplus \alpha \xrightarrow{\overline{\alpha}+\beta}_{mrev} \psi^{-1}(\psi(s) \oplus \overline{\alpha}) \oplus \beta .$$

All we need to show now is that:

$$\psi^{-1}(\psi(s) \oplus \overline{\alpha}) \oplus \alpha = s$$
$$\psi(\psi^{-1}(\psi(s) \oplus \overline{\alpha}) \oplus \beta)) = \psi(s) \oplus (\overline{\alpha} + \beta) ,$$

which clearly is the case. $\qquad\square$

Corollary 3. *Let r be a home state of a* WFST-*system STS. Then STS^{mrev} is solvable iff STS and $\overline{STS}_r$ are solvable.*

The above corollary and the proof of the last theorem provide a method for *constructing* a PT-net implementing mixed step reversibility provided that one can synthesise PT-nets for two step transition systems using, e.g., theory of regions [1, 10].

We have obtained a method for checking the feasibility of mixed reversability. This is indeed useful, in view of Proposition 6. Moreover, for set transition systems the result extends to direct reversibility.

Theorem 2. *Let r be a home state of a well-formed set transition system STS. Then STS^{rev} is solvable iff STS and $\overline{STS}_r$ are solvable.*

Proof. ($\Longrightarrow$) Let $STS^{rev} \simeq_\psi CRG_N$ and T be the set of transitions of N. Then $STS \simeq_\psi CRG_{N|_T}$ and $\overline{STS}_r \simeq_\psi CRG_{N'}$, where N' is $N|_{\overline{T}}$ with the initial marking set to $\psi(r)$.

($\Longleftarrow$) Follows from Theorem 1 and Corollary 2. $\qquad\square$

6 From Sequential Reversibility to Step Reversibility

Checking the feasibility of step reversibility and then constructing a suitable PT-net can be difficult. Our next result shows that in certain cases one can carry out this task more easily, if we are given a net that simultaneously solves the original transition system, overapproximates its reversed version that contains only spikes, and underapproximates its mixed reversed version.

Theorem 3. *Let $N = (P, T \cup \overline{T}, F, M_0)$ be a* PT-*net, and $STS = (S, T, \rightarrow, s_0)$ be a* WFST-*system such that:*

$$(STS^{spike})^{rev} \lhd CRG_N \lhd STS^{mrev} \tag{4}$$

$$STS \simeq CRG_{N|_T} . \tag{5}$$

Then STS^{mrev} is solvable. Moreover, if STS is a set transition system, then STS^{rev} is solvable.

Proof. The states as well as the initial states of $(STS^{spike})^{rev}$, STS^{mrev}, and STS are all the same; moreover, $((STS^{spike})^{rev}|_T)^{seq} = (STS^{mrev}|_T)^{seq} = STS^{seq}$. Similarly, the initial states of CRG_N and $CRG_{N|_T}$ are the same and we have $(CRG_N)|_T = CRG_{N|_T}$.

Moreover, all transition systems in (4) and (5) satisfy *FD & REA & SEQ*, and there is a bijection ψ such that:

$$(STS^{spike})^{rev} \lhd_\psi CRG_N \lhd_{\psi^{-1}} STS^{mrev} \text{ and } STS \simeq_\psi CRG_{N|_T} . \tag{6}$$

By (4) and *SEQ* of step transition systems and reachability graphs and the fact that we may assume that each $t \in T$ appears in the labels of the arcs of STS, we have for any $t \in T$:

$$reach_N = reach_{N|_T} \text{ and } eff_N(t) = -eff_N(\bar{t}) . \tag{7}$$

We first show that it can be assumed that, for all $t \in T$:

$$pre_N(\bar{t}) \geq post_N(t) \text{ and } post_N(\bar{t}) \geq pre_N(t) . \tag{8}$$

Indeed, suppose that $F(p, \bar{t}) < F(t, p)$. We then modify F to become F' which is the same as F except that $F'(p, \bar{t}) = F(t, p)$ and $F'(\bar{t}, p) = F(p, t)$. Let N' be the resulting PT-net. Clearly, $eff_N(x) = eff_{N'}(x)$, for every $x \in T \cup \bar{T}$.

After this modification—which does not affect transitions in T—(5) is still satisfied after taking N' to play the role of N. However, the satisfaction of (4) is not so immediate. But the modification can only restrict the enabling of steps, and the enabling of transitions other than $\bar{t}$ is unchanged. Thus

$$CRG_{N'} \blacktriangleleft CRG_N \lhd STS^{mrev} .$$

Hence, if (4) does not hold with N' playing the role of N, then there is $M \in reach_{N'} \subseteq reach_N$ and $k \geq 1$ such that:

$$M \xrightarrow{\bar{t}^k}_{CRG_N} M' \text{ and } \neg M \xrightarrow{\bar{t}^k}_{CRG_{N'}} . \tag{9}$$

By (4) and the first part of (9), we have:

$$M' \xrightarrow{t^k}_{CRG_N} M \quad \text{and so} \quad M(p) \geq F(t^k, p) . \tag{10}$$

By construction, the only reason for the second part of (9) to hold is that $F'(p, \bar{t}^k) > M(p)$. Thus, by $F'(p, \bar{t}^k) = F(t^k, p)$, we obtain $F(t^k, p) > M(p)$, yielding a contradiction with (10).

We can apply the above modification as many times as needed, finally concluding that (8) can be assumed to hold for N as any modification does not invalidate the conditions captured by (8) that were got by the previous modifications.

We next show that STS^{mrev} is solvable, after constructing a PT-net $\widetilde{N} = (\widetilde{P}, T \cup \bar{T}, \widetilde{F}, \widetilde{M_0})$, in the following way (Fig. 5):

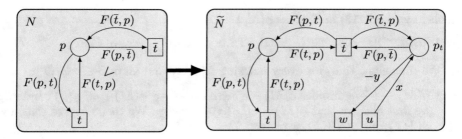

Fig. 5. Constructing place p_t in the proof of Theorem 3, where $x = \textit{eff}_N(u)(p) > 0$ and $y = \textit{eff}_N(w)(p) \leq 0$. Note that $u, w \in T \cup \overline{T} \setminus \{\overline{t}\}$.

- $\widetilde{P} = \bigcup_{p \in P} P_p$, where, for every $p \in P$,[5]

$$P_p = \{p\} \cup \{p_t \mid t \in T \wedge F(p, \overline{t}) > F(t, p)\} \text{ and } \widetilde{M_0}(P_p) = \{M_0(p)\} .$$

- The connections in $\widetilde{N}$ are set as follows, where $p \in P$ and $u \in T \cup \overline{T} \setminus \{\overline{t}\}$:
  - $\widetilde{F}(p, \overline{t}) = F(t, p)$ and $\widetilde{F}(\overline{t}, p) = F(p, t)$.
  - $\widetilde{F}(p_t, \overline{t}) = F(p, \overline{t})$ and $\widetilde{F}(\overline{t}, p_t) = F(\overline{t}, p)$.
  - $\textit{eff}_N(u)(p) > 0$ implies $\widetilde{F}(p_t, u) = 0$ and $\widetilde{F}(u, p_t) = \textit{eff}_N(u)(p)$.
  - $\textit{eff}_N(u)(p) \leq 0$ implies $\widetilde{F}(u, p_t) = 0$ and $\widetilde{F}(p_t, u) = -\textit{eff}_N(u)(p)$.
  - $\widetilde{F}$ on $(P \times T) \cup (T \times P)$ is as F unless it has been set explicitly above.

In what follows, for every marking M of N, we use $\phi(M)$ to denote the marking of $\widetilde{N}$ such that $\phi(M)(P_p) = \{M(p)\}$, for every $p \in P$. Hence $\phi(M_0) = \widetilde{M_0}$.

We now present a number of straightforward properties of $\widetilde{N}$. We first observe that, by (8), for all $t \in T$, $u \in T \cup \overline{T}$, and $p \in P$,

$$\begin{array}{ll} pre_{\widetilde{N}}(\overline{t}) \geq post_{\widetilde{N}}(t) & \textit{eff}_{\widetilde{N}}(t) = -\textit{eff}_{\widetilde{N}}(\overline{t}) \\ post_{\widetilde{N}}(\overline{t}) \geq pre_{\widetilde{N}}(t) & \textit{eff}_{\widetilde{N}}(u)(P_p) = \{\textit{eff}_N(u)(p)\} . \end{array} \tag{11}$$

Therefore, for every marking M of N and every $\kappa \in \textit{mult}(T \cup \overline{T})$ such that $M + \textit{eff}_N(\kappa) \geq \varnothing$,

$$\phi(M) + \textit{eff}_{\widetilde{N}}(\kappa) = \phi(M + \textit{eff}_N(\kappa)) . \tag{12}$$

The construction does not affect the enabling of steps involving just one transition as well as steps α over T since $p_t \in P_p$ cannot disable α if it is not also disabled by p. Hence, for all markings M of N, $u \in T \cup \overline{T}$, $k \geq 1$, and $\alpha \in \textit{mult}(T)$,

$$M[u^k\rangle_N \iff \phi(M)[u^k\rangle_{\widetilde{N}} \text{ and } M[\alpha\rangle_N \iff \phi(M)[\alpha\rangle_{\widetilde{N}} . \tag{13}$$

---

[5] Intuitively, each $p_t \in P_p$ is a (suitably adjusted) copy of p.

Hence, by (6, 12, 13) and $\widetilde{M_0} \doteq \phi(M_0)$,

$$(STS^{spike})^{rev} \vartriangleleft_{\phi \circ \psi} CRG_{\widetilde{N}} \text{ and } STS \simeq_{\phi \circ \psi} CRG_{\widetilde{N}|_T} \simeq_{\phi^{-1}} CRG_{N|_T} . \qquad (14)$$

We then show that, for every marking M of $\widetilde{N}$ and all $\alpha, \beta \in mult(T)$:

(A) $\phi(M) \xrightarrow{\overline{\alpha}+\beta}_{\widetilde{N}}$ implies $\phi(M) - eff_{\widetilde{N}}(\alpha) \xrightarrow{\alpha+\beta}_{\widetilde{N}} \phi(M) + eff_{\widetilde{N}}(\beta)$. Indeed, we first observe that $\phi(M) - eff_{\widetilde{N}}(\alpha) \in reach_{\widetilde{N}}$. We then observe that, by $\phi(M) \geq pre_{\widetilde{N}}(\overline{\alpha} + \beta)$, we have:

$$\begin{aligned}
\phi(M) - eff_{\widetilde{N}}(\alpha) \geq & \quad pre_{\widetilde{N}}(\overline{\alpha} + \beta) - eff_{\widetilde{N}}(\alpha) \\
= & \quad pre_{\widetilde{N}}(\overline{\alpha}) + pre_{\widetilde{N}}(\beta) - post_{\widetilde{N}}(\alpha) + pre_{\widetilde{N}}(\alpha) \\
\geq_{by(11)} & \quad pre_{\widetilde{N}}(\alpha + \beta) .
\end{aligned}$$

Hence $\alpha + \beta$ is enabled at $\phi(M) - eff_{\widetilde{N}}(\alpha)$, and *(A)* holds as we have:

$$\phi(M) - eff_{\widetilde{N}}(\alpha) + eff_{\widetilde{N}}(\alpha + \beta) = \phi(M) + eff_{\widetilde{N}}(\beta) .$$

(B) $\phi(M) \xrightarrow{\alpha+\beta}_{\widetilde{N}}$ implies $\phi(M) + eff_{\widetilde{N}}(\alpha) \xrightarrow{\overline{\alpha}+\beta}_{\widetilde{N}} \phi(M) + eff_{\widetilde{N}}(\beta)$. Indeed, by *SEQ* of reachability graphs, $\phi(M) \xrightarrow{\alpha}_{\widetilde{N}} \phi(M) + eff_{\widetilde{N}}(\alpha) = M'$. Suppose that $M' \xrightarrow{\overline{\alpha}+\beta}_{\widetilde{N}}$ does not hold. Then there is $q \in \widetilde{P}$ such that

$$\widetilde{F}(q, \overline{\alpha} + \beta) > M'(q). \qquad (15)$$

Moreover, $\phi(M)(q) \geq \widetilde{F}(q, \alpha+\beta)$ and $M'(q) = \phi(M)(q) - \widetilde{F}(q, \alpha) + \widetilde{F}(\alpha, q)$. Hence:

$$\begin{aligned}
\widetilde{F}(q, \overline{\alpha} + \beta) & > \phi(M)(q) - \widetilde{F}(q, \alpha) + \widetilde{F}(\alpha, q) \\
& \geq \widetilde{F}(q, \alpha + \beta) - \widetilde{F}(q, \alpha) + \widetilde{F}(\alpha, q) ,
\end{aligned}$$

and so, by erasing $\widetilde{F}(q, \beta)$ from both sides of inequality (as $\widetilde{F}(q, \overline{\alpha} + \beta) = \widetilde{F}(q, \overline{\alpha}) + \widetilde{F}(q, \beta)$ and $\widetilde{F}(q, \alpha + \beta) = \widetilde{F}(q, \alpha) + \widetilde{F}(q, \beta)$), $\widetilde{F}(q, \overline{\alpha}) > \widetilde{F}(\alpha, q)$. Thus there is $t \in \alpha$ and such that $\widetilde{F}(q, \overline{t}) > \widetilde{F}(t, q)$ and so, by the definition of $\widetilde{N}$, $q = p_t$, for some $p \in P$. Now, it follows from the construction of $\widetilde{N}$, there are $\alpha_0, \alpha_1, \beta_0, \beta_1$ and $k \geq 1$ such that $\alpha = t^k + \alpha_0 + \alpha_1$ and $\beta = \beta_0 + \beta_1$ and $t \not\in \alpha_0 + \alpha_1$ and, for $x = \alpha, \beta$, we have:

$$\begin{aligned}
\widetilde{F}(x_1, p_t) = \widetilde{F}(p_t, x_0) = 0 = \widetilde{F}(p_t, \overline{x}_1) = \widetilde{F}(\overline{x}_0, p_t) \\
\widetilde{F}(p_t, \overline{x}_0) = \widetilde{F}(x_0, p_t) \qquad \widetilde{F}(p_t, x_1) = \widetilde{F}(\overline{x}_1, p_t) .
\end{aligned}$$

By *SEQ* of reachability graphs,

$$\phi(M) \xrightarrow{\alpha_1+\beta_1}_{\widetilde{N}} \phi(M) + eff_{\widetilde{N}}(\alpha_1 + \beta_1) \xrightarrow{t^k}_{\widetilde{N}} \phi(M) + eff_{\widetilde{N}}(\alpha_1 + \beta_1 + t^k) .$$

Thus, by (14), $\phi(M) + eff_{\widetilde{N}}(\alpha_1 + \beta_1 + t^k) \xrightarrow{\overline{t}^k}_{\widetilde{N}} \phi(M) + eff_{\widetilde{N}}(\alpha_1 + \beta_1)$, and so

$$\begin{aligned}
& \phi(M)(p_t) + eff_{\widetilde{N}}(\alpha_1 + \beta_1 + t^k)(p_t) \\
& = \phi(M)(p_t) + eff_{\widetilde{N}}(t^k)(p_t) + eff_{\widetilde{N}}(\alpha_1 + \beta_1)(p_t) \\
& = \phi(M)(p_t) + eff_{\widetilde{N}}(t^k)(p_t) - F(p_t, \alpha_1 + \beta_1) + F(\alpha_1 + \beta_1, p_t) \\
& = \phi(M)(p_t) + eff_{\widetilde{N}}(t^k, p_t) - F(p_t, \alpha_1 + \beta_1) \geq \widetilde{F}(p_t, \overline{t}^k) .
\end{aligned} \qquad (16)$$

We therefore have: $M'(p_t) = M(p_t) + \textit{eff}_{\widetilde{N}}(t^k)(p_t) - \widetilde{F}(p_t, \alpha_1) + \widetilde{F}(\alpha_0, p_t) \geq_{by(16)} \widetilde{F}(p_t, \overline{t}^k) + \widetilde{F}(p_t, \beta_1) + \widetilde{F}(\alpha_0, p_t) = \widetilde{F}(p_t, \overline{t}^k) + \widetilde{F}(p_t, \beta_1) + \widetilde{F}(p_t, \overline{\alpha}_0) = \widetilde{F}(p_t, \overline{\alpha}) + \widetilde{F}(p_t, \beta) = \widetilde{F}(p_t, \overline{\alpha} + \beta)$,

yielding a contradiction with (15). Thus $M' \xrightarrow{\overline{\alpha}+\beta}_{\widetilde{N}}$ holds. By $M' \xrightarrow{\overline{\alpha}+\beta}_{\widetilde{N}}$ and *(B)* we have:

$$M' + \textit{eff}_{\widetilde{N}}(\overline{\alpha} + \beta) = \phi(M) + \textit{eff}_{\widetilde{N}}(\alpha) + \textit{eff}_{\widetilde{N}}(\overline{\alpha} + \beta) = \phi(M) + \textit{eff}_{\widetilde{N}}(\beta) .$$

We now conclude, by (14), *(A)*, and *(B)*, that $STS^{mrev} \simeq_{\phi \circ \psi} CRG_{\widetilde{N}}$.

Finally, if all the steps labelling the arcs of STS are sets, then we can construct a new net $\widetilde{N}'$, adding to $\widetilde{N}$ a fresh set of places $P' = \{p_{tu} \mid t \in T \land u \in \overline{T}\}$, where each p_{tu} is such that $\widetilde{M}_0(p_{tu}) = 1$ and has exactly the following connections $\widetilde{F}(t, p_{tu}) = \widetilde{F}(p_{tu}, t) = \widetilde{F}(u, p_{tu}) = \widetilde{F}(p_{tu}, u) = 1$.

Such places ensure that each step enabled at a reachable marking of $\widetilde{N}$ is a subset of T or a subset of $\overline{T}$. Moreover, the enabling of such steps is not affected by adding P', so that in this case we get indeed $STS^{rev} \simeq CRG_{\widetilde{N}'}$.   □

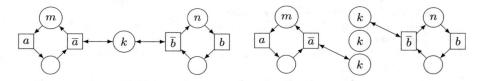

Fig. 6. Net $N_{n,m}$ with $k = \max(m, n)$ and $m, n \geq 1$ (left), and the same net after applying the construction from Theorem 3 (right).

Example 3. Figure 6 depicts a family $N_{n,m}$ of PT-nets which satisfy the assumptions of the last theorem. We clearly have $CRG_{N_{n,m}} \not\simeq STS^{mrev}$, where STS is the reachability graph of the net obtained from $N_{n,m}$ by deleting $\overline{a}$ and $\overline{b}$. However, the construction from the proof of Theorem 3 yields the PT-net $CRG_{\widetilde{N}_{n,m}}$ satisfying $CRG_{\widetilde{N}_{n,m}} \simeq STS^{mrev}$.

7   Concluding Remarks

In this paper we conducted what is to the best of our knowledge the first study of reversibility in the P/T-net model, when the *step semantics*, based on executing steps (multisets) of actions rather than single actions is considered, thus capturing *real parallelism*. In a quite more abstract setting, the (partial) reversal of steps, thus generating *mixed steps* possibly containing both original and reversed events, has been previously studied in [18], but now we are seeing here when and how the reversal can be really done in a concrete operational framework, as Petri Nets are.

Among the topics for future research we would single out an investigation of the impact of allowing multiple reverses of a given action (*splitting reverses*). Such an idea has already been applied in the case of sequential transition systems, making some non-reversible transition system reversible.

Acknowledgement. This research was supported by Cost Action IC1405. The first author was partially supported by the Spanish project TRACES (TIN2015-67522-C3-3), and by Comunidad de Madrid as part of the program S2018/TCS-4339 (BLOQUES-CM) co-funded by EIE Funds of the European Union.

References

1. Badouel, E., Darondeau, P.: Theory of regions. In: Reisig, W., Rozenberg, G. (eds.) ACPN 1996. LNCS, vol. 1491, pp. 529–586. Springer, Heidelberg (1998). https://doi.org/10.1007/3-540-65306-6_22
2. Barylska, K., Best, E., Erofeev, E., Mikulski, Ł., Piątkowski, M.: Conditions for Petri net solvable binary words. ToPNoC **11**, 137–159 (2016)
3. Barylska, K., Erofeev, E., Koutny, M., Mikulski, Ł., Piątkowski, M.: Reversing transitions in bounded Petri nets. Fundamenta Informaticae **157**, 341–357 (2018)
4. Barylska, K., Koutny, M., Mikulski, Ł., Piątkowski, M.: Reversible computation vs. reversibility in Petri nets. Sci. Comput. Program. **151**, 48–60 (2018)
5. Cardelli, L., Laneve, C.: Reversible structures. In: Proceedings of CMSB 2011, pp. 131–140 (2011)
6. Cohen, M.E.: Systems for financial and electronic commerce, 3 September 2013. US Patent 8,527,406
7. Danos, V., Krivine, J.: Reversible communicating systems. In: Gardner, P., Yoshida, N. (eds.) CONCUR 2004. LNCS, vol. 3170, pp. 292–307. Springer, Heidelberg (2004). https://doi.org/10.1007/978-3-540-28644-8_19
8. Danos, V., Krivine, J.: Transactions in RCCS. In: Abadi, M., de Alfaro, L. (eds.) CONCUR 2005. LNCS, vol. 3653, pp. 398–412. Springer, Heidelberg (2005). https://doi.org/10.1007/11539452_31
9. Danos, V., Krivine, J., Sobocinski, P.: General reversibility. Electr. Notes Theor. Comp. Sci. **175**(3), 75–86 (2007)
10. Darondeau, P., Koutny, M., Pietkiewicz-Koutny, M., Yakovlev, A.: Synthesis of nets with step firing policies. Fundam. Inform. **94**(3–4), 275–303 (2009)
11. de Frutos Escrig, D., Koutny, M., Mikulski, Ł.: An efficient characterization of petri net solvable binary words. In: Khomenko, V., Roux, O.H. (eds.) PETRI NETS 2018. LNCS, vol. 10877, pp. 207–226. Springer, Cham (2018). https://doi.org/10.1007/978-3-319-91268-4_11
12. Erofeev, E., Barylska, K., Mikulski, Ł., Piątkowski, M.: Generating all minimal Petri net unsolvable binary words. In: Proceedings of PSC 2016, pp. 33–46 (2016)
13. Esparza, J., Nielsen, M.: Decidability issues for Petri nets. BRICS Report Series **1**(8), 1–25 (1994)
14. Hujsa, T., Delosme, J.-M., Munier-Kordon, A.: On the reversibility of live equal-conflict Petri nets. In: Devillers, R., Valmari, A. (eds.) PETRI NETS 2015. LNCS, vol. 9115, pp. 234–253. Springer, Cham (2015). https://doi.org/10.1007/978-3-319-19488-2_12
15. Lanese, I., Mezzina, C.A., Stefani, J.-B.: Reversing higher-order Pi. In: Gastin, P., Laroussinie, F. (eds.) CONCUR 2010. LNCS, vol. 6269, pp. 478–493. Springer, Heidelberg (2010). https://doi.org/10.1007/978-3-642-15375-4_33

16. Özkan, H.A., Aybar, A.: A reversibility enforcement approach for Petri nets using invariants. WSEAS Trans. Syst. **7**, 672–681 (2008)
17. Phillips, I., Ulidowski, I.: Reversing algebraic process calculi. J. Logical Algebraic Program. **73**(1–2), 70–96 (2007)
18. Phillips, I., Ulidowski, I.: Reversibility and asymmetric conflict in event structures. J. Logical Algebraic Meth. Program. **84**(6), 781–805 (2015)
19. Reisig, W.: Understanding Petri Nets - Modeling Techniques, Analysis Methods, Case Studies. Springer, Heidelberg (2013). https://doi.org/10.1007/978-3-642-33278-4
20. De Vos, A.: Reversible Computing: Fundamentals, Quantum Computing, and Applications. Wiley (2011)

On Interval Semantics of Inhibitor and Activator Nets

Ryszard Janicki$^{(\boxtimes)}$

Department of Computing and Software, McMaster University,
Hamilton, Ontario L8S 4K1, Canada
janicki@mcmaster.ca

Abstract. An interval operational semantics - in a form of interval sequences and step sequences - is introduced for elementary activator nets, and a relationship between inhibitor and activator nets is discussed. It is known that inhibitor and activator nets can simulate themselves for both standard firing sequence semantics and firing step sequence semantics. This paper shows that inhibitor and activator nets are not equivalent with respect to interval sequence and interval step sequence semantics, however, in some sense, they might be interpreted as equivalent with respect to pure interval order operational semantics.

Keywords: Interval order · Inhibitor net · Activator net · Interval sequence · Operational semantics

1 Introduction

Inhibitor nets are Petri nets enriched with inhibitor arcs and activator nets are Petri nets enriched with activator arcs. This paper deals with elementary inhibitor nets and elementary activator nets.

Inhibitor arcs allow a transition to check for an *absence* of a token. They have been introduced in [2] to solve a synchronization problem not expressible in classical Petri nets. In principle they allow 'test for zero', an operator the standard Petri nets do not have (cf. [25]). *Activator arcs* (also called 'read', or 'contextual' arcs [3,24]), formally introduced in [16,24], are conceptually orthogonal to the inhibitor arcs, they allow a transition to check for a *presence* of a token.

Elementary inhibitor nets [16], or just inhibitor nets in this paper, are very simple. They are just classical *elementary nets* of [26,29], i.e. one-safe place-transition nets without self-loops (cf. [7]), extended with inhibitor arcs. Nevertheless they can easily express complex behaviours involving 'not later than' cases [3,16,20,21], priorities, various versions of simultaneities, etc. [14,18,31].

Similarly *elementary nets with activator arcs* [16] are just classical elementary nets extended with activator arcs.

Partially supported by Discovery NSERC Grant of Canada.

S. Donatelli and S. Haar (Eds.): PETRI NETS 2019, LNCS 11522, pp. 192–212, 2019.
https://doi.org/10.1007/978-3-030-21571-2_12

However the expressiveness of elementary nets with inhibitor arcs is often misunderstood and misinterpreted. While for most known models each elementary net with inhibitor arcs can always equivalently be represented by an appropriate elementary net with activator arcs [16,21], the activator arcs *can not* always be simulated by self-loops. If only firing sequences, i.e. languages, generated by nets are concerned, then both inhibitor and activator elementary nets can be represented by equivalent one-safe nets with self-loops. However this is absolutely not true if simultaneous executions, for instance steps, are allowed (cf. [3,16,31]).

It is widely believed that each elementary net with inhibitor arcs can always equivalently be represented by an appropriate elementary net with *activator* arcs. The idea is that an *inhibitor arc* which tests whether a place is empty, can be simulated by an *activator arc* which tests whether its *complement place* (cf. [29]) is not empty. While this is true for plain firing sequence semantics and firing step sequence semantics (cf. [16,21]), it is absolutely false for interval order operational semantics expressed for example in terms of interval sequences [18] or ST-traces [30,31].

Usually an operational or observational semantics of concurrent systems is defined pretty straightforward, either in terms of sequences (often also called 'traces', cf. [11,23]), i.e., total orders, or step sequences, i.e., stratified orders [3,16,21,31]. It has been known for long time that any execution that can be observed by a single observer must be an interval order [15,31,32], so the most precise operational semantics ought to be defined in terms of interval orders. The interval orders do not have a natural sequential representation (such a representation for total and stratified orders is respectively provided by plain sequences and step sequences). To represent interval orders with sequences one might use sequences of the beginnings and endings of events involved, or sequences of appropriate maximal antichains [8,9,15]. The former approach leads to the concept of ST-traces [30], that were used in [31] to define partial semantics of activator nets, and interval sequences [19], that were used in [18] to define interval order observational semantics for inhibitor nets. The latter approach was used in the model of operational semantics proposed in [13,17]. When comparing operational semantics based on ST-traces with the one based on interval sequences, it was pointed out in [18] that when interval operational semantics is concerned, activator nets and inhibitor nets might not be equivalent, however this issue was not elaborated.

In this paper we define interval operational semantics of activator nets by using interval sequences to represent interval orders, in the same way as it was done in [18] and in [17]. It is assumed that transitions have a beginning and an end and a system state consists of a marking together with some transitions that have started, but have not finished yet [13,18,30,31]. The sequences of beginnings and ends are represented by interval sequences, which generate appropriate interval orders.

We will show that with respect to this semantic the activator nets are different from inhibitor nets. For activator nets such semantics can be regarded as not sound (in a sense of [17]) and it is inconsistent with the step sequence semantics of [15,20]. Moreover there are some sets of interval sequences that can be generated by inhibitor nets but cannot by any activator net.

We will argue that operational semantics defined in terms of interval step sequences might make sense and such semantics is introduced for activator nets. Such semantics is still not sound but it is consistent with the step sequence semantics of [15,20].

However, if we are only interested in interval orders generated by nets, not their particular representations, inhibitor and activator nets can be considered as equivalent.

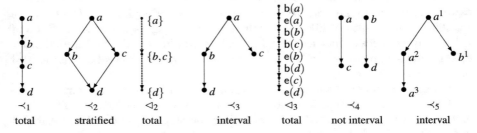

Fig. 1. Examples of partial orders represented as Hasse diagrams (cf. [9]). Note that order $\prec_1$, being total, is uniquely represented by a sequence $abcd$, the stratified order $\prec_2$ is uniquely represented by a step sequence $\{a\}\{b,c\}\{d\}$, and the interval order $\prec_3$ is (*not* uniquely) represented by a sequence that represents $\lhd_3$, i.e. $b(a)e(a)b(b)b(c)e(b)b(d)e(c)e(d)$. The interval order $\prec_5$ is built from *enumerated* events.

2   Partial Orders and Sequences

We first recall some concepts and results that will be used throughout the paper.

2.1   Partial Orders

A (*strict*) *partial order* is a pair $po = (X, \prec)$ such that X is a set and is $\prec$ an irreflexive and transitive binary relation on X. Moreover,

- $\frown$ is a binary *incomparability* relation comprising all pairs (a, b) of distinct elements of X such that $a \not\prec b$ and $b \not\prec a$.
- *po* is *total* if $\frown$ is empty.
- *po* is *stratified* if $\frown$ is transitive, i.e. $\frown \cup\, id_X$ is an equivalence relation.
- *po* is *interval* if $a \prec c$ and $b \prec d$ implies $a \prec d$ or $b \prec c$, for all $a, b, c, d \in X$.
- *po* is *discrete* if for all $a, b \in X$, both $\{c \mid a \frown c\}$ and $\{c \mid a \prec c \prec b\}$ are finite.

Clearly all finite partial orders are discrete. Figure 1 illustrates the above definitions. We will often write $\prec_{po}$ and $\frown_{po}$ instead of $\prec$ and $\frown$, and also, if X is known, we will just write $\prec$ instead of $(X, \prec)$.

For interval orders, the name and intuition follow from Fishburn's Theorem [8], which has two versions, the original one and that for discrete partial orders. The original version is usually formulated as follows.

Theorem 1 ([8,9]). *A countable partial order $(X, \prec)$ is interval if and only if there is a total order $(Y, \lhd)$ and two mappings, $\mathsf{b}, \mathsf{e} : X \to Y$, such that, for all $a, b \in X$:*

1. $\mathsf{b}(a) \lhd \mathsf{e}(a)$,
2. $a \prec b \iff \mathsf{e}(a) \lhd \mathsf{b}(b)$. ◇

Usually, $\mathsf{b}(a)$ is interpreted as the 'beginning', and $\mathsf{e}(a)$ as the 'end', of *interval a*.

Note that the above formulation does *not* assume that either of the mappings b and e is injective. The constructions from the (different) proofs provided in [8,9] do not result in injective mappings. The original formulation of [8] assumes that Y is the set of real numbers and $\lhd$ is just the total ordering of real numbers.

For discrete orders, a stronger version of Theorem 1 has been provided in [15].

Theorem 2 ([15]). *A discrete partial order $(X, \prec)$ is interval if and only if there is a discrete total order $(Y, \lhd)$ and two injective mappings, $\mathsf{b}, \mathsf{e} : X \to Y$, such that, for all $a, b \in X$, we have $\mathsf{b}(X) \cap \mathsf{e}(Y) = \emptyset$ and*

1. $\mathsf{b}(a) \lhd \mathsf{e}(a)$,
2. $a \prec b \iff \mathsf{e}(a) \lhd \mathsf{b}(b)$. ◇

Both versions will explicitly be used in this paper.

2.2   Sequences and Step Sequences

To build sequences we will use four kinds of basic elements: $\Sigma = \{a, b, \dots\}$ are *events*; $\widehat{\Sigma} = \{a^i \mid a \in \Sigma, i \geq 1\}$ are *enumerated* events; $\mathscr{B} = \{B_\alpha \mid \alpha \in \Sigma \cup \widehat{\Sigma}\}$ are the *beginnings* of events; and $\mathscr{E} = \{E_\alpha \mid \alpha \in \Sigma \cup \widehat{\Sigma}\}$ are the *endings* of events. Moreover, $\mathsf{ev}(B_\alpha) = \mathsf{ev}(E_\alpha) = \alpha$, $\mathscr{B}_X = \{B_\alpha \mid \alpha \in X\}$, $\mathscr{E}_X = \{E_\alpha \mid \alpha \in X\}$, and $\mathscr{B}\mathscr{E}_X = \mathscr{B}_X \cup \mathscr{E}_X$, for all $\alpha \in \Sigma \cup \widehat{\Sigma}$ and $X \subseteq \Sigma \cup \widehat{\Sigma}$. We will use various finite sequences over these four sets of basic elements as well as finite sequences of sets of the basic elements. Sequences of sets will be called '*step sequences*'.

For a sequence (step sequence) $x = \kappa_1 \dots \kappa_k$, we assume the following:

- $\mathsf{alph}(x)$ is the set of all basic elements occurring in x, and we denote $\alpha \in x$, for every $\alpha \in \mathsf{alph}(x)$. For example, $\mathsf{alph}(aba) = \{a, b\}$ and $\mathsf{alph}(\{a^1, b^1\}\{a^2\}) = \{a^1, a^2, b^1\}$.
- $\mathsf{ev}(x) = \mathsf{ev}(\mathsf{alph}(x))$, provided that x is over $\mathscr{B} \cup \mathscr{E}$. For example, $\mathsf{ev}(B_a B_b E_a) = \{a, b\}$.

- x is *singular* if no basic element occurs in it more than once.
- if x is singular, then $\mathsf{ord}(x)$ is the partial order with the domain $\mathsf{alph}(x)$ and such that $\alpha \prec_{\mathsf{ord}(x)} \beta$ if α precedes β in x. The order $\mathsf{ord}(x)$ is total if x is a sequence, and the order $\mathsf{ord}(x)$ is stratified if x is a step sequence. For example, $a \prec_{\mathsf{ord}(abc)} c$ and $a \frown_{\mathsf{ord}(\{a,b\}\{c,d\})} b$.
- For a set A of basic elements, $x \cap A$ denotes the *projection* of x onto A, which is obtained from x by erasing all the elements not belonging to A. For example, $eadbca \cap \{a, b\} = aba$ and $ab \cap \{c\} = \varepsilon$.
- If x is a sequence or step sequence over Σ, then its *enumerated representation* $\mathsf{enum}(x)$ is obtained from x by replacing each i-th occurrence of a by a^i. For example, $\mathsf{enum}(abbaba) = a^1 b^1 b^2 a^1 b^3 a^2$, $\mathsf{enum}(\{a, b\}\{a, b, c\}\{a, c\}) = \{a^1, b^1\}\{a^2, b^2, c^1\}\{a^3, c^2\}$.
- If x is a sequence or step sequence over $\mathscr{BE}_\Sigma$, then its *enumerated representation* $\mathsf{enum}(x)$ is obtained from x by replacing each i-th occurrence of B_a by B_{a^i}, and each i-th occurrence of E_a by E_{a^i}. For example, $\mathsf{enum}(E_a B_b E_a) = E_{a^1} B_{b^1} E_{a^2}$.
- For every sequence or step sequence x over Σ or $\mathscr{BE}_\Sigma$, $\mathsf{enumord}(x)$ is a partial order defined as $\mathsf{enumord}(x) = \mathsf{ord}(\mathsf{enum}(x))$. If x is a sequence then $\mathsf{enumord}(x)$ is total and if x is a step sequence then $\mathsf{enumord}(x)$ is stratified.

2.3   Interval Sequences

Interval sequences, proposed and discussed in [18,19], are close to much older ST-traces [30]. The substantial difference is that ST-traces were defined for Petri nets, whereas interval sequences do not assume any system model.

Definition 3. *A sequence x over $\mathscr{BE}_\Sigma$ is interval if $x \cap \mathscr{BE}_{\{a\}} \in (B_a E_a)^*$, for every $a \in \Sigma$. A sequence x over $\mathscr{BE}_{\widehat{\Sigma}}$ is interval if $x \cap \mathscr{BE}_{\widehat{\{a\}}} = B_{a^1} E_{a^1} \dots B_{a^k} E_{a^k}$ ($k \geq 0$), for every $a \in \Sigma$. All interval sequences are denoted by* IntSeq *(or* $\mathsf{IntSeq}(\mathscr{BE}_\Sigma)$*).* ◇

For example, $B_a B_b E_b E_a B_a E_a$ and $B_{a^1} B_{b^1} E_{b^1} E_{a^1} B_{a^2} E_{a^2}$ are interval sequence, but $B_a B_b E_b E_a E_a B_a$ and $B_{a^1} B_{b^1} E_{b^1} E_{a^2} B_{a^2} E_{a^1}$ are not.

If x is a singular interval sequence over $\mathscr{BE}_\Sigma$, then we have $x \cap \{B_a, E_a\} = B_a E_a$, for every $a \in \mathsf{ev}(x)$. All interval sequences over $\mathscr{BE}_{\widehat{\Sigma}}$ are singular, and if x is an interval sequence over Σ, then $\mathsf{enum}(x)$ is an interval sequence over $\mathscr{BE}_{\widehat{\Sigma}}$.

A singular interval sequence x *generates* an interval order given by

$$\mathsf{intord}(x) = (\mathsf{ev}(x), \{(\alpha, \beta) \mid \alpha, \beta \in \mathsf{ev}(x) \wedge E_\alpha \prec_{\mathsf{ord}(x)} B_\beta\}).$$

Also,

$$\mathsf{enumintord}(x) = \mathsf{intord}(\mathsf{enum}(x))$$

is the *enumerated* interval order generated by an interval sequence x over $\mathscr{BE}_\Sigma$.

For example, Fig. 1 depicts two interval orders generated in this way, $po_3 = (\{a, b, c, d\}, \prec_3) = \mathsf{intord}(B_aE_aB_bB_cE_bB_dE_cE_d)$ and $po_5 = (\{a^1, a^2, a^3, b^1\}, \prec_5) = \mathsf{enumintord}(B_aE_aB_aB_bE_aB_aE_bE_a)$. In the latter case $\Sigma = \{a, b\}$.

We would like to point out that the interval sequence representation of interval orders requires Theorem 2.

The interval sequences generating the same interval orders can be characterized similarly as Mazurkiewicz Traces are defined [22].

Let $\sim$ be a symmetric binary relation on sequences over $\mathcal{BE}$ such that $z \sim z'$ if z and z' can be decomposed as $z = x\tau\tau'w$ and $z' = x\tau'\tau w$, for some $\tau, \tau' \in \mathcal{B}$ or $\tau, \tau' \in \mathcal{E}$. Moreover, let $\approx$ be an equivalence relation defined as the transitive and reflexive closure of $\sim$. The equivalence class of $\approx$ containing z will be denoted by $[x]_\approx$.

Proposition 4. *[17] Let z be an interval sequence, w be an interval sequence over $\mathcal{BE}_\Sigma$, and x be a singular interval sequence.*

1. If $z \sim z'$, then z' is an interval sequence.
2. $w \sim w'$ if and only if $\mathsf{enum}(w) \sim \mathsf{enum}(w')$.
3. $[x]_\approx = \{z \in \mathsf{IntSeq} \mid \mathsf{intord}(z) = \mathsf{intord}(x)\}$.
4. $[w]_\approx = \{z \in \mathsf{IntSeq} \mid \mathsf{enumintord}(z) = \mathsf{enumintord}(w)\}$.   ◇

To simplify our notation, from now on we assume that
- for each sequence or step sequence z over Σ: $\mathsf{enumord}(z) = (\mathsf{alph}(z), \vartriangleleft_z)$.
- for each interval sequence x over $\mathcal{BE}_\Sigma$: $\mathsf{enumintord}(x) = (\mathsf{ev}(x), \blacktriangleleft_x)$.

2.4   Sound Interval Sequence Operational Semantics

Let $\mathscr{M}$ be a model of concurrent system (i.e. Petri net, process algebra expressions, some automaton, etc.) that allows to define its operational semantics in terms of interval sequences, and $\mathsf{issem}(\mathscr{M})$ be the set of interval sequences that describes the operational semantics of $\mathscr{M}$. If $\mathscr{M}$ is for example an inhibitor net of [18] or [13], or a safe net with arc-based conditions of [17] then $\mathsf{issem}(\mathscr{M})$ is just a set of all *interval firing sequences* generated by such nets.

Since each interval sequence x uniquely defines an interval order $\mathsf{enumintord}(x)$, we may define the interval order operational semantics

$$\mathsf{IOSEM}(\mathscr{M}) = \{\mathsf{enumintord}(x) \mid x \in \mathsf{issem}(\mathscr{M})\}.$$

The above definition looks as uncontroversial, however it is uncontroversial only when it could be ensured that any interval sequence generating an interval partial order in $\mathsf{IOSEM}(\mathscr{M})$ is a firing interval sequence from $\mathsf{issem}(\mathscr{M})$; in other words, that the following holds:

$$\{z \in \mathsf{IntSeq} \mid \mathsf{enumintord}(z) \in \mathsf{IOSEM}(\mathscr{M})\} = \mathsf{issem}(\mathscr{M}),$$

or, equivalently:

$$\bigcup_{x \in \mathsf{issem}(\mathscr{M})} [x]_{\approx} = \mathsf{issem}(\mathscr{M}).$$

The models satisfying the above equations will be said to have a *sound interval sequence operational semantics.*

If an interval sequence operational semantics of $\mathscr{M}$ is not sound, $\mathsf{issem}(\mathscr{M})$ may still be valid concept, but we have to use $\mathsf{IOSEM}(\mathscr{M})$ very carefully, as it may not be a valid construction.

3   Inhibitor and Activator Petri Nets

An *inhibitor* net is a tuple $IN = (P, T, F, I, m_0)$, where P is a set of *places*, T is a set of *transitions*, P and T are disjoint, $F \subseteq (P \times T) \cup (T \times P)$ is a *flow relation*, $I \subseteq P \times T$ is a set of *inhibitor arcs*, and $m_0 \subseteq P$ is the *initial marking*. An inhibitor arc $(p, e) \in I$ means that e can be enabled only if p *is not marked*. In diagrams (p, e) is indicated by an edge with a small circle at the end. Any set of places $m \subseteq P$ is called a *marking*.

Similarly, and *activator* net is a tuple $AN = (P, T, F, A, m_0)$, where P, T, F, m_0 are as for IA, and $A \subseteq P \times T$ is a set of *activator arcs*. An activator arc $(p, e) \in I$ means that e can be enabled only if p *is marked*. In diagrams (p, e) is indicated by an edge with a small bullet at the end.

In both cases, the quadruple (P, T, F, m_0) is just a plain elementary net [26, 29].

For every $x \in P \cup T$, the set $^\bullet x = \{y \mid (y, x) \in F\}$ denotes the *input* nodes of x and the set $x^\bullet = \{y \mid (x, y) \in F\}$ denotes the *output* nodes of x. The set $x^\circ = \{y \mid (x, y) \in I \cup I^{-1}\}$ is the set of nodes connected by an inhibitor arc to x, while $x^\odot = \{y \mid (x, y) \in A \cup A^{-1}\}$ is the set of nodes connected by an activator arc to x. The dot-notation extends to sets in the natural way, e.g. the set $X^\bullet$ comprises all outputs of the nodes in X, and similarly for '$^\circ$' and '$^\odot$'. We assume that for every $t \in T$, both $^\bullet t$ and $t^\bullet$ are non-empty and disjoint. These requirements do not always appear in the literature, but following [26, 29] we use them for two reasons. Firstly because they are quite natural, and secondly because they allow us to avoid many unnecessary technicalities (cf. [29]). Additionally, both of $^\bullet t$ and $t^\bullet$ must have an empty intersection with t° (or $t^\odot$). Figure 3 shows two examples of elementary inhibitor nets, N and N^1.

The classical *firing sequence* semantics for an *inhibitor net* $IN = (P, T, F, I, m_0)$ is defined as follows:

- A transition t is *enabled* at marking m if $^\bullet t \subseteq m$ and $(t^\bullet \cup t^\circ) \cap m = \emptyset$.
- An enabled t can *occur* leading to a new marking $m' = (m \setminus {}^\bullet t) \cup t^\bullet$, which is denoted by $m[t\rangle m'$, or $m[t\rangle_{IN} m'$.

– A *firing sequence* from marking m to marking m' is a sequence of transitions $t_1 \ldots t_k$ $(k \geq 0)$ for which there are markings $m = m_0, \ldots, m_k = m'$ such that $m_{i-1}[t_i\rangle m_i$, for every $1 \leq i \leq k$. This is denoted by $m[t_1 \ldots t_k\rangle m'$ and $t_1 \ldots t_k \in \mathsf{FS}_{IN}(m \rightsquigarrow m')$.

Similarly we can define *firing step sequence* semantics as follows:

– A set $A \subseteq T$ is a *step* if for all distinct $t, r \in A$, we have $(t^\bullet \cup {}^\bullet t) \cap (r^\bullet \cup {}^\bullet r) = \emptyset$.
– A step A is *enabled* at marking m if ${}^\bullet A \subseteq m$ and $(A^\bullet \cup A^\circ) \cap m = \emptyset$.
– An enabled A can *occur* leading to a new marking $m' = (m \setminus {}^\bullet A) \cup A^\bullet$, which is denoted by $m[A\rangle m'$, or $m[A\rangle_{IN} m'$.
– A *firing step sequence* from marking m to marking m' is a sequence of transitions $A_1 \ldots A_k$ $(k \geq 0)$ for which there are markings $m = m_0, \ldots, m_k = m'$ such that $m_{i-1}[A_i\rangle m_i$, for every $1 \leq i \leq k$.
This is denoted by $m[A_1 \ldots A_k\rangle m'$ and $A_1 \ldots A_k \in \mathsf{FSS}_{IN}(m \rightsquigarrow m')$.
– The set of *stratified orders* leading from marking m to marking m' is defined as $\mathsf{SO}_{IN}(m \rightsquigarrow m') = \{\mathsf{enumord}(x) \mid x \in \mathsf{FSS}_{IN}(m \rightsquigarrow m')\}$.

The firing sequence semantics and firing step sequence semantics for an activator net $AN = (P, T, F, A, m_0)$ are defined in almost identical way, the only differences are:

– a transition t is *enabled* at marking m if ${}^\bullet t \cup t^\odot \subseteq m$ and $t^\bullet \cap m = \emptyset$; and
– a step A is *enabled* at marking m if ${}^\bullet A \cup A^\odot \subseteq m$ and $A^\bullet \cap m = \emptyset$.

Relationships between inhibitor nets and activator nets are often presented in terms of complementary places [16, 20].

We say that places $p, q \in P$ are *complementary* [4, 10] (p is a complement of q and vice versa) if $p \neq q$, ${}^\bullet p = q^\bullet$ and $p^\bullet = {}^\bullet q$, and $|m_0 \cap \{p, q\}| = 1$. If p and q are complementary, we will write $p = \widetilde{q}, q = \widetilde{p}$, and clearly $p = \widetilde{\widetilde{p}}, q = \widetilde{\widetilde{q}}$. We may assume that *every inhibitor (activator) place has its complement*, i.e. $(p, t) \in I \implies \widetilde{p} \in P$ ($(p, t) \in A \implies \widetilde{p} \in P$). If it does not, we can always add it, as it *does not change the net behaviour* (cf. [10, 16, 20, 26]). The elementary inhibitor (activator) nets with this property are called *complement closed*.

We may always assume that an inhibitor (activator) net IN is complement closed, if not we can replace it by INC, constructed as shown in Fig. 2, which has the same behavioural properties as IN. Similarly for activator nets, so the words 'complement closed' will be omitted.

An inhibitor net $IN = (P, T, F, I, m_0)$ and an activator net $AN = (P, T, F, A, m_0)$ are *complement equivalent*, written $IN \stackrel{\mathsf{cpl}}{=} AN$, if $A = \{(p, t) \mid (\widetilde{p}, t) \in I\}$.

For example the nets INC and AIN from Fig. 2 are complement equivalent, i.e. $INC \stackrel{\mathsf{cpl}}{=} AIN$.

The following result states that, as far as firing sequences and firing step sequences are concerned, each elementary net with an inhibitor arc can be simulated by an appropriate elementary net with activator arcs and vice versa.

Proposition 5 ([16,21]). *Let $IN = (P, T, F, I, m_0)$ be an inhibitor net, and let $AN = (P, T, F, A, m_0)$ be an activator net such that $IN \overset{\mathsf{cpl}}{\equiv} AN$. For each $m, m' \subseteq P$, $t \in T$ and $A \subseteq T$, we have:*

1. *$m[t\rangle_{IN} m' \iff m[t\rangle_{AN} m'$,*
2. *$m[A\rangle_{IN} m' \iff m[A\rangle_{AN} m'$.*   ◇

In principle, Proposition 5 states that if $IN \overset{\mathsf{cpl}}{\equiv} AN$ then IN and AN can be considered as *bisimilar* [23] with respect to both firing sequence and step firing sequence. Moreover, directly from Proposition 5 we can get the following result.

Proposition 6. *When an inhibitor net IN and an activator net AN satisfy $IN \overset{\mathsf{cpl}}{\equiv} AN$, then for each $m, m' \subseteq P$:*

1. *$\mathsf{FS}_{IN}(m \rightsquigarrow m') = \mathsf{FS}_{AN}(m \rightsquigarrow m')$, and*
2. *$\mathsf{FSS}_{IN}(m \rightsquigarrow m') = \mathsf{FSS}_{AN}(m \rightsquigarrow m')$.*   ◇

We would like to point out that despite Propositions 5 and 6, not all related properties of inhibitor and activator nets are always the same. For example the absence of a token, unlike the presence of a token, may not be tested in some circumstances [16, 20].

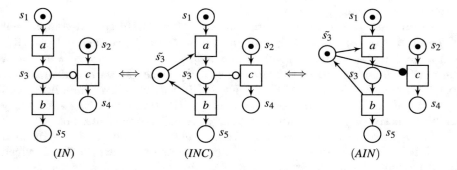

Fig. 2. The net IN is an inhibitor net but it is not complement closed. Adding the place $\tilde{s}_3$ makes it complement closed and transforms it into the net INC. The net AIN was derived from INC by replacing the inhibitor arc (s_3, c) with the activator arc $(\tilde{s}_3, c)$ (see Proposition 5). All three nets generate exactly the same set of firing sequences and firing step sequences (but *not* interval firing sequences!).

4   Interval Sequence Semantics of Inhibitor Nets

If transitions have a beginning and an end, one cannot adequately describe a system state by a marking alone. We need marking plus some transitions that have started, but have not finished yet. One way of describing such system state is the concept of ST-marking, proposed in [30] and explored among others in [31]. Another, slightly simpler for elementary inhibitor nets, way is to use the model recently proposed in [18]. The idea of this approach is briefly illustrated in Fig. 3. If inhibitor arcs are not involved, to represent transitions by their beginnings and ends we might just replace each transition $\boxed{t}$ by the net $\boxed{B_t} \!\rightarrow\! \bigcirc\!t\!\rightarrow\!\boxed{E_t}$, as proposed for example in [5] for nets with priorities or in [33] for timed Petri nets. Each inhibitor arc must be replaced by two when transformation is made and this construction is explained in detail in [18][1]. Within this model, the interval order semantics of the inhibitor net IN in Fig. 3 is fully represented by the firing sequence semantics of the inhibitor net IN^1. Assuming that we can 'hold a token' in transitions and holding a token in c overlap with holding tokens in a and b, the net IN can generate the interval order $\prec_{IN}^{int}$, which is represented for example by an interval sequence $B_a B_c E_a B_b E_b E_c$ which is a firing sequence of the net IN^1.

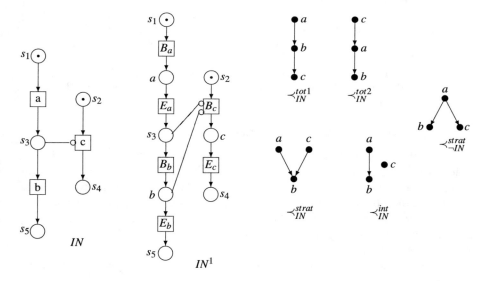

Fig. 3. Interval orders semantics of an inhibitor net. All runs are assumed to start from $\{s_1, s_2\}$ and end at $\{s_4, s_5\}$. The stratified order $\prec_{\neg IN}^{strat}$ is *not* generated by neither IN nor IN^1.

The *interval firing semantics* of an inhibitor net $IN = (P, T, F, I, m_0)$ is defined as follows:

– $\mathcal{BE}_T$ are $\mathcal{BE}$-*transitions*. Intuitively, B_a and E_a are the 'beginning' and 'end' of execution of a.

---

[1] When the inhibitor arc (b, B_c) is removed from the net IN^1 of Fig. 3, the interval sequence $B_a E_a B_b B_c E_b E_c$ is a proper firing sequence of this new N^1 and $\blacktriangleleft_{B_a E_a B_b B_c E_b E_c} = \prec_{\neg IN}^{strat}$, but the stratified order $\prec_{\neg IN}^{strat}$ is not expected to be a behaviour generated by the net IN. See [18] for details.

- We say that a set $m \subseteq P \cup T$ is an *extended marking* if: $m \cap (^\bullet m \cup m^\bullet) = \emptyset$. Intuitively, $a \in m$ means that transition $a \in T$ is being executed in the system state represented by m.
- For each $a \in T$ we define:

$$^\bullet B_a = {}^\bullet a, \qquad B_a^\bullet = \{a\},$$
$$^\bullet E_a = \{a\}, \qquad E_a^\bullet = a^\bullet,$$
$$B_a^\circ = a^\circ \cup (a^\circ)^\bullet, \qquad E_a^\circ = \emptyset.$$

- A $\mathscr{BE}$-transition $\tau \in \mathscr{BE}_T$ is *enabled* at *extended marking* $m \subseteq P \cup T$ if

$$^\bullet \tau \subseteq m \text{ and } (\tau^\bullet \cup \tau^\circ) \cap m = \emptyset.$$

- A $\mathscr{BE}$-transition τ enabled at m can *occur* leading to a *new extended marking*

$$m' = (m \setminus {}^\bullet \tau) \cup \tau^\bullet,$$

which is denoted by: $m[\![\tau\rangle\!\rangle m'$.
- A *firing interval sequence* from extended marking m to extended marking m' is a sequence of $\mathscr{BE}$-transitions $\tau_1 \ldots \tau_k$ $(k \geq 0)$ for which there are extended markings $m = m_0, \ldots, m_k = m'$ such that $m_{i-1}[\![\tau_i\rangle\!\rangle m_i$, for every $1 \leq i \leq k$. This is denoted by $m[\![\tau_1 \ldots \tau_k\rangle\!\rangle m'$ and $\tau_1 \ldots \tau_k \in \mathsf{FIS}_{IN}(m \rightsquigarrow m')$.
- An extended marking m is *reachable* if $\mathsf{FIS}_{IN}(m_0 \rightsquigarrow m) \neq \emptyset$.
- A set of *interval orders* leading from extended marking m to extended marking m' is given by

$$\mathsf{IO}_{IN}(m \rightsquigarrow m') = \{\mathsf{enumintord}(x) \mid x \in \mathsf{FIS}_{IN}(m \rightsquigarrow m')\}.$$

It is not immediately obvious that the definition of $\mathsf{FIS}_{IN}(m \rightsquigarrow m')$ is even valid, as this requires to show that each element of $\mathsf{FIS}_{IN}(m \rightsquigarrow m')$ is an interval sequence. Moreover all total order representations of a given interval order are considered equivalent and none is preferred, hence, to have interval firing sequences consistent with interval orders they represent, if $x \in \mathsf{FIS}_{IN}(m \rightsquigarrow m')$, then $\mathsf{intord}(x) = \mathsf{intord}(y)$ should imply $y \in \mathsf{FIS}_{IN}(m \rightsquigarrow m')$. In Sect. 2.4, the latter property is called 'soundness'.

The following two results from [18] show that both two above properties are satisfied.

Proposition 7. ([18]). *For all markings $m, m' \subseteq P$, we have:*

1. $\mathsf{FIS}_{IN}(m \rightsquigarrow m') \subseteq \mathsf{IntSeq}(\mathscr{BE}^*)$,
2. $\{x \in \mathsf{IntSeq} \mid \mathsf{enumintord}(x) \in \mathsf{IO}_{IN}(m \rightsquigarrow m')\} \subseteq \mathsf{FIS}_{IN}(m \rightsquigarrow m')$. $\diamond$

Moreover the interval sequence semantics and the step sequence semantics are consistent for inhibitor nets.

Proposition 8. ([18]). *For all markings $m, m' \subseteq P$:*

$$\mathsf{SO}_{IN}(m \rightsquigarrow m') = \mathsf{IO}_{IN}(m \rightsquigarrow m') \cap \mathsf{STRAT},$$

where STRAT is the set of all finite stratified orders. $\diamond$

Unfortunately, for activator nets the situation is entirely different.

5   Interval Sequence Semantics of Activator Nets

One of the simplest possible activator nets is the net AN_0 presented below. Its standard transformation into inhibitor net, after some natural simplifications, results in the net IN_0, and interval sequence semantics of IN_0 is represented by firing sequence semantics of $\widehat{IN_0}$.

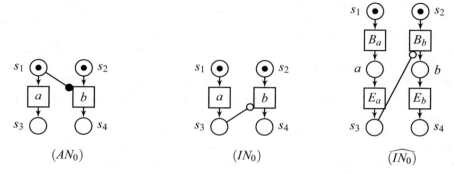

$$(AN_0) \qquad\qquad (IN_0) \qquad\qquad (\widehat{IN_0})$$

Assume $m = \{s_3, s_4\}$. Clearly $\mathsf{FS}_{AN_0}(m_0 \rightsquigarrow m) = \mathsf{FS}_{IN_0}(m_0 \rightsquigarrow m) = \{ba\}$, $\mathsf{FSS}_{AN_0}(m_0 \rightsquigarrow m) = \mathsf{FSS}_{IN_0}(m_0 \rightsquigarrow m) = \{\{b\}\{a\}, \{a, b\}\}$, and $\mathsf{FIS}_{IN_0}(m_0 \rightsquigarrow m) = \mathsf{FS}_{\widehat{IN_0}}(m_0 \rightsquigarrow m) = \{B_b E_b B_a E_a, B_a B_b E_a E_b, B_a B_b E_b E_a, B_b B_a E_a E_b, B_b B_a E_b E_a\}$. We also have $\blacktriangleleft_{B_a B_b E_a E_b} = \blacktriangleleft_{B_a B_b E_b E_a} = \blacktriangleleft_{B_b B_a E_a E_b} = \blacktriangleleft_{B_b B_a E_b E_a}$, and $\mathsf{IO}_{IN_1}(m_0 \rightsquigarrow m) = \{\blacktriangleleft_{B_b E_b B_a E_a}, \blacktriangleleft_{B_a B_b E_a E_b}\}$, where

$$\blacktriangleleft_{B_b E_b B_a E_a} = \begin{array}{c} \bullet\ b \\ \Big| \\ \bullet\ a \end{array} \qquad \blacktriangleleft_{B_a B_b E_a E_b} = \begin{array}{cc} \bullet & \bullet \\ a & b \end{array}$$

The standard transformation of inhibitor nets into activator ones applied to $\widehat{IN_0}$ results in the net $\widehat{AIN_0}$ below (which is clearly different from $\widehat{AN_0}$):

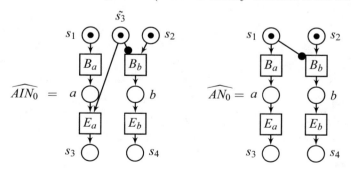

Of course we have $\mathsf{FIS}_{IN_0}(m_0 \rightsquigarrow m) = \mathsf{FS}_{\widehat{IN_0}}(m_0 \rightsquigarrow m) = \mathsf{FS}_{\widehat{AIN_0}}(m_0 \rightsquigarrow m)$, but can we really claim $\mathsf{FIS}_{AN_0}(m_0 \rightsquigarrow m) = \mathsf{FS}_{\widehat{AIN_0}}(m_0 \rightsquigarrow m)$, no matter how $\mathsf{FIS}_{AN_0}(m' \rightsquigarrow m'')$ is defined? The problematic interval sequences are $B_a B_b E_a E_b$ and $B_a B_b E_b E_a$, or, more precisely, their common prefix $B_a B_b$.

Consider 'holding a token' process for the net IN_0. One may start execution of a, hold a token in a and start b when still holding a token in a. It is possible since b can start its execution until the token is placed in s_3. In the net $\widehat{IN_0}$ holding a token in a, or executing a, is modeled by having a token in the place a and B_b can be fired unless there is a token in s_3, so firing interval sequences may start from the sequence $B_a B_b$.

Now consider 'holding a token' process for the net AN_0. Intuitively starting an execution of a should *remove* the token from s_1, so b *cannot start* when the token is held in a! Hence any interval sequence that start with $B_a B_b$ is *not* a proper model of any execution of the net AN_0. The opposite case, starting with b, and beginning execution of a when holding a token in b is unproblematic, so some executions may start with $B_b B_a$. Such a behaviour is rightly modeled by the net $\widehat{AN_0}$ where we have $\mathsf{FS}_{\widehat{AN_0}}(m_0 \rightsquigarrow m) = \{B_b E_b B_a E_a, B_b B_a E_a E_b, B_b B_a E_b E_a\} \subsetneq \mathsf{FS}_{\widehat{AIN_0}}(m_0 \rightsquigarrow m)$.

This means that the *adequate interval sequence semantics of the activator net AN_0 is rather given by the plain firing sequence semantics of $\widehat{AN_0}$*. Therefore we have to define the interval sequence semantics for activator nets in such a way that $\mathsf{FIS}_{AN_0}(m_0 \rightsquigarrow m) = \mathsf{FS}_{\widehat{AN_0}}(m_0 \rightsquigarrow m)$. However, in such a case we have $\mathsf{FIS}_{AN_0}(m_0 \rightsquigarrow m) \subsetneq \mathsf{FIS}_{IN_0}(m_0 \rightsquigarrow m)$. We can define $\mathsf{IO}_{AN}(m \rightsquigarrow m')$ for activator nets in the same fashion as for inhibitor nets, and then we have $\mathsf{IO}_{AN_0}(m_0 \rightsquigarrow m) = \{\mathsf{enumintord}(x) \mid x \in \mathsf{FIS}_{AN_0}(m \rightsquigarrow m')\} = \{\blacktriangleleft_{B_b E_b B_a E_a}, \blacktriangleleft_{B_b B_a E_a E_b}\}$. But this means that $\mathsf{IO}_{AN_0}(m_0 \rightsquigarrow m) = \mathsf{IO}_{IN_0}(m_0 \rightsquigarrow m)$, despite the fact that $\mathsf{FIS}_{AN_0}(m_0 \rightsquigarrow m) \neq \mathsf{FIS}_{IN_0}(m_0 \rightsquigarrow m)$, so such semantics is definitely not sound in the sense of Sect. 2.4. We will discuss this issue in detail later.

The case orthogonal to the net AN_0 is the net AN_1 presented below. Its standard transformation into inhibitor net, after some natural simplifications, results in the net IN_1, and interval sequence semantics of IN_1 is represented by firing sequence semantics of $\widehat{IN_1}$.

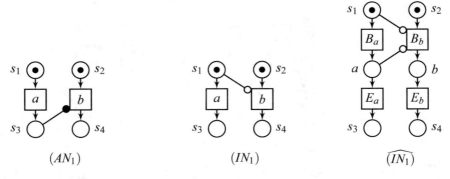

(AN_1) (IN_1) $\widehat{(IN_1)}$

Again assume $m = \{s_3, s_4\}$. Clearly $\mathsf{FS}_{AN_1}(m_0 \rightsquigarrow m) = \mathsf{FS}_{IN_1}(m_0 \rightsquigarrow m) = \{ab\}$, $\mathsf{FSS}_{AN_1}(m_0 \rightsquigarrow m) = \mathsf{FSS}_{IN_1}(m_0 \rightsquigarrow m) = \{\{a\}\{b\}\}$, and $\mathsf{FIS}_{IN_1}(m_0 \rightsquigarrow m) = \mathsf{FS}_{\widehat{IN_1}}(m_0 \rightsquigarrow m) = \{B_a E_a B_b E_b\}$.

The standard transformation of inhibitor nets into activator ones applied to $\widehat{IN_1}$ results in the net $\widehat{AIN_1}$ below, which could easily be simplified to the more natural net $\widehat{AN_1}$ on the right.

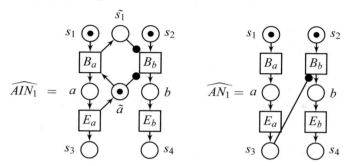

This case is not problematic as we have $\mathsf{FIS}_{IN_1}(m_0 \rightsquigarrow m) = \mathsf{FS}_{\widehat{IN_1}}(m_0 \rightsquigarrow m) = \mathsf{FS}_{\widehat{AIN_1}}(m_0 \rightsquigarrow m) = \mathsf{FS}_{\widehat{AN_1}}(m_0 \rightsquigarrow m) = \{B_a E_a B_b E_b\}$. We can then consider the firing sequence semantics of $\widehat{AN_1}$ as the interval order semantics of AN_1, i.e. to assume $\mathsf{FIS}_{AN_1}(m_0 \rightsquigarrow m) = \mathsf{FS}_{\widehat{AN_1}}(m_0 \rightsquigarrow m)$.

We can now define formally an *interval sequence semantics* of *activator nets*. Let $AN = (P, T, F, A, m_0)$ be an activator net. We define its interval sequence semantics by repeating the definition of interval sequence semantics for inhibitor nets with replacing the rules involving inhibitor arcs by:

- For each $a \in T$ we define: $B_a^{\odot} = a^{\odot}$ and $E_a^{\odot} = \emptyset$.
- A $\mathscr{BE}$-transition $\tau \in \mathscr{BE}_T$ is *enabled* at *extended marking* $m \subseteq P \cup T$ if

$$^{\bullet}\tau \cup \tau^{\odot} \subseteq m \text{ and } \tau^{\bullet} \cap m = \emptyset.$$

We will show that an equivalent of Proposition 7(1) also holds for activator nets, so interval firing sequence semantics is valid.

Proposition 9. *For all markings $m, m' \subseteq P$, we have:* $\mathsf{FIS}_{AN}(m \rightsquigarrow m') \subseteq \mathsf{IntSeq}(\mathscr{BE}^*)$.

Proof. It suffice to show that if $m[\![x\rangle\!\rangle m'$, then $x \in \mathsf{IntSeq}(\mathscr{BE}^*)$, i.e. for each $a \in T, x \cap \mathscr{BE}_{\{a\}} \in (B_a E_a)^*$. Let $x = x_1 B_a x_2$ and $m[\![x_1 B_a\rangle\!\rangle m''$. Since $B_a^{\bullet} = \{a\}$, $a \in m''$. We also have: for any $m_a \subseteq P \cup T$, if $a \in m_a$, then B_a is not enabled in m_a, so the only way to remove a from m_a is to fire E_a (as $^{\bullet}E_a = \{a\}$). Hence we must have $x = x_1 B_a y E_a z$ where $y \cap \mathscr{BE}_{\{a\}} = \varepsilon$. $\square$

However an equivalent of Proposition 7(2) may not hold. Consider the net AN_0 discussed above. We have $\mathsf{IO}_{AN_0}(m_0 \rightsquigarrow m) = \{\blacktriangleleft_{B_b E_b B_a E_a}, \blacktriangleleft_{B_b B_a E_a E_b}\}$, so $\{x \in \mathsf{IntSeq} \mid \mathsf{enumintord}(x) \in \mathsf{IO}_{AN_0}(m_0 \rightsquigarrow m)\} = \{B_b E_b B_a E_a, B_a B_b E_a E_b, B_a B_b E_b E_a, B_b B_a E_a E_b, B_b B_a E_b E_a\}$, and $\mathsf{FIS}_{AN_0}(m_0 \rightsquigarrow m) = \{B_b E_b B_a E_a, B_b B_a E_a E_b, B_b B_a E_b E_a\}$, so an equivalent of Proposition 7(2) does not hold for the activator net AN_0.

Moreover we have:

Corollary 10. *An activator net* $AN = (P, T, A, m_0)$ *such that*

$$\mathsf{FIS}_{AN}(m_0 \leadsto m) = \{B_b E_b B_a E_a, B_a B_b E_a E_b, B_a B_b E_b E_a, B_b B_a E_a E_b, B_b B_a E_b E_a\},$$

for some $m \subseteq P$, *does not exist.*

Proof. By inspection as we may restrict our search to the nets with $T = \{a, b\}$. □

An interval sequence semantics of activator nets, as defined above, is *not consistent* with their step sequence semantics. Consider the nets presented below.

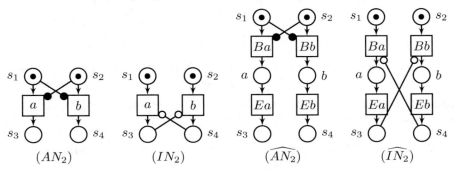

(AN_2) $\qquad$ (IN_2) $\qquad$ $(\widehat{AN_2})$ $\qquad$ $(\widehat{IN_2})$

Assume $m = \{s_3, s_4\}$. We have $\mathsf{FSS}_{AN_2}(m_0 \leadsto m) = \mathsf{FSS}_{IN_2}(m_0 \leadsto m) = \{\{a, b\}\}$, $\mathsf{FIS}_{IN_2}(m_0 \leadsto m) = \{B_a B_b E_a E_b, B_b B_a E_a E_b, B_a B_b E_b E_a, B_b B_a E_b E_a\}$, but $\mathsf{FIS}_{AN_2}(m_0 \leadsto m) = \emptyset$. Hence $\mathsf{SO}_{AN_2}(m_0 \leadsto m) = \mathsf{SO}_{IN_2}(m_0 \leadsto m) = \{\lhd_{\{a,b\}}\}$, $\mathsf{IO}_{IN_2}(m_0 \leadsto m) = \{\blacktriangleleft_{B_a B_b E_a E_b}\}$, where $\lhd_{\{a,b\}} = \blacktriangleleft_{B_a B_b E_a E_b} = a \frown b$, but $\mathsf{IO}_{AN_2}(m_0 \leadsto m) = \emptyset$, so $\mathsf{SO}_{AN_2}(m_0 \leadsto m) \neq \mathsf{IO}_{AN_2}(m_0 \leadsto m) \cap \mathsf{STRAT}$.

This means an equivalent of Proposition 8 does not hold, i.e. *the step sequence semantics of activator nets and the interval sequence semantics of activator nets described above are not consistent.*

However, standard firing sequence semantics and and interval sequence semantics are consistent.

Lemma 11. *For every two* $m, m' \subseteq P$, *then for each* $t \in T$,

$$m[t\rangle m' \iff m[\![B_t E_t\rangle\!\rangle m'.$$

Proof. Since $^\bullet B_t = {}^\bullet t$, $B_t^\odot = t^\odot$, and $^\bullet t \cup t^\odot \subseteq m$, t is enabled at m if and only if B_t is enabled at m.
($\Rightarrow$) If $m[t\rangle m'$ then $m' = (m \setminus {}^\bullet t) \cup t^\bullet$. Let $m[\![B_t\rangle\!\rangle m_B$, i.e. $m_B = (m \setminus {}^\bullet B_t) \cup B_t^\bullet = (m \setminus {}^\bullet t) \cup \{t\}$. Hence E_t is enabled at m_B. Let $m_B[\![E_t\rangle\!\rangle m_E$, i.e. $m_E = (m_B \setminus {}^\bullet E_t) \cup E_t^\bullet = (((m \setminus {}^\bullet t) \cup \{t\}) \setminus \{t\}) \cup t^\bullet = (m \setminus {}^\bullet t) \cup t^\bullet = m'$. Hence $m[\![B_t\rangle\!\rangle m_B[\![E_t\rangle\!\rangle m'$, i.e. $m[\![B_t E_t\rangle\!\rangle m'$.
($\Leftarrow$) If $m[\![B_t E_t\rangle\!\rangle m'$ then by the same reasoning as in the proof of ($\Rightarrow$) we can show that $m' = (m \setminus {}^\bullet t) \cup t^\bullet$. Hence $m[t\rangle m'$. □

In case of activator nets, to get an interval semantics consistent with step sequence semantics one needs to introduce and use interval *step* sequences.

6 Interval Step Sequence Semantics of Activator Nets

When considering instantaneous (or zero time) events we often have to decide if simultaneous execution of them has to be taken into account. It is often assumed that the simultaneous execution of instantaneous events is not allowed or it is irrelevant. Validity of such assumption depends in part on the model of time that is used.

If we assume standard continues real time, then from the time-energy uncertainty relations (cf. [6, 12, 27])

$$\Delta t \Delta E \geq \frac{\hbar}{2\pi},$$

where t denotes time, E denotes energy and $\hbar$ is Planck's constant, we must conclude that simultaneous execution of instantaneous events is unobservable as it would require infinite energy ($\Delta t = 0$). For discrete time, often assumed in computation theory, we have $\Delta t > 0$ so the assumption that we can observe simultaneity of instantaneous events might be valid. However, if interval sequence operational semantics (or its equivalent) is sound, i.e.

$$\{z \in \mathsf{IntSeq} \mid \mathsf{enumintord}(z) \in \mathsf{IOSEM}(\mathscr{M})\} = \mathsf{issem}(\mathscr{M}),$$

then simultaneous execution of instantaneous events is irrelevant as each step $\{B_a, B_b\}$ is equivalently represented by sequences $B_a B_b$ and $B_b B_a$, each step $\{E_a, E_b\}$ is equivalently represented by sequences $E_a E_b$ and $E_b E_a$, and each step $\{B_a, E_b\}$ is equivalently represented by the sequence $B_a E_b$; and all these rules can easily be extended to bigger steps. Hence, in terms of interval orders, interval sequence semantics and interval step sequence semantics would be equivalent, so why use the more complex one?

The situation is much different when interval sequence operational semantics is not sound, as for instance the one defined in this paper for activator nets. In such cases some kind of interval step sequence semantics might be very useful.

Definition 12. *A step sequence x over $\mathscr{BE}_\Sigma$ is interval if $x \cap \mathscr{BE}_{\{a\}} \in (B_a E_a)^*$, for every $a \in \Sigma$. A step sequence x over $\mathscr{BE}_{\widehat{\Sigma}}$ is interval if $x \cap \mathscr{BE}_{\widehat{\{a\}}} = B_{a^1} E_{a^1} \ldots B_{a^k} E_{a^k}$ ($k \geq 0$), for every $a \in \Sigma$. All interval step sequences are denoted by $\mathsf{IntStSeq}$ (or $\mathsf{IntStSeq}(\mathscr{BE}_\Sigma)$).* ◇

Similarly as for interval sequences, a singular interval *step* sequence x *generates* an interval order given by

$$\mathsf{intord}(x) = (\mathsf{ev}(x), \{(\alpha, \beta) \mid \alpha, \beta \in \mathsf{ev}(x) \wedge E_\alpha \prec_{\mathsf{ord}(x)} B_\beta\}).$$

Also,

$$\mathsf{enumintord}(x) = \mathsf{intord}(\mathsf{enum}(x))$$

is the *enumerated* interval order generated by an interval *step* sequence x over $\mathscr{BE}_\Sigma$.

In this case we use Theorem 1 to justify the interval step sequence representation of interval orders.

Let $AN = (P, T, F, A, m_0)$ be an activator net. We define its *interval step sequence* semantics almost exactly in the same way as we defined its interval sequence semantics. The only (natural) differences are the following:

- A set $A \subseteq \mathscr{BE}_T$ is a *step* if for all distinct $\tau_1, \tau_2 \in A$, we have $(\tau_1^\bullet \cup {}^\bullet\tau_1) \cap (\tau_2^\bullet \cup {}^\bullet\tau_2) = \emptyset$.
- a step $A \subseteq \mathscr{BE}_T$ is *enabled* at *extended marking* $m \subseteq P \cup T$ if ${}^\bullet A \cup A^\odot \subseteq m$ and $A^\bullet \cap m = \emptyset$.
- An enabled A can *occur* leading to a new marking $m' = (m \setminus {}^\bullet A) \cup A^\bullet$, which is denoted by $m[\![A\rangle\!\rangle m'$, or $m[\![A\rangle\!\rangle_{AN} m'$.
- The set of *interval step sequences* leading from marking m to m' is denoted by $\mathsf{FISS}_{AN}(m_0 \rightsquigarrow m)$, and the set of *interval orders* leading from marking m to m' is defined as $\mathsf{IOS}_{AN}(m \rightsquigarrow m') = \{\mathsf{enumintord}(x) \mid x \in \mathsf{FISS}_{AN}(m \rightsquigarrow m')\}$.

Interval step sequence semantics can be defined for the inhibitor nets as well in an almost identical manner. However Proposition 7(2) makes its redundant as it can be simulated in full by plain interval sequence semantics[2].

For our net AN_2, with $m = \{s_3, s_4\}$, we have $\mathsf{FISS}_{AN_2}(m_0 \rightsquigarrow m) = \mathsf{FSS}_{\widehat{AN_2}}(m_0 \rightsquigarrow m)$, i.e. $\mathsf{FISS}_{AN_2}(m_0 \rightsquigarrow m) = \{\{B_a, B_b\}\{E_a, E_b\}, \{B_a, B_b\}\{E_a\} \{E_b\}, \{B_a, B_b\}\{E_b\}\{E_a\}\}$. Hence, $\mathsf{SO}_{AN_2}(m_0 \rightsquigarrow m) = \{\lhd_{\{a,b\}}\}$, $\mathsf{IOS}_{AN_2}(m_0 \rightsquigarrow m) = \{\blacktriangleleft_{\{B_a, B_b\}\{E_a, E_b\}}\}$, where $\lhd_{\{a,b\}} = \blacktriangleleft_{\{B_a, B_b\}\{E_a, E_b\}} = a \frown b$, so $\mathsf{SO}_{AN_2}(m_0 \rightsquigarrow m) = \mathsf{IOS}_{AN_2}(m_0 \rightsquigarrow m)$.

We will show that this relationship is also true in general, but first we prove an equivalent of Proposition 9.

Proposition 13.
For all markings $m, m' \subseteq P$, we have: $\mathsf{FISS}_{AN}(m \rightsquigarrow m') \subseteq \mathsf{IntStSeq}(\mathscr{BE}^*)$.

Proof. It suffice to show that if $m[\![x\rangle\!\rangle m'$, then $x \in \mathsf{IntStSeq}(\mathscr{BE}^*)$, i.e. for each $a \in T$, $x \cap \mathscr{BE}_{\{a\}} \in (B_a E_a)^*$. Let $x = x_1 A x_2$, $B_a \in A$ and $m[\![x_1 A\rangle\!\rangle m''$. Since $B_a^\bullet = \{a\}$, $a \in m''$. We also have: for any $m_a \subseteq P \cup T$, if $a \in m_a$ and $B_a \in A'$, then A' is not enabled in m_a, so the only way to remove a from m_a is to fire A'' containing E_a (as ${}^\bullet E_a = \{a\}$). Hence we must have $x = x_1 A' y A'' z$ where $B_a \in A'$, $E_a \in A''$ and $y \cap \mathscr{BE}_{\{a\}} = \varepsilon$. $\quad\square$

Hence interval step sequence semantics, we have proposed, is valid. However an equivalent of Proposition 7(2) still may not hold for interval step sequence semantics of activator nets. Consider the net AN_2 discussed above. We have $\mathsf{IO}_{AN_2}(m_0 \rightsquigarrow m) = \{\blacktriangleleft_{\{B_a, B_b\}\{E_a, E_b\}}\}$, so for example the interval step sequence $\{B_b\}\{E_b\}\{B_a\}\{E_a\}$ belongs to $\{x \in \mathsf{IntStSeq} \mid \mathsf{enumintord}(x) \in \mathsf{IO}_{AN_2}(m_0 \rightsquigarrow m)\}$. However $\mathsf{FISS}_{AN_2}(m_0 \rightsquigarrow m) = \{\{B_a, B_b\}$

---

[2] It can easily be shown that $m[\![\{B_{a_1}, \ldots, B_{a_k}\}\rangle\!\rangle_{IN} m' \iff m[\![B_{a_{i_1}}\rangle\!\rangle_{IN} \ldots [\![B_{a_{i_k}}\rangle\!\rangle_{IN} m'$ for any permutation $l_1, \ldots, l_k$ of $1, \ldots, k$, which is not true for $[\![\ldots\rangle\!\rangle_{AN}$.

$\{E_a, E_b\}, \{B_a, B_b\}\{E_a\}\{E_b\}, \{B_a, B_b\}\{E_b\}\{E_a\}\}$, i.e. $\{B_b\}\{E_b\}\{B_a\}\{E_a\} \notin$ $\mathsf{FISS}_{AN_2}(m_0 \rightsquigarrow m)$. Hence an equivalent of Proposition 7(2) does not hold for the activator net AN_2 and interval step sequence semantics, so this semantics is also *not sound* in the sense of Sect. 2.4.

For every step $A \subseteq T$, let $B_A = \{B_a \mid a \in A\}$ and $E_A = \{E_a \mid a \in A\}$. We may now formulate and prove the result corresponding to Lemma 11 for interval firing sequences.

Lemma 14. *For every two markings* $m, m' \subseteq P$ *and every step* $A \subseteq T$,

$$m[A\rangle m' \iff m[\![B_A E_A\rangle\!\rangle m').$$

Proof. Since A is a step then for all $t, r \in A$ we have $(t^\bullet \cup {}^\bullet t) \cap (r^\bullet \cup {}^\bullet r) = \emptyset$, which also means $(E_t^\bullet \cup {}^\bullet B_t) \cap (E_r^\bullet \cup {}^\bullet B_r) = \emptyset$. Since ${}^\bullet B_A = {}^\bullet A$, $B_A^\bullet = A^\odot$, and ${}^\bullet A \cup A^\odot \subseteq m$, A is enabled at m if and only if B_A is enabled at m.
($\Rightarrow$) If $m[A\rangle m'$ then $m' = (m \setminus {}^\bullet A) \cup A^\bullet$. Let $m[\![B_A\rangle\!\rangle m_B$, i.e. $m_B = (m \setminus {}^\bullet B_A) \cup B_A^\bullet = (m \setminus {}^\bullet A) \cup A$. Hence E_A is enabled at m_B. Let $m_B[\![E_A\rangle\!\rangle m_E$, i.e. $m_E = (m_B \setminus {}^\bullet E_A) \cup E_A^\bullet = (((m \setminus {}^\bullet A) \cup A) \setminus A) \cup A^\bullet = (m \setminus {}^\bullet A) \cup A^\bullet = m'$. Hence $m[\![B_A\rangle\!\rangle m_B[\![E_A\rangle\!\rangle m'$, i.e. $m[\![B_A E_A\rangle\!\rangle m'$.
($\Leftarrow$) If $m[\![B_A E_A\rangle\!\rangle m'$ then by the same reasoning as in the proof of ($\Rightarrow$) we can show that $m' = (m \setminus {}^\bullet A) \cup A^\bullet$. Hence $m[A\rangle m'$. □

We can now show that the *interval step sequence semantics and the step sequence semantics are consistent for activator nets.*

Proposition 15. *For all markings* $m, m' \subseteq P$:

$$\mathsf{SO}_{AN}(m \rightsquigarrow m') = \mathsf{IOS}_{AN}(m \rightsquigarrow m') \cap \mathsf{STRAT},$$

where STRAT *is the set of all finite stratified orders.*

Proof. Directly from Lemma 14. □

However, if we are interested in interval orders only, not a way they are generated, then inhibitor nets and activator nets can be regarded as equivalent.

Proposition 16. *When an inhibitor net* IN *and an activator net* AN *satisfy* $IN \overset{\mathrm{cpl}}{\equiv} AN$, *then for each* $m, m' \subseteq P$:

$$\mathsf{IO}_{IN}(m \rightsquigarrow m') = \mathsf{IOS}_{AN}(m \rightsquigarrow m').$$

Proof. (sketch) Assume $IN = (P_{IN}, T_{IN}, F, I, m_0)$ and $AN = (P_{AN}, T_{AN}, F, I, m_0)$. Of course $P_{IN} = P_{AN}$ and $T_{IN} = T_{AN}$ but such disticntion will make our proof easier to formulate. Additionally, suppose that $m[\![\{E_{a_1}, \ldots, E_{a_k}\}\rangle\!\rangle_{AN} m'$, for $m, m' \subseteq P_{AN} \cup T_{AN}$. Then for any permutation $i_1, \ldots, i_k$ of $1, \ldots, k$ we clearly have $m[\![\{E_{a_{i_1}}\}\rangle\!\rangle_{AN} \ldots [\![\{E_{a_{i_k}}\}\rangle\!\rangle_{AN} m'$, and vice versa.

($\Rightarrow$) Suppose that we have $m[\![B_{a_1}\rangle\!\rangle_{IN} \ldots [\![B_{a_k}\rangle\!\rangle_{IN} m'[\![E_a\rangle\!\rangle_{IN} m''$ for some $m, m', m'' \subseteq P_{IN} \cup T_{IN}$. Define $A = \{a_1, \ldots, a_k\}$. Hence we have $A \cup \{a\} \subseteq m'$, ${}^\bullet B_A \subseteq m$, $(B_A^\bullet \cup B_A^\circ) \cap m = \emptyset$, $m' = (m \setminus {}^\bullet B_A) \cup B_A^\bullet$ and $m'' = (m' \setminus \{a\}) \cup a^\bullet$. Since $IN \overset{\text{cpl}}{\equiv} AN$ then we also have ${}^\bullet B_A \cup B_A^\odot \subseteq m$. Hence $m[\![B_A\rangle\!\rangle_{AN} m'[\![E_a\rangle\!\rangle m''$. From this it follows $\mathsf{IO}_{IN}(m \rightsquigarrow m') \subseteq \mathsf{IOS}_{AN}(m \rightsquigarrow m')$.

($\Leftarrow$) Suppose we have $m[\![B_A\rangle\!\rangle_{AN} m'[\![E_a\rangle\!\rangle_{AN} m''$ for some $m, m', m'' \subseteq P_{AN} \cup T_{AN}$ and $A = \{a_1, \ldots, a_k\} \subseteq T_{AN}$. Hence we have $A \cup \{a\} \subseteq m'$, ${}^\bullet B_A \cup B_Z^\odot \subseteq m$, $B_A^\bullet \cap m = \emptyset$, $m' = (m \setminus {}^\bullet B_A) \cup B_A^\bullet$ and $m'' = (m' \setminus \{a\}) \cup a^\bullet$. Let $a_{i_1}, \ldots, a_{i_k}$ be any permutation of the elements of A. From Proposition 7(2) we conclude that $m[\![B_{a_{i_1}}\rangle\!\rangle_{IN} \ldots [\![B_{a_{i_k}}\rangle\!\rangle_{IN} m'[\![E_a\rangle\!\rangle_{IN} m''$, which implies $\mathsf{IOS}_{AN}(m \rightsquigarrow m') \subseteq \mathsf{IO}_{IN}(m \rightsquigarrow m')$.   □

As the nets IN_2 and AN_2 indicate we *cannot* replace $\mathsf{IOS}_{AN}(m \rightsquigarrow m')$ with $\mathsf{IO}_{AN}(m \rightsquigarrow m')$ in Proposition 16. We might say (up to a few behaviour preserving simplifications) that $IN_2 \overset{\text{cpl}}{\equiv} AN_2$, but $\mathsf{IO}_{IN_2}(m_0 \rightsquigarrow \{s_3, s_4\}) = \mathsf{IOS}_{AN_2}(m_0 \rightsquigarrow \{s_3, s_4\}) = \{\lhd_{\{a,b\}}\}$ while $\mathsf{IO}_{AN_2}(m_0 \rightsquigarrow \{s_3, s_4\}) = \emptyset$.

7   Final Comments

Two interval order type semantics have been proposed for elementary activator nets, one uses interval sequences and Theorem 2 and another uses interval step sequences and original Fishburn result, i.e. Theorem 1. Both semantics were analyzed compared with interval sequence semantics of inhibitor nets as proposed in [18] and with classical step sequence semantics of [16, 20]. We have shown that for both the interval sequence and the interval step sequence semantics are not sound in the sense of [17, 18], and that the interval sequence semantics of activator nets is inconsistent with its traditional step sequence semantics. On the other hand, the interval sequence semantics of inhibitor nets is sound and consistent with their step semantics as shown in [18].

To show that there is a kind of interval semantics for activator nets that is consistent with its step sequence semantics we have introduced the concept of interval step sequences. Because the intervals sequence semantics of inhibitor nets is sound, interval step sequences are redundant and irrelevant for inhibitor nets, however for activator nets the interval step sequence semantics is richer than interval sequence semantics and it is consistent with the step sequence semantics.

We have also shown that inhibitor nets can produce sets of firing sequences that cannot be produced by any activator net. Despite all of this, when we concentrate on interval orders only and abstract away intervals that generate them, inhibitor and activator nets might be treated as equivalent (cf. Proposition 16).

The results of this paper emphasize the difference between *interval semantics*, as in [1, 28], and *interval order semantics*, as in [13, 18, 30, 31]. Interval orders are partial orders so for different a and b we have either $a \prec b$ or $a \frown b$, i.e. only two possible relationships; while for two intervals a and b, we might have up to seven relationships: *a before b, a equal b, a meets b, a overlaps b, a during b, a starts b* and *a finishes b* (cf. [1]). Interval order semantics is usually simpler, but not necessarily equivalent to interval semantics.

When a model of concurrent system $\mathcal{M}$ is *sound* as defined in Sect. 2.4, then interval orders are a good abstractions of appropriate intervals; and interval order semantics and interval semantics might be considered as equivalent. When $\mathcal{M}$ is *not sound*, we have to be very careful when using interval orders as some of their interval representations may be invalid. Interval semantics in terms of interval sequences or interval step sequences may still be well defined and valid, but interval orders may not.

Acknowledgment. The author gratefully acknowledges three anonymous referees for their helpful comments.

References

1. Allen, J.F.: Maintaining knowledge about temporal intervals. Commun. ACM **26**, 832–843 (1983)
2. Agerwala, T., Flynn, M.: Comments on capabilities, limitations and "correctness" of Petri nets. Comput. Archit. News **4**(2), 81–86 (1973)
3. Baldan, P., Busi, N., Corradini, A., Pinna, G.M.: Domain and event structure semantics for Petri nets with read and inhibitor arcs. Theor. Comp. Science **323**, 129–189 (2004)
4. Best, E., Devillers, R.: Sequential and concurrent behaviour in Petri net theory. Theor. Comput. Sci. **55**, 97–136 (1987)
5. Best, E., Koutny, M.: Petri net semantics of priority systems. Theor. Comput. Sci. **94**, 141–158 (1992)
6. Busch, P.: The time-energy uncertainty relation. In: Muga, G., Mayato, S., Egusquiza, I. (eds.) Time in Quantum Mechanics. Lecture Notes in Physics, vol. 734. Springer, Heidelberg (2008). https://doi.org/10.1007/978-3-540-73473-4_3
7. Desel, J., Reisig, W.: Place/transition Petri nets. In: Reisig, W., Rozenberg, G. (eds.) ACPN 1996. LNCS, vol. 1491, pp. 122–173. Springer, Heidelberg (1998). https://doi.org/10.1007/3-540-65306-6_15
8. Fishburn, P.C.: Intransitive indifference with unequal indifference intervals. J. Math. Psychol. **7**, 144–149 (1970)
9. Fishburn, P.C.: Interval Orders and Interval Graphs. John Wiley, New York (1985)
10. Goltz, U., Reisig, W.: The non-sequential behaviour of Petri nets. Inf. Control **57**(2), 125–147 (1983)
11. Hoare, C.A.R.: Communicating sequential processes. Commun. ACM **26**(1), 100–106 (1983)
12. Jammer, M.: The Philosophy of Quantum Mechanics. John Wiley, New York (1974)
13. Janicki, R.: Modeling operational semantics with interval orders represented by sequences of antichains. In: Khomenko, V., Roux, O.H. (eds.) PETRI NETS 2018. LNCS, vol. 10877, pp. 251–271. Springer, Cham (2018). https://doi.org/10.1007/978-3-319-91268-4_13
14. Janicki, R., Kleijn, J., Koutny, M., Mikulski, L.: Step traces. Acta Inform. **53**, 35–65 (2016)
15. Janicki, R., Koutny, M.: Structure of concurrency. Theor. Comput. Sci. **112**, 5–52 (1993)

16. Janicki, R., Koutny, M.: Semantics of inhibitor nets. Inf. Comput. **123**(1), 1–16 (1995)
17. Janicki, R., Koutny, M.: Operational semantics, interval orders and sequences of antichains. Fundamenta Informaticae (2019, to appear)
18. Janicki, R., Yin, X.: Modeling concurrency with interval orders. Inf. Comput. **253**, 78–108 (2017)
19. Janicki, R., Yin, X., Zubkova, N.: Modeling interval order structures with partially commutative monoids. In: Koutny, M., Ulidowski, I. (eds.) CONCUR 2012. LNCS, vol. 7454, pp. 425–439. Springer, Heidelberg (2012). https://doi.org/10.1007/978-3-642-32940-1_30
20. Kleijn, J., Koutny, M.: Process semantics of general inhibitor nets. Inf. Comput. **190**, 18–69 (2004)
21. Kleijn, J., Koutny, M.: Formal languages and concurrent behaviour. Stud. Comput. Intell. **113**, 125–182 (2008)
22. Mazurkiewicz, A.: Introduction to Trace Theory. In: Diekert V., Rozenberg, G. (eds.) The Book of Traces, pp. 3–42. World Scientific, Singapore (1995)
23. Milner, R.: Communication and Concurrency. Prentice-Hall, Upper Saddle River (1989)
24. Montanari, U., Rossi, F.: Contextual nets. Acta Informatica **32**(6), 545–596 (1995)
25. Murata, T.: Petri nets: properties, analysis and applications. Proc. IEEE **77**(4), 541–579 (1989)
26. Nielsen, M., Rozenberg, G., Thiagarajan, P.S.: Behavioural notions for elementary net systems. Distrib. Comput. **4**, 45–57 (1990)
27. Petri, C.A.: Nets, time and space. Theor. Comput. Sci. **153**, 3–48 (1996)
28. Popova-Zeugmann, L., Pelz, E.: Algebraical characterisation of interval-timed Petri nets with discrete delays. Fundamenta Informaticae **120**(3–4), 341–357 (2012)
29. Rozenberg, G., Engelfriet, J.: Elementary net systems. In: Reisig, W., Rozenberg, G. (eds.) ACPN 1996. LNCS, vol. 1491, pp. 12–121. Springer, Heidelberg (1998). https://doi.org/10.1007/3-540-65306-6_14
30. van Glabbeek, R., Vaandrager, F.: Petri net models for algebraic theories of concurrency. In: de Bakker, J.W., Nijman, A.J., Treleaven, P.C. (eds.) PARLE 1987. LNCS, vol. 259, pp. 224–242. Springer, Heidelberg (1987). https://doi.org/10.1007/3-540-17945-3_13
31. Vogler, W.: Partial order semantics and read arcs. Theor. Comput. Sci. **286**(1), 33–63 (2002)
32. Wiener, N.: A contribution to the theory of relative position. Proc. Camb. Philos. Soc. **17**, 441–449 (1914)
33. Zuberek, W.M.: Timed Petri nets and preliminary performance evaluation. In: Proceedings of the 7-th Annual Symposium on Computer Architecture, La Baule, France, pp. 89–96 (1980)

Reversing Unbounded Petri Nets

Łukasz Mikulski[1] and Ivan Lanese[2(✉)]

[1] Folco Team, Nicolaus Copernicus University, Torun, Poland
lukasz.mikulski@mat.umk.pl
[2] Focus Team, University of Bologna/INRIA, Bologna, Italy
ivan.lanese@gmail.com

Abstract. In Petri nets, computation is performed by executing transitions. An effect-reverse of a given transition b is a transition that, when executed, undoes the effect of b. A transition b is reversible if it is possible to add enough effect-reverses of b so to always being able to undo its effect, without changing the set of reachable markings.

This paper studies the *transition reversibility problem*: in a given Petri net, is a given transition b reversible? We show that, contrarily to what happens for the subclass of bounded Petri nets, the transition reversibility problem is in general *undecidable*. We show, however, that the same problem is *decidable in relevant subclasses* beyond bounded Petri nets, notably including all Petri nets which are cyclic, that is where the initial marking is reachable from any reachable marking. We finally show that some non-reversible Petri nets can be restructured, in particular by adding new places, so to make them reversible, while preserving their behaviour.

Keywords: Petri nets · Reverse transition · Reversibility

1 Introduction

Reversible computation, a computational paradigm where any action can be undone, is attracting interest due to its applications in fields as different as low-power computing [24], simulation [9], robotics [27] and debugging [29]. Reversible computation has been explored in different settings, including digital circuits [32], programming languages [26,33], process calculi [12], and Turing machines [5]. In this paper, we focus on reversible computation in the setting of Petri nets.

In the early years of investigation of Petri nets (the seventies), a notion of *local reversibility* [8,19], requiring each transition to have a *reverse transition*, was used. The reverse b^- of a transition b undoes the effect of b in any marking (i.e., in any state) reachable by executing b. That is, after having executed b

This work has been partially supported by COST Action IC1405 on Reversible Computation - extending horizons of computing. The second author has been also partially supported by French ANR project DCore ANR-18-CE25-0007.

S. Donatelli and S. Haar (Eds.): PETRI NETS 2019, LNCS 11522, pp. 213–233, 2019.
https://doi.org/10.1007/978-3-030-21571-2_13

from some marking M it is always possible to execute b^-, and this leads back to marking M. Local reversibility is close to the definition currently used in, e.g., process calculi [12], programming languages [26,33] and Turing machines [5]. As time passed, a notion of *global reversibility* [15], requiring the initial state to be reachable by any reachable state of the net, attracted more interest because of its applications in controllability enforcing [23,31] and reachability testing [28]. The notion of local reversibility occurred for some time under the name of symmetric Petri nets [15], however, this name later changed the meaning to denote other forms of state space symmetry [10,21]. In this paper, with reversibility we mean a form of local reversibility (detailed below), and following [7] we write *cyclicity* to refer to the other notion. In order to relate reversibility and cyclicity, one would like to understand whether a Petri net can be made cyclic by adding reverse transitions or, more generally, whether reverse transitions can be added.

This problem was first tackled in [2] for the restricted class of bounded Petri nets. They provided the main insight below: the main issue is that adding reverse transitions *must not change the behaviour of the net*, which was defined as the set of reachable markings. However, this happens if the reverse of a transition b can trigger also in a marking not reachable by executing b as last transition, hence reverse transitions cannot always be added. However, in bounded Petri nets one can always add a *complete set of effect-reverses*. An effect-reverse of a transition b is a transition that, when executed, has the same effect as the reverse of b. However, an effect-reverse may not be enabled in all the markings the reverse is, hence adding one effect-reverse is in general not enough. A set of effect-reverses able to reverse a given transition b in all the markings where the reverse b^- can do it is called complete.

Hence, following [2], we define the *transition reversibility problem* as follows: in a given Petri net, can we add a complete set of effect-reverses for a given transition b without changing the set of reachable markings? We say a net is reversible if the answer to the transition reversibility problem is positive for each transition. The approach in [2] cannot be easily generalised to cope with unbounded Petri nets. The problem is hard in the unbounded case (indeed, we will show it to be undecidable) since adding even a single reverse (or effect-reverse) transition can have a great, and not easily characterisable, impact on the net. Indeed, the problem MESTR (Marking Equality with Single Transition Reverse) of establishing whether adding a single reverse (or effect-reverse) transition to a given net changes the set of reachable markings is, in general, undecidable [3]. One can, however, try to add effect-reverses, hoping to get a complete set of them without needing the ones for which the MESTR problem is undecidable.

In this paper, we tackle the transition reversibility problem in nets which are not necessarily bounded. We propose an approach based on identifying pairs of markings which forbid to add the effect-reverses of a specific transition b (Sect. 3). We call such pairs *b-problematic*. We show that a transition b is reversible iff there is no b-problematic pair (Corollary 1). We then study relevant properties of b-problematic pairs, including decidability and complexity issues (Sect. 4). In particular, we show that the existence of b-problematic pairs is *undecidable*

(Theorem 3), which is surprising since for a given transition b the set of minimal b-problematic pairs is finite (Proposition 2), and checking whether a given pair of markings is b-problematic is decidable (Corollary 2).

Given that the problem is undecidable, we identify relevant subclasses of Petri nets where the problem becomes decidable (Sect. 5). We show, in particular, that cyclicity implies reversibility (Corollary 3), which in our opinion provides a novel link between the two notions.

In order to have more reversible nets, we study whether a net can be restructured so to make it reversible while preserving its behaviour, in the sense described below (Sect. 6). First, we show that some nets, but not all, can be made reversible by extending them with new places, while preserving the requirements and effect of transitions on existing ones. Second, we consider whether the reversed behaviour of a net (obtained by considering the net as a labelled transition system and then adding reverse transitions) could be obtained as a behaviour of any Petri net, possibly completely different from the starting one. Surprisingly, this is possible only for Petri nets that can be made reversible by just extending them with new places (Theorem 5).

2   Background

In this section we introduce the notions needed for our developments. While these are largely standard, we mainly follow the presentation from [2,4].

The Group $\mathbb{Z}^X$ and the Monoid $\mathbb{N}^X$

The set of all integers is denoted by $\mathbb{Z}$, while the set of non-negative integers by $\mathbb{N}$. Given a set X, the cardinality (number of elements) of X is denoted by $|X|$, the powerset (set of all subsets) by 2^X. Multisets over X are members of $\mathbb{N}^X$, i.e., functions from X into $\mathbb{N}$. We extend the notion for all integers in an intuitive way obtaining mappings from X into $\mathbb{Z}$. If the set X is finite, mappings from X into $\mathbb{Z}$ (as well as $\mathbb{N}$) will be represented by vectors of $\mathbb{Z}^{|X|}$, written as $[x_1, \ldots, x_{|X|}]$ (assuming a fixed ordering of the set X). Given a function f we represent its domain restriction to a set X (subset of its domain) as $f\!\downarrow_X$.

The group $\mathbb{Z}^X$, for a set X, is the set of mappings from X into $\mathbb{Z}$ with componentwise addition $+$ (note that $\mathbb{N}^X$ with $+$ is a monoid). If $Y, Z \in \mathbb{Z}^X$ then $(Y + Z)(x) = Y(x) + Z(x)$ for every $x \in X$, while for $A, B \subseteq \mathbb{Z}^X$ we have $A + B = \{Y + Z \mid Y \in A \wedge Z \in B\}$. We define subtraction, denoted by $-$, analogously. The star operation is defined as $Y^* = \bigcup\{Y_i \mid i \in \mathbb{N}\}$, where Y_0 is a constant function equal 0 for every argument, denoted by $\boldsymbol{0}$, and $Y_{i+1} = Y_i + Y$. Rational subsets of $\mathbb{N}^X$ are subsets built from atoms (single elements of $\mathbb{N}^X$) with the use of finitely many operations of union $\cup$, addition $+$ and star *. The partial order $\leq$ (both on mappings and tuples) is understood componentwise, while $<$ means $\leq$ and $\neq$. Given $A \subseteq \mathbb{Z}^X$, $\min(A)$ is the set of minimal elements in A. For tuples over X we define $first : X^n \rightarrow X$ by $first((x_1, x_2, \ldots, x_n)) = x_1$.

Transition Systems

A *labelled transition system* (or, simply, *lts*) is a tuple $TS = (S, T, \rightarrow, s_0)$ with a set of *states* S, a finite set of *labels* T, a set of *arcs* $\rightarrow \subseteq (S \times T \times S)$, and an *initial state* $s_0 \in S$. We draw an lts as a graph with states as nodes, and labelled edges defined by arcs. A label a is *fireable* at $s \in S$, denoted by $s[a\rangle$, if $(s, a, s') \in \to$, for some $s' \in S$. A state s' is *reachable from* s through the execution of a sequence of transitions $\sigma \in T^*$, written $s[\sigma\rangle s'$, if there is a directed path from s to s' whose arcs are labelled consecutively by σ. A state s' is *reachable* if it is reachable from the initial state s_0. The set of states reachable from s is denoted by $[s\rangle$. A state s' is reachable by a (also said a-reachable) if it is reachable via a sequence of transitions having a as last element. A sequence $\sigma \in T^*$ is *fireable*, from a state s, denoted by $s[\sigma\rangle$, if there is some state s' such that $s[\sigma\rangle s'$. A labelled transition system $TS = (S, T, \to, s_0)$ is called *finite* if the set S is finite.

We assume that for each $a \in T$, the set of arcs labelled by a is nonempty.

Let $TS_1 = (S_1, T, \to_1, s_{0_1})$ and $TS_2 = (S_2, T, \to_2, s_{0_2})$ be ltss. A total function $\zeta \colon S_1 \to S_2$ is a *homomorphism* from TS_1 to TS_2 if $\zeta(s_{0_1}) = s_{0_2}$ and $(s, a, s') \in \to_1 \Leftrightarrow (\zeta(s), a, \zeta(s')) \in \to_2$, for all $s, s' \in S_1$, $a \in T$. TS_1 and TS_2 are *isomorphic* if ζ is a bijection.

Petri Nets

A *Place/Transition Petri net* (or, simply, *net*) is a tuple $N = (P, T, F, M_0)$, where P is a finite set of *places*, T is a finite set of *transitions*, F is the *flow function* $F \colon ((P \times T) \cup (T \times P)) \to \mathbb{N}$ specifying the arc weights, and M_0 is the *initial marking*. Markings are mappings $M \colon P \to \mathbb{N}$.

Petri nets admit a natural graphical representation (see, e.g., net N_1 in Fig. 1 of Sect. 3). Nodes represent places and transitions, arcs represent the weight function (we drop the weight if it is 1). Places are indicated by circles, and transitions by boxes. Markings are depicted by tokens inside places.

The *effect* of a transition a on a place p is $\mathit{eff}_p(a) = F(a, p) - F(p, a)$. The *(total) effect* of transition $a \in T$ is a mapping $\mathit{eff}(a) \colon P \to \mathbb{Z}$, where $\mathit{eff}(a)(p) = \mathit{eff}_p(a)$ for every $p \in P$. For a transition $a \in T$ we define two mappings: en_a, called *entries*, and ex_a, called *exits*, as follows: $en_a, ex_a \colon P \to \mathbb{N}$ and $en_a(p) = F(p, a)$ as well as $ex_a(p) = F(a, p)$ for every $p \in P$. A transition $a \in T$ is *enabled* at a marking M, denoted by $M[a\rangle$, if $M \geq en_a$. The firing of a at marking M leads to M', denoted by $M[a\rangle M'$, if $M[a\rangle$ and $M' = M + \mathit{eff}(a)$ (note that there is no upper limit to the number of tokens that a place can hold). The notions of enabledness and firing, $M[\sigma\rangle$ and $M[\sigma\rangle M'$, are extended in the usual way to sequences $\sigma \in T^*$, and $[M\rangle$ denotes the set of all markings reachable from M. We assume that each transition is enabled in at least one reachable marking. It is easy to observe that transition enabledness is *monotonic*: if a transition a is enabled at marking M and $M \leq M'$, then a is also enabled at M'.

Note that markings as well as entries and exits of a transition are multisets over P (mappings from P to $\mathbb{N}$), while total effect is a mapping from P to $\mathbb{Z}$, hence we represent all of them as vectors (after fixing an order on P).

The *reachability graph* of a Petri net $N = (P, T, F, M_0)$ is defined as the lts

$$RG(N) = ([M_0\rangle, T, \{(M, a, M') \mid M, M' \in [M_0\rangle \wedge M[a\rangle M'\}, M_0).$$

Intuitively, the reachability graph has reachable markings as states and firings as arcs. If a labelled transition system TS is isomorphic to the reachability graph of a Petri net N, then we say that N *solves* TS, and TS is *synthesisable* to N. A Petri net $N = (P, T, F, M_0)$ is *bounded* if $[M_0\rangle$ is finite (hence its reachability graph is a finite lts), otherwise the net is *unbounded*. A place $p \in P$ is bounded if there exists $n_p \in \mathbb{N}$ such that $M(p) < n_p$ for every $M \in [M_0\rangle$, otherwise the place is *unbounded*. The set of all bounded places of the net N is denoted by *bound*(N). Note that every place of a bounded net is bounded, while in each unbounded net there exists at least one unbounded place.

We now define reverses of transitions, and effect-reverses of transitions.

Definition 1 (transition reverse and effect-reverse). *The* reverse *of a transition $a \in T$ in a net $N = (P, T, F, M_0)$ is the transition a^- such that for each $p \in P$ we have $F(p, a^-) = F(a, p)$ and $F(a^-, p) = F(p, a)$. An* effect-reverse *of a transition $a \in T$ is any transition a^{-e} such that $eff(a^{-e}) = -eff(a)$.*

A minimum effect-reverse *(that is, an effect-reverse without self-loops) of a transition $a \in T$ is a transition a^{-e} such that for each $p \in P$ we have $en_{a^{-e}}(p) = -eff_p(a^{-e})$ and $ex_{a^{-e}}(p) = 0$ if $eff_p(a^{-e}) \leq 0$, and $en_{a^{-e}}(p) = 0$ and $ex_{a^{-e}}(p) = eff_p(a^{-e})$ otherwise.*

Notably, the reverse of a transition is also an effect-reverse, but not every effect-reverse is a reverse. Furthermore, a reverse of a transition a is able to reverse the transition a in any marking reachable by a, while an effect-reverse may do it only in some of these markings. A set of effect-reverses for transition a is *complete* if it includes enough effect-reverses to reverse a at any marking reachable by a.

We now define the notions of reversibility and cyclicity we are interested in.

Definition 2 (reversibility). *A transition b is reversible in a net N if it is possible to add to N a complete set of effect-reverses of b without changing the set of reachable markings. A net N is reversible if all its transitions are reversible.*

Definition 3 (cyclicity). *Let $N = (P, T, F, M_0)$ be a Petri net. A marking M reachable in N is called a home state if it is reachable from any other marking reachable in N. A net N is cyclic if M_0 is a home state.*

3   Problematic Pairs

In this paper, we tackle the transition reversibility problem, that is we want to decide whether, in a given Petri net N (possibly unbounded), we can add a complete set of effect-reverses of a given transition b without changing the set of reachable markings. In other words, if N is the original net and N' the one obtained by adding the complete set of effect-reverses, then their reachability graphs $RG(N)$ and $RG(N')$ differ only for the presence of reverse transitions in $RG(N')$. In particular, no new markings are reachable in N', hence $RG(N)$ and $RG(N')$ have the same set of states.

We remark that the net obtained by adding complete sets of effect-reverses for each transition (without changing the set of reachable markings) is by construction reversible and also trivially cyclic, hence understanding the transition reversibility problem is the key for understanding also reversibility and cyclicity.

In this section we show that the transition reversibility problem for transition b is equivalent to deciding the absence of particular pairs of markings, that we call b-*problematic*, introduced below. This characterisation will help us in solving the transition reversibility problem for relevant classes of Petri nets.

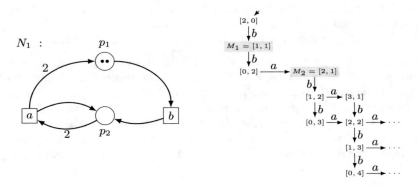

Fig. 1. A Petri net N_1 and part of its reachability graph (transition system).

Let us start by considering the sample net N_1 in Fig. 1. Net N_1 has two unbounded places, hence the theory from [2], valid for bounded nets, does not apply. Furthermore, adding a complete set of effect-reverses for transition b in net N_1 changes the set of reachable markings. Intuitively, the reason is that the net contains two markings M_1 and M_2 such that in M_1 at least one effect-reverse should trigger, in M_2 it must not (since M_2 is not reachable by b), but M_2 is greater than M_1. Hence, by monotonicity, we have a contradiction.

We now introduce the notion of b-problematic pair to formalise the intuition.

Definition 4 (b-problematic pair). *Let $N = (P, T, F, M_0)$ be a net, and $b \in T$ a transition. A pair (M_1, M_2) of markings reachable in N is b-problematic if $M_1 < M_2$, $\exists_{\sigma \in T^*} M_0[\sigma b\rangle M_1$ and $\forall_{\rho \in T^*} \neg M_0[\rho b\rangle M_2$.*[1]

The set of all b-problematic pairs in N is denoted by $\mathbb{P}_b(N)$.

We say a pair is problematic if it is b-problematic for some transition b.

Pair $([1, 1], [2, 1])$ is b-problematic in net N_1 in Fig. 1. Intuitively, an effect-reverse of b should trigger in marking $[1, 1]$ (reachable by b) but not in marking $[2, 1]$ (not reachable by b), but monotonicity forbids this.

The notion of problematic pair has been also implicitly used by [7], to define the notion of complete Petri net and study the inverse of a Petri net. Completeness allows one to block reverse transitions in all markings that do not occur as second component of a problematic pair.

We can now prove that b-problematic pairs forbid to reverse transition b.

---

[1] Note that there can be two reasons for $\neg M_0[\rho b\rangle M_2$. Either ρb is not enabled at M_0 or $M_0[\rho b\rangle M$ but $M \neq M_2$.

Proposition 1. *Let $N = (P, T, F, M_0)$ be a Petri net and $b \in T$ a transition. If there exists in N a b-problematic pair, then b is not reversible in N.*

Proof. Let (M_1, M_2) be the b-problematic pair. A complete set of effect-reverses should include at least one effect-reverse b^{-e} triggering in M_1. By monotonicity b^{-e} triggers also in M_2, but, since M_2 is not reachable by b, adding b^{-e} changes the set of reachable markings. This proves the thesis. □

The result above is independent of the number of effect-reverses in the complete set, hence even considering an infinite set of effect-reverses would not help.

We have shown that the existence of b-problematic pairs implies that b is not reversible. We now prove the opposite implication, namely that absence of b-problematic pairs implies reversibility of b.

Theorem 1. *Let $N = (P, T, F, M_0)$ be a Petri net and $b \in T$ a transition. If no b-problematic pair exists in N, then b is reversible in N.*

Proof. Take a reachable marking M. If it is greater than a b-reachable marking then it is b-reachable (otherwise we would obtain a b-problematic pair). Take the set bR of b-reachable markings. Let $\min(bR)$ be the set of all minimal elements in bR. They are all incomparable, hence by Dickson's Lemma [14] the set $\min(bR)$ is finite. Let us consider the set of effect-reverses of b composed of the effect-reverses b^{-e} of b such that $en_{b^{-e}} \in \min(bR)$. Note that we have one such effect-reverse for each marking in $\min(bR)$. By monotonicity, we have at least one effect-reverse triggering at each marking in bR, and no effect-reverse triggering outside bR, hence this is a complete set of effect-reverses (that do not change the set of reachable markings) as desired. □

The result above is an unbounded version of the procedure presented in the proof of [2, Theorem 4.3] for bounded nets. By combining the two results above we get the following corollary, showing that b-problematic pairs provide an equivalent formulation of the transition reversibility problem.

Corollary 1. *Let $N = (P, T, F, M_0)$ be a Petri net and $b \in T$ a transition. The transition b is reversible in N iff no b-problematic pair exists in N.*

The formulation in terms of b-problematic pairs raises the following question:

– Is it decidable whether a b-problematic pair exists in a given net?

In Sect. 4 we answer negatively the question above, hence the transition reversibility problem is in general undecidable. This raises additional questions:

1. Can we decide whether a given pair of markings is b-problematic for a net?
2. Can we find relevant classes of nets where the transition reversibility problem is decidable?
3. Can we transform a net into a reversible one while preserving its behaviour?

We will answer the questions above in Sects. 4, 5 and 6, respectively.

4   Undecidability of the Existence of b-problematic Pairs

The main result of this section is the undecidability of the existence of b-problematic pairs and, as a consequence, of the transition reversibility problem.

Before proving our main result we show, however, that a given net has finitely many minimal b-problematic pairs, and that one can decide (indeed it is equivalent to the Reachability Problem) whether a given pair of markings is b-problematic. These results combined seem to hint at the decidability of the transition reversibility problem. However, this is not the case.

We start by proving that there are finitely many minimal b-problematic pairs.

Proposition 2. *Let $N = (P, T, F, M_0)$ be a net, and $b \in T$ a transition. There exist finitely many minimal b-problematic pairs in N.*

Proof. A b-problematic pair $([x_1, \ldots, x_n], [y_1, \ldots, y_m])$ can be seen as a tuple $(x_1, \ldots, x_n, y_1, \ldots, y_m)$, hence the result follows from Dickson's Lemma [14]. □

We now show by reduction to the Reachability Problem that one can decide whether a given pair of markings is b-problematic.

Lemma 1. *One can reduce the problem of checking whether a given pair of markings is b-problematic to the Reachability Problem.*

Proof. Let (M_1, M_2) be the given pair of markings. One has to check that they are reachable and that the marking $M_1 - \mathit{eff}(b)$ is reachable, and $M_2 - \mathit{eff}(b)$ is not, which are four instances of reachability. □

On the other hand, we can reduce the Reachability Problem to checking whether a given pair of markings is b-problematic.

Theorem 2. *One can reduce the Reachability Problem to the problem of checking whether a given pair of markings is b-problematic.*

Proof. The proof is by construction. The construction is depicted in Fig. 2. Let $N = (P, T, W, M_0)$ be a net and M a marking of this net. We will build a new net N' and a pair of markings (M_1, M_2) in it such that (M_1, M_2) is b-problematic in N' if and only if M is reachable in N.

We define net N' as $(P \cup \{q_1, q_2, q_3\}, T \cup \{b, c\}, W', M_0')$. We assume that q_1, q_2, q_3, b, c are fresh objects. For the vector representation of markings in N' we fix the order of places as follows: first the places from N (in the order they have in N), then places q_1, q_2, q_3 in order. We set $M_0' = [x_1, \ldots, x_n, 1, 0, 0]$ with $[x_1, \ldots, x_n] = M_0$. W' extends W as follows. Place q_1 is connected by a self-loop with every transition $a \in T$, and a preplace of the two new transitions b, c. Place q_2 is a postplace for both new transitions, while q_3 is a postplace for c only. Moreover, c and b take from any place $p \in P$ a number of tokens equal to, respectively, $M_0(p)$ and $M(p)$.

In the constructed net we can always reach marking $M_2 = [0, \ldots, 0, 1, 1]$ (by executing c in M_0') but never reach it by executing b (since only c adds a token to place q_3, but b and c are in conflict because of place q_1). We can also reach marking $M_1 = [0, \ldots, 0, 1, 0]$ by executing b iff M is reachable in N. Indeed, if b triggers in some marking $M_3 > M$ then some tokens are left in the places from N, and they are never consumed since b consumes the token in q_1, which is needed to perform any further transition. Thus, the pair (M_1, M_2) is b-problematic in N' if and only if M is reachable in N.

This proves that we can use a decision procedure checking whether a pair of markings is b-problematic in order to solve the Reachability Problem. $\qquad\square$

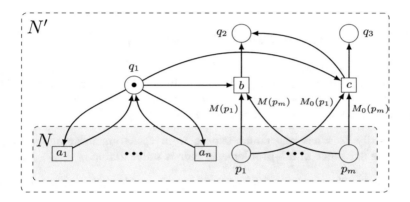

Fig. 2. Reducing reachability to checking whether a pair of markings is b-problematic.

We can combine the two results above to give a precise characterisation of the complexity of checking whether a given pair of markings is b-problematic.

Corollary 2. *The problem of deciding whether a given pair of markings is b-problematic and the Reachability Problem are equivalent.*

As a consequence, known bounds for the complexity of the Reachability Problem apply to the problem of deciding whether a given pair of reachable markings is b-problematic as well. For instance, we know that it is not elementary [11].

At this stage it looks like there should be a procedure to construct the finite set $\min(\mathbb{P}_b(N))$. Indeed, we can perform an additional step in this direction: we can compute a (finite) over-approximation $\min(\mathcal{E}_b(N) + \{\mathit{eff}(b)\})$ (we remind that $\mathcal{E}_b(N)$ is the set of markings of N where b is enabled) of the (finite) set $\min(\mathit{first}(\mathbb{P}_b(N)))$ of the minimals of the markings that may occur as first component in a b-problematic pair. Both the set $\min(\mathit{first}(\mathbb{P}_b(N)))$ and its over-approximation are finite, but deciding membership and browsing the over-approximation is easy, while even the emptiness problem is actually undecidable for the set $\min(\mathit{first}(\mathbb{P}_b(N)))$.

We first need a lemma on monotonicity properties of b-problematic pairs.

Lemma 2. *Let* $N = (P, T, F, M_0)$ *be a net,* $b \in T$ *a transition, and* (M_1, M_2) *a b-problematic pair.*

- *If* $M_3 < M_1$ *and* $\exists_{\sigma \in T^*} M_0[\sigma b\rangle M_3$ *then* (M_3, M_2) *is a b-problematic pair.*
- *If* $M_2 < M_4$, M_4 *is reachable and* $\forall_{\rho \in T^*} \neg M_0[\rho b\rangle M_4$ *then* (M_1, M_4) *is a b-problematic pair.*

Proof. Directly from Definition 4, using the transitivity of $<$.   □

We now prove the correctness of our over-approximation.

Lemma 3. *Let* $N = (P, T, F, M_0)$ *be a net,* $b \in T$ *a transition. Then:*

$$\min(\mathit{first}(\mathbb{P}_b(N))) \subseteq \min(\mathcal{E}_b(N) + \{\mathit{eff}(b)\})$$

Proof. Take a b-problematic pair (M_1, M_2). By definition M_1 is an element of $\mathcal{E}_b(N) + \{\mathit{eff}(b)\}$. Assume it is not minimal. Then there exists $M_3 \in \min(\mathcal{E}_b(N) + \{\mathit{eff}(b)\})$ such that $M_3 < M_1$ and so, by Lemma 2, (M_3, M_2) is also a b-problematic pair, hence $M_1 \notin \min(\mathit{first}(\mathbb{P}_b(N)))$.   □

Contrarily to what one could expect from these encouraging preliminary results, the existence of a b-problematic pair is undecidable.

Theorem 3. *The problem of the existence of a b-problematic pair is undecidable.*

Proof. Assume towards a contradiction that we can decide the existence of a b-problematic pair in a net N. Consider the decision problem MESTR from [3] – *Are the reachability sets of two given nets N and N', where N' is obtained from N by adding the single reverse b^- of a given transition b, equal?* We show below that we can reduce MESTR to checking the existence of a b-problematic pair.

Consider the following procedure. If N cannot be reversed by a set of effect-reverses then it cannot be reversed by a single reverse b^-. Hence, if there exists a b-problematic pair in N, then the answer for MESTR is negative.

Otherwise we construct the set of markings

$$X = (\mathbb{N}^P \setminus (\min(\mathcal{E}_b(N) + \mathit{eff}(b)) + \mathbb{N}^P)) \cap (\mathit{ex}_b + \mathbb{N}^P).$$

X consists of markings $\hat{M}$ (reachable or not) such that:

1. b^- triggers in $\hat{M}$;
2. $\hat{M}$ is not b-reachable;

Condition 1 holds thanks to $(\mathit{ex}_b + \mathbb{N}^P)$. Condition 2 holds since $\hat{M}$ is not in the set $(\min(\mathcal{E}_b(N) + \mathit{eff}(b)) + \mathbb{N}^P)$.

We show now that, if there exists no b-problematic pair in N, then X contains all reachable markings which are not b-reachable. Assume towards a contradiction that $M \notin X$ is reachable in N but not reachable by b. Then $M \notin \min(\mathcal{E}_b(N) + \mathit{eff}(b))$ (because it is not reachable by b) and there exists

$M' \in \min(\mathcal{E}_b(N) + \mathit{eff}(b))$ which is smaller than M (because otherwise $M \in X$). Hence (M', M) is a problematic pair which we excluded.

As a result, if there exists no b-problematic pair in N, it is enough to ask, whether there is any reachable marking in X. If one such marking exists then the answer for MESTR is positive, otherwise it is negative. It remains to show that (i) one can construct the set X and (ii) check whether it contains any marking reachable in N.

To prove part (i) we utilise the fact that $(\min(\mathcal{E}_b(N) + \mathit{eff}(b))$ is finite and can be easily computed using Dickson's lemma and the coverability set (see [16]) of N. Moreover, $\{ex_b\}$ is a singleton while $\mathbb{N}^P$ is the set of all multisets over P. Summing up, $\mathbb{N}^P$, $(\min(\mathcal{E}_b(N) + \mathit{eff}(b))$, and $ex_b + \mathbb{N}^P$ are rational subsets of $\mathbb{N}^P$ and one can provide rational expressions for them. Since, by [18], rational subsets of $\mathbb{N}^P$ form an effective Boolean algebra (see also [1] for more details), we can construct (giving a rational expression) the set X.

To prove part (ii) we use the results from [20], showing that the emptiness problem for intersection of the set of reachable markings and any rational set of markings given by a rational expression is decidable for Petri nets.

Summing up, if the existence of a b-problematic pair is decidable, then also MESTR is decidable, which is not true. Hence, the existence of a b-problematic pair is undecidable. $\square$

5   Decidable Subclasses

We have shown that in general the transition reversibility problem is undecidable, as well as the existence of a b-problematic pair. We already know from [2] that the problem becomes decidable in the class of bounded nets. We show below a few other classes of Petri nets (not necessarily bounded) where the problem is decidable. In particular, we show that the transition reversibility problem is decidable, and indeed complete sets of effect-reverses always exist, for cyclic nets.

Note that cyclicity does not ensure that a net actually has effect-reverses for all transitions. However, we show that complete sets of effect-reverses can always be added without changing the set of reachable markings in such nets. This result links cyclicity to reversibility. We prove a more general result, and then show the one above to be an instance.

Proposition 3. *Let $N = (P, T, F, M_0)$ be a net and $b \in T$ one of its transitions. If all the reachable markings enabling b are home states then there exists no b-problematic pair in N.*

Proof. The proof follows the schema in Fig. 3. Assume towards a contradiction that $M_1 < M_2$ is a b-problematic pair. Let M be the marking such that $M[b\rangle M_1$ (reachable since M_1 is reachable by b). By the assumption, M is a home state, hence there exists a path σ from M_1 to M. Applying σb in M_2 (possible by monotonicity) leads to M_2 back, with b as the last transition, hence (M_1, M_2) is not b-problematic against the hypothesis. $\square$

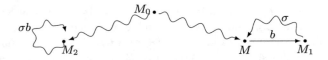

Fig. 3. Idea of the proof of Proposition 3.

Proposition 4. *Let N be a cyclic net and $b \in T$ one of its transitions. There exists no b-problematic pair in N.*

Proof. Directly from Proposition 3, since in a cyclic net every reachable marking is a home state.   □

Corollary 3. *Each cyclic net is reversible.*

The inverse implication does not hold, since one has to actually add effect-reverses to obtain a cyclic net. Moreover, an alternative proof of the result above can be based on the semi-linearity of reachability set for cyclic nets [7].

We present below another class of nets where reversibility is decidable.

Proposition 5. *Take a net N. If the set of reachable markings which are not home states is finite then one can decide whether N is reversible or not.*

Proof. First, we can construct the set of all markings which are not home states (finite by assumption) as follows. We do not discuss whether finiteness is decidable or not. By [13] one can verify whether a given marking is a home state. Moreover, every marking reachable from a home state is a home state. Hence, we can find all non home states as follows: explore all states starting from the initial one and stop exploring a branch every time you find a home state.

Consider now the construction in Theorem 1. We will decide whether an arbitrary effect-reverse (triggering in a marking M_1) of some transition b changes the set of reachable markings. This happens only if there exists a marking $M_2 > M_1$ which is not b-reachable. We will consider a few cases, distinguished by whether M_2 and M_1 are home states or not.

If both of them are home states then consider the net N' obtained by changing the initial marking to any home state. N' has precisely all home states of N (called a home space [6]), including M_1 and M_2, as reachability set and is cyclic. By Proposition 4 (M_1, M_2) is not b-problematic in N', hence M_2 must be b-reachable. Thus it is b-reachable also in N against the hypothesis, hence this case can never happen.

If M_2 is not a home state (M_1 may be home state or not), then thanks to Corollary 2 we can check all the combinations of a marking in $\min(\mathcal{E}_b(N) + \{eff(b)\})$ (which is a finite over-approximation of the possible first elements thanks to Lemma 3) with a marking which is not a home state (the corresponding set is finite and can be constructed as discussed at the beginning of the proof).

Assume M_2 is a home state, but M_1 is not. Consider the set of all minimal home states larger than M_1. This is finite from Dickson's Lemma [14]. We can construct it using the computed set of reachable markings which are not home states and the coverability set [16] which helps to decide whether there will be any reachable marking larger than the one considered. This way, we can browse a tree of all markings larger than M_1, and cut the branch whenever we find a home state or the considered marking is not coverable.

Suppose that M_2 is not in this set, then there is a home state M_3, which is smaller than M_2 but larger than M_1. If M_3 is b-reachable then (M_3, M_2) is a problematic pair, which is impossible, since both of them are home states. Otherwise, (M_1, M_3) is a problematic pair and the net is not reversible. □

6   Removing Problematic Pairs

We know from Proposition 1 that we cannot reverse a net N with problematic pairs. In this section we investigate whether a net N that is not reversible can be made reversible by modifying it but preserving its behaviour. First, we consider adding new places to the net, but preserving its behaviour. Second, we allow to completely change the structure of the net, again preserving its behaviour. The second case is a generalisation of the first one, but, surprisingly, they turn out to work well for the same set of nets.

We need to define what exactly "preserving its behaviour" means. We cannot require the reachability graphs to be the same, as done in previous sections, since the set of places changes. We could require them to be isomorphic, but it is enough to have a homomorphism from the restructured net to the original one. Indeed, this weaker condition is enough to prove our results.

If only new places are added, one may try to simply require transitions to have their usual behaviour on pre-existing places. This approach is safe, i.e. by restricting the attention to old places, no new marking appears, and no new transition between markings becomes enabled. Yet, there is the risk to disable transitions, and, as a consequence, to make unreachable some markings which were previously reachable.

We define below the notion of extension of a net N, which requires (1) that no new behaviours will appear, and (2) that previous behaviours are preserved.

Definition 5 (extension of a net). *Let $N = (P, T, F, M_0)$ be a net. A net $N' = (P \cup P', T, F', M_0')$ is an extension of N if $M_0' \downarrow_P = M_0$ and:*

1. *$F'(p, a) = F(p, a)$ and $F'(a, p) = F(a, p)$ for each $p \in P$ and $a \in T$;*
2. *for each reachable marking M in N and each reachable marking M' in N' such that $M' \downarrow_P = M$ and each transition b, b is enabled in M' iff it is enabled in M.*

Notably, from each extension N' of a net N there is a homomorphism from $RG(N')$ to $RG(N)$, as shown by the following lemma.

Lemma 4. *Let $N = (P, T, F, M_0)$ be a net and $N' = (P \cup P', T, F', M_0')$ an extension of N. Then there is a homomorphism from $RG(N')$ to $RG(N)$.*

Proof. The homomorphism is defined by function $\downarrow_P \colon RG(N') \to RG(N)$. The condition on the initial state is satisfied by construction. Let us consider the other condition.

Let us take two markings M_1', M_2' in $RG(N')$ and a transition t such that $M_1'[t\rangle M_2'$. Transition t is enabled in $M_1' \downarrow_P$ too, thanks to condition 2 in Definition 5, and $M_1' \downarrow_P [t\rangle M_2' \downarrow_P$ thanks to condition 1 as desired. By induction on the length of the shortest path from M_0' to M_1' we can show that $M_1' \downarrow_P$ is reachable in N, thus $M_1' \downarrow_P$ and $M_2' \downarrow_P$ are in $RG(N)$.

Let us now take two markings M_1, M_2 in $RG(N)$ and a transition t such that $M_1[t\rangle M_2$. Let M_1' be such that $M_1' \downarrow_P = M_1$. Transition t is enabled in M_1 too, thanks to condition 2 in Definition 5, and $M_1'[t\rangle M_2'$ for some M_2' such that $M_2' \downarrow_P = M_2$ thanks to condition 1 as desired. By induction on the length of the shortest path from M_0 to M_1 we can show that M_1' is reachable in N', thus M_1' and M_2' are in $RG(N')$. This completes the proof. $\qquad\square$

Checking condition 2 of the definition of extension above is equivalent to solving the problem of state space equality, which is in general undecidable [20]. However, it is decidable in many relevant cases. In particular, complementary nets, based on the idea presented in [30], are a special case of extension of a net. Intuitively, a complementary net has one more place p' for each bounded place p in the original net, with a number of tokens such that at each time the sum of tokens in p and p' is equal to the bound of tokens for p.

Definition 6 (complementary net). *Let $N = (P, T, F, M_0)$ be a net. The complementary net for N is a net $N' = (P \cup P', T, F', M_0')$ constructed by adding a complement place p' for every bounded place $p \in P$ obtaining $P' = \{p' \mid p \in bound(N)\}$. For all $M \in [M_0\rangle$ and $p \in bound(N)$, we define $\widehat{M} \in \mathbb{N}^{P \cup P'}$, such that $\widehat{M}(p) = M(p)$ and $\widehat{M}(p') = (\max_{M'' \in [M_0\rangle} M''(p)) - M(p)$, setting $M_0' = \widehat{M_0}$. We also set $F'(p', a) = max(F(a, p) - F(p, a), 0)$ and $F'(a, p') = max(F(p, a) - F(a, p), 0)$, for all $p' \in P'$ and $a \in T$. Furthermore, $F'(p, a) = F(p, a)$ and $F'(a, p) = F(a, p)$ for all $p \in P$ and $a \in T$.*

Complementary nets are a special case of extension of nets.

Lemma 5. *Let N be a net and N' the complementary net for N. Then N' is an extension of N.*

Proof. Structural conditions hold by construction, while the complementary places do not change the sets of transitions enabled in reachable markings. $\qquad\square$

Another example of extension just adds trap places which only collect tokens. Such a place can be used to compute the number of times a given transition fires.

Extending a net N to make it reversible is a very powerful technique. Indeed, we will show below that if there is any reversible net N' with a homomorphism from $RG(N')$ to $RG(N)$, then there is also a reversible extension of N.

Theorem 4. *Let $N = (P, T, F, M_0)$ be a net. If there is a reversible net $N' = (P', T', F', M_0')$ and a homomorphism ζ' from $RG(N')$ to $RG(N)$ then there is a reversible extension N'' of N.*

Proof. Note that $T = T'$. Let $N'' = (P \cup P', T, F \cup F', M'_0 \cup M_0)$ be the simple union of N and N' synchronised on the set of transitions. All the markings in $RG(N'')$ are of the form $M' \cup M$ with $\zeta'(M') = M$. The proof is a simple induction on the length of the shorter derivation from $M'_0 \cup M_0$ to $M' \cup M$.

We show below that N'' is an extension of N. The condition 1 holds by the construction. Let us check condition 2. A transition t enabled in a marking $M' \cup M$ of N'' is also trivially enabled in marking M of N. Viceversa, a transition t enabled in marking M of N is also enabled in any marking $M' \cup M$ of N''. Indeed, since there is a homomorphism from $RG(N')$ to $RG(N)$, t is enabled in M'. Being enabled in both M' and M it is enabled also in $M' \cup M$.

We now show that $RG(N'')$ and $RG(N')$ are isomorphic. We define the isomorphism as $\zeta''(M' \cup M) = M'$. Existence of corresponding transitions follows as above. We also note that ζ'' is bijective since $M = \zeta'(M')$, hence $RG(N'')$ and $RG(N')$ are isomorphic.

We now have to show that N'' is reversible. Assume towards a contradiction that it is not, hence from Corollary 1 it has a problematic pair $(M'_1 \cup M_1, M'_2 \cup M_2)$. $M'_1 \cup M_1 < M'_2 \cup M_2$ implies $M'_1 < M'_2$, and from the isomorphism M'_1 is reachable by b while M'_2 is not, hence (M'_1, M'_2) is problematic in N', against the hypothesis that M' is reversible. This concludes the proof. $\square$

The previous result shows that if there is a reversible net with the same behaviour as a given net, then the given net also has a reversible extension.

We can also instantiate the result above in terms of the solvability of the reversed transition system of a given net.

Definition 7 (Reversed transition system). *The reversed transition system of net N is obtained by taking the reachability graph $RG(N)$ of net N and by adding for each arc (M_1, a, M_2) in $RG(N)$ a new arc (M_2, a^-, M_1).*

If the reversed transition system is synthesisable, then it can also be solved by an extension of the original net.

Theorem 5. *Let $N = (P, T, F, M_0)$ be a net and TS its reversed transition system. If TS is synthesisable, then N has a reversible extension.*

Proof. Let $N' = (P', T', F', M'_0)$ be a solution of TS. By definition $RG(N')$ without reverse transitions and $RG(N)$ are isomorphic. Using the technique in the proof of Theorem 4 we can show that the simple union of N' without reverse transitions and N synchronised on transitions is an extension of N. Furthermore, one can add reverse transitions by synchronising the reverse in N with the reverse in N', and this does not change the set of reachable markings. Indeed, an additional marking would be reachable in N' too, against the hypothesis. $\square$

Hence the questions "Is the reversed transition system of a net N synthesisable?" and "Can we find a reversible extension of a net N?" are equivalent. From now on we will concentrate on the second formulation.

Thanks to Corollary 1, this second formulation is also equivalent to "Can we find an extension N' of a net N such that, for each transition b, net N' has no b-problematic pair?". Naturally, the answer to this question depends on the structure of the set of b-problematic pairs in N.

We discuss below the answer to this question in various classes of nets. For instance, we can give a positive answer to the question if for each problematic pair (M_1, M_2) in N there is at least a bounded place p such that the number of tokens in p in M_1 and in M_2 is different. Indeed, such pairs are no more b-problematic in the complementary net, which is thus reversible.

However, this is not the case for all nets. In particular, there is no reversible extension of the Petri Net in N_1 in Fig. 1, as shown by the result below.

Example 1. In order to show that there is no reversible extension N' of the net N_1 in Fig. 1 we show that any extension of N_1 has at least one b-problematic pair. We show, in particular, that the pair of markings $([1,1], [2,1])$ remains b-problematic in any extension. Assume this is not the case. Then one of the new places, let us call it p, should have less tokens in M_2 than in M_1. In particular, this should happen if we go to M_1 via b and to M_2 via bba. Hence, the effect of ba on p should be negative. This is not possible, since we have an infinite path $(ba)^*$, which would be disabled against the hypothesis.

As a consequence of Theorem 5, the reversed transition system of N_1 is not synthesisable. We can generalise the example above as follows:

Lemma 6. *If a net N has a b-problematic pair (M_1, M_2) such that M_2 is reachable from M_1 by σ, and there is a marking $M \in [M_0\rangle$ where σ^ω is enabled then there is no reversible extension N' of N.*

Unfortunately, the above result together with the construction of complementary nets do not cover the whole spectrum of possible behaviours. We give below two examples of nets that do not satisfy the premises of Lemma 6 and where b-problematic pairs cannot be removed using complementary nets. However, in Example 2 there is an extension where b can be reversed, while this is not the case in Example 3.

Example 2. Consider the net N_2 in Fig. 4. There is only one minimal b-problematic pair, composed of two markings $M_1 = [0,1,0]$ and $M_2 = [1,2,0]$ (emphasised in grey). This pair does not satisfy the premises of Lemma 6. Moreover, all places in this net are unbounded and we cannot use the complementary net construction. However, one can consider adding a forth place p_b which is a postplace of transition b (with weight 1). After that, since $M_1(p_b) = 1$ and $M_2(p_b) = 0$, the two considered markings no longer form a b-problematic pair. Note that, however, markings $[1,2,0]$ and $[2,3,0]$ form an a-problematic pair that satisfies the premises of Lemma 6 and hence the whole net cannot be reversed.

Example 3. Consider net N_3 in Fig. 5. All places in N_3 are unbounded, and the computation $c(caba)^\omega$ is enabled in N_3, hence additional places that count the number of executions of each transition are unbounded as well. Therefore we cannot use the idea described in Example 2, nor the complementary net.

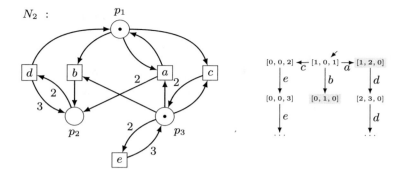

Fig. 4. Net N_2 and part of its reachability graph.

We show now that Lemma 6 cannot be used either. Suppose that there is a b-problematic pair (M_1, M_2) with M_2 reachable from M_1. This means that the computation $M_0[\sigma\rangle M_x[b\rangle M_1[\rho\rangle M_2$ is enabled in N_3. Note that for every reachable marking M: (i) if $M[bc\rangle$ then $M[cb\rangle$ and (ii) if $M(p_3) > 1$ then $M[ba\rangle$ implies $M[ab\rangle$, (iii) if $M(p_1) > 3$ then $M[bac\rangle$ implies $M[cab\rangle$. Moreover if $M(p_3) \leq 1$ then a is not enabled at M while if $M(p_1) \leq 2$ then ac is not enabled at M. Let $\rho'b\sigma'$ be such that $M_x[\rho'\rangle M_z[b\rangle M_y[\sigma'\rangle M_2$ and σ' is the shortest possible. By rearrangement (i) described above, σ' starts with a (otherwise we can move b forward and σ' is not the shortest possible). In order for a to be enabled we need $M_y(p_3) \geq 2$, but if $M_y(p_3) > 2$ we could swap a and b against the hypothesis that σ' is minimal. Hence, the only possibility is $M_y(p_3) = 2$. Thus aa is not enabled. Hence, σ' is a or it starts with ac. Let us consider the second case. Since ac is enabled we have $M_y(p_1) > 2$. By rearrangement (iii) we have $M_z(p_1) \leq 3$ which implies $M_y(p_1) \leq 1$, otherwise σ' would not be minimal, hence this case can never happen. Thus, $\sigma' = a$, $M_y(p_3) = 2$, $M_2(p_3) = 1$. As a consequence, $M_x(p_3) < M_1(p_3) \leq M_2(p_3) = 1$ (as M_1 and M_2 form a problematic pair) and the only possibility is $M_x(p_3) = 0$. The only reachable marking with no tokens in p_3 is the initial marking $[4, 0, 0]$. Thus, $M_1 = [2, 0, 1]$ which is a contradiction, since the only marking reachable from $[2, 0, 1]$ is b-reachable. Hence, Lemma 6 cannot be used.

Fortunately, we can reuse the reasoning from Example 1. We have to show that the pair of states marked in grey remains b-problematic for every extension of N_3. Assume that there exists an extension of N_3 for which markings M_1' and M_2' corresponding to M_1 and M_2 do not form a b-problematic pair. This means that there is a new place p such that $M_1'(p) > M_2'(p)$. In particular, $\mathit{eff}_p(b) > \mathit{eff}_p(c) + \mathit{eff}_p(b) + \mathit{eff}_p(a)$. Hence the effect of ca is negative. This is not possible since we have an infinite path $c(ca)^*$, which would be disabled, against Definition 5. Hence no reversible extension exists.

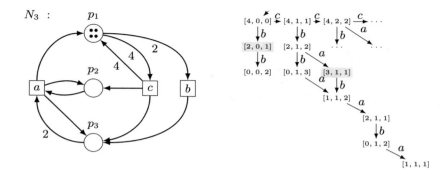

Fig. 5. Unsolvable net N_3 and part of its reachability graph.

7   Conclusions

In the paper we presented an approach to equip a possibly unbounded Petri net with a set of effect-reverses for each transition without changing the set of reachable markings. We have shown that, contrarily to the bounded case, this is not always possible. We introduced the notion of b-problematic pair of markings, which makes the analysis of the net easier.

We have shown, in particular, that a net can be reversed iff it has no b-problematic pairs. Furthermore, we have shown that sometimes b-problematic pairs can be removed by extending the net, and that, if the labelled transition system of the reverse net is synthesisable, then this can always be done.

However, our techniques cannot cover the whole class of Petri nets, since the undecidability of the existence of at least one b-problematic pair remains. This result might surprise, since there exist only finitely many minimal b-problematic pairs, and one can easily compute a finite over-approximation of the set containing all the first components of such minimal b-problematic pairs.

The particular case above shows that Dickson's Lemma guarantees finiteness of a set of minimals, but not the decidability of its emptiness. In order to use Dickson's Lemma constructively to compute all the elements in this finite set, we need a procedure deciding whether there exists any element larger than a given one. But this is just another formulation of the emptiness problem we want to solve using Dickson's Lemma constructively. In our opinion, this is the main reason of the counter-intuitiveness of some facts presented in this paper.

As future work one can try to reduce the gap between the sets of nets for which we can and we cannot add effect-reverses. Also, exploiting results of [22] on the undecidability of reachability sets for nets with 5 unbounded places, one may try bound the number of unbounded places needed to prove undecidability of the transition reversibility problem.

Furthermore, the relation between the results above and reversibility in other models should be explored. As already mentioned, reversibility is the notion normally used in process calculi [12], programming languages [26,33] and Turing machines [5]. The Janus approach [33] obtains reversibility without using history information, as we do in Sect. 5. This approach requires a carefully crafted language, e.g., assignments, conditional and loops in Janus are nonstandard. Ensuring reversibility in existing models (from Turing machines [5] and CCS [12] to Erlang [26]) normally requires history information, and indeed in Sect. 6 additional places are used to keep such history information. However, while most of the approaches use dedicated constructs to store history information, here we add history information within the model, and this explains why this is not always possible. This is always possible in Turing machines [5], which are however sequential, while Petri nets are concurrent, and more expressive than Petri nets. The only result in the concurrency literature we are aware of showing that history information can be coded inside the model is the mapping of reversible higher-order pi-calculus into higher-order pi-calculus in [25], which however completely changes the structure of the system, while here we only add new places preserving the original backbone of the system. Indeed, our result is close to [17] where reversibility for distributed Erlang programs is obtained via monitoring, since both approaches feature a distributed state and a minimal interference with the original system. Yet the approach in [17] requires a known and well-behaved communication structure ensured by choreographies. Summarising, the results in this paper can help in answering general questions about reversibility, such as "Which kinds of systems can be reversed without history information?" and "Which kinds of systems can be reversed using only history information modelled inside the original language and preserving the structure of the system?"

References

1. Badouel, E., Bernardinello, L., Darondeau, P.: Petri Net Synthesis. Springer, Heidelberg (2015). https://doi.org/10.1007/978-3-662-47967-4
2. Barylska, K., Erofeev, E., Koutny, M., Mikulski, Ł., Piątkowski, M.: Reversing transitions in bounded Petri nets. Fundamenta Informaticae **157**(4), 341–357 (2018)
3. Barylska, K., Koutny, M., Mikulski, Ł., Piątkowski, M.: Reversible computation vs. reversibility in Petri nets. Sci. Comput. Program. **151**, 48–60 (2018)
4. Barylska, K., Mikulski, Ł.: On decidability of persistence notions. In: 24th Workshop on Concurrency, Specification and Programming, pp. 44–56 (2015)
5. Bennett, C.H.: Logical reversibility of computation. IBM J. Res. Dev. **17**(6), 525–532 (1973)
6. Best, E., Esparza, J.: Existence of home states in Petri nets is decidable. Inf. Process. Lett. **116**(6), 423–427 (2016)
7. Bouziane, Z., Finkel, A.: Cyclic Petri net reachability sets are semi-linear effectively constructible. In: Infinity, ENTCS, pp. 15–24. Elsevier (1997)
8. Cardoza, E., Lipton, R., Meyer, A.: Exponential space complete problems for Petri nets and commutative semigroups (preliminary report). In: Proceedings of STOC 1976, pp. 50–54. ACM (1976)

9. Carothers, C.D., Perumalla, K.S., Fujimoto, R.: Efficient optimistic parallel simulations using reverse computation. ACM TOMACS **9**(3), 224–253 (1999)
10. Colange, M., Baarir, S., Kordon, F., Thierry-Mieg, Y.: Crocodile: a symbolic/symbolic tool for the analysis of symmetric nets with bag. In: Kristensen, L.M., Petrucci, L. (eds.) PETRI NETS 2011. LNCS, vol. 6709, pp. 338–347. Springer, Heidelberg (2011). https://doi.org/10.1007/978-3-642-21834-7_20
11. Czerwiński, W., Lasota, S., Lazic, R., Leroux, J., Mazowiecki, F.: The reachability problem for Petri nets is not elementary. arXiv preprint arXiv:1809.07115 (2018)
12. Danos, V., Krivine, J.: Reversible communicating systems. In: Gardner, P., Yoshida, N. (eds.) CONCUR 2004. LNCS, vol. 3170, pp. 292–307. Springer, Heidelberg (2004). https://doi.org/10.1007/978-3-540-28644-8_19
13. de Frutos Escrig, D., Johnen, C.: Decidability of home space property. Université de Paris-Sud. Centre d'Orsay. LRI (1989)
14. Dickson, L.E.: Finiteness of the odd perfect and primitive abundant numbers with n distinct prime factors. Am. J. Math. **35**(4), 413–422 (1913)
15. Esparza, J., Nielsen, M.: Decidability issues for Petri nets. BRICS Report Series 1(8) (1994)
16. Finkel, A.: The minimal coverability graph for Petri nets. In: Rozenberg, G. (ed.) ICATPN 1991. LNCS, vol. 674, pp. 210–243. Springer, Heidelberg (1993). https://doi.org/10.1007/3-540-56689-9_45
17. Francalanza, A., Mezzina, C.A., Tuosto, E.: Reversible choreographies via monitoring in Erlang. In: Bonomi, S., Rivière, E. (eds.) DAIS 2018. LNCS, vol. 10853, pp. 75–92. Springer, Cham (2018). https://doi.org/10.1007/978-3-319-93767-0_6
18. Ginsburg, S., Spanier, E.: Bounded ALGOL-like languages. Trans. Am. Math. Soc. **113**(2), 333–368 (1964)
19. Hack, M.: Petri nets and commutative semigroups. Technical Report CSN 18, MIT Laboratory for Computer Science, Project MAC (1974)
20. Hack, M.: Decidability questions for Petri nets. Ph.D. thesis, MIT (1976)
21. Haddad, S., Kordon, F., Petrucci, L., Pradat-Peyre, J.-F., Treves, L.: Efficient state-based analysis by introducing bags in Petri nets color domains. In: ACC, pp. 5018–5025. IEEE (2009)
22. Jancar, P.: Undecidability of bisimilarity for Petri nets and some related problems. Theor. Comput. Sci. **148**(2), 281–301 (1995)
23. Kezić, D., Perić, N., Petrović, I.: An algorithm for deadlock prevention based on iterative siphon control of Petri net. Automatika: časopis za automatiku, mjerenje, elektroniku, računarstvo i komunikacije **47**(1–2), 19–30 (2006)
24. Landauer, R.: Irreversibility and heat generated in the computing process. IBM J. Res. Dev. **5**, 183–191 (1961)
25. Lanese, I., Mezzina, C.A., Schmitt, A., Stefani, J.-B.: Controlling reversibility in higher-order Pi. In: Katoen, J.-P., König, B. (eds.) CONCUR 2011. LNCS, vol. 6901, pp. 297–311. Springer, Heidelberg (2011). https://doi.org/10.1007/978-3-642-23217-6_20
26. Lanese, I., Nishida, N., Palacios, A., Vidal, G.: A theory of reversibility for Erlang. J. Log. Algebr. Methods Program. **100**, 71–97 (2018)
27. Laursen, J.S., Schultz, U.P., Ellekilde, L.: Automatic error recovery in robot assembly operations using reverse execution. In: IROS, pp. 1785–1792. IEEE (2015)
28. Leroux, J.: Vector addition system reversible reachability problem. In: Katoen, J.-P., König, B. (eds.) CONCUR 2011. LNCS, vol. 6901, pp. 327–341. Springer, Heidelberg (2011). https://doi.org/10.1007/978-3-642-23217-6_22

29. McNellis, J., Mola, J., Sykes, K.: Time travel debugging: root causing bugs in commercial scale software. CppCon talk (2017). https://www.youtube.com/watch?v=l1YJTg_A914
30. Murata, T.: Petri nets: properties, analysis and applications. Proc. IEEE **77**(4), 541–580 (1989)
31. Reveliotis, S.A., Choi, J.Y.: Designing reversibility-enforcing supervisors of polynomial complexity for bounded Petri nets through the theory of regions. In: Donatelli, S., Thiagarajan, P.S. (eds.) ICATPN 2006. LNCS, vol. 4024, pp. 322–341. Springer, Heidelberg (2006)
32. Shende, V.V., Prasad, A.K., Markov, I.L., Hayes, J.P.: Synthesis of reversible logic circuits. IEEE Trans. CAD Integr. Circ. Syst. **22**(6), 710–722 (2003)
33. Yokoyama, T., Glück, R.: A reversible programming language and its invertible self-interpreter. In: PEPM, pp. 144–153. ACM Press (2007)

Concurrent Processes

Generalized Alignment-Based Trace Clustering of Process Behavior

Mathilde Boltenhagen[1]($\boxtimes$), Thomas Chatain[1]($\boxtimes$), and Josep Carmona[2]($\boxtimes$)

[1] LSV, CNRS, ENS Paris-Saclay, Inria, Université Paris-Saclay, Cachan, France
{boltenhagen,chatain}@lsv.fr
[2] Universitat Politècnica de Catalunya, Barcelona, Spain
jcarmona@cs.upc.edu

Abstract. Process mining techniques use event logs containing real process executions in order to mine, align and extend process models. The partition of an event log into trace variants facilitates the understanding and analysis of traces, so it is a common pre-processing in process mining environments. Trace clustering automates this partition; traditionally it has been applied without taking into consideration the availability of a process model. In this paper we extend our previous work on process model based trace clustering, by allowing cluster centroids to have a complex structure, that can range from a partial order, down to a subnet of the initial process model. This way, the new clustering framework presented in this paper is able to cluster together traces that are distant only due to concurrency or loop constructs in process models. We show the complexity analysis of the different instantiations of the trace clustering framework, and have implemented it in a prototype tool that has been tested on different datasets.

1 Introduction

Process Mining is becoming an essential discipline to cope with the tons of process data arising in organizations [1]. Now an organization can use some of the available commercial tools to elicit and streamline its processes, so that its decisions are based on the evidences found in the data. In any of these existing software tools, the notion of *trace variant* is fundamental: it denotes a singular sequential execution of the process from start to end. All observed traces that correspond to the same permutation of activities (although the other data attributes, e.g., the customer name, are different), are included into the same trace variant. When trace variants are found, stakeholders then analyze them in order to find out possible incoherences between observed and modeled behavior [2].

In reality, however, the previous flow for analyzing process data is not as ideal as one may think. First, *event logs* that contain the process data stored by an organization, can contain noise, a phenomenon that affects the capability of identifying the right trace variant. Second, processes are not static entities in organizations, but instead evolve over time, which implies also a drift on

© Springer Nature Switzerland AG 2019
S. Donatelli and S. Haar (Eds.): PETRI NETS 2019, LNCS 11522, pp. 237–257, 2019.
https://doi.org/10.1007/978-3-030-21571-2_14

the number and type of trace variants. Third, processes describing concurrent behavior will tend to separate in different trace variants different interleavings of the same Mazurkiewicz trace, although perhaps the analysis for all these traces should be the same. The same applies in case of loops, where often it is not necessary to separate traces that only differ in the number of loop iterations.

In this paper we present a novel clustering technique that is able to tackle the aforementioned situations. Intuitively, the idea is to cluster the event log in a way that traces in the same cluster can be very distant when considered as words, but actually they correspond to the same trace variant when concurrency and loop behavior is disregarded. We build upon a technique presented in a recent paper [3], which assumes that a process model exists. This assumption is realistic in many contexts, e.g., in *Process-Aware Information Systems* (PAIS), process models are often available [4].

We extend the technique in [3] by allowing clustering centroids to now be partial-orders or even subnets of the process model. We present properties that relate them, and show how the subnet case can still be encoding in a SAT instance. Correspondingly, we adapt the notion of inter- and intra-cluster distance and spot quality criteria, so that a characterization of optimal clustering can be defined by users.

We see a great potential on the techniques presented in this paper: first, the techniques proposed can help into simplifying the analysis of event logs, by enabling a better (more abstract) characterization of trace variants. Second, the novel concurrency-and-loop-aware trace clustering proposed is significantly more robust than the ones found in the literature, and tends to avoid redundancy between different clusters.

Related Work. Several techniques have been proposed in the last decade for trace clustering [5–11]. They can be partitioned into *vector space approaches* [5, 7], *context aware approaches* [8,9] and *model-based approaches* [6,10,11]. All the aforementioned clustering algorithms consider only the event log as input, and use different internal representations for producing the clusters. In contrast, in a recent paper [3], we presented a different view on clustering event log traces, by assuming that a process model exists. All the aforementioned techniques do not allow concurrency or loop behavior.

The use of an explicit characterization of concurrency has been considered recently in process discovery: the works in [12,13] show how to improve the discovery of a process model by folding the initial unfolding that satisfies the independence relations given as inputs. In the area of conformance checking, the same phenomena has been observed: the work in [14] assumes traces are represented as partial order, thus allowing again an explicit characterization of concurrency in the problem formalization.

Perhaps the works more similar to the one of this paper are [15,16], where a transition system representing the event log is clustered, so that a set of simpler process models is generated. Tailored state-based properties that guarantee certain Petri net classes are used to guide the clustering, whereas in this work the computation of subnets is unrestricted.

Our work is also related to [17] which clusters events and detects deviation. However, our work focuses on an existing model and the results may consider different directions like repairs while [17] gives a pre-processing of data.

Complexity of our works is related to [18] which demonstrates several methods for distance between automata. Even for dynamic functions, the complexity have been proved PSPACE-complete and is even more complex for Petri nets which is formalism used in this paper.

Organization of the Paper. In the next section we provide an example of the main contributions of the paper. Then, preliminaries are given in Sect. 3, and Sect. 4 defines the quality criteria for trace clustering. In Sects. 5 and 6 we present the two main clustering perspectives proposed in this paper, and the complexity analysis of the problem of computing an optimal clustering is reported in Sect. 7. Finally, in Sect. 8 we provide an evaluation of the prototype implementation of the techniques of this paper over several event logs. Section 9 summarizes this paper and provides futures research lines.

2   A Motivating Example

In [3], we introduced the idea of a trace clustering technique based on a known process model. Each group of traces is related to a trace variant, corresponding to a *full run* of the model (Definition 2), which serves as centroid. The traces in the cluster must all be sufficiently close to the centroid. This allows to identify executions of the model which reproduce typical observed traces, and also to isolate deviant log traces which are too far from what the model describes. In Fig. 1, we present an example of alignment-based clustering.

Our definitions deal with the distance between log traces and the centroid of their cluster. Since these are usually presented as words over an alphabet of actions, a notion of distance on words is used, typically Levenshtein's edit distance. Sometimes, concurrency and loop behavior is not important to differentiate two traces of a business process, as illustrated by the following example.

Example 1. Model of Fig. 1 describes the behaviors of users rating an app. First, users start the form (s). They give either a good (g) or a bad (b) mark attached to a comment (c) or a file (f). Bad ratings get apologies (a), a silent transition (τ) enables to avoid them. Finally, users can donate to the developers of the app (d). The company may be interested in grouping users by behavior, to visualize the differences; for instance which profiles provide bad marks. Trace clusterings of Figs. 1, 2 and 3 has been created, for a maximal distance of alignment to 1.

The order of concurrent actions, like writing a comment before or after giving the rating, does not need to matter to distinguish behaviors in this process. In the alignment-based trace clustering from [3], $\langle s, f, b, a \rangle$ and $\langle s, b, f, a \rangle$, of unhappy customers who uploaded a file, differ only on concurrent actions, and are separated in different clusters. In contrast, Fig. 2 shows a new trace clustering approach where concurrency is disregarded, with the consequence that the two

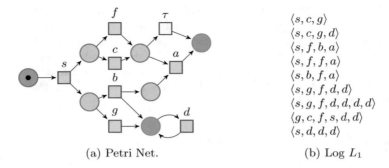

$\langle s, c, g \rangle$
$\langle s, c, g, d \rangle$
$\langle s, f, b, a \rangle$
$\langle s, f, f, a \rangle$
$\langle s, b, f, a \rangle$
$\langle s, g, f, d, d \rangle$
$\langle s, g, f, d, d, d, d \rangle$
$\langle g, c, f, s, d, d \rangle$
$\langle s, d, d, d \rangle$

(a) Petri Net.   (b) Log L_1

Centroids	Traces	Distance
$\langle s, c, \tau, g \rangle$	$\langle s, c, g \rangle$	0
	$\langle s, c, g, d \rangle$	1
$\langle s, b, f, a \rangle$	$\langle s, b, f, a \rangle$	0
	$\langle s, f, f, a \rangle$	1
$\langle s, f, b, a \rangle$	$\langle s, f, b, a \rangle$	0
$\langle s, g, f, \tau, d, d \rangle$	$\langle s, g, f, d, d \rangle$	0
$\langle s, g, f, \tau, d, d, d, d \rangle$	$\langle s, g, f, d, d, d, d \rangle$	0
non-clustered	$\langle g, c, f, s, d, d \rangle$	NA
	$\langle s, d, d, d \rangle$	NA

(c) Clusters

Fig. 1. Alignment-based Trace Clustering (ATC).

Centroids	Traces	Distance
	$\langle s, c, g \rangle$	0
	$\langle s, c, g, d \rangle$	1
	$\langle s, b, f, a \rangle$	0
	$\langle s, f, b, a \rangle$	0
	$\langle s, f, f, a \rangle$	1
	$\langle s, g, f, d, d \rangle$	0
	$\langle s, g, f, d, d, d, d \rangle$	0
non-clustered	$\langle g, c, f, s, d, d \rangle$	NA
	$\langle s, d, d, d \rangle$	NA

Fig. 2. Alignment and Partial Order based Trace Clustering (APOTC).

previous traces now belong to the same cluster. Underneath, the method uses partial-order runs (called processes) of the model instead of sequential runs, shown on the first column of Fig. 2.

Furthermore, donating twice or four times to the developers of the app represent very close behaviors and accordingly, similar profiles. Hence traces $\langle s, g, f, d, d \rangle$ and $\langle s, g, f, d, d, d, d \rangle$ should then be clustered in the same group. This is why we propose yet another trace clustering technique that allows for repetitive behavior in the same cluster, that uses subnets of the process model. Then the two traces below belong to a unique cluster, shown in Fig. 3.

Furthermore, aligning traces to the model is fundamental to avoid clustering (highly) deviant traces. For instance, the log trace $\langle g, c, f, s, d, d \rangle$ is left non-clustered in our work, but would be clustered with $\langle s, g, f, d, d \rangle$ for a trace clustering based only on the log. We compared the results of clusterings to [11] which grouped data by attributes frequency and occurrences, e.g. the activity names. Those traces are then groups with fitting traces.

3   Preliminaries

3.1   Process Models and Trace Clustering

We assume process models are described as Petri nets [19]. Formally:

Definition 1 (Process Model (Labeled Petri Net)). *A Process Model defined by a* labeled Petri net system *(or simply* Petri net*) is a tuple* $N = \langle P, T, F, m_0, m_f, \Sigma, \lambda \rangle$, *where* P *is the set of places,* T *is the set of transitions (with* $P \cap T = \emptyset$*),* $F \subseteq (P \times T) \cup (T \times P)$ *is the flow relation,* m_0 *is the initial marking,* m_f *is the final marking,* Σ *is an alphabet of actions and* $\lambda : T \to \Sigma \cup \{\tau\}$ *labels every transition by an action or as silent.*

Semantics. The semantics of Petri nets is given in term of *firing sequences*. Given a node $x \in P \cup T$, we define its pre-set $^{\bullet}x \overset{\text{def}}{=} \{y \in P \cup T \mid (y, x) \in F\}$ and its post-set $x^{\bullet} \overset{\text{def}}{=} \{y \in P \cup T \mid (x, y) \in F\}$. A marking is an assignment of a non-negative integer to each place. A transition t is *enabled* in a marking m when all places in $^{\bullet}t$ are marked. When a transition t is enabled, it can *fire* by removing a token from each place in $^{\bullet}t$ and putting a token to each place in $t^{\bullet}$. A marking m' is *reachable* from m if there is a sequence of firings $\langle t_1 \ldots t_n \rangle$ that transforms m into m', denoted by $m[t_1 \ldots t_n\rangle m'$.

The set of reachable markings from m_0 is denoted by $[m_0\rangle$. A Petri net is *k-bounded* if no marking in $[m_0\rangle$ assigns more than k tokens to any place. A Petri net is *safe* if it is 1-bounded. In this paper we assume safe Petri nets.

Definition 2 (Full Run). *A firing sequence* $u = \langle t_1 \ldots t_n \rangle$ *such that* $m_0[u\rangle m_f$ *is called a* full run *of* N. *We denote by* $Runs(N)$ *the set of full runs of* N.

Centroids	Traces	Distance
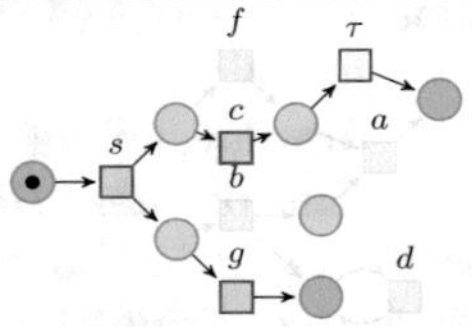	$\langle s, c, g \rangle$	0
	$\langle s, c, g, d \rangle$	1
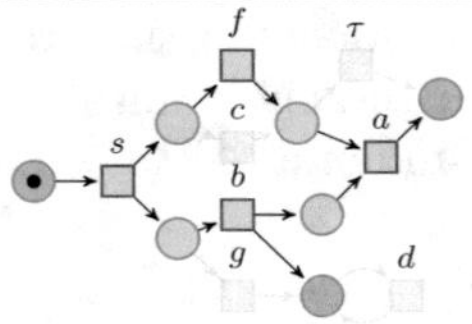	$\langle s, b, f, a \rangle$	0
	$\langle s, f, b, a \rangle$	0
	$\langle s, f, f, a \rangle$	1
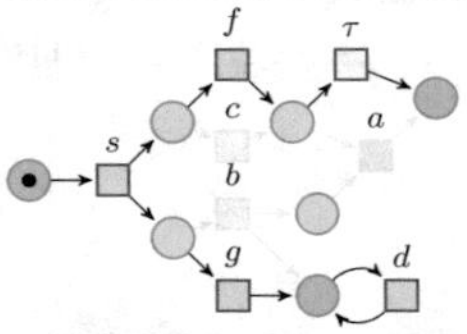	$\langle s, g, f, d, d \rangle$	0
	$\langle s, g, f, d, d, d, d \rangle$	0
non-clustered	$\langle g, c, f, s, d, d \rangle$	NA
	$\langle s, d, d, d \rangle$	NA

Fig. 3. Alignment and Model Subnet based Trace Clustering (AMSTC).

Given a full run $u = \langle t_1 \ldots t_n \rangle \in \mathit{Runs}(N)$, the sequence of actions $\lambda(u) \stackrel{\text{def}}{=} \langle \lambda(t_1) \ldots \lambda(t_n) \rangle$ is called a *(model) trace of* N. When the labeling function λ is injective, like in the model of Fig. 1, we sometimes identify the transition t with its label $\lambda(t)$. Then, full runs coincide with model traces. Examples for the model of Fig. 1 are $\langle s, c, g \rangle$, $\langle s, f, b, a \rangle$, $\langle s, f, g, d, d, d \rangle$.

Definition 3 (Log). *A* log *over an alphabet* Σ *is a finite set of words* $\sigma \in \Sigma^*$, *called* log traces.

Figure 1c shows log traces of recorded behaviors.

Definition 4 (Trace Clustering). *Given a log* L, *a trace clustering over* L *is a partition over a (possibly proper) subset of the traces in* L.

Figures 1 and 2 show two different examples of trace clustering, with 5 and 4 clusters respectively.

Alignment-based trace clustering is a particular form of trace clustering: it relies on a model N of the observed system. The idea of alignment-based trace clustering is to explicit the relation between log traces and full runs of N. Concretely, each cluster of log traces will be assigned a full run u of N, presented as the centroid of the cluster. Hence, traces in the same cluster are not only similar among them, but they are related to a run of the model, which together validates a part of the model and explains the observed log traces.

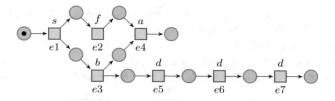

Fig. 4. Example of process of the Petri net in Fig. 1

Definition 5 (Alignment-based Trace Clustering (ATC) [3]). *For a log L and a Petri net $N = \langle P, T, F, m_0, m_f, \Sigma, \lambda \rangle$, an alignment-based trace clustering of L w.r.t. N is a tuple $C = \langle \{u_1 \ldots u_n\}, \chi \rangle$ where $u_1 \ldots u_n$ ($n \in \mathbb{N}$) are full runs of N which serve as centroids for the clusters and $\chi : L \to \{\text{nc}, u_1 \ldots u_n\}$ maps log traces either to the centroid of its cluster $\chi(\sigma)$, or to none of the clusters, denoted by* nc.

Each set $\chi^{-1}(u_i)$, for $i \in \{1 \ldots n\}$, defines the cluster whose centroid is u_i. The set $\chi^{-1}(\text{nc})$ contains the traces which are left non-clustered.

Figure 1 shows a clustering of the traces of a log L_1 based on a model N. For this cluster, $\chi^{-1}(\langle s, g, c, \tau \rangle)$ is also the set of sequences: $\{\langle s, c, g \rangle, \langle s, g, c, d \rangle\}$. We remark that the traces $\langle g, c, f, s, d, d \rangle$ and $\langle s, d, d, d \rangle$ have not been classified for this clustering.

3.2 Partial-Order Semantics

In full runs of a process model, transition occurrences are totally ordered. However transitions can be handled in different orders for the same process in case of concurrency. In the model of Fig. 1 traces $\langle s, b, f, a \rangle$ and $\langle s, f, b, a \rangle$ follow the same process but differ by the order of the transitions.

They can however be seen as two linearizations of a common representation based on partial-order runs which represents a *process*.

Definition 6 (Partial-Order Representation of Runs: Process). *A (non-branching) process $\mathcal{P}$ of a Petri Net $N = \langle P, T, F, m_0, m_f, \Sigma, \lambda \rangle$ is a tuple $\mathcal{P} = \langle B, E, G, B_0, B_f, h \rangle$ where:*

- *(B, E, G, B_0, B_f) is a non-branching, finite, acyclic Petri Net, i.e.*
  - *its causality relation G^+ is acyclic, and*
  - *it has no forward and no backward branchings:*

$$\forall b \in B \quad \exists! e \in E \cup \{\bot\} \quad b \in e^\bullet$$
$$\forall b \in B \quad \exists! e \in E \cup \{\top\} \quad b \in {}^\bullet e$$

where $\bot$ and $\top$ are virtual events satisfying $\bot^\bullet \stackrel{\text{def}}{=} B_0$ and ${}^\bullet\top \stackrel{\text{def}}{=} B_f$.

– $h : (B \cup E) \to (P \cup T)$ *is a function that maps the non-branching process* $\mathcal{P}$ *in the Petri Net* N *with the following relations:*
  - $h(B) \subseteq P$ *and* $h(E) \subseteq T$
  - $\forall e \in E, h_{|\bullet e}$ *is a bijection between* $\bullet e$ *and* $\bullet h(e)$, *same reasoning for* $h_{|e\bullet}$
  - $h_{|B_0}$ *is a bijection between* B_0 *and* m_0, *likewise for* $h_{|B_f}$

Figure 4 shows a process of the Petri Net in Fig. 1. Each event e_i corresponds to a transition of N (for instance $h(e_2) = f$).

Definition 7 (Process representation of a full run). *Every full run* u *of a (safe) model* N *induces a process of* N. *This process is unique up to isomorphism [20] and is denoted by* $\Pi(u)$.

In general, a process represents several full runs, which differ only by the ordering of concurrent actions. For instance, both sequences $\langle s, f, b, a, d, d, d \rangle$ and $\langle s, b, d, d, d, f, a \rangle$ induce the process of the Fig. 4.

We write $Runs(\mathcal{P})$ for the set of full runs of the process $\mathcal{P}$. For every full run $\langle e_1 \ldots e_n \rangle$ of a process $\mathcal{P}$ of a Petri net N, the sequence $u \overset{\text{def}}{=} \langle h(e_1) \ldots h(e_n) \rangle \in T^*$ is called a *linearization* of $\mathcal{P}$. Every linearization of $\mathcal{P}$ is a full run of N.

3.3 Distances Between Log and Model Traces

A key element in this work, and in Process Mining in general, is to align log traces to full runs of the model. In this work, in particular, we target good alignment between every log trace $\sigma \in L$ and the centroid of its cluster $u = \chi(\sigma)$. By the labeling λ of transitions of the model, full runs are mapped to words over the alphabet Σ of actions, called model traces, and the quality of the alignment between σ and u can be quantified as the distance $dist(\sigma, \lambda(u))$, where $dist$ is a distance between finite words over Σ. In this paper, we use Levenshtein's edit distance, which is usually considered appropriate in Process Mining.

Definition 8 (Levenshtein's edit distance). *Levenshtein's edit distance* $dist(w1, w2)$ *between two words* w_1 *and* $w_2 \in \Sigma^*$ *is the minimal number of edits needed to transform* w_1 *to* w_2. *Editions can be substitutions to a letter by another one, deletions or additions of a letter in words.*

We will abuse notations, and write $dist(\sigma, u)$ for $dist(\sigma, \lambda(u))$, and $dist(u_1, u_2)$ for $dist(\lambda(u_1), \lambda(u_2))$. For example, the full run $\langle s, g, c \rangle$ and the log trace $\langle s, g, d, c \rangle$ have only one difference: the addition of d. They are at distance 1.

4 Quality Criteria for Trace Clustering

Figure 5 shows two alignment-based trace clusterings of a new log L_2, based on the model N of Fig. 1. The two clusterings have been created for the same model and log, and contain different centroids.

Clustering	Centroids u	Traces σ	$dist(\sigma, u)$	Quality criteria
$\mathcal{C}_1$	$\langle s, c, g, \tau \rangle$	$\langle s, f, g \rangle$	1	$d(\mathcal{C}_1) = 2$
		$\langle s, c, g \rangle$	0	
		$\langle s, g, c \rangle$	2	
		$\langle s, c, g, d \rangle$	1	$\Delta(\mathcal{C}_1) = 12$
	$\langle s, b, f, a \rangle$	$\langle s, f, b, a \rangle$	2	
		$\langle s, c, b, a \rangle$	2	$n(\mathcal{C}_1) = 3$
		$\langle s, b, f, a \rangle$	0	
		$\langle s, b, c, a \rangle$	1	$\check{c}(\mathcal{C}_1) = 0.83$
	$\langle s, f, \tau, g, d, d \rangle$	$\langle s, f, g, d \rangle$	1	
		$\langle s, f, g, d, d, d, d \rangle$	2	$\Phi(\mathcal{C}_1) = 3$
	non-clustered	$\langle s, f, a, a, a \rangle$	NA	
		$\langle s, f, b, d, d \rangle$	NA	
$\mathcal{C}_2$	$\langle s, f, g, \tau \rangle$	$\langle s, f, g \rangle$	0	
		$\langle s, c, g, d \rangle$	2	$d(\mathcal{C}_1) = 2$
		$\langle s, f, g, d \rangle$	1	
	$\langle s, f, b, a \rangle$	$\langle s, f, b, a \rangle$	0	$\Delta(\mathcal{C}_2) = 14$
		$\langle s, f, a, a, a \rangle$	2	
		$\langle s, c, b, a \rangle$	1	$n(\mathcal{C}_2) = 4$
		$\langle s, b, f, a \rangle$	2	
		$\langle s, b, c, a \rangle$	2	$\check{c}(\mathcal{C}_2) = 1.0$
		$\langle s, f, b, d, d \rangle$	2	
	$\langle s, f, g, \tau, d, d, d, d \rangle$	$\langle s, f, g, d, d, d, d \rangle$	0	$\Phi(\mathcal{C}_2) = 2$
	$\langle s, g, c, \tau \rangle$	$\langle s, c, g \rangle$	2	
		$\langle s, g, c \rangle$	0	

Fig. 5. Two possible clusterings for the same set of trace logs.

In this section, we provide criteria which contribute in the qualification of a good clustering. We have identified the following criteria:

- $d(\mathcal{C})$, *maximum distance between a trace and the centroid of its cluster*: this criterion, defined by $\max_{\sigma \in L \setminus \chi^{-1}(\mathrm{nc})} dist(\sigma, \chi(\sigma))$, will be minimized to increase the fit of the centroids to their traces. In case of a log containing noise, a small distance may induce many non-clustered traces.
- $\Delta(\mathcal{C})$, *sum of distances* : the sum $\Delta(\mathcal{C}) \overset{\text{def}}{=} \sum_{\sigma \in L \setminus \chi^{-1}(\mathrm{nc})} dist(\sigma, \chi(\sigma))$ can be seen as a variant or a refinement of the previous criterion $d(\mathcal{C})$. It will also be minimized in order to get the most representative centroids.
- $n(\mathcal{C})$, *number of clusters* : The number of clusters provides an interesting perspective, which is analogous to the number of trace variants of a process model, but in this case from the log perspective.
- $\check{c}(\mathcal{C})$, *ratio of clustered traces:* this ratio, defined as $\check{c}(\mathcal{C}) \overset{\text{def}}{=} \frac{|L| - |\chi^{-1}(\mathrm{nc})|}{|L|}$, is close to 1 for a process model that covers most of the behavior of the log. $\check{c}(\mathcal{C})$ also highlights the ratio of distant traces, i.e. traces that deviate from the model, for a given maximum distance $d(\mathcal{C})$.
- $\Phi(\mathcal{C})$, *inter-cluster distance* : the distance between the centroids is also an important parameter. For an ATC $\mathcal{C} = \langle \{u_1 \dots u_n\}, \chi \rangle$, the inter-cluster

distance is defined as $\Phi(\mathcal{C}) \stackrel{\text{def}}{=} \min_{i \neq j} dist(u_i, u_j)$. A larger distance involves distant clusters, this is why this parameter should be maximized in order to prevent overlay between the clusters.

Most of the detailed criteria come from the Data Mining domain [21,22]. Other measures like the Dunn [23], which compares distances between items that share or not a cluster, and the Silhouette [24], that computes if items is close enough to their clusters instead of the others, help the user to analyze its clustering. As usual when multiple parameters are taken into account, there will not exist in general a unique clustering optimizing all the criteria together. Instead, every clustering problem should consider a good balance between the parameters to optimize. Our tool, which is described in Sect. 8, returns the optimal clustering for a given pre-defined setting.

Example 2. Figure 5 shows two ATCs of the model N of Fig. 1 and a new set of log traces L_2. The results differ on the parameters to optimize. The first one minimize the inter-cluster distance $\Phi(\mathcal{C}_1)$ for a given distance between the trace and the centroid $d(\mathcal{C}_1)$ to 2. However, some traces are left non-clustered which do not appear in the second clustering. In contrast, the centroids are closer $(\Phi(\mathcal{C}_2) < \Phi(\mathcal{C}_1))$ and the number of clusters is larger $(n(\mathcal{C}_1) < n(\mathcal{C}_2))$.

5   Fitting Centroids to Concurrency

The aim of ATC is to group traces which are similar to a full run of the model. In this section, we want to go further and allow one to cluster together traces which differ only by the order of execution of transitions which are presented as concurrent in the model. In the ATC of Fig. 1, traces $\langle s, b, f, a \rangle$ and $\langle s, f, b, a \rangle$ are clustered separately, and since no model trace is at distance ≤ 1 to both of them, every ATC $\mathcal{C}$ which would cluster them together would have an inter-cluster $d(\mathcal{C}) > 1$. Yet, $\langle s, b, f, a \rangle$ and $\langle s, f, b, a \rangle$ are perfectly aligned with two different interleavings of *the same* execution of the model, if one understands "execution" as process like in Definition 6. The following definition of trace clustering, precisely uses processes as cluster centroids.

Definition 9 (Alignment and Partial Order based Trace Clustering (APOTC)). *As full run clustering, an alignment and partial order based trace clustering, of a log L and a Petri net $N = \langle P, T, F, m_0, m_f, \Sigma, \lambda \rangle$, is a tuple $\mathcal{C} = \langle \{\mathcal{P}_1 \ldots \mathcal{P}_n\}, \chi \rangle$ where $\mathcal{P}_1 \ldots \mathcal{P}_n$ $(n \in \mathbb{N})$ are processes of N which serve as centroids for the clusters and $\chi : L \rightarrow \{\text{nc}, \mathcal{P}_1 \ldots \mathcal{P}_n\}$ maps log traces either to the centroid of its cluster $\chi(\sigma)$, or to none of the clusters, denoted by* nc.

5.1   Quality Criteria for APOTC

All the quality criteria of ATC are considered in APOTC, but they need to be redefined now that centroids are processes. Indeed we need to compare log traces to processes, which represent (finite) sets of full runs. Naturally, the distance between a model trace σ and a process $\mathcal{P}$ will be defined as the distance to its closest linearization of $\mathcal{P}$.

Definition 10. *We define the distance $dist(\sigma, \mathcal{P})$ (abusing notation dist again) between a trace σ and a process $\mathcal{P}$ as $dist(\sigma, \mathcal{P}) \stackrel{def}{=} \min_{u \in Runs(\mathcal{P})} dist(\sigma, u)$.*

This allows us to define $d(\mathcal{C})$ and $\Delta(\mathcal{C})$ for APOTC like for ATC, respectively as $\max_{\sigma \in L \setminus \chi^{-1}(\mathbf{nc})} dist(\sigma, \chi(\sigma))$ and $\sum_{\sigma \in L \setminus \chi^{-1}(\mathbf{nc})} dist(\sigma, \chi(\sigma))$ with $\chi(\sigma)$ are processes.

The inter-cluster distance of an APOTC $\mathcal{C} = \langle \{\mathcal{P}_1 \ldots \mathcal{P}_n\}, \chi \rangle$ is also defined as for ATC, as the minimum distance between two centroids: $\Phi(\mathcal{C}) = \min_{i \neq j} dist(\mathcal{P}_i, \mathcal{P}_j)$, using the appropriate notion of distance between processes:

Definition 11. *The distance between two processes is the minimal distance between their linearizations: $dist(\mathcal{P}, \mathcal{P}') \stackrel{def}{=} \min_{\substack{u \in Runs(\mathcal{P}) \\ u' \in Runs(\mathcal{P}')}} dist(u, u')$.*

Example 3. Figure 2 shows an APOTC of the model and log of Fig. 1. Traces $\langle s, b, f, a \rangle$ and $\langle s, f, b, a \rangle$ can now be clustered together, yielding a smaller $n(\mathcal{C})$ for equivalent $\Delta(\mathcal{C})$ and $d(\mathcal{C})$.

5.2   Relating APOTC to ATC

Any ATC can be casted as an APOTC. All the full runs centroids of an ATC, which are sequential executions, can be represented as processes using Definition 7. The following theorem explains how this transformation affects the quality criteria of the clusterings.

Theorem 1. *For any ATC $\mathcal{C}_u = \langle \{u_1 \ldots u_n\}, \chi_u \rangle$, we define $\forall i \in \{1 \ldots n\}$ $\mathcal{P}_i \stackrel{def}{=} \Pi(u_i)$ and $\chi_{\mathcal{P}} \stackrel{def}{=} \Pi \circ \chi_u$ (by convention $\Pi(\mathbf{nc}) = \mathbf{nc}$) inducing $\mathcal{C}_{\mathcal{P}} = \langle \{\mathcal{P}_1 \ldots \mathcal{P}_n\}, \chi_{\mathcal{P}} \rangle$ its corresponding APOTC of the same process model N and the same log L. The distances below follow the properties:*

1. *$d(\mathcal{C}_u) \geq d(\mathcal{C}_{\mathcal{P}})$ and $\Delta(\mathcal{C}_u) \geq \Delta(\mathcal{C}_{\mathcal{P}})$ with equality if the model is sequential*
2. *$\Phi(\mathcal{C}_u) \geq \Phi(\mathcal{C}_{\mathcal{P}})$ with equality if the model is sequential*
3. *$n(\mathcal{C}_u) = n(\mathcal{C}_{\mathcal{P}})$ and $\check{c}(\mathcal{C}_u) = \check{c}(\mathcal{C}_{\mathcal{P}})$*

Proof. We first observe that the obtained set $\{\mathcal{P}_1 \ldots \mathcal{P}_n\}$ is by Definition 7 a set of subnets of N and $\chi_{\mathcal{P}}$ maps every clustered log traces to a subnet and non-clustered log traces to $\mathbf{nc}$. Then $\mathcal{C}_{\mathcal{P}} = \langle \{\mathcal{P}_1 \ldots \mathcal{P}_n\}, \chi_{\mathcal{P}} \rangle$ is indeed an APOTC.

1. Every trace σ of L is either clustered ($\chi_u(\sigma) = u_i$, $i \in \{1 \ldots n\}$) or non-clustered ($\chi_u(\sigma) = \mathbf{nc}$). The maximum distance between traces and centroids $d(\mathcal{C}_u)$ depends only on clustered traces: $\forall \sigma \in L_{\setminus \chi_u^{-1}(nc)}$ $dist(\sigma, \chi_u(\sigma)) \leq d(\mathcal{C}_u)$. By Definition 7 $\chi_u(\sigma) \in Runs(\chi_{\mathcal{P}}(\sigma))$. Then for any clustered trace σ, we have $d(\mathcal{C}_{\mathcal{P}}) \leq dist(\sigma, \chi_{\mathcal{P}}(\sigma)) \leq dist(\sigma, \chi_u(\sigma)) \leq d(\mathcal{C}_u)$ with equality if the model is sequential (no other run in $Runs(\chi_{\mathcal{P}}(\sigma))$). Furthermore, $\Delta(\mathcal{C})$ is the sum of the distances: $\Delta(\mathcal{C}_{\mathcal{P}}) \leq \Delta(\mathcal{C}_u)$.

2. Let u_i and u_j, $i, j \in \{1 \ldots n\}$, be two centroids of the ATC. The corresponding processes of those centroids are defined by $\mathcal{P}_i = \Pi(u_i)$ and $\mathcal{P}_j = \Pi(u_j)$ and $u_i \in Runs(\mathcal{P}_i)$ and $u_j \in Runs(\mathcal{P}_j)$. This implies $dist(\mathcal{P}_i, \mathcal{P}_j) \leq dist(u_i, u_j)$ with equality if the model is sequential (no other run in the processes). Consequently $\Phi(\mathcal{C}_\mathcal{P}) \leq \min_{i \neq j} dist(u_i, u_j) = \Phi(\mathcal{C}_u)$ with equality if the model is sequential.
3. This is immediate by definition of $\chi_\mathcal{P}$. □

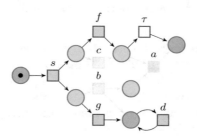

Fig. 6. A subnet of the Petri Net in Fig. 1. Only transitions s, g, f, τ, d are kept. Transitions in light gray do not belong to the subnet.

As a summary, casting an ATC to an APOTC improves the distances between traces and centroids; in contrast, the resulting APOTC may get a lower (i.e. poorer) inter-cluster distance than the ATC. The number of clusters and ratio of clustered traces are preserved.

This means that clusters that were distant in the ATC may become closer in the APOTC, which appears negative when seen from the perspective of good clusterings presenting distant clusters. But, in the other hand, clusters that become closer will typically be those that one precisely wanted to merge because they represent different interleavings of processes. This is exactly what happens in Example 3. Merging clusters then results in a lower number of clusters $n(\mathcal{C})$, which also helps to get a human understandable clustering and facilitates the analysis of the results by decision makers.

Example 4. When casting the ATC of Fig. 1 to an APOTC, the clusters with centroid $\langle s, b, f, d, a \rangle$ and $\langle s, f, d, a \rangle$ become two clusters with the same process as centroid. This leads to an inter-cluster distance $\Phi(\mathcal{C}_\mathcal{P}) = 0$ for the APOTC. But, after merging these two clusters, one gets the better APOTC presented in Fig. 2.

6   Fitting Centroids to Concurrency and Repetitive Behavior

In Fig. 2, we show that APOTC separates process arising from traces corresponding to different number of loop iterations, e.g., the traces $\langle s, g, f, d, d \rangle$ and $\langle s, g, f, d, d, d, d \rangle$. The issue is due of the finite size of runs of processes. Indeed process centroids are partial order runs which do not allow loops and infinite sequences of events. To overcome this limitation, we introduce subnets of models.

Definition 12 (Subnet of Petri net). *A* subnet *of a Petri net* $N = \langle P, T, F, m_0, m_f, \Sigma, \lambda \rangle$ *is a Petri net* $\langle P, T', F_{|T'}, m_0, m_f, \Sigma_{|T'}, \lambda \rangle$ *with* $T' \subseteq T$, *and* $F_{T'} \stackrel{\text{def}}{=} F \cap (P \times T' \cup T' \times P)$.

Figure 6 presents a subnet of the model of Fig. 1. Observe that our definition of subnets, based on selecting transitions, restricts the semantics of the net and cannot produce new behaviors. Formally:

Lemma 1. *Every full run (resp. process) of a subnet of a Petri net N, is a full run (resp. process) of N.*

We now formalize AMSTC, which consider subnets as centroids:

Definition 13 (Alignment and Model Subnet-based Trace Clustering (AMSTC)). *For a log L and a Petri net $N = \langle P, T, F, m_0, m_f, \Sigma, \lambda \rangle$, an alignment and model subnet trace clustering, of L w.r.t. N is a tuple $\mathcal{C} = \langle \{\mathcal{N}_1 \ldots \mathcal{N}_n\}, \chi \rangle$ where $\mathcal{N}_1 \ldots \mathcal{N}_n$ are subnets of N which serve as centroids for the clusters and $\chi : L \rightarrow \{\text{nc}, \mathcal{N}_1 \ldots \mathcal{N}_n\}$ maps log traces either to the centroid of its cluster $\chi(\sigma)$, or to none of the clusters, denoted by* nc.

Figure 3 shows an AMSTC of log traces based on the model of Fig. 1.

6.1 Quality Criteria for AMSTC

All the previous criteria for ATC and APOTC also apply to AMSTC, but the notion of distance needs to be adapted again. The quality criteria $d(\mathcal{C})$ and $\Delta(\mathcal{C})$ rely on the distance between log traces and the centroids of their clusters. Here, centroids are subnets, and the distance is defined as $dist(\sigma, \mathcal{N}) \stackrel{\text{def}}{=} \min_{u \in Runs(\mathcal{N})} dist(\sigma, u)$. Computing this distance corresponds to aligning the trace to the model.

The inter-cluster distance, $\Phi(\mathcal{C})$ is the minimal distance between two subnet centroids, now defined as: $dist(\mathcal{N}, \mathcal{N}') \stackrel{\text{def}}{=} \min_{\substack{u \in Runs(\mathcal{N}) \\ u' \in Runs(\mathcal{N}')}} dist(u, u')$.

Example 5. Figure 3 shows an AMSTC of the Petri Net and the log traces of Fig. 1 for $d(\mathcal{C}) = 1$. Then the traces $\langle s, g, f, d, d \rangle$ and $\langle s, g, f, d, d, d, d \rangle$ are grouped in the same cluster.

Intra-cluster Distance. If applied unrestricted, AMSTC can use as centroids, subnets with branchings and loops, and then cluster together very different log traces. The intra-cluster distance aims at controlling this aspect. For instance, taking as centroid the complete net of Fig. 1 would not yield a satisfactory AMSTC. Instead, traces in the same cluster should be similar and represent a generalized notion of trace variant. This criterion is quantified by the *intra-cluster distance* $\Theta(\mathcal{C})$. Clusterings with low $\Theta(\mathcal{C})$ will be preferred.

– $\Theta(\mathcal{C})$, *the intra-cluster distance*: Before defining the intra-cluster distance of a clustering $\mathcal{C}$, we focus on each of its centroids separately: for every centroid $\mathcal{N}_k$, define

$$\Theta'(\mathcal{N}_k) \stackrel{\text{def}}{=} \sup_{\mathcal{P},\mathcal{P}' \in Proc(\mathcal{N}_k)} \frac{dist(\mathcal{P},\mathcal{P}')}{(1+\epsilon)^{max(|\mathcal{P}|,|\mathcal{P}'|)}}$$

where $|\mathcal{P}|$ denotes the number of events in $\mathcal{P}$, and $\epsilon > 0$ is a parameter set by the user in order to limit (more or less) the influence of long processes. Indeed, when the subnet $\mathcal{N}_k$ has loops, it has infinitely many processes, arbitrary large, which yields arbitrary large distance to the smaller processes. Yet, such subnets may be relevant, as illustrated by Example 6. This is why our definition penalizes more for distances between small processes.

Finally, the intra-cluster distance of a clustering $\mathcal{C} = \langle\{\mathcal{N}_1 \ldots \mathcal{N}_n\}, \chi\rangle$ is:

$$\Theta(\mathcal{C}) \stackrel{\text{def}}{=} \max_k \Theta'(\mathcal{N}_k)$$

Example 6. The AMSTC of Fig. 3 has only one centroid with a loop. Because of the loop d, this centroid has infinitely many processes. With $\epsilon = 0.1$, the intra-cluster distance of this AMSTC is bounded by 2.39, increasing ϵ to 0.5 returns $\Theta(\mathcal{C}) = 0.12$ which penalizes significantly less the loop in the subnet centroid.

6.2   Relating AMSTC to APOTC

Every APOTC $\mathcal{C}_\mathcal{P}$ induces an AMSTC whose subnet centroids are subnets are defined according to the process centroids of $\mathcal{C}_\mathcal{P}$.

Definition 14 (Subnet induced by a process). *Every process* $\mathcal{P} = (B, E, G, B_0, B_f, h)$ *of a model* $N = \langle P, T, F, m_0, m_f, \Sigma, \lambda\rangle$, *induces a subnet of* N *defined by* $\Psi(\mathcal{P}) \stackrel{\text{def}}{=} (P, h(E), G_{|h(E)}, h(B_0), h(B_f))$.

Figure 6 shows the subnet corresponding to the last process of Fig. 2. All the transitions in the process belong to the subnet, and the loop fits traces with arbitrary repetition of activity d (for instance $\langle s, f, g, d, d\rangle$ and $\langle s, f, g, d, d, d, d\rangle$).

The following theorem relates APOTC and the induced AMSTC, analogously to Theorem 1 for ATC and APOTC.

Theorem 2. *For any APOTC* $\mathcal{C}_\mathcal{P} = \langle\{\mathcal{P}_1 \ldots \mathcal{P}_n\}, \chi_{\mathcal{P}_i}\rangle$, *we define* $\forall i \in \{1 \ldots n\}$ $\mathcal{N}_i \stackrel{\text{def}}{=} \Psi(\mathcal{P}_i)$ *and* $\chi_{\mathcal{N}_\mathcal{P}} \stackrel{\text{def}}{=} \Psi \circ \chi_\mathcal{P}$ *(by convention* $\Psi(\text{nc}) = \text{nc})$ *inducing* $\mathcal{C}_{\mathcal{N}_\mathcal{P}} = \langle\{\mathcal{N}_1 \ldots \mathcal{N}_n\}, \chi_{\mathcal{N}_\mathcal{P}}\rangle$ *its corresponding AMSTC of the same process model* N *and the same log* L. *The distances below follow the properties:*

1. $d(\mathcal{C}_\mathcal{P}) \geq d(\mathcal{C}_{\mathcal{N}_\mathcal{P}})$ *and* $\Delta(\mathcal{C}_\mathcal{P}) \geq \Delta(\mathcal{C}_{\mathcal{N}_\mathcal{P}})$ *with equality if the model is acyclic*
2. $\Phi(\mathcal{C}_\mathcal{P}) \geq \Phi(\mathcal{C}_{\mathcal{N}_\mathcal{P}})$ *with equality if the model is acyclic.*
3. $n(\mathcal{C}_\mathcal{P}) = n(\mathcal{C}_{\mathcal{N}_\mathcal{P}})$ *and* $\check{c}(\mathcal{C}_\mathcal{P}) = \check{c}(\mathcal{C}_{\mathcal{N}_\mathcal{P}})$

Proof. The correspondence APOTC-AMSTC is similar to the correspondance ATC-APOTC demonstrated in Theorem 1. When properties exactly coincide between ATC and APOTC for sequential models, same results are found between APOTC and AMSTC for acyclic models: without loops, every subnet has a single process and the distances are preserved. □

As subnets may allow infinite runs, the sum of differences $\Delta(\mathcal{C})$ between log traces and centroids may decrease in an expansion from APOTC to AMSTC. Unfortunately, the inter-cluster distance $\Phi(\mathcal{C})$ can also be lower.

When AMSTCs Meet APOTCs. Observe that, in our definition of AMSTC, only the behavior of the subnets is considered. Hence, nothing penalizes a clustering for having dead transitions in a cluster, i.e., transitions which do not participate in any full run of the subnet. Intuitively, this situation is not satisfactory since we expect the subnets to give information about the part of the net which really participates in the observed traces. By the way, notice that the subnets induced by processes following Definition 14 never have any dead transition. These subnets also have another property: they all have at least one full run. Let us call *fair* an AMSTC in which every centroid has these two properties. The following theorem establishes a relation between APOTCs and fair AMSTCs.

Theorem 3. *For a log L and an acyclic and trace-deterministic[1] model N, the transformation defined in Theorem 2 establishes a bijection from the set of APOTC to the set of fair AMSTCs $\mathcal{C}$ with intra-cluster distance $\Theta(\mathcal{C}) = 0$.*

Proof. Since N is acyclic, for every process $\mathcal{P}$ of N, the subnet induced by $\mathcal{P}$ has no other process than $\mathcal{P}$ itself. This proves that any AMSTC $\mathcal{C}$ obtained from an APOTC has intra-cluster distance $\Theta(\mathcal{C}) = 0$. It is also fair as we noticed earlier.

Now, every centroid $\mathcal{N}_i$ of a fair AMSTC $\mathcal{C}$ with $\Theta(\mathcal{C}) = 0$ has a single process (call it $\mathcal{P}_i$): indeed, since the model is trace-deterministic, every subnet centroid in $\mathcal{C}$ having two different processes would lead to $\Theta(\mathcal{C}) > 0$. This establishes a bijection between the centroids of fair AMSCs with intra-cluster distance 0, and the processes of N, which serve as centroids in APOTCs. This bijection between centroids induces naturally our bijection between APOTCs and AMSTCs. □

In summary, AMSTC handles both concurrency and repetitive behavior, and under some situations behaves similarly to APOTC.

7   Complexity of Alignment-Based Trace Clusterings

For a log L and a model N, one is typically interested in computing a trace clustering (ATC, APOTC or AMSTC) $\mathcal{C}$ of L w.r.t. N of sufficient quality, i.e. satisfying some constraints on the quality criteria $d(\mathcal{C})$, $\Delta(\mathcal{C})$, $n(\mathcal{C})$, $\check{c}(\mathcal{C})$... We

---

[1] N is *trace-deterministic* if the mapping $u \in Runs(N) \mapsto \lambda(u) \in \Sigma^*$ is injective.

will see that, at least from a theoretical point of view, the complexity lies already in the existence of a clustering, and the specification of many quality constraints does not change the complexity.

For a non-empty log L and a model N, there exists an ATC C of L w.r.t. N having $\check{c}(C) > 0$ (i.e. such that at least one trace is clustered), iff N has a full run. Indeed, when no constraint is given about the quality criteria $d(C)$, $\Delta(C)$, $n(C)$, $\Phi(C)\ldots$, any full run of N can serve as centroid, and any log trace can be assigned to any cluster. The same holds for APOTC, where centroids are processes of N, since N has a process iff N has a full run; it holds again for AMSTC, taking into account the constraint that any subnet used as centroid should have a full run, or the stronger constraint that the subnet should not have any dead transition, as discussed in Sect. 6.2.

Now, deciding if a model has a full run u, corresponds to checking reachability of the final marking. The problem of reachability in Petri nets is known to be decidable, but non-elementary [25], and still PSPACE-complete for safe Petri nets. But the complexity trivially drops to NP-complete[2] if a bound l is given (with l an integer coded in unary) on the length of u.

In practice, relevant clusterings will not use very long full runs (or processes for APOTC) as centroids. Also for AMSTC, no very long full run will be considered in the computation of $d(C)$, $\Delta(C)$ or $\Phi(C)$. Typically, a bound l on the length of the full runs can be assumed, for instance 2 times the length of the longer log trace. Let us call l-bounded a trace clustering satisfying this constraint.

Theorem 4. *The problem of deciding, for a log L, a model N, an integer bound l, integers $d_{\max}$, $\Delta_{\max}$, $n_{\max}$ and a rational number $\check{c}_{\min}$, the existence of a l-bounded ATC (respectively APOTC, AMSTC) C of L w.r.t. N, having $d(C) \leq d_{\max}$, $\Delta(C) \leq \Delta_{\max}$, $n(C) \leq n_{\max}$ and $\check{c}(C) \geq \check{c}_{\min}$, is NP-complete.*

Proof. As observed earlier, the problem is NP-hard even with the only constraint that at least one trace is clustered (i.e. $\check{c}(C) > 0$, or equivalently $\check{c}(C) \geq \frac{1}{|L|}$). It remains to show that it is in NP: indeed, if there exists a (l-bounded) clustering, there exists one with no more that $|L|$ clusters (forgetting empty clusters cannot weaken the quality criteria); and, by assumption, the size of centroids (defined as $|\sigma|$ for ATC, $|\mathcal{P}|$ for APOTC, number of transitions in the subnet for AMSTC) is bounded by l. So, it is possible to guess a clustering C in polynomial time. For APOTC and AMSTC, one can also guess in P time the full run $u \in Runs(\chi(\sigma))$, for every clustered trace σ, which will achieve the $dist(\sigma, \chi(\sigma))$. Now, checking that C satisfies the constraints, only requires to compute Levenshtein's edit distances and minima over sets of polynomial size. This can be done in P time.   □

For ATC, the problem remains in NP with an additional constraint on the inter-cluster distance ($\Phi(C) \geq \Phi_{\min}$) because the inter-cluster-distance can be computed in P time.

---

[2] NP-hardness can be obtained by reduction from the problem of reachability in a safe acyclic Petri net, known to be NP-complete [26,27].

On the other hand, incorporating new constraints like bounds on $\Phi(\mathcal{C})$ for APOTC or AMSTC, or on the intra-cluster distance $\Theta(\mathcal{C})$, may increase the complexity. The principle of the algorithm remains: guess non-deterministically a clustering, then check if it satisfies the constraints. Hence, the complexity depends on the complexity of the algorithm used as an oracle to check, given a log, a model and a clustering $\mathcal{C}$, if $\mathcal{C}$ satisfies the constraints. Precisely, if there exists such an oracle algorithm in some complexity class A, then the l-bounded trace clustering problem is in NP^A. For instance, for APOTC, checking if $\Phi(\mathcal{C}) \geq \Phi_{\min}$ is in NP; in consequence, the trace clustering problem with such constraint is in $\mathrm{NP}^{\mathrm{NP}}$. We get the same result for constraints on the intra-cluster distance $(\Theta(\mathcal{C}) \geq \Theta_{\min})$ for AMSTC.

8   SAT Encoding and Experimentation

The NP-completeness established in Theorem 4 for our trace clustering problems suggests to encode them as SAT problems. For each clustering problem, our tool DARKSIDER constructs a pseudo-Boolean[3] formula and calls a solver (currently MINISAT+ [28]). Every solution to the formula is interpreted as a trace clustering. This is already what we did for a version of ATC presented in our previous paper [3], to which we refer the reader for a more detailed description of the SAT encoding. Here, we present the main ideas for the encoding of AMSTC, which is done in the same spirit[4].

The assignment of log traces to clusters in an AMSTC $\mathcal{C} = \langle \{\mathcal{N}_1 \ldots \mathcal{N}_n\}, \chi \rangle$ is encoded using variables $(\chi_{\sigma k})_{\sigma \in L, \ k=1\ldots n}$ meaning that $\chi(\sigma) = \mathcal{N}_k$. Variables $(c_{kt})_{k=1\ldots n, \ t \in T}$ code the fact that transition t appears in subnet $\mathcal{N}_k$.

In order to encode that a sequence $u = \langle t_1 \ldots t_n \rangle$ is a full run of N (or of a subnet $\mathcal{N}_k$), we use a set of Boolean variables:

- $\tau_{i,t}$ for $i = 1 \ldots n, \ t \in T$: means that transition $t_i = t$; and
- $m_{i,p}$ for $i = 0 \ldots n, \ p \in P$: means that place p is marked in marking m_i reached after firing $\langle t_1 \ldots t_i \rangle$ (remind that we consider only safe nets, therefore the $m_{i,p}$ are Boolean variables).

They are involved in constraints like, for instance:

- Initial marking: $\left(\bigwedge_{p \in m_0} m_{0,p} \right) \wedge \left(\bigwedge_{p \in P \setminus m_0} \neg m_{0,p} \right)$
- Transitions are enabled when they fire: $\bigwedge_{i=1}^{n} \bigwedge_{t \in T} (\tau_{i,t} \implies \bigwedge_{p \in \bullet t} m_{i-1,p})$.

Finally, variables $\delta_{\sigma i}$ are used to detect and count the mismatches between a (clustered) log trace $\sigma \in L$ and the (closest) execution of the subnet $\chi(\sigma)$ which serves as centroid for its cluster. These variables are needed to encode

---

[3] Pseudo-Boolean constraints are generalizations of Boolean constraints. They allow one to specify constant bounds on the number of variables which can/must be assigned to true among a set V of variables.

[4] Due to AMSTC being the most general trace clustering, our experiments focus on this method.

the constraints $d(\mathcal{C}) \leq d_{\max}$ and $\Delta(\mathcal{C}) \leq \Delta_{\max}$, or to construct a minimization objective for the solver when one wants to find AMSTC which minimize these quantities.

As explained in Sect. 7, dealing with constraints about the inter-cluster and intra-cluster distance pushes our AMSTC problem out of the NP complexity class. Concretely, this means that such constraints are not adapted for a SAT encoding. Instead, we use an approximation of the inter-cluster distance, by bounding the number of common transitions between centroids, and the intra-cluster distance, by bounding the number of transition per centroids.

Table 1. Experimental results for the computation of ATC and AMSTC with our tool DARKSIDER, obtained on a virtual machine with CPU Intel®Core i5-530U-1.8 GHz*2 and 5.2 GB RAM. AO (almost optimal) indicates experiments where we stopped the SAT solver before it finds an *optimal* solution; this way, we got much better execution times and still very satisfactory solutions.

Model			$|L|$	Clustering	Formulas size		Execution Time (sec)	$d(\mathcal{C})$	$n(\mathcal{C})$	$\Phi(\mathcal{C})$	$\check{c}(\mathcal{C})$
Reference	$|T|$	$|P|$			Variables	Constraints					
Fig. 1	8	7	12	ATC	13854	26457	0.66	2	3	3	1.0
Fig. 1	8	7	12	ATC	13854	26457	0.50	0	3	2	0.25
Fig. 1	8	7	12	AMSTC	27306	51897	1.52	2	2	2	1.0
Fig. 1	8	7	12	AMSTC	27306	51897	1.14	0	3	1	0.83
Fig. 1	8	7	300	ATC	348800	641530	1252.53	2	3	2	0.91
Fig. 1	8	7	300	AMSTC	868592	1641844	1449.14	2	2	4	0.99
subnet of M1 [29]	17	14	12	ATC	98924	218568	9.18	2	3	1	0.38
subnet of M1 [29]	17	14	12	AMSTC	191876	418462	15.76	2	2	5	0.54
subnet of M1 [29]	17	14	100	ATC	433491	925574	124.57	2	3	3	0.80
subnet of M1 [29]	17	14	100	AMSTC	1185839	2573419	5659.75	2	2	5	0.95
subnet of M1 [29]	17	14	100	AMSTC	1185839	2573419	AO:100.16	2	2	5	0.95
M1 [29]	40	40	12	ATC	297176	678664	108.30	3	2	5	0.41
M1 [29]	40	40	12	AMSTC	713769	1656046	88.27	3	2	5	0.5
M1 [29]	40	40	100	ATC	692247	1385046	AO: 309.50	3	3	3	0.80
M1 [29]	40	40	100	AMSTC	4978835	11559283	AO: 230.12	3	3	3	0.83

8.1   Experimental Results

Our tool DARKSIDER[5] implements the computation of ATCs and AMSTCs and optimizes the inter-cluster distance and the proportion of clustered traces.

We firstly experimented the clusterings of the model in Fig. 1 and the logs of Fig. 5. We increased the log size and the model size to observe the limits of finding optimal solutions of our alignment-based trace clustering problems. Log traces are slightly noisy, to show different kinds of solutions. As the computation of $\Delta(\mathcal{C})$ takes time due to the numerous combinations for the pseudo-SAT encoding, this criterion has been removed for testing various sizes entries. Likewise, for the complexity reasons explained in Sect. 7, our tool does not implement the optimization of the inter-cluster distance $\Phi(\mathcal{C})$; it simply computes it a posteriori.

---

[5] https://github.com/BoltMaud/darksider.

Table 1 shows experimental results. Our encoding automatically gets the minimal number of required clusters which is usually a parameter to set [22]. Notice that our tool computes *optimal* clusterings for a given setting w.r.t. the inter-cluster distance. Of course, this quest of optimality is very expensive in computation time. This is why, for the larger problems, we stopped the solver after it found solutions that we considered close to the optimal. These experiments are labeled AO in Table 1. For the model with 17 transitions and 14 places and 100 traces, this dramatically decreased the execution time from 5659.75 to 100.16 s. Furthermore, for larger logs and larger models, like 100 traces and the model M1 of [29], we stopped the computation of the optimum, however we got almost optimal results with a ratio of clustered traces to 0.83, where the optimum would be 0.95.

9   Conclusion and Future Work

In this work, we have investigated novel alignment-based trace clustering techniques, that generalize the notion of centroid in different directions: concurrency and repetitive behavior. The paper proposes quality criteria for characterizing trace clustering, and adapts them for each one of the new instantiations proposed. Also, the situations where the two different instantiations collapse are described formally. Furthermore, a complexity analysis on the different instantiations of the methods is reported. The approach has been implemented and tested over some datasets, showing that it can be applied in practice.

As future work we have many avenues to follow: first, we plan to investigate the SAT encoding and interaction with the SAT solver, so that a better performance can be attained. Second, we plan to explore applications of the theory of this paper; we see several possibilities in different process mining sub-domains, ranging from concept drift, down to predictive monitoring.

Acknowledgments. This work has been supported by Farman institute at ENS Paris-Saclay and by MINECO and FEDER funds under grant TIN2017-86727-C2-1-R.

References

1. van der Aalst, W.M.P.: Process Mining - Data Science in Action, 2nd edn. Springer, Heidelberg (2016). https://doi.org/10.1007/978-3-662-49851-4
2. Carmona, J., van Dongen, B.F., Solti, A., Weidlich, M.: Conformance Checking - Relating Processes and Models. Springer, Cham (2018). https://doi.org/10.1007/978-3-319-99414-7
3. Chatain, T., Carmona, J., van Dongen, B.: Alignment-based trace clustering. In: Mayr, H.C., Guizzardi, G., Ma, H., Pastor, O. (eds.) ER 2017. LNCS, vol. 10650, pp. 295–308. Springer, Cham (2017). https://doi.org/10.1007/978-3-319-69904-2_24
4. Dumas, M., van der Aalst, W.M.P., ter Hofstede, A.H.M.: Process-Aware Information Systems: Bridging People and Software Through Process Technology. Wiley, New York (2005)

5. Greco, G., Guzzo, A., Pontieri, L., Saccà, D.: Discovering expressive process models by clustering log traces. IEEE Trans. Knowl. Data Eng. **18**(8), 1010–1027 (2006)
6. Ferreira, D., Zacarias, M., Malheiros, M., Ferreira, P.: Approaching process mining with sequence clustering: experiments and findings. In: Alonso, G., Dadam, P., Rosemann, M. (eds.) BPM 2007. LNCS, vol. 4714, pp. 360–374. Springer, Heidelberg (2007). https://doi.org/10.1007/978-3-540-75183-0_26
7. Song, M., Günther, C.W., van der Aalst, W.M.P.: Trace clustering in process mining. In: Ardagna, D., Mecella, M., Yang, J. (eds.) BPM 2008. LNBIP, vol. 17, pp. 109–120. Springer, Heidelberg (2009). https://doi.org/10.1007/978-3-642-00328-8_11
8. Bose, R.P.J.C., van der Aalst, W.M.P.: Context aware trace clustering: Towards improving process mining results. In: Proceedings of the SIAM International Conference on Data Mining, SDM 2009, pp. 401–412 (2009)
9. Bose, R.P.J.C., van der Aalst, W.M.P.: Trace clustering based on conserved patterns: towards achieving better process models. In: Rinderle-Ma, S., Sadiq, S., Leymann, F. (eds.) BPM 2009. LNBIP, vol. 43, pp. 170–181. Springer, Heidelberg (2010). https://doi.org/10.1007/978-3-642-12186-9_16
10. Weerdt, J.D., vanden Broucke, S.K.L.M., Vanthienen, J., Baesens, B.: Active trace clustering for improved process discovery. IEEE Trans. Knowl. Data Eng. **25**(12), 2708–2720 (2013)
11. Hompes, B., Buijs, J., van der Aalst, W., Dixit, P., Buurman, H.: Discovering deviating cases and process variants using trace clustering. In: Proceedings of the 27th Benelux Conference on Artificial Intelligence (BNAIC 2015) (2015)
12. Ponce-de-León, H., Rodríguez, C., Carmona, J., Heljanko, K., Haar, S.: Unfolding-based process discovery. In: Finkbeiner, B., Pu, G., Zhang, L. (eds.) ATVA 2015. LNCS, vol. 9364, pp. 31–47. Springer, Cham (2015). https://doi.org/10.1007/978-3-319-24953-7_4
13. Ponce de León, H., Rodríguez, C., Carmona, J.: POD - a tool for process discovery using partial orders and independence information. In: Proceedings of the BPM Demo Session 2015 Co-located with the 13th International Conference on Business Process Management (BPM 2015), pp. 100–104 (2015)
14. Lu, X., Fahland, D., van der Aalst, W.M.P.: Conformance checking based on partially ordered event data. In: Fournier, F., Mendling, J. (eds.) BPM 2014. LNBIP, vol. 202, pp. 75–88. Springer, Cham (2015). https://doi.org/10.1007/978-3-319-15895-2_7
15. de San Pedro, J., Cortadella, J.: Mining structured Petri nets for the visualization of process behavior. In: Proceedings of the 31st Annual ACM Symposium on Applied Computing, pp. 839–846 (2016)
16. Mokhov, A., Cortadella, J., de Gennaro, A.: Process windows. In: 17th International Conference on Application of Concurrency to System Design, ACSD 2017, pp. 86–95 (2017)
17. Lu, X., Fahland, D., van den Biggelaar, F.J.H.M., van der Aalst, W.M.P.: Detecting deviating behaviors without models. In: Reichert, M., Reijers, H.A. (eds.) BPM 2015. LNBIP, vol. 256, pp. 126–139. Springer, Cham (2016). https://doi.org/10.1007/978-3-319-42887-1_11
18. Benedikt, M., Puppis, G., Riveros, C.: Regular repair of specifications. In: Proceedings of the 26th Annual IEEE Symposium on Logic in Computer Science. LICS 2011, pp. 335–344 (2011)
19. Murata, T.: Petri nets: properties, analysis and applications. Proc. IEEE **77**(4), 541–574 (1989)

20. Engelfriet, J.: Branching processes of Petri nets. Acta Informatica **28**(6), 575–591 (1991)
21. Gonzalez, T.F.: Clustering to minimize the maximum intercluster distance. Theor. Comput. Sci. **38**, 293–306 (1985)
22. Berkhin, P.: A survey of clustering data mining techniques. In: Kogan, J., Nicholas, C., Teboulle, M. (eds.) Grouping Multidimensional Data, pp. 25–71. Springer, Heidelberg (2006). https://doi.org/10.1007/3-540-28349-8_2
23. Dunn, J.C.: A fuzzy relative of the ISODATA process and its use in detecting compact well-separated clusters. J. Cybern. **3**(3), 32–57 (1973). https://doi.org/10.1080/01969727308546046
24. Rousseeuw, P.J.: Silhouettes: a graphical aid to the interpretation and validation of cluster analysis. J. Comput. Appl. Math. **20**, 53–65 (1987)
25. Czerwinski, W., Lasota, S., Lazic, R., Leroux, J., Mazowiecki, F.: The reachability problem for Petri nets is not elementary (extended abstract). CoRR abs/1809.07115 (2018)
26. Stewart, I.A.: Reachability in some classes of acyclic Petri nets. Fundam. Inform. **23**(1), 91–100 (1995)
27. Cheng, A., Esparza, J., Palsberg, J.: Complexity results for 1-safe nets. Theor. Comput. Sci. **147**(1&2), 117–136 (1995)
28. Eén, N., Sörensson, N.: Translating pseudo-boolean constraints into SAT. JSAT **2**(1–4), 1–26 (2006)
29. Taymouri, F., Carmona, J.: Model and event log reductions to boost the computation of alignments. In: Ceravolo, P., Guetl, C., Rinderle-Ma, S. (eds.) SIMPDA 2016. LNBIP, vol. 307, pp. 1–21. Springer, Cham (2018). https://doi.org/10.1007/978-3-319-74161-1_1

Finding Complex Process-Structures
by Exploiting the Token-Game

Lisa Luise Mannel[✉] and Wil M. P. van der Aalst

Process and Data Science (PADS), RWTH Aachen University, Aachen, Germany
{mannel,wvdaalst}@pads.rwth-aachen.de

Abstract. In process discovery, the goal is to find, for a given event log, the model describing the underlying process. While process models can be represented in a variety of ways, in this paper we focus on the representation by Petri nets. Using an approach inspired by language-based regions, we start with a Petri net without any places, and then insert the maximal set of places considered fitting with respect to the behavior described by the log. Traversing and evaluating the whole set of all possible places is not feasible since their number is exponential in the number of activities. Therefore, we propose a strategy to drastically prune this search space to a small number of candidates, while still ensuring that all fitting places are found. This allows us to derive complex model structures that other discovery algorithms fail to discover. In contrast to traditional region-based approaches this new technique can handle infrequent behavior and therefore also noisy real-life event data. The drastic decrease of computation time achieved by our pruning strategy, as well as our noise handling capability, is demonstrated and evaluated by performing various experiments.

Keywords: Process discovery · Petri nets · Language-based regions

1 Introduction

More and more processes executed in companies are supported by information systems which store each event executed in the context of a so-called *event log*. For each event, such an event log typically describes a name identifying the executed activity, the respective case specifying the execution instance of the process, the time when the event was observed, and often other data related to the activity and/or process instance. An example event log is shown in Fig. 1.

In the context of process mining, many algorithms and software tools have been developed to utilize the data contained in event logs: in *conformance checking*, the goal is to determine whether the behaviors given by a process model and event log comply. In *process enhancement*, existing process models are improved. Finally, in *process discovery*, a process model is constructed aiming to reflect the

© Springer Nature Switzerland AG 2019
S. Donatelli and S. Haar (Eds.): PETRI NETS 2019, LNCS 11522, pp. 258–278, 2019.
https://doi.org/10.1007/978-3-030-21571-2_15

behavior defined by the given event log: the observed events are put into relation to each other, preconditions, choices, concurrency, etc. are discovered, and brought together in a model, e.g. a Petri net.

Case ID	Activity	Time Stamp	Resource
1	▶	01.01.2019	R0
1	A	02.01.2019	R1
2	▶	07.01.2019	R0
1	C	11.03.2019	R2
2	B	01.05.2019	R4
2	C	07.07.2019	R2
2	C	08.07.2019	R2
2	C	11.07.2019	R2
1	C	26.08.2019	R2
1	D	27.09.2019	R3
1	■	29.09.2019	R0
2	E	07.12.2019	R4
2	■	08.12.2019	R0
		...	

Real-life processes usually have a start and end, and therefore it is reasonable to assume a designated start activity ▶ to be executed at the beginning of each process instance, as well as a corresponding end activity ■. For a (simplified) process of package delivery by the postal service, this could, for example, be *package registration* and *confirmation of delivery*. Each delivery process would be corresponding to a case, with possible activities like *attempt delivery* or *relocate package*. Possible resources could be the car used for delivery or the employee processing the package.

Fig. 1. Excerpt of an example event log. The two visible cases correspond to the activity sequences $\langle \blacktriangleright, A, C, C, D, \blacksquare \rangle$ and $\langle \blacktriangleright, B, C, C, C, E, \blacksquare \rangle$.

Process discovery is non-trivial for a variety of reasons. The behavior recorded in an event log cannot be assumed to be complete, since behavior allowed by the process specification might simply not have happened yet. Additionally, real-life event logs often contain rare patterns, either due to infrequent behavior or due to logging errors. Especially the latter type should not be taken into account when discovering the process model, but finding a balance between filtering out noise and at the same time keeping all desired information is often a non-trivial task. Ideally, a discovered model should be able to produce the behavior contained within the event log, not allow for behavior that was not observed, represent all dependencies between the events and at the same time be uncomplicated enough to be understood by a human interpreter. It is rarely possible to fulfill all these requirements simultaneously. Based on the capabilities and focus of the used algorithm, the discovered models can vary greatly. Often, there is no one and only true model, but instead, a trade-off between the aspects noted above has to be found. In Sect. 2 we give an overview of work related to our paper. For a detailed introduction to process mining we refer the interested reader to [1].

In this paper we suggest an algorithm inspired by language-based regions, that guarantees to find a model defining the minimal language containing the input language [2]. Due to our assumptions, the usually infinite set of all possible places is finite. In contrast to prominent discovery approaches based on language-based regions (see Sect. 2), we do not use integer linear programming to find the subset of fitting places. Instead, we replay the event log on candidate places to evaluate whether they are fitting. We achieve that by playing the token game for each trace in the log, and then utilizing the results to skip uninteresting sections of the search space as suggested in [3]. In contrast to the brute-force approach evaluating every single candidate place, our technique drastically increases the efficiency of candidate evaluation by combining this skipping of candidates with a smart candidate traversal strategy, while still providing the same guarantees. Additionally, our algorithm lends itself to apply efficient noise-filtering, as well as other user-definable constraints on the set of fitting places. As a final step, we suggest to post-process the discovered set of fitting places, thereby reducing the complexity of the resulting model and making it interpretable by humans. Altogether, our approach has the potential to combine the capability of discovering complex model structures, typical for region-based approaches, with the ability to handle noise and simplify the model according to user-definable constraints. We illustrate the capabilities of our algorithm by providing results of measuring its decrease in computation time compared to the brute-force approach, testing its noise-handling abilities, and illustrating the rediscovery of complex models.

An overview of related work is given in the next section. In the remainder of this paper we provide a detailed description, formalization and discussion of our discovery approach. In Sect. 3 basic notations and definitions are given. We present a detailed motivation and overview of our approach in Sect. 4. Section 5 provides an extensive explanation and formalization. In Sect. 6, we briefly discuss our implementation, including some tweaks and optimizations that can be used to further improve our approach. A comparison to existing discovery algorithms is given in Sect. 7 together with results and evaluation of testing. Finally, we conclude the paper with a summary and suggestion of future work in Sect. 8.

2   Related Work

Process discovery algorithms make use of a variety of formal and informal representations to model the behavior they extract from a given event log. However, the basic idea is similar: based on the event data given by an event log, the different event types are coordinated and ordered using some kind of connection between them. In this paper, we focus on the formal representation by Petri nets, where the event types correspond to transitions and the coordinating connections correspond to places. However, our ideas can be adapted to other representations as well. Discovery algorithms that produce a formal process model can provide desirable formal guarantees for their discovered models, for example, the ability to replay each sequence of events contained in the log, or the ability to re-discover a model based on a sufficient description of its behavior.

As noted above in process discovery there are several, often conflicting quality criteria for a discovered model. To decrease computation time and the complexity of the found Petri net, many existing discovery algorithms further abstract from a given log to another representation, containing only a fraction of the original information, based on which a formal model is created. These algorithms can rediscover only subclasses of Petri nets, and often the resulting model does not allow for the log to be fully replayed, or allows for much more than the log suggests. Examples are the Alpha Miner variants [4] and the Inductive Mining family [5]. Other miners based on heuristics, like genetic algorithms or Heuristic Miner [6] cannot provide guarantees at all.

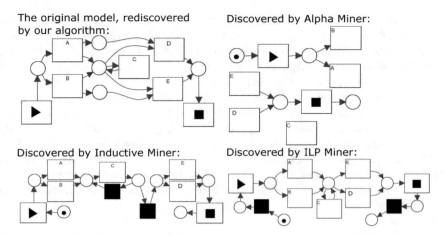

Fig. 2. There exists a variety of discovery algorithms that are able to mine a process model based on the log $[\langle \blacktriangleright, A, C, C, D\blacksquare\rangle, \langle \blacktriangleright, B, C, C, C, E\blacksquare\rangle]$ from Fig. 1. As illustrated, the resulting Petri nets differ significantly between the algorithms. In particular, the places connecting A to D and B to E, which ensure that the first choice implies the second choice, are rarely discovered by existing algorithms.

Due to omitting part of the information contained in the log, the miners described above are not able to discover complex model structures, most prominently non-free choice constructs. The task of creating a Petri net that precisely corresponds to a given description of its behavior is known as the synthesis problem and closely related to region theory [7]. Traditionally, this description was given in form of a transition system, which was then transformed into a Petri net using state-based region theory [8]. The approach has been adapted for process discovery by developing algorithms, e.g. FSM Miner, that generate a transition system based on the log, which is then transformed into a Petri net [9,10]. Other approaches use language-based region theory, where the given description is a language, rather than a transition system [11–13]. An event log can be directly interpreted as a language. Therefore, language-based regions can be applied directly to synthesize a Petri net from a given event log [2]. Here the basic strategy is to start with a Petri net that has one transition for each activity type contained in the log. Then all places that allow replaying the log are added. The result is a Petri net, that defines a language which is a minimal superset of the input language. Currently such algorithms, most prominently ILP Miner, are based on integer linear programming [14–16]. However, available implementations make use of an abstraction of the log to increase performance, thus losing their ability to find all possible places.

In contrast to most other discovery algorithms, region-based approaches guarantee that the model can replay the complete event log, and at the same time does not allow for much different behavior. In particular, complex model structures like non-free choice constructs are reliably discovered. On the downside, region-based discovery algorithms often lead to complex process models that are impossible to understand for the typically human interpreter. They are also known to expose severe issues with respect to low-frequent behavior often contained in real-life event logs. Finally, finding all the fitting places out of all possible places tends to be extremely time-consuming.

To illustrate the differences between existing discovery algorithms, in Fig. 2 we show the results of selected discovery algorithms applied to the log shown in Fig. 1. The original model, that produced the log, can be rediscovered by the approach suggested in this paper. Alpha Miner, Inductive Miner and ILP Miner cannot discover this model because they are restricted to mine only for certain structures and/or a subset of possible places. In particular, the implication of the second choice (C orD) by the first choice (A or B) is not discovered by any of these algorithms.

3   Basic Notation, Event Logs and Process Models

Throughout our paper we will use the following notations and definitions: A set, e.g. $\{a, b, c\}$, does not contain any element more than once, while a multiset, e.g. $[a, a, b, a] = [a^3, b]$, may contain multiples of the same element. By $\mathbb{P}(X)$ we refer to the power set of the set X, and $\mathbb{M}(X)$ is the set of all multisets over this set. In contrast to sets and multisets, where the order of elements is irrelevant, in sequences the elements are given in a certain order, e.g. $\langle a, b, a, b \rangle \neq \langle a, a, b, b \rangle$. We refer to the i'th element of a sequence σ by $\sigma(i)$. The size of a set, multiset or sequence X, that is $|X|$, is defined to be the number of elements in X.

We define activities, traces, and logs as usual, except that we require each trace to begin with a designated start activity and end with a designated end activity. Since process models, in general, have a start and end, this is a reasonable assumption. It implies, that in any discovered model all places are intermediate places that are not part of an initial or final marking. Thus, every candidate place we consider during execution of our algorithm, does not contain any initial tokens. This greatly simplifies the presentation of our work. Note, that any log can easily be transformed accordingly.

Definition 1 (Activity, Trace, Log). *Let $\mathcal{A}$ be the universe of all possible activities (actions, events, operations, ...), let $\blacktriangleright \in \mathcal{A}$ be a designated start activity and let $\blacksquare \in \mathcal{A}$ be a designated end activity. A trace is a sequence containing $\blacktriangleright$ as the first element, $\blacksquare$ as the last element and in-between elements of $\mathcal{A} \backslash \{\blacktriangleright, \blacksquare\}$. Let $\mathcal{T}$ be the set of all such traces. A log $L \subseteq \mathbb{M}(\mathcal{T})$ is a multiset of traces.*

In the following definition of Petri nets, note that we require the set of transitions to correspond to a subset of the universe of activities. Therefore our Petri nets are free of silent or duplicate transitions. In combination with not having to deal with markings, this results in a finite set of candidate places. Also note, that we do not use weighted arcs, and can therefore assume the arcs to be implicitly defined by the sets of ingoing and outgoing transitions of the places. This definition of a subset of Petri nets, natural with respect to our discovery algorithm, removes a lot of notational overhead (Fig. 3).

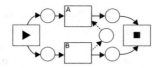

Fig. 3. Example of a Petri net $N = (A, \mathcal{P})$ with transitions $A = \{a, b, \blacktriangleright, \blacksquare\}$ and places $\mathcal{P} = \{(\{\blacktriangleright\}|\{a\}), (\{\blacktriangleright\}|\{b\}), (\{a\}|\{\blacksquare\}), (\{b\}|\{\blacksquare\})\}$. The behavior of N is the set of fitting traces $\{\langle \blacktriangleright, a, b, \blacksquare \rangle, \langle \blacktriangleright, b, a, \blacksquare \rangle\}$. A possible place $(\{b\}|\{a\})$ is underfed with respect to the trace $\langle \blacktriangleright, a, b, \blacksquare \rangle$.

Definition 2 (Petri nets). *A Petri net is a pair $N = (A, \mathcal{P})$, where $A \subseteq \mathcal{A}$ is the set of transitions, and $\mathcal{P} \subseteq \{(I|O) \mid I \subseteq A \wedge I \neq \emptyset \wedge O \subseteq A \wedge O \neq \emptyset\}$ is the set of places. We call I the set of* ingoing *activities of a place and O the set of* outgoing *activities.*

Places, that are not able to perfectly replay the given log, can be unfitting in two ways: If at some point during replay there is no token at the place, but the log requires the consumption of a token anyway, we call the place *underfed*. If at the end of a replayed trace there is at least one token left at the place, we call the place *overfed*. This categorization has been extensively discussed in [3] and is the key to our efficient candidate traversal: as detailed in Sect. 4, by evaluating one place to be underfed (overfed) we can determine a whole set of places to be underfed (overfed), without even looking at them.

Definition 3 (Overfed/Underfed/Fitting Places, see [3]). *Let $N = (A, \mathcal{P})$ be a Petri net, let $p = (I|O) \in \mathcal{P}$ be a place, and let σ be a trace. With respect to the given trace σ, p is called*

- *underfed, denoted by $\triangledown_\sigma(p)$, if and only if $\exists k \in \{1, 2, ..., |\sigma|\}$ such that $|\{i \mid i \in \{1, 2, ...k-1\} \wedge \sigma(i) \in I\}| < |\{i \mid i \in \{1, 2, ...k\} \wedge \sigma(i) \in O\}|$,*
- *overfed, denoted by $\triangle_\sigma(p)$, if and only if $|\{i \mid i \in \{1, 2, ...|\sigma|\} \wedge \sigma(i) \in I\}| > |\{i \mid i \in \{1, 2, ...|\sigma|\} \wedge \sigma(i) \in O\}|$,*
- *fitting, denoted by $\square_\sigma(p)$, if and only if not $\triangledown_\sigma(p)$ and not $\triangle_\sigma(p)$.*

Note, that a place can be underfed and overfed at the same time.

Definition 4 (Behavior of a Petri net). *We define the* behavior *of the Petri net $(A, \mathcal{P})$ to be the set of all fitting traces, that is $\{\sigma \in \mathcal{T} \mid \forall p \in \mathcal{P}: \square_\sigma(p)\}$.*

4   Algorithmic Framework

As input our algorithm takes a log L, that can be directly interpreted as a language, and a parameter $\tau \in [0, 1]$. This parameter τ in principle determines the fraction of the log that needs to be replayable by a fitting place, and is essential for our noise handling strategy. We provide details on this threshold at the end of this section.

Inspired by language-based regions, the basic strategy of our approach is to begin with a Petri net, whose transitions correspond to the activity types used in the given log. From the finite set of unmarked, intermediate places $\mathcal{P}_{\text{all}}$ we want to select a subset $\mathcal{P}_{\text{final}}$, such that the language defined by the resulting net defines the minimal language containing the input language, while, for human readability, using only a minimal number of places to do so. Note, that by filtering the log for noise the definition of the desired language becomes less rigorous, since the allegedly noisy parts of the input language are ignored, and thus the aforementioned property can no longer be guaranteed.

We achieve this aim by applying several steps detailed in Sect. 5. First of all, we observe that the set $\mathcal{P}_{\text{all}}$ contains several places that can never be part of a solution, independently of the given log. By ignoring these places we reduce our set of candidates to $\mathcal{P}_{\text{cand}} \subseteq \mathcal{P}_{\text{all}}$. Next, we apply the main step of our approach: while utilizing the token-game to skip large parts of the candidate space, we actually evaluate only a subset of candidates, $\mathcal{P}_{\text{visited}} \subseteq \mathcal{P}_{\text{cand}}$. We can guarantee that the set of fitting places is a subset of these evaluated places, that is $\mathcal{P}_{\text{fit}} \subseteq \mathcal{P}_{\text{visited}}$. Finally, aiming to achieve a model that is interpretable by human beings, we reduce this set of fitting places to a set of final places $\mathcal{P}_{\text{final}} \subseteq \mathcal{P}_{\text{fit}}$ by removing superfluous places.

The main challenge of our approach lies in the size of the candidate space: there are $|\mathcal{P}_{\text{all}}| = |\mathbb{P}(A) \backslash \emptyset \times \mathbb{P}(A) \backslash \emptyset| \approx (2^{|A|})^2$ possible places to be considered. Keeping all of them in memory and even more replaying the log for this exponentially large number of candidates will quickly become infeasible, even for comparably small numbers of activities. Reducing the set $\mathcal{P}_{\text{all}}$ to $\mathcal{P}_{\text{cand}}$ is by far not a sufficient improvement.

Towards a solution to this performance issue, we propose an idea allowing us to drastically reduce the amount of traversed candidate places, while still providing the complete set $\mathcal{P}_{\text{fit}}$ as outcome. The monotonicity results on Petri net places introduced in [3] form the basis of our approach. Intuitively, if a candidate place $p_1 = (I_1|O_1)$ is underfed with respect to some trace σ, then at some point during the replay of σ there are not enough tokens in p_1. By adding another outgoing arc to p connecting it to some transition $a \notin O_1$ we certainly do not increase the number of tokens in the place and therefore the resulting place $p_2 = (I_1|O_1 \cup \{a\})$ must be underfed as well. Thus, by evaluating p_1 to be underfed we can infer that all candidates $(I_1|O_2)$ with $O_1 \subseteq O_2$ are underfed as well, without having to evaluate them. A similar reasoning can be applied to overfed places. This is formalized in Lemma 1. For more details, we refer the reader to the original paper [3].

Lemma 1 (Monotonicity Results (see [3])). *Let $p_1 = (I_1|O_1)$ be a place and let σ be a trace. If $\triangledown_\sigma(p_1)$, then $\triangledown_\sigma(p_2)$ for all $p_2 = (I_2|O_2)$ with $I_1 \supseteq I_2$ and $O_1 \subseteq O_2$. If $\triangle_\sigma(p_1)$, then $\triangle_\sigma(p_2)$ for all $p_2 = (I_2|O_2)$ with $I_1 \subseteq I_2$ and $O_1 \supseteq O_2$.*

As detailed in [3], these monotonicity results allow us to determine a whole set of places to be unfitting by evaluating a single unfitting candidate. Combining this idea with the candidate traversal strategy presented in Sect. 5 allows us to skip most unfitting places when traversing the candidate space, without missing any other, possibly interesting place. It is important to note that we do not simply skip the replay of the log, but actually do not traverse these places at all. This greatly increases the performance of our algorithm.

Setting our algorithm apart from other region-based approaches is its ability to directly integrate noise filtering. When evaluating a visited place, we refer to a user-definable parameter τ as detailed in the following definition:

Definition 5 (Fitness with Respect to a Threshold). *With respect to a given log L and threshold τ, we consider a place $p =(I|O)$ to be*

- *fitting, that is $\Box_L^\tau(p)$, if and only if* $\frac{|\{\sigma \in L \mid \Box_\sigma(p) \wedge \sigma \cap (I \cup O) \neq \emptyset\}|}{|\{\sigma \in L \mid \exists a \in \sigma : a \in (I \cup O)\}|} \geq \tau,$
- *underfed, that is $\triangledown_L^\tau(p)$, if and only if* $\frac{|\{\sigma \in L \mid \triangledown_\sigma(p)\}|}{|\{\sigma \in L \mid \exists a \in \sigma : a \in (I \cup O)\}|} > (1 - \tau),$
- *overfed, that is $\triangle_L^\tau(p)$, if and only if* $\frac{|\{\sigma \in L \mid \triangle_\sigma(p)\}|}{|\{\sigma \in L \mid \exists a \in \sigma : a \in (I \cup O)\}|} > (1 - \tau).$

Intuitively, a place p is fitting/underfed/overfed/ with respect to L and τ, if it is underfed/overfed/fitting with respect to a certain fraction of traces in L that involve the activities of p. This fraction is determined by the threshold τ. By defining the value of τ, the user of our algorithm can choose to ignore a fraction of traces when evaluating the places, making the result much more robust with respect to infrequent behavior. If $\tau = 1$, then all places are required to be perfectly fitting. In [3] it is shown that Lemma 1 can be extended to the use of such a threshold. Despite our slightly modified definition, their proof remains valid. The impact of different values of τ will be investigated in Sect. 7.

5   Computing a Desirable Subset of Places

As input, our algorithm expects a log L, and a user-definable parameter $\tau \in [0, 1]$. The activities contained in L, $A \subseteq \mathcal{A}$, define the set of transitions of the Petri net we want to discover. These define a finite set of unmarked, intermediate places $\mathcal{P}_{\mathrm{all}}$, that we could insert, as the starting point of our algorithm.

5.1   Pre-pruning of Useless Places

Within the set $\mathcal{P}_{\mathrm{all}}$, there are many candidates that are guaranteed to be unfitting, independently of the given log. These are all places $(I|O)$, with $\blacktriangleright \in O$ or with $\blacksquare \in I$ for designated start and end activities $\blacktriangleright$, $\blacksquare$. By completely excluding such places from our search space, we remain with a much smaller set of candidates $\mathcal{P}_{\mathrm{cand}} \subseteq \mathcal{P}_{\mathrm{all}}$: For a set of activities A the number of candidates is bounded by $2^{|A|} \times 2^{|A|}$. By removing all places with $\blacktriangleright \in O$ or $\blacksquare \in I$, we effectively decrease the size of the activity set A by one for each, incoming and outgoing activities. The new bound on the number of candidates is $2^{|A|} \times 2^{|A|}$, thus reducing its size by 25%.

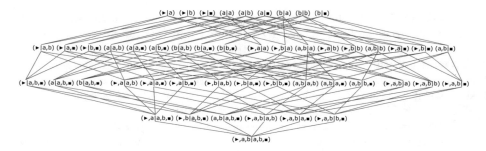

Fig. 4. Example of the candidate space based on four activities ▶, a, b and ■. The horizontal levels and colored edges indicate the relations between the activity sets: each candidate place $(I_1|O_1)$ is connected to a place $(I_2|O_2)$ from the level above by a blue line if $I_1 = I_2$ and $O_1 \supseteq O_2$, and by a red line if $I_1 \supseteq I_2$ and $O_1 = O_2$.

5.2   Developing an Efficient Candidate Traversal Strategy

We illustrate our idea with the help of a running example, where $A = \{▶, a, b, ■\}$ is the given set of activities. The corresponding pre-pruned candidate space is visualized in Fig. 4. The organization and coloring are chosen to clarify the relations between the candidates, which we are going to utilize to further prune the candidate space.

Our strategy for traversing the set of candidates is the key to the effective utilization of the described monotonicity results. Since we cannot efficiently store and retrieve an exponential amount of candidate places, we need a scheme to deterministically compute the next place we want to evaluate based on a limited set of information we keep in storage. This scheme should at the same time guarantee that all fitting places are visited, each place is visited only a limited number of times (preferably at most once), and we are able to prune the search space by employing the results obtained by previous place evaluations. In the following, we are going to develop such a candidate traversal scheme.

5.3   Organization of the Candidate Space as Trees

We organize the candidate space in a set of trees, as described in Definition 6. An example is shown in Fig. 5. Note, that there are many ways to organize the candidates in such a structure and this example shows merely one of these possibilities.

Let $A \subseteq \mathcal{A}$ be the given set of activities and let $>_i$ and $>_o$ be two total orderings on A. In the remainder, we assume all sets of ingoing activities of a place to be ordered lexicographically according to $>_i$, and sets of outgoing activities according to $>_o$. Possible strategies of computing such orderings are noted in Sect. 6.

Definition 6 (Complete Candidate Tree). *A* complete candidate tree *is a pair* $CT = (N, F)$ *with* $N = \{(I|O) \mid I \subseteq A\backslash\{\blacksquare\}, O \subseteq A\backslash\{\blacktriangleright\}, I \neq \emptyset, O \neq \emptyset\}$. *We have that* $F = F_{red} \cup F_{blue}$, *with*

$$F_{red} = \{((I_1|O_1), (I_2|O_2)) \in N \times N \mid |O_2| = 1, O_1 = O_2,$$

$$\exists a \in I_1 \colon (I_2 \cup \{a\} = I_1 \wedge \forall a' \in I_2 \colon a' <_i a)\} \; \textbf{\textit{(red edges)}}$$

$$F_{blue} = \{((I_1|O_1), (I_2|O_2)) \in N \times N \mid I_1 = I_2,$$

$$\exists a \in O_1 \colon (O_2 \cup \{a\} = O_1 \wedge \forall a' \in O_2 \colon a' <_o a)\} \; \textbf{\textit{(blue edges)}}.$$

If $((I_1|O_1), (I_2|O_2)) \in F$, *we call the candidate* $(I_1|O_1)$ *the* child *of its* parent $(I_2|O_2)$. *A candidate* $(I|O)$ *with* $|I| = |O| = 1$ *is called a* base root.

Note, that there is a certain asymmetry in the definition of F_{red} and F_{blue}: red edges connect only places which have exactly one outgoing transition, while for blue edges the number of outgoing transitions is unlimited. This is necessary to obtain the collection of trees we are aiming for, and which is further investigated below: if we did not restrict one of the edge types in this way, the resulting structure would contain cycles. If we restricted both types of edges, many candidates would not be connected to a base root at all. However, the choice of restricted edge type, red or blue, is arbitrary.

In the following we show that each candidate is organized into exactly one tree, that can be identified by its unique base root. Therefore, the number of connected trees contained in one structure CT as described in Definition 6 is exactly the number of base roots. In our running example (Fig. 5) there are 9 such trees.

Lemma 2. *The structure CT described by Definition 6 organizes the candidate space into a set of trees rooted in the base roots, where every candidate is connected to exactly one base root.*

Proof (Lemma 2). Let $CT = (N, F)$ be the structure defined in Definition 6. We show that every candidate $(I|O) \in N$ has a unique parent, and, by induction on the number of activities of a candidate, that each candidate is the descendant of exactly one of the base roots. This implies that there are no cycles and the structure is indeed a set of connected trees rooted in the base roots.

If $|I \cup O| = 2$ then p is a base root that cannot have any parents and the claim holds. Now assume that the claim holds for all candidates with $|I \cup O| = n$.

Consider a candidate $p_1 = (I_1|O_1)$ with $|I_1 \cup O_1| = n + 1$. Let $p_2 = (I_2|O_2)$ be any potential parent of of p_1. We distinguish two cases:

Case $|O_1| = 1$: This implies $I_1 = \{a_1, a_2, ..., a_{n-1}, a_n\}$. We have that $(p_1, p_2) \notin F_{\text{blue}}$, because, otherwise, we would have $O_2 = \emptyset$. If $(p_1, p_2) \in F_{\text{red}}$, then by definition $O_1 = O_2$ and $I_1 = I_1\backslash\{a_n\}$.

Case $|O_1| \geq 2$: This implies $O_1 = \{a_1, a_2, ..., a_{k-1}, a_k\}$, for some $k \in [2, n-1]$. We have that $(p_1, p_2) \notin F_{\text{red}}$, because red edges require $|O_1| = 1$. If $(p_1, p_2) \in F_{\text{blue}}$, then by definition we have that $I_1 = I_2$ and $O_2 = O_1\backslash\{a_k\}$.

In both cases, p_2 is fully defined based on p_1 and therefore the unique parent. Since $|I_2 \cup O_2| = n$, the claim holds for p_2 and thus for p_1 as well. By induction, the claim holds for all candidates in N. $\square$

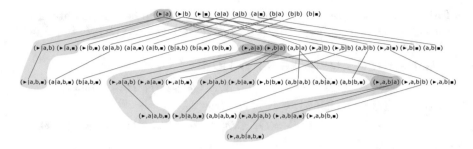

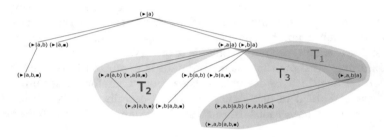

Fig. 5. Example of a complete candidate tree based on $A = \{\blacktriangleright, a, b, \blacksquare\}$. In reference to Lemma 3, an example subtree of red edges is marked by a red background, while all subtrees of blue edges attached to it are marked by a blue background. Together they form the whole tree rooted in the base root ($\{\blacktriangleright\}|\{a\}$).

Fig. 6. We illustrate our tree cutting strategy using the tree rooted in the base root ($\{\blacktriangleright\}|\{a\}$) from our running example (Fig. 5). For the example place ($\{\blacktriangleright, a\}|\{a\}$) we highlight the blue-edged subtree (T_2) and red-edged subtree (T_1) referred to in Lemma 3. In reference to Lemma 4 we indicate the subtrees attached by blue edges, which coincides with the blue-edged subtree (T_2), and the subtrees attached by red edges (T_3). We emphasize that $T_1 \neq T_3$.

5.4   Tree Traversal and Pruning of Candidate Space

Definition 7 (Tree Traversal Strategy). *Let L be a log, τ threshold, and A the set of activities contained in L. Let $CT = (N, F)$ be the complete candidate tree based on A, with $N = \mathcal{P}_{cand}$. We traverse one connected tree after the other, by using a depth-first strategy starting on each base root of CT.*

Let $p = (I|O)$ be a visited place. If p is fitting, i.e. $\square_L^\tau(p)$, we add it to $\mathcal{P}_{fit}$. If p is underfed ($\triangledown_L^\tau(p)$), we skip all subtrees attached to p by a blue edge. If p is overfed ($\triangle_L^\tau(p)$) and $O = \{a\}$ with $\nexists a' \in A: a' >_o a$, we skip all subtrees attached to p by a red edge. After skipping a subtree this way, we proceed with our traversal as if we had already visited those places.

Formalized in Theorem 1, we show that this strategy allows us to skip sets of unfitting places and at the same time guarantee, that no place in $\mathcal{P}_{fit}$ is missed.

Lemma 3. *Let L, τ be a given log and threshold, $CT = (N, F)$ a complete candidate tree and $p_1 \in N$ any candidate. If $\bigtriangledown_L^\tau(p_1)$ $(\bigtriangleup_L^\tau(p_1))$, then for all candidates $p_2 \in N$ within the blue-edged (red-edged) subtree rooted in p_1, we have that $\bigtriangledown_L^\tau(p_2)$ $(\bigtriangleup_L^\tau(p_2))$.*

Proof (Lemma 3). Let $p_1 = (I_1|O_1) \in N$ be a candidate with $\bigtriangledown_L^\tau(p_1)$ $(\bigtriangleup_L^\tau(p_1))$. Consider any descendant $p_2 = (I_2|O_2)$ attached to p_1 via a path of blue (red) edges. From the definition of blue (red) edges (see Definition 6) we infer that $I_1 = I_2$ and $O_1 \subseteq O_2$ $(O_1 = O_2$ and $I_1 \subseteq I_2)$. Then from the monotonicity results presented Lemma 1 it follows that $\bigtriangledown_L^\tau(p_2)$ $(\bigtriangleup_L^\tau(p_2))$. $\qquad\square$

According to Lemma 3, if a place p is underfed, then blue-edged subtree rooted in p contains only underfed places, and if p is overfed, then the red-edged subtree rooted in p contains only overfed places. In the following, we try to extend these results to subtrees rooted in p in general. We will show that for subtrees attached to p by blue edges this is possible in a straight-forward way. Unfortunately, due to the asymmetry of the definition of the edges, the same argument does not always apply to subtrees attached by red edges: here it is entirely possible that there are blue edges within the corresponding subtree, and "overfedness" of p cannot be necessarily extended to the contained places. This is illustrated in Fig. 6.

Lemma 4. *Let L, τ be a given log and threshold, $CT = (N, F)$ a complete candidate tree and $p_1 = (I_1|O_1) \in N$ any candidate. If $\bigtriangledown_L^\tau(p_1)$, then for all candidates p_2 within the subtrees attached to p_1 with a blue edge we have $\bigtriangledown_L^\tau(p_2)$. If $\bigtriangleup_L^\tau(p_1)$ and $O_1 = \{a\}$ with $\nexists a' \in A: a' >_o a$, then for all candidates p_2 within the subtrees attached to p_1 we have $\bigtriangleup_L^\tau(p_2)$.*

Proof (Lemma 4). Assume $\bigtriangledown_L^\tau(p_1)$. Due to the definition of blue edges, it holds for each child $p' = (I'|O')$ in a subtree attached to p_1 by such an edge, that $|O'| > |O_1|$, and thus in particular $|O'| \geq 2$. By definition, all descendants of p' have at least two outgoing activities. This implies that there is no red edge in the whole subtree, since they require by definition at most one outgoing activity. Thus, by Lemma 3, for every place p_2 in a subtree rooted in such a p' we have $\bigtriangledown_L^\tau(p_2)$, and the first claim holds.

Now assume $\bigtriangleup_L^\tau(p_1)$ and $O_1 = \{a\}$ with $\nexists a' \in A: a' >_o a$. Due to the definition of red edges, it holds for p_1 and each descendant $p' = (I'|O')$ of p_1 reachable by red edges, that $O' = \{a\}$. This implies that there is no blue edge in the whole subtree rooted in p_1, since they require the existence of an activity a' with $a <_o a'$. Thus, by Lemma 3, for every place p_2 in the subtree rooted in p_1, we have $\bigtriangleup_L^\tau(p_2)$. $\qquad\square$

Theorem 1. *Given a log L, a threshold τ and the complete candidate tree $CT = (N, F)$ with $N = \mathcal{P}_{cand}$, let $\mathcal{P}_{fit}$ be the set of fitting places with respect to L and τ, that is $\{p \in N \mid \Box_L^\tau(p)\}$. Let $\mathcal{P}_{visited}$ be the set of places visited by our tree traversal strategy described in Definition 7. It holds that $\mathcal{P}_{fit} \subseteq \mathcal{P}_{visited} \subseteq \mathcal{P}_{cand}$.*

Proof (Theorem 1). As proven in Lemma 2 every candidate is contained in exactly one tree rooted in a base root and is thus visited exactly once by standard depth-first search. Thus $\mathcal{P}_{\text{visited}} \subseteq \mathcal{P}_{\text{cand}}$.

Using depth-first search for tree traversal guarantees that for any visited candidate p, we visit the complete subtree rooted in p, before proceeding to another subtree. Thus skipping a subtree does not influence the traversal of other subtrees. If p is underfed/overfed, we can apply Lemma 4 to guarantee that all places contained in the skipped subtrees are underfed/overfed as well. Thus no fitting places are skipped, and we have that $\mathcal{P}_{\text{fit}} \subseteq \mathcal{P}_{\text{visited}} \subseteq \mathcal{P}_{\text{cand}}$. □

As mentioned earlier, we cannot store the complete candidate tree due to its exponential size, and thus the challenge of the tree traversal lies in the deterministic computation of the next candidate based on limited information only. In our algorithm, this information is only the last candidate and its fitness status. We define a total ordering on the set of places easily computable based on the given orderings $>_i, >_o$. We can use this ordering to deterministically select the next subtree to traverse based on the current candidate. The fitness status is used to decide whether to actually traverse the selected subtree, or skip it and select the next one.

While the theoretical worst-case scenario still requires traversing the full candidate space, we have achieved a drastic increase in performance in practical applications. This is presented in detail in Sect. 7.

5.5   Evaluation of Potentially Fitting Candidates

For each place visited during candidate traversal, we need to determine its fitness status. Fitting places are added to the set of fitting places $\mathcal{P}_{\text{fit}}$, which will be the input for the post-processing step (see Sect. 5.6). Overfed and underfed places are not added, instead this fitness status can be used in the context of candidate traversal to skip sets of unfitting places.

To determine the fitness status of a place $p = (I|O)$, we use token-based replay. We replay every trace $\sigma \in L$ on the place p: for each activity $a \in \sigma$, from first to last, if $a \in O$ we decrement the number of tokens at the place by one. Then, if $a \in I$ we increment the number of tokens by one. If the number of tokens becomes negative we consider the place to be underfed with respect to this trace, that is $\triangledown_\sigma(p)$. Otherwise, if the trace has been fully replayed and the number of tokens is larger than zero, we consider the place to be overfed, that is $\triangle_\sigma(p)$. Note that the place can be underfed and overfed at the same time. Based on the replay results of the single traces and the user-definable threshold τ (see Definition 5), we evaluate the fitness status of the place with respect to the whole log.

5.6   Post-processing

In the previous step, a subset of places $\mathcal{P}_{\text{fit}}$ was computed. These are exactly the places considered to be fitting with respect to the given log L and threshold τ.

However, many of them can be removed without changing the behavior of the Petri net implicitly defined by $\mathcal{P}_{\text{fit}}$. Since the process model we return is likely to be interpreted by a human being, such places are undesirable. These places are called *implicit* or sometimes *redundant* and have been studied extensively in the context of Petri nets [17–21]. In the post-processing step, we find and remove those undesirable places by solving an Integer Linear Programming Problem as suggested for example in [21]. The resulting set of places $\mathcal{P}_{\text{final}} \subseteq \mathcal{P}_{\text{fit}}$ is finally inserted into the Petri net that forms the output of our algorithm.

6   Implementation

We implemented our algorithm as a plug-in for ProM [22] named `eST-Miner` using Java. There are many ways in which our idea of organizing the candidate space into a tree structure can be optimized for example with respect to certain types of models, logs or towards a certain goal. Other ideas on how to improve performance can be found in [3].

   In our implementation, we investigated the impact of the orderings of the ingoing and outgoing activity sets ($>_i, >_o$ in Sect. 5) on the fraction of cut off places. They determine the position of candidates within our tree structure. If these orderings are such that underfed/overfed places are positioned close to the root, this leads to large subtrees and thus many places being cut off due to monotonicity results. In Sect. 7, we present first results of testing different activity orderings and evaluate their impact.

7   Testing Results and Evaluation

In this section, we present the results of testing our algorithm on various data sets. We use a selection of artificial log-model pairs to demonstrate our ability to rediscover complex structures and deal with noise. The efficiency of our search space pruning technique and the resulting increase in speed are evaluated using artificial logs as well as real-life logs. An overview of these logs is given in Table 1.

Rediscoverability of Models: In Fig. 7, a simple model is shown, which includes non-free choice constructs: the places $(\{a\}|\{e\})$ and $(\{b\}|\{f\})$ are not part of any directly follows relation, and are therefore not discovered by most existing algorithms that provide formal guarantees. Well-known discovery techniques like Alpha-Miner variants [4] or the Inductive Mining family [5] fail at such tasks. Attempts to extend the capabilities of Alpha Miner to discover such places [4] have been only partially successful. These approaches may result in process models that cannot replay the event log. In some cases, the model may have no fitting traces due to livelocks or deadlocks. More complex structures involving non-free choice constructs, like the model depicted in Fig. 8, are difficult to mine and not rediscovered by most algorithms [4].

   In contrast to existing non-region-based algorithms, our approach guarantees that all fitting places are found, and thus we are able to rediscover every model

Table 1. List of logs used for evaluation. The upper part lists real-life logs while the lower part shows artificial logs. Logs are referred to by their abbreviations. The `Sepsis` log has been reduced by removing all traces that occur at most twice. The log `HP2018` has not yet been published. The much smaller 2017 version can be found at [23]. We use a reduced version with all single trace variants removed. The log `Teleclaims*` results from removing the natural start and end activity activities in the log `Teleclaims` and removing the 15% less common traces. The `Artificial1` log contains a single trace $\langle a, b, c, d, e, f \rangle$. The log `Artificial2` is a log generated based on a random model containing a loop, XOR-Split and silent transition.

Log name	Abbreviation	Activities	Trace variants	Reference
Sepsis-mod	Sepsis	11	27	[24]
HelpDesk2018SiavAnon-mod	HD2018	11	595	(see caption)
Road Traffic Fine Management	RTFM	11	231	[25]
Teleclaims	Teleclaims	11	12	[26]
Teleclaims-mod	Teleclaims*	9	12	[26]
repairexample	Repair	12	77	[26]
running-example	RunEx	8	6	[26]
MyLog1	Artificial1	6	1	(see caption)
MyLog2	Artificial2	7	78	(see caption)
N7	a++CE	7	3	[4]

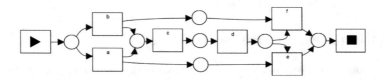

Fig. 7. The shown model can be rediscovered by our algorithm. Since (a, e) and (b, f) are not part of any directly follows relation, most other discovery algorithms fail to discover the corresponding places [4].

that is free of duplicate and silent transitions, assuming that the log is complete (i.e., non-fitting places can be identified). In particular, we can rediscover both models shown in Figs. 7 and 8.

Dealing with Noise: By setting the threshold τ to 1 we require every fitting place, and thus the whole discovered Petri net, to be able to perfectly replay every trace of the log. However, event logs often contain some noise, be it due to exceptional behavior or due to logging errors. Often, we want to ignore such irregularities within the log when searching for the underlying model. Therefore, we suggest using the parameter τ as a noise filtering technique utilizing the internal working of our algorithm. In contrast to approaches modifying the whole event log, this allows us to filter for each place individually: based on the portion

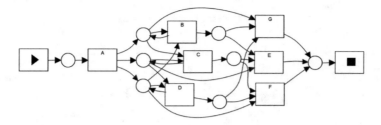

Fig. 8. The Petri net shown in this Figure is an example of a model rediscoverable by our algorithm, that existing non-region-based algorithms fail to find [4]. The only free choice is whether to do B, C or D after doing A. The rest of the trace is implicitly determined by this choice.

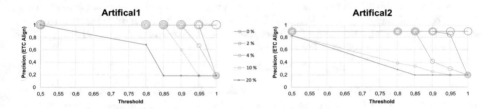

Fig. 9. For various levels of noise (0, 2, 4, 10, 20%) introduced randomly on logs with 1000 traces each, we present the precision results (ETC Align) resulting from applying our algorithm with different values for the threshold τ. Lower values of τ have a positive impact on precision, and all tested models could be rediscovered for certain values of τ.

of the log relevant for the current place, we can ignore a certain fraction of irregular behavior specified by τ without losing other valuable information.

We test our implementation using different values for τ on logs modified to contain several levels of noise. We run our algorithm with different values for τ on logs with 1000 traces and added random noise of 2, 4, 10 and 20%. The resulting model is tested for precision with respect to the original log using the ETC Align Precision Metric implemented in ProM [22]. As shown in Fig. 9, a lower threshold τ, in general, leads to increasing precision of the discovered model. In fact, for adequate values of τ the original models could be rediscovered. Thus, by choosing adequate values for the threshold τ, our algorithm is able to handle reasonable levels of noise within the given event log.

Measuring Performance: The main contribution of our approach lies in our strategy of computing the set of fitting places $\mathcal{P}_{\text{fit}}$. The post-processing step, where implicit/redundant places are removed, has not been optimized and constitutes an application of existing results. Therefore our performance evaluation focuses on the first step. Based on several real-life logs as well as artificial logs we investigate two measures of performance: first, the absolute computation time needed

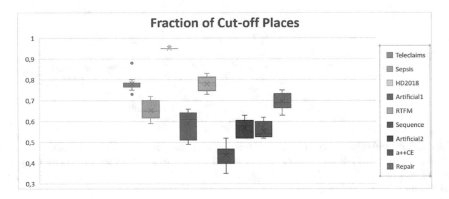

Fig. 10. Depending on the given log and activity orderings, the fraction of places skipped by our algorithm can vary greatly. In this figure, we present box plots showing the fraction of cut-off places $\mathcal{P}_{\text{skipped}}$ for 10 sample runs of our algorithm on different logs, given as the fraction of the complete candidate space ($\mathcal{P}_{\text{skipped}}/\mathcal{P}_{\text{cand}}$). The threshold τ has been set to 1 for all runs.

to discover $\mathcal{P}_{\text{fit}}$, compared to the time needed by a brute force approach traversing the whole set of candidates ($\mathcal{P}_{\text{cand}}$), and second, the fraction of places that were cut off, that is $\mathcal{P}_{\text{skipped}}/\mathcal{P}_{\text{cand}}$. For each log, we performed 10 runs of our algorithm using two mutually independent random activity orderings for in- and outgoing activity sets to survey the influence on these performance measures.

Fraction of Skipped Places: The fraction of cut-off places can vary greatly between different event logs, but also within several runs of our algorithm on the same event log, depending on the chosen activity orderings for in- and outgoing activities. In Fig. 10, we present the results for several logs, based on 10 runs of our algorithm for each. Interestingly, the fraction of cut-off places is highest for the real-life event logs RTFM, HD2018, Sepsis and Teleclaims. For HD2018 it goes as high as 96% of the candidate places, that were never visited. The reason for this could be the more restrictive nature of these large and complex logs, resulting in a smaller set of fitting places $\mathcal{P}_{\text{fit}}$, and thus more possibilities to cut off branches. The lowest results are obtained for the artificial Artificial1 log. Here we could confirm the expectation stated in Sect. 6, that an ordering of high average index first for ingoing activities and lo average index first for outgoing activities, leads to significantly more places being cut off than using the same ordering for both activity sets.

Comparison to the Brute Force Approach: We evaluate the increase in performance of computing $\mathcal{P}_{\text{fit}}$ using our algorithm pruning the candidate space, in comparison to the brute force approach traversing the candidate space without any pruning. We choose three real-life event logs, RTFM, HD2018 and Sepsis and perform 10 runs of our algorithm on each. In Fig. 11 the results of these tests are presented: we compare the time needed by the brute force approach to the minimum, maximum and average time needed for each of the three logs. As is to be expected, the impact of applying our technique is most significant for large logs, where replaying the log on a place takes a long time, and thus cutting off places has a big effect. This is the case for the RTFM and HP2018 logs. The Sepsis log, on the other hand, has shorter and fewer traces, reflected in a smaller difference in the time needed by our algorithm compared to the brute force approach.

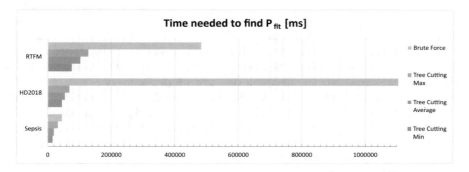

Fig. 11. Minimal, maximal and average time in milliseconds needed by 10 sample runs of our tree cutting algorithm on three real-life logs, with $\tau = 1$, compared to the time needed by the brute force approach traversing the complete candidate space.

8   Conclusion

We have introduced a process discovery algorithm inspired by language-based regions to find a model fitting with respect to a given event log and threshold τ. In contrast to non-region-based approaches our algorithm is able to discover complex structures, most prominently implicit dependencies expressed by non-free choice constructs. In particular, for $\tau = 1$, we can guarantee that the set of fitting places we discover defines a Petri net, such that the language of this net is the minimal language containing the input language (given by the event log). Our candidate traversal strategy allows us to skip large parts of the search space, giving our algorithm a huge performance boost over the brute force approach.

A well-known disadvantage of applying region theory is, that in the context of infrequent behavior within the log, the resulting models tend to be overfitting, not reflecting the core of the underlying process. We can avoid this issue, since our approach lends itself to using the threshold τ as an intuitive noise control mechanics, utilizing the internal workings of our algorithm.

We can see several possibilities for future research based on the presented ideas. The most important contribution of our work is the reduction of the search space to merely a fraction of its original size: we organize all candidate places into trees and are able to cut off subtrees, that are known to contain only unfitting places. Our approach would strongly benefit from any strategy allowing for more subtrees to be cut off or pre-pruning of the candidate space. Note, that within our work we focus on formal strategies that provide the guarantee that all fitting places are discovered. For practical applications it is important to develop heuristic techniques to increase the number of skipped places. Compared to many other approaches this is relatively easy.

New insights can be gained from further testing and evaluating different activity orderings or tree traversal schemes. By developing heuristics on how to choose orderings, based on certain characteristics of the given log, one can optimize the number of skipped candidate places without losing formal guarantees. These guarantees are no longer given, when applying heuristic approaches that allow for fast identification and skipping of subtrees that are likely to be uninteresting. However, for practical applications, the increase in performance is likely to justify the loss of a few fitting places. Alternatively, by evaluating the more interesting subtrees first, the user can be shown a preliminary result, while the remaining candidates are evaluated in the background. User-definable or heuristically derived combinations of ingoing and outgoing activity sets can either be cut off during traversal, or excluded from the search space from the very beginning. Finally, note that our approach allows for each subtree to be evaluated independently, and thus lends itself to increase performance by implementing it in a parallelized manner.

A major issue of our algorithm is the inability to deal with silent and duplicate activities. There exist approaches to identify such activities, either as a general preprocessing step [27] or tailored towards a particular algorithm [28,29]. The applicability of such strategies to our approach remains for future investigation.

We emphasize that our idea is applicable to all process mining related to Petri net definable models, and therefore we see potential not only in our discovery algorithm itself, but also in the combination with, and enhancement of, existing and future approaches.

Acknowledgments. We thank the Alexander von Humboldt (AvH) Stiftung for supporting our research. We thank Alessandro Berti for his support with implementing the eST-Miner, and Sebastiaan J. van Zelst for reviewing this paper.

References

1. van der Aalst, W.: Process Mining: Data Science in Action, 2nd edn. Springer, Heidelberg (2016). https://doi.org/10.1007/978-3-662-49851-4
2. Bergenthum, R., Desel, J., Lorenz, R., Mauser, S.: Process mining based on regions of languages. In: Alonso, G., Dadam, P., Rosemann, M. (eds.) BPM 2007. LNCS, vol. 4714, pp. 375–383. Springer, Heidelberg (2007). https://doi.org/10.1007/978-3-540-75183-0_27
3. van der Aalst, W.: Discovering the "glue" connecting activities - exploiting monotonicity to learn places faster. In: It's All About Coordination - Essays to Celebrate the Lifelong Scientific Achievements of Farhad Arbab, pp. 1–20 (2018)
4. Wen, L., van der Aalst, W., Wang, J., Sun, J.: Mining process models with non-free-choice constructs. Data Min. Knowl. Discov. **15**(2), 145–180 (2007)

5. Leemans, S.J.J., Fahland, D., van der Aalst, W.M.P.: Discovering block-structured process models from event logs - a constructive approach. In: Colom, J.-M., Desel, J. (eds.) PETRI NETS 2013. LNCS, vol. 7927, pp. 311–329. Springer, Heidelberg (2013). https://doi.org/10.1007/978-3-642-38697-8_17
6. Weijters, A., van der Aalst, W.: Rediscovering workflow models from event-based data using little thumb. Integr. Comput.-Aided Eng. **10**(2), 151–162 (2003)
7. Badouel, E., Bernardinello, L., Darondeau, P.: Petri Net Synthesis. TTCSAES. Springer, Heidelberg (2015). https://doi.org/10.1007/978-3-662-47967-4
8. Ehrenfeucht, A., Rozenberg, G.: Partial (set) 2-structures. Acta Informatica **27**(4), 343–368 (1990)
9. Carmona, J., Cortadella, J., Kishinevsky, M.: A region-based algorithm for discovering petri nets from event logs. In: Dumas, M., Reichert, M., Shan, M.-C. (eds.) BPM 2008. LNCS, vol. 5240, pp. 358–373. Springer, Heidelberg (2008). https://doi.org/10.1007/978-3-540-85758-7_26
10. van der Aalst, W., Rubin, V., Verbeek, H., van Dongen, B., Kindler, E., Günther, C.W.: Process mining: a two-step approach to balance between underfitting and overfitting. Softw. Syst. Model. **9**(1), 87 (2008)
11. Lorenz, R., Mauser, S., Juhás, G.: How to synthesize nets from languages: A survey. In: Proceedings of the 39th Conference on Winter Simulation: 40 Years! The Best is Yet to Come, WSC 2007, pp. 637–647. IEEE Press, Piscataway (2007)
12. Darondeau, P.: Deriving unbounded Petri nets from formal languages. In: Sangiorgi, D., de Simone, R. (eds.) CONCUR 1998. LNCS, vol. 1466, pp. 533–548. Springer, Heidelberg (1998). https://doi.org/10.1007/BFb0055646
13. Bergenthum, R., Desel, J., Lorenz, R., Mauser, S.: Synthesis of Petri nets from finite partial languages. Fundam. Inf. **88**(4), 437–468 (2008)
14. van der Werf, J.M.E.M., van Dongen, B.F., Hurkens, C.A.J., Serebrenik, A.: Process discovery using integer linear programming. In: van Hee, K.M., Valk, R. (eds.) PETRI NETS 2008. LNCS, vol. 5062, pp. 368–387. Springer, Heidelberg (2008). https://doi.org/10.1007/978-3-540-68746-7_24
15. van Zelst, S.J., van Dongen, B.F., van der Aalst, W.M.P.: Avoiding over-fitting in ILP-based process discovery. In: Motahari-Nezhad, H.R., Recker, J., Weidlich, M. (eds.) BPM 2015. LNCS, vol. 9253, pp. 163–171. Springer, Cham (2015). https://doi.org/10.1007/978-3-319-23063-4_10
16. van Zelst, S., van Dongen, B., van der Aalst, W.: ILP-based process discovery using hybrid regions. In: ATAED@Petri Nets/ACSD (2015)
17. Berthelot, G.: Checking properties of nets using transformation. In: Rozenberg, G. (ed.) APN 1985. LNCS, vol. 222, pp. 19–40. Springer, Berlin (1986). https://doi.org/10.1007/BFb0016204
18. Berthelot, G.: Transformations and decompositions of nets. In: Brauer, W., Reisig, W., Rozenberg, G. (eds.) ACPN 1986. LNCS, vol. 254, pp. 359–376. Springer, Heidelberg (1987). https://doi.org/10.1007/978-3-540-47919-2_13
19. Colom, J.M., Silva, M.: Improving the linearly based characterization of P/T nets. In: Rozenberg, G. (ed.) ICATPN 1989. LNCS, vol. 483, pp. 113–145. Springer, Heidelberg (1991). https://doi.org/10.1007/3-540-53863-1_23
20. Garcia-Valles, F., Colom, J.: Implicit places in net systems. In: Proceedings 8th International Workshop on Petri Nets and Performance Models, pp. 104–113 (1999)
21. Berthomieu, B., Botlan, D.L., Dal-Zilio, S.: Petri net reductions for counting markings. CoRR abs/1807.02973 (2018)

22. van Dongen, B.F., de Medeiros, A.K.A., Verbeek, H.M.W., Weijters, A.J.M.M., van der Aalst, W.M.P.: The ProM framework: a new era in process mining tool support. In: Ciardo, G., Darondeau, P. (eds.) ICATPN 2005. LNCS, vol. 3536, pp. 444–454. Springer, Heidelberg (2005). https://doi.org/10.1007/11494744_25
23. Polato, M.: Dataset belonging to the help desk log of an Italian company (2017)
24. Mannhardt, F.: Sepsis cases - event log (2016)
25. De Leoni, M., Mannhardt, F.: Road traffic fine management process (2015)
26. van der Aalst, W.M.P.: Event Logs and Models Used in Process Mining: Data Science in Action (2016). http://www.processmining.org/event_logs_and_models_used_in_book
27. Lu, X., Fahland, D., van den Biggelaar, F.J.H.M., van der Aalst, W.M.P.: Handling duplicated tasks in process discovery by refining event labels. In: La Rosa, M., Loos, P., Pastor, O. (eds.) BPM 2016. LNCS, vol. 9850, pp. 90–107. Springer, Cham (2016). https://doi.org/10.1007/978-3-319-45348-4_6
28. Li, J., Liu, D., Yang, B.: Process mining: extending α-algorithm to mine duplicate tasks in process logs. In: Chang, K.C.-C., et al. (eds.) APWeb/WAIM -2007. LNCS, vol. 4537, pp. 396–407. Springer, Heidelberg (2007). https://doi.org/10.1007/978-3-540-72909-9_43
29. Wen, L., Wang, J., Sun, J.: Mining invisible tasks from event logs. In: Dong, G., Lin, X., Wang, W., Yang, Y., Yu, J.X. (eds.) APWeb/WAIM -2007. LNCS, vol. 4505, pp. 358–365. Springer, Heidelberg (2007). https://doi.org/10.1007/978-3-540-72524-4_38

Concurrent Programming
from PSEUCO to Petri

Felix Freiberger[1,2(✉)] and Holger Hermanns[1]

[1] Saarland University, Saarland Informatics Campus, Saarbrücken, Germany
freiberger@depend.uni-saarland.de
[2] Saarbrücken Graduate School of Computer Science,
Saarland Informatics Campus, Saarbrücken, Germany

Abstract. The growing importance of concurrent programming has
made practical concurrent software development become a cornerstone
of many computer science curricula. Since a few years, a sound bridge
from concurrency theory to concurrence practice is available in the form
of PSEUCO, a light-weight programming language featuring both message
passing and shared memory concurrency. That language is at the core of
an award-winning lecture at Saarland Informatics Campus. This paper
presents a novel two-step semantic mapping from PSEUCO programs to
colored Petri nets, developed for the sake of further strengthening the
educational concept behind PSEUCO. The approach is fully integrated in
PSEUCO.COM, our open-source teaching tool for PSEUCO, empowering students to interact with the Petri-net-based semantics of PSEUCO. In addition, we present a source-level exploration tool for PSEUCO, also based on
this semantics, that gives users an IDE-like debugging experience while
enabling full control over the nondeterminism inherent in their programs.
The debugger is also part of PSEUCO.COM, allowing students to access it
without any set-up.

Keywords: Concurrency · Education · Colored Petri nets ·
Programming · Semantics

1 Introduction

Over the past decades, concurrent computation has grown tremendously in
importance within computer science. This concerns both the theoretical modeling of concurrent systems with formalisms like Petri nets as well as the practical
development of concurrent programs in real-world programming languages. The
latter has made modern concurrent programming an integral part of computer
science education. The ACM curricula recommendations [1] advocate a strong
educational component on "Parallel and Distributed Computing" and stipulate
that "Communication and coordination among processes is rooted in the message
passing and shared memory models of computing and such algorithmic concepts
as atomicity, consensus, and conditional waiting".

© Springer Nature Switzerland AG 2019
S. Donatelli and S. Haar (Eds.): PETRI NETS 2019, LNCS 11522, pp. 279–297, 2019.
https://doi.org/10.1007/978-3-030-21571-2_16

Listing 1. A message passing PSEUCO program. Expressions having the form c <! x (lines 11, 18 and 21) send the value of x on channel c. Expressions having the form <? c (lines 4, 19 and 22) receive a value from channel c.

```
1  void factorial(intchan c) {
2      int z, j, n;
3      while (true) {
4          z = <? c; // receive input
5
6          n = 1;
7          for (j = z; j > 0 ; j--) {
8              n= n*j;
9          }
10
11         c <! n; // send result
12     };
13 }
14
15 mainAgent {
16     intchan cc;
17     agent a = start(factorial(cc));
18     cc <! 3;
19     int mid = <? cc;
20     println("3! evaluates to " + mid + ".");
21     cc <! mid;
22     println("(3!)! evaluates to " + (<? cc) + ".");
23 }
```

At Saarland University in Saarbrücken, Germany, this is addressed since 2005 in the mandatory Bachelor-level *Concurrent Programming* lecture [7], which in 2013 was awarded with the German "Preis des Fakultätentages Informatik" for its "innovative concept combining classical process calculi with practical programming challenges". At its core, the lecture revolves around PSEUCO, a light-weight programming language featuring both message passing and shared memory concurrency concepts, with its message passing syntax being inspired by Go [10]. It includes support for data structures with condition synchronization. Listing 1 shows a small message passing PSEUCO program that computes the factorial of the factorial of 3.

PSEUCO is overarching the lecture topics and bridges from a theoretical part – introducing process calculi in the form of Milner's CCS [16] – to a practical part dealing with Go and Java. The latter is effectively accomplished by providing a transpiler from PSEUCO to Java source code which can either be output for inspection or immediately be compiled to Java byte code and executed.

To facilitate foundational reasoning about and analysis of concurrent programs, PSEUCO also has a formal semantics mapping to value-passing CCS. A corresponding PSEUCO-to-CCS-compiler, plus tools to facilitate analysis of the resulting transition system, are provided to students as part of an IDE called

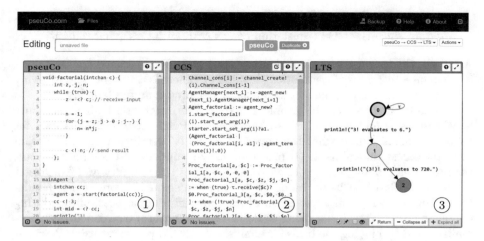

Fig. 1. PSEUCO.COM showing Listing 1 and the CCS-based semantics.

PSEUCO.COM [4]. It is based on web technologies to ensure it can be accessed easily by students. It requires no setup, updates automatically and works in all modern browsers including on tablets. All computations are performed on the client side (except for some advanced analysis tasks such as checks related to memory models) so the tool works offline after initial use. The Concurrent Programming lecture also uses PSEUCO.COM for the theoretical part as an IDE for CCS terms. Figure 1 shows a screenshot of PSEUCO.COM showing the program from Listing 1 ①, a fraction of its corresponding CCS term ② and the resulting LTS ③, minimized up to observational congruence.

Over the years, both PSEUCO and PSEUCO.COM have been subject to continuous improvement and have become an integral part of the annual lecture editions. While being an overall success, room for disruptive improvement has become apparent. Pragmatically speaking, this concerns readability of the compiler output and debugging. Conceptually speaking, it concerns a structured and concurrency-preserving and hence more faithful formal semantics.

- Students typically have a very hard time understanding the CCS terms produced by the compiler because the latter uses several low-level hacks. Among them:
  - Control flow is resolved into goto-style [5] spaghetti-code in CCS syntax.
  - The CCS terms generated contain many static helper constructs, such as an **AgentManager**, responsible for assigning unique ids to agents.
  - Synchronous (handshaking) and asynchronous (buffered) communication channels are internally distinguished by mapping them to negative, respectively positive integer identifiers. Every time a channel is used, the CCS term branches depending on the sign of the channel identifier.
- PSEUCO.COM lacks debugging support. It can show the LTS induced by the CCS term (induced by the PSEUCO program), but this does not provide an adequate debugging experience. The LTS is notoriously large, very often too large to grasp – Listing 1 already induces an LTS with 48 states.

– The students lack a feeling of the true concurrency inherent to a PSEUCO program since the CCS translation comes with an interleaving semantics.

All in all, the lesson the students learn is thus typically restricted to "OK, there is a formal – but messy – semantics", instead of "There is a natural way of giving a formal concurrency semantics to a concurrent programming language".

A deeper analysis of the problems led to the insight that all these problems can be overcome by instead providing a Petri-net-based semantics for PSEUCO. This is indeed what the paper develops. It describes a novel extension to PSEUCO.COM that aims at (i) providing easy-to-understand compiler output, (ii) providing a debugging experience that matches the usability and feature set of classic IDEs while being based on a complete semantics preserving nondeterminism, allowing full exploration of all nondeterministic possibilities during debugging, (iii) exposing students to Petri nets as a natural true concurrency formalism, and (iv) laying the basis for analysis of PSEUCO programs using Petri net techniques. At the core of this work is a two-level formalization of the semantics in terms of colored Petri nets.

Related Work. Higher-level Petri nets are an attractive base for the formal semantics of programming languages or process calculi. Among the pioneering works, $B(PN)^2$ [3] has been proposed as a concurrent programming notation geared towards Petri nets. A compositional semantics maps $B(PN)^2$ to M-Nets, a Petri net dialect specifically designed as a vehicle for giving semantics to concurrent programming languages [2]. Just like our approach, M-Nets are based on *colored* Petri nets. They support CCS-style composition which is coherent with their unfolding operation. This is orthogonal to the approach presented here, which does not provide composition operations, but focuses on providing a Petri-net-based semantics for a programming language closely resembling traditional imperative programming, together with tool support for use in teaching. PETRUCHIO [15] is a tool-supported approach that focuses on dynamically changing communication structures in Petri nets, especially rooted in the π-calculus [17]. It comes with a translation of the latter into Petri nets, so as to enable analysis with net verification tools. More recently, Nested-Unit Petri Nets [9] have been proposed as an extension of (uncolored) Petri nets, to be used when mapping compositional, process calculi-inspired programs to Petri nets. The addition of units allows more efficient storage of markings, speeding up analysis in the presence of appropriately-defined units.

Organization of the Paper. Section 2 reviews the main features of the PSEUCO programming language. Section 3 introduces colored Petri nets and a JavaScript library for handling them. Section 4 introduces C^P_PN, a higher-level Petri net notation that is used as an intermediate step, its implementation and a translation from PSEUCO to C^P_PN. Section 5 details how this translation and a debugger based on it are included in the PSEUCO.COM web application. Finally, Section 6 concludes this paper.

2   PSEUCO in a Nutshell

To set the stage for what follows, this section reviews the most important aspects of the PSEUCO language design, closely following its presentation in [4].

Mainstream programming is nowadays dominated by imperative programming languages. PSEUCO is an imperative language featuring a heavily simplified Java-like look and feel paired with language concepts inspired by the Go programming language [10]. It also has similarities with Holzmann's Promela language [11].

A very simplistic PSEUCO example is depicted in Listing 2. This program implements concurrent counting. A shared integer, n, is initialized to 10. The procedure countdown() decrements this counter five times. The mainAgent, which is run when the program is started, starts a second agent that runs countdown() before calling countdown() itself. After both agents have executed this procedure, the mainAgent prints the final value of n. To ensure mutually exclusive access to the shared variable, a globally defined lock named guard_n is used within the countdown() procedure.

PSEUCO also provides native support for message passing concurrency. An example is presented in Listing 1. An agent running the procedure factorial interacts via a channel with the mainAgent. In a nutshell, factorial computes the factorial of a number received from channel c and reports the result on the same channel c. This channel is declared locally in line 16 and passed as a parameter of factorial. Its type intchan indicates that it accepts integers and is unbuffered, meaning that it induces a handshake between the agents sending to (via <!) and receiving from (via <?) it. PSEUCO also has channels that can hold

---

Listing 2. Shared memory concurrent counting in PSEUCO.

```
1  int n = 10;
2  lock guard_n;
3
4  void countdown() {
5      for (int i = 5; i >= 1; i--) {
6          lock(guard_n);
7          n--;
8          unlock(guard_n);
9      }
10 }
11
12 mainAgent {
13     agent a = start(countdown());
14     countdown();
15     join(a);
16     println("The value is " + n);
17 }
```

Listing 3. Replacement for lines 4 to 11 in Listing 1.

```
1  select {
2      case <? t: {
3          return;
4      }
5      case z = <? c: { // lines 6 to 11 identical to Listing 1
6          n = 1;
7          for (j = z; j > 0 ; j--) {
8              n= n*j;
9          }
10
11         c <! n; // send result
12     }
13 };
```

strings or Booleans. After starting the agent, the `mainAgent` feeds the number 3 into the channel `cc` and then waits for results to be sent back to him. The result is returned back to the `factorial` agent. After the second round, the main agent prints the result.

This program does not terminate the `factorial` agent. Explicit termination can be achieved by applying three changes. First, the expansion uses a new channel declared by inserting `boolchan2 t;` before line 1. This channel is a FIFO buffer which can hold up to 2 Booleans. Second, the main agent is instructed to send a message on that channel at the end of its execution by inserting `t <! true;` after line 22. Finally, the `factorial` agent may now receive a message on two different channels (`t` and `c`) and therefore a `select-case` statement is used to specify dedicated reactions by replacing lines 4 to 11 with Listing 3. In case any message on `t` is received, the agent immediately terminates. Otherwise, it proceeds as previously. PSEUCO has borrowed the `select-case` concept from Go [10]. A `select` statement consist of several `cases`. Except for `default` cases, each case has a guard and a statement. The guard contains exactly one send (`<!`) or receive operation (`<?`). At runtime, a case can be selected only if the message passing operation of the guard is possible, i.e. if the channel can be read or be written to, respectively. One of those cases is selected nondeterministically and its guard and statement are processed. A `default` case can always be selected. If there are multiple cases that can be selected, one of them is selected nondeterministically.

These examples give an impression of the features provided by PSEUCO, all of which are given semantics by translation to CCS. In addition, PSEUCO supports arrays, `structs` and `monitors` with condition synchronization, however, these can be viewed as syntactic sugar and are not considered in this paper.

3   A Library for Colored Petri Nets

PSEUCO is an imperative programming language, and as such any PSEUCO program operates on variables. In this context, colored Petri nets [12] offer a clear advantage over basic Petri nets as a semantic model for the language.

Colored Petri Nets. We generally follow the definitions from [13] and assume any syntax for expressions where *Exp* is the set of expressions, *Types* is a set of types and *Vars* is a set of variables. Let $\mathbb{B} \in$ *Types* be the set of Booleans. Let *Values* $:= \bigcup_{t \in Types} t$ be the set of all values. We use *Type* : *Vars* $\rightarrow$ *Types* to express the type of a variable and *Type* : *Exp* $\rightarrow$ *Types* for the type of an expression. For a set of variables *Vars*, let *Type*(*Vars*) denote the set of types $\{Type(v) \,|\, v \in Vars\}$. We assume a function *Var* : *Exp* $\rightarrow$ *Vars* that returns the variables in an expression. A *binding* b is a function b : *Vars* $\rightarrow$ *Values* such that $\forall v \in$ *Vars* : $b(v) \in Type(v)$. Let *Bindings* be the set of bindings. Lastly, we assume an evaluation function *eval* : *Exp* $\times$ *Bindings* $\rightarrow$ *Values* such that $\forall e \in$ *Exp* : $\forall b \in$ *Bindings* : $eval(e, b) \in Type(e)$. When evaluating closed expressions, we omit the second argument to *eval*. Let X_{MS} be the set of all multisets over X.

Definition 1 (Colored Petri net). *A colored Petri net is a tuple CPN $=$ $(\Sigma, P, T, A, N, C, G, E, I)$ satisfying the requirements below:*

(i) Σ *is a finite set of non-empty types, called* color sets.
(ii) P, T *and* A *are pairwise disjoint sets of* places, transitions *and* arcs.
(iii) $N : A \rightarrow P \times T \cup T \times P$ *is a* node *function.*
(iv) $C : P \rightarrow \Sigma$ *is a* color *function.*
(v) $G : T \rightarrow Exp$ *is a* guard *function such that*
  $\forall t \in T : Type(G(t)) = \mathbb{B} \wedge Type(Var(G(t))) \subseteq \Sigma.$
(vi) $E : A \rightarrow Exp$ *is an* arc expression *function such that*
  $\forall a \in A : Type(E(a)) = C(p(a))_{MS} \wedge Type(Var(E(a))) \subseteq \Sigma$
  where $p(a)$ is the place of $N(a)$.
(vii) $I : P \rightarrow Exp$ *is an* initialization *function such that*
  $\forall p \in P : Type(I(p)) = C(p)_{MS}$ *and* $\forall p \in P : Var(I(p)) = \emptyset,$
  i.e. all expressions returned by I are closed.

CPN in JavaScript. Colored Petri nets will serve as a semantic model for PSEUCO, and tool support for experiencing and exploring this semantics is at the core of our educational approach. For this purpose, we provide a JavaScript library to express and evaluate colored Petri nets, which we will call `colored-petri-nets`. The library implements support for the concepts needed to materialize Definition 1 and its semantic underpinning by introducing a syntax for arc expressions, implementing a data structure for Petri nets and providing an algorithm for finding enabled steps. In addition, for nets where only a finite number of values are reachable, it allows converting the colored Petri net into a basic Petri net. For simplicity, the library enforces some restrictions and simplifications:

1. In Definition 1, arc expressions evaluate to a multiset of colors. Our expression syntax only allows expressing single colors, which are always treated as singleton multisets. Multiple tokens can be consumed or produced by the use of multiple arcs.
2. Usually, colored Petri nets are defined over an arbitrary set Σ of color sets (or types). Our implementation, however, only supports a single type, $\Sigma = \{\mathbb{V}\}$, with $\mathbb{V}$ containing numbers, booleans, arrays and objects (where keys are strings and values are valid colors).
3. In full generality, both arcs from places to transitions ("incoming" arcs) and arcs from transitions to places ("outgoing" arcs) are inscribed with the same kind of expression. However, allowing arbitrary expressions on incoming arcs complicates computing the set of enabled markings because the binding are to be guessed or deduced. CPN tools [13] solve this problem by using a sophisticated algorithm [14] to compute the values of variables that can be deduced and by restricting unbounded variables to *small color sets*. For reasons of simplicity, we instead use a restricted pattern syntax on incoming arcs. When given token colors to read, these patterns evaluate to partial bindings. If they are compatible, they combine to the single binding under which the guard and the outgoing arcs' expressions are evaluated, assuming no unbound variables remain. While this restricts a single token read to return only one specific binding, nondeterminism can be retrieved by allowing expressions with inherent nondeterminism or by duplicating transitions.

The restricted pattern syntax for incoming arcs is inspired by JavaScript [6], especially by valid left-hand sides of JavaScript assignments. Various extensions and modifications aim to provide a more complete set of matching capabilities. The constructs in Table 1 can be used in patterns.

For outgoing arcs, the syntax and semantics of expressions are also based on a fragment of JavaScript, but feature some additions. These mostly serve the purpose of increasing the expressiveness without needing to allow procedural code. Just as in traditional JavaScript, we support (i) conditionals (x ? 42 : 1337); (ii) logical *or* (||) and *and* (&&); (iii) equality (==) and inequality (!=) checks; (iv) numerical comparison (>, <, >= and <=); (v) division-free basic arithmetic (+, - and *); (vi) boolean negation (!); (vii) property access (point.coordinates.x), including the length property of arrays; (viii) grouping with (and); (ix) integer and boolean literals; (x) variables; (xi) array literals ([4, 5, 6]); and (xii) object literals ({ a: 1, b: 2 }), including ES6-style shorthand notation ({ a, b }). In addition, support is provided for (xiii) array concatenation via the new @ operator; (xiv) spreads in object literals ({ a: 1, ...x, b: 2, ...y, a: 2 }) which copy in all keys and values from another object, using the rightmost value if a key is duplicated; and (xv) an evaluation function (eval(a + b, vars)) that evaluates a subexpression in a separate environment that is passed in object form as the second argument.

The JavaScript library colored-petri-nets is freely available at https://dgit.cs.uni-saarland.de/pseuco/colored-petri-nets.

Table 1. Constructs allowed in `colored-petri-nets` patterns.

Name	Example	Description
Variable	`x`	Matches anything and binds the value
Wildcard	`_`	Matches anything and drops the value
Array patterns	`[a, b, c]`	Matches any array of matching length and recursively matches the components
Empty slots in array patterns	`[a, , c]`	Act as a wildcard
Spread in array pattern	`[hd, ...tl]`	Matches the remainder of the array, only allowed in the last slot
Object pattern	`{ a, b: c }`	Matches objects having all required keys and matches the values to the pattern, if one is specified, or binds it to a variable named like the key
Spread in object pattern	`{ one, two, ...rest }`	Matches all unmentioned keys into a new object, only allowed in the last slot

4   Augmenting CPN for Concurrent Programming

As discussed in Section 1, there are three problems impeding readability of the
CCS terms produced by the current PSEUCo to CCS compiler: (i) hard-to-follow
program flow, (ii) an abundance of static helper constructs and (iii) the insertion
of runtime logic into user code (e.g. for message passing). While the switch to
Petri nets (and using a graphical representation for them) inherently improves
on problem (i), without additional care, issues (ii) and (iii) would resurface.

For example, when considering message passing in PSEUCo, Petri nets are
obviously capable of expressing both synchronous and asynchronous message
passing channels. However, since channel variables can be dynamically reas-
signed, static analysis cannot always determine whether a channel variable refers
to a synchronous or an asynchronous channel, or whether two channel variables
in different agents could refer to the same channel. Therefore, a naïve implementation would be bound to introducing a central storage place for the contents
of all asynchronous channels. Each use of a channel would then have to perform a run-time check to determine the type of channel, and in the case of an
asynchronous channel proceed by synchronizing with the central storage place.
Such constructs blow up the resulting Petri net and impede readability. They
also hinder graph layout by their introduction of highly interconnected places. In
addition, these constructs are not specific to PSEUCo, so we would like to make
them reusable for compiling other programming languages to Petri nets.

To this end, we introduce an abstraction layer between PSEUCo and colored
Petri nets, called *colored program Petri nets* (C^P_PN). This is a high-level notation

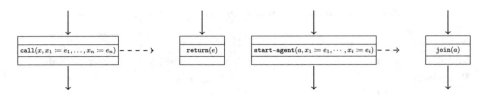

Fig. 2. Calling and agent management.

based on colored Petri nets tailored to concurrent programs using shared memory and/or message passing concurrency. As $C^P_P N$ are meant as an easily visualizable and reusable intermediate form between an imperative concurrent programming language and Petri nets, we do not define an executable semantics for them, but instead provide a translational semantics to ordinary colored Petri nets.

4.1 $C^P_P N$ Overview

At a glance, a $C^P_P N$ is very similar to a colored Petri net. However, in $C^P_P Ns$, tokens are strictly associated with *agents* (or *threads*) on the program level, and token colors can be viewed as object valuations together representing the states of local variables of the agent. Agents have an identity that becomes relevant when waiting for an agent's termination or when handling reentrant locks. On the $C^P_P N$ level, this is echoed by regular (i.e. CPN-typical) transitions being restricted to always consume and produce exactly one token, and the initial marking being constrained to contain a single token. In return, $C^P_P N$ includes *command transitions* to handle (i) procedure calls, (ii) agent creation and management, (iii) message passing, (iv) global variables and (v) mutexes.

Calling and Agent Management (see Figure 2). For both `call` and `start-agent`, the x_i are local variables representing the arguments of the called procedure and the e_i are expressions representing their values. For `call`, x is the variable that captures the return value. Similarly, for `start-agent`, a captures the identity of the newly started agent. This allows waiting for the termination of the agent with a `join(a)` transition. For `return`, e is an expression representing the return value.

Note that `call` and `start-agent` have two outgoing arcs, one of which is dashed. The dashed arc represents the called procedure or the newly started agent, while the solid arc represents the caller.

Message Passing Support (see Figure 3). `init-chan` creates a new channel of capacity c, assigned to variable x. Sending and receiving messages is handled by `send`, `receive` and `default` transitions. Any place that has an outgoing arc to such a transition may not have an outgoing arc to other transitions. The `default` transition is always allowed and allows bailing out of a place that has message passing transitions.

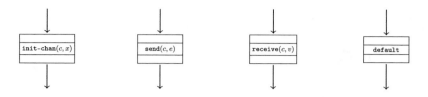

Fig. 3. Message passing support: Channel creation and sending/receiving messages.

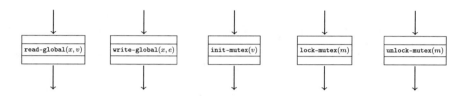

Fig. 4. Shared memory support: Global variables and mutexes.

Shared Memory Support (see Figure 4). C^P_PNs support the use of global variables that all agents can access. These variables cannot be used directly but can be accessed using two dedicated commands: `read-global`(x, v) copies the value of the global variable x to the local variable v, and `write-global`(x, e) writes the result of evaluating the expression e to x. Coordination is possible through the use of *mutexes* (or *locks*). They are initialized with `init-mutex`, which saves their identity in variable v. Then, they can be used with `lock-mutex` and `unlock-mutex`.

Example. Figure 5 depicts a C^P_PN for the PSEUCo program listed in Listing 1. In the visual representation, the three lines labeling the transition bodies indicate the name of the transition, which kind of command transition it is and the guard (which defaults to *true*).

Syntax of C^P_PN. The ensemble of structures enabled by C^P_PN is as follows. Let *GlobVars* be a set of identifiers for *global variables*. We define a set *Cmds* of *commands* that can be associated with transitions to make them command transitions. For example, the transition `factorial-send` in Figure 5 has the command `send`(c, n) associated with it, indicating that the result of evaluating the expression n is sent over the channel returned by evaluating the expression c. Similarly, we define commands for all types of command transitions appearing in Figures 2, 3, and 4. Let *MPCommCmds* $\subset$ *Cmds* be the set of possible message passing commands, i.e. sending, receiving or `default` transitions. Similarly, *StartAgentCmds*, *CallCmds* and *ReturnCmds* refer to the corresponding respective subsets of *Cmds*.

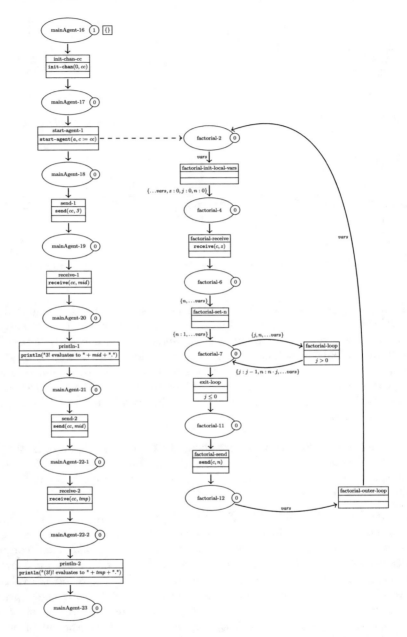

Fig. 5. A $C^B_P N$ for Listing 1. The notation adheres to certain simplifications that are introduced in Section 4.2.

Definition 2 (C^P_PN). A C^P_PN is a tuple $CPN = (\Sigma, P, T, A, Cmd, S, N, C, G, E, I)$ satisfying the requirements below:

(i) $(\Sigma, P, T, A, N, C, G, E, I)$ is a colored Petri net.

(ii) $Cmd : T \to Cmds \cup \{\bot\}$ is a command function.

(iii) $S : A \to \mathbb{B}$ indicates for each arc whether it is starting a new procedure or agent.

(iv) The arc expression function E only returns expressions that always evaluate to singleton multisets:

$$\forall a \in A : \forall b \in Bindings : |eval(E(a), b)| = 1.$$

(v) The initialization function specifies a single token:

$$\exists p \in P : |eval(I(p))| = 1 \wedge \forall p' \in P \setminus \{p\} : |eval(I(p))| = 0.$$

(vi) Let $post(x) := \{y \,|\, \exists a \in A : N(a) = (x, y)\}$ denote the set of successors and $pre(x) := \{y \,|\, \exists a \in A : N(a) = (y, x)\}$ denote the set of predecessors of a place or transition. For each transition, the number of incoming and outgoing arcs must be correct, in the following sense:

1. $\forall t \in T : |pre(t)| = 1$

2. $\forall t \in T : |post(t)| = \begin{cases} 2 & \text{if } Cmd(t) \in StartAgentCmds \cup CallCmds \\ 0 & \text{if } Cmd(t) \in ReturnCmds \\ 1 & \text{otherwise} \end{cases}$

For call and start transitions, exactly one arc must be marked as starting:

3. $\forall t \in T : |\{a \,|\, a \in post(t) \wedge S(a) = true\}| = 1$

(vii) Let $T_{MPComm} := \{t \in T \,|\, Cmd(t) \in MPCommCmds\}$ be the set of message passing communication transitions. If a transition has an outgoing arc to any transition in T_{MPComm}, all outgoing arcs must lead to such transitions, i.e. $\forall p \in P : post(p) \cap T_{MPComm} \neq \emptyset \implies post(p) \subseteq T_{MPComm}$.

Relative to Definition 1, we can note the following differences:

- There is a command function (see (ii)) assigning commands to transitions. If $Cmd(T) = \bot$, we call T a *Petri transition*, otherwise, it is a *command transition*.
- The arc expression function must yield expressions that always return a single token. This, together with condition (vi), ensures that Petri transitions must consume and produce exactly one token at all times.
- The initial marking is now restricted to contain a single token.
- Condition (vi) and the new function S (see (iii)) ensure command transitions have the correct number of incoming and outgoing arcs, and `call` and `start-agent` transitions have one outgoing arc designated as the starting arc, represented by a dashed line in the graphical representation.
- An additional restriction, item (vii), ensures that any place that has an outgoing arc to a message passing command transition can only have arcs to such transitions.

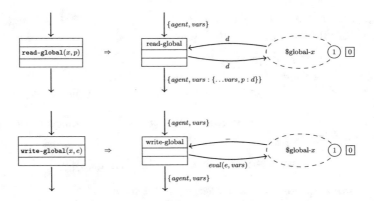

Fig. 6. Unfolding of shared memory command transitions. Each `read-global` and `write-global` command transition is replaced by the construct above, creating a new place global-x (initialized with a token of color 0) whenever a global variable named x is seen for the first time.

4.2 Translation From $C^P_P N$ to CPN

As mentioned previously, $C^P_P N$s do not posses an executable semantics, but instead are translated to regular colored Petri nets. At its core, the construction unfolds the command transitions into the structures corresponding to their intended functionality.

1. All arc expressions and the initial marking are updated by adding an `agent` property that contains an agent id and the current recursion depth, initially both 0. This enables e.g. `join` transitions to recognize agents and allows matching the token of a returning procedure to its caller. The color sets of all places are adjusted accordingly.
2. A fixed set of management places is added to handle id generation for locks, channels and agents, manage agent termination and store channel contents and lock states.
3. For each global variable, a management place is added to store its value.
4. Each command transition is replaced with a specific construct implementing the command functionality, typically by synchronizing with one or more management places. For example, Figure 6 shows the unfolding for shared memory command transitions. For message passing command transitions, these replacement constructs in addition can synchronize with each other to implement handshaking over synchronous channels. Similarly, each pair of `call` and `return` command transitions causes a linking transition to be inserted that handles returns from that specific `return` transition to that `call` transition.

The various details of the needed constructions are too verbose to include them in all detail here due to lack of space. A majority of these constructions result in a linear increase in the size of the net, similar to Figure 6. However, overall, the size of the resulting CPN is quadratic in the number of `call`, `return`,

send and **receive** transitions due to the additional transitions needed to handle handshaking and returns.

Implementation. The JavaScript library `colored-petri-nets` introduced in Section 3 has full support for C^P_PN and for the translation to colored Petri nets. On top of the restrictions and simplifications that apply to plain colored Petri nets (see Section 3), two additional simplifications are added: First, command transitions are not allowed to have a guard different from *true*. Second, arcs belonging to command transitions do not have arc inscriptions. Therefore, command transitions are not allowed to change the token color except in the way dictated by their command. For command transitions where arc inscriptions seem necessary, e.g. when passing arguments to a called procedure or newly started agent, the behavior is instead controlled by additional parameters within the command. The notation used in Figures 2, 3, 4, 5, and 6 matches these simplifications.

4.3   From PSEUCO to C^P_PN

As per the design goals of C^P_PN, compiling PSEUCO programs to C^P_PN is rather straightforward. This task is taken care of by `pseuco-cpn-compiler`, a JavaScript-based implementation of such a compiler. The compiler starts with a net consisting of a single place, then simply traverses the abstract syntax tree of the input program, processing children in reverse order while building up the net from bottom to top. This direction is advantageous because it simplifies building a *source map*, indicating which program statement each place belongs to, as the compiler creates the place representing the program state *before* a certain statement while processing that statement. The JavaScript library `pseuco-cpn-compiler` is freely available at https://dgit.cs.uni-saarland.de/pseuco/pseuco-cpn-compiler. It currently supports the array- and structure-free subset of PSEUCO. In conjunction with the `colored-petri-nets` library, it allows compiling PSEUCO programs into regular colored Petri nets or, for PSEUCO programs with a bounded state space, basic Petri nets.

5   PSEUCO.COM: An Educational Tool Backed by Petri Nets

As previously discussed, the main motivation of this work has been to enhance PSEUCO.COM by providing a more easily digestible semantics of PSEUCO programs and by providing IDE-like debugging capabilities. This section details the result of these efforts.

Colored Petri Nets in PSEUCO.COM. To integrate the `colored-petri-nets` library and `pseuco-cpn-compiler` into the educational tool PSEUCO.COM, an appealing way is needed to visualize Petri nets. For the purpose of working with labeled transition systems, PSEUCO.COM does already employ a force-directed

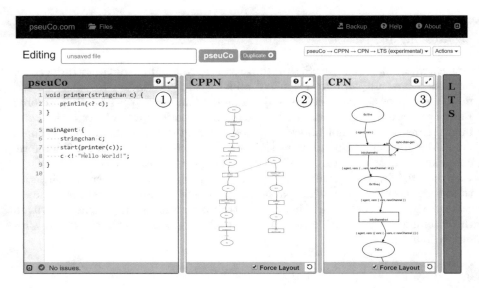

Fig. 7. PSEUCO.COM showing a sample PSEUCO program and its Petri net semantics.

graph layout system allowing the user to interactively explore a transition system by expanding or collapsing states (i.e. showing and hiding their successors). Petri nets incur less need for interactive exploration as they tend to stay small even for PSEUCO programs resulting in thousands of LTS states using the existing CCS-based compiler. Still, the integration of Petri nets into PSEUCO.COM uses a force-based graph layout, so as to allow users to influence the layout of the net by dragging and dropping, similarly to what is supported for LTS. In addition to standard forces like electrical charges and spring forces along arcs, PSEUCO.COM employs custom forces to orient graphs. They ensure that regular arcs typically point downwards, while called agents and procedures are positioned horizontally. This is demonstrated in Figure 7, showing a sample PSEUCO program ① and two graphs describing the corresponding C^P_PN ② and a fragment of its unfolding ③. To speed up convergence, the force layout is initialized by node positions precomputed with `dagre` which employs an algorithm similar to Graphviz [8]. The force layout can be disabled to fall back to the static layout provided by `dagre`.

Debugging PSEUCO *Programs in* PSEUCO.COM. The CCS-based semantics employed by PSEUCO.COM so far has been of little help for students seeking to understand the behavior of their programs. At its core, this problem is rooted in the difficulty of mapping a state of the resulting LTS to the state of the underlying PSEUCO program, a process that requires parsing convoluted CCS terms and an understanding of the internal low-level hacks used by the PSEUCO to CCS compiler. While switching to the Petri-net-based semantics alleviates this by replacing CCS terms with markings in a fixed net, significantly improving readability, a partial understanding of the compiler's internals is still required. To resolve this, we present the PSEUCO.COM debugging feature. While designed to be usable even without an understanding of Petri nets, it is based upon the semantics described above and backed by the C^P_PN-based PSEUCO compiler included in PSEUCO.COM.

In its core, the debugger is a tool to explore the marking graph of the underlying Petri net. However, instead of showing the current marking directly, the debugger translates the marking into PSEUCO terminology. Figure 8 shows the debugger in action, demonstrating its main features: It allows the user to

- inspect the console output ①,
- see running agents and their local variables and call stack ②,
- identify the statement an agent is currently executing ③,
- see global variables and the state of asynchronous channels and locks (not present in the example),
- see which agents are currently waiting for message passing synchronization to happen ④,
- single-step agents ⑤, manually resolving nondeterminism if present,
- automatically execute single agents ⑥ as long as their behavior is deterministic,
- automatically execute the whole program ⑦, resolving nondeterminism randomly,
- set breakpoints ⑧ to interrupt automatic execution and to
- return to any previous state of the program ⑨.

All of this functionality is rooted in the linkage between the PSEUCO program and the Petri net levels, mentioned above. The compiler, `pseuco-cpn-compiler`, annotates its output C^P_PN with a source map allowing the debugger to link elements of the net to the original program. When converting the C^P_PN to a colored Petri net, the `colored-petri-nets` module preserves these annotations and generates additional metadata identifying the newly introduced places.

When inspecting a marking, the debugger uses the agent id and recursion depth embedded in the token colors (see Section 4.2) to identify running agents, their local variables and stack frames. The source map information of an agent tokens' place identifies that agent's position in the source code. Global variables and their values are identified by looking for places added by the replacement rule introduced in Figure 6 and tokens stored in them. Asynchronous channel contents and in-progress handshaking events are handled similarly.

In summary, this approach allows PSEUCO.COM to present a debugger that supports a similar feature set and user experience than a traditional IDE. Being built upon on the complete Petri-net-based semantics, however, allows preserving the full nondeterministic behavior of the program, ensuring that every execution of the program that can occur in practice can not only be reproduced but also specifically chosen in the debugger.

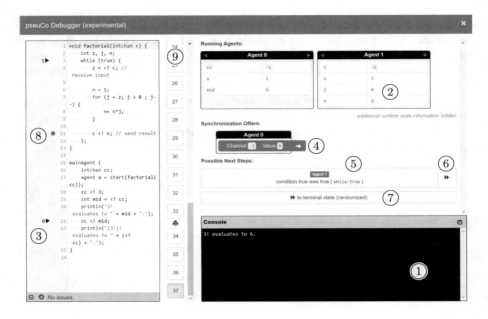

Fig. 8. PSEUCO.COM in a debugging session for the program from Listing 1.

6   Conclusion and Future Work

This paper has presented a translational semantics for PSEUCO to Petri nets tailored to the use in education. Its intermediate step, the higher-level Petri net formalism C^P_PN, is geared towards easy visualization of program semantics and reusability for other programming languages. The transpiler is highly integrated into PSEUCO.COM, providing easy access to it to students and teachers and extending PSEUCO.COM with a structured and concurrency-preserving semantics and a nondeterminism-preserving debugging facility. The transpiler and its underlying colored Petri nets implementation are available as open-source JavaScript libraries.

There is obvious room for improvement in these tools, most importantly expanding the transpiler to support the full feature set of PSEUCO, which includes arrays, data structures and monitors with condition synchronization. Doing so cleanly requires the introduction of additional command transitions in C^P_PN.

The Petri net extension is brand-new and has as such not been used in a lecture edition so far. This is planned for summer 2019, together with a shift in the theoretical course focus, now embracing the Petri net perspective on concurrency.

Acknowledgments. We would like to thank Bernd Finkbeiner for supervising an earlier phase of this project. Michaela Klauck provided helpful feedback during development of the debugging feature of PSEUCO.COM. This work was partially supported by the ERC Advanced Investigators Grant 695614 (POWVER) and by the Deutsche Forschungsgemeinschaft (DFG, German Research Foundation) – project number 389792660 – TRR 248 (see https://perspicuous-computing.science).

References

1. Computer Science Curricula 2013. ACM Press and IEEE Computer Society Press (2013). https://doi.org/10.1145/2534860
2. Best, E., Fraczak, W., Hopkins, R.P., Klaudel, H., Pelz, E.: M-nets: an algebra of high-level Petri nets, with an application to the semantics of concurrent programming languages. Acta Inf. **35**(10), 813–857 (1998). https://doi.org/10.1007/s002360050144
3. Best, E., Hopkins, R.P.: $B(PN)^2$ — a basic Petri net programming notation. In: Bode, A., Reeve, M., Wolf, G. (eds.) PARLE 1993. LNCS, vol. 694, pp. 379–390. Springer, Heidelberg (1993). https://doi.org/10.1007/3-540-56891-3_30
4. Biewer, S., Freiberger, F., Held, P.L., Hermanns, H.: Teaching academic concurrency to amazing students. In: Aceto, L., Bacci, G., Bacci, G., Ingólfsdóttir, A., Legay, A., Mardare, R. (eds.) Models, Algorithms, Logics and Tools. LNCS, vol. 10460, pp. 170–195. Springer, Cham (2017). https://doi.org/10.1007/978-3-319-63121-9_9
5. Dijkstra, E.W.: Letters to the editor: go to statement considered harmful. Commun. ACM **11**(3), 147–148 (1968). https://doi.org/10.1145/362929.362947
6. ECMA International: ECMAScript© 2015 Language Specification, 6th edn., June 2015. Standard ECMA-262
7. Eisentraut, C., Hermanns, H.: Teaching concurrency concepts to freshmen. Trans. Petri Nets Other Models Concurr. **1**, 35–53 (2008). https://doi.org/10.1007/978-3-540-89287-8_3
8. Gansner, E.R., Koutsofios, E., North, S.C., Vo, K.: A technique for drawing directed graphs. IEEE Trans. Softw. Eng. **19**(3), 214–230 (1993). https://doi.org/10.1109/32.221135
9. Garavel, H.: Nested-unit Petri nets. J. Log. Algebr. Methods Program. **104**, 60–85 (2019). https://doi.org/10.1016/j.jlamp.2018.11.005, http://www.sciencedirect.com/science/article/pii/S2352220817302018
10. The Go Programming Language Specification. http://golang.org/ref/spec
11. Holzmann, G.: The Spin Model Checker – Primer and Reference Manual, 1st edn. Addison-Wesley Professional, Boston (2003)
12. Jensen, K.: Coloured Petri nets and the invariant-method. Theor. Comput. Sci. **14**, 317–336 (1981). https://doi.org/10.1016/0304-3975(81)90049-9
13. Jensen, K.: Coloured Petri Nets - Basic Concepts, Analysis Methods and Practical Use - Volume 1. EATCS Monographs on Theoretical Computer Science. Springer, Heidelberg (1992). https://doi.org/10.1007/978-3-662-06289-0
14. Kristensen, L.M., Christensen, S.: Implementing coloured Petri nets using a functional programming language. High. Order Symb. Comput. **17**(3), 207–243 (2004). https://doi.org/10.1023/B:LISP.0000029445.29210.ca
15. Meyer, R., Strazny, T.: Petruchio: from dynamic networks to nets. In: Touili, T., Cook, B., Jackson, P. (eds.) CAV 2010. LNCS, vol. 6174, pp. 175–179. Springer, Heidelberg (2010). https://doi.org/10.1007/978-3-642-14295-6_19
16. Milner, R. (ed.): A Calculus of Communicating Systems. LNCS, vol. 92. Springer, Heidelberg (1980). https://doi.org/10.1007/3-540-10235-3
17. Milner, R.: Communicating and Mobile Systems - The Pi-calculus. Cambridge University Press, New York (1999)

Algorithmic Aspects

Improving Saturation Efficiency
with Implicit Relations

Shruti Biswal$^{(\boxtimes)}$ and Andrew S. Miner

Department of Computer Science, Iowa State University, Ames, IA 50010, USA
{sbiswal,asminer}@iastate.edu

Abstract. Decision diagrams are a well-established data structure for reachability set generation and model checking of high-level models such as Petri nets, due to their versatility and the availability of efficient algorithms for their construction. Using a decision diagram to represent the transition relation of each event of the high-level model, the saturation algorithm can be used to construct a decision diagram representing all states reachable from an initial set of states, via the occurrence of zero or more events. A difficulty arises in practice for models whose state variable bounds are unknown, as the transition relations cannot be constructed before the bounds are known. Previously, on-the-fly approaches have constructed the transition relations along with the reachability set during the saturation procedure. This can affect performance, as the transition relation decision diagrams must be rebuilt, and compute-table entries may need to be discarded, as the size of each state variable increases. In this paper, we introduce a different approach based on an implicit and unchanging representation for the transition relations, thereby avoiding the need to reconstruct the transition relations and discard compute-table entries. We modify the saturation algorithm to use this new representation, and demonstrate its effectiveness with experiments on several benchmark models.

Keywords: Petri nets · Decision diagram · Saturation ·
Reachability set generation

1 Introduction

High-level formalisms can be used to model complex discrete-state systems. The generation of the reachable state space, or *reachability set*, for such systems is an essential step for different kinds of studies. Formal verification techniques, such as model checking, may require the entire state space of a system to verify that some or all states satisfy certain properties, such as the presence of a safety property at all states. However, the reachability set of a system can be extremely large due to the state explosion problem, making the generation of the reachability set difficult.

© Springer Nature Switzerland AG 2019
S. Donatelli and S. Haar (Eds.): PETRI NETS 2019, LNCS 11522, pp. 301–320, 2019.
https://doi.org/10.1007/978-3-030-21571-2_17

Present-day *symbolic* techniques usually outperform the traditional *explicit* techniques used for reachability set generation. The *saturation* [7] algorithm is one such symbolic strategy for reachability set generation. An efficient implementation [14] of the saturation algorithm uses *multi-valued decision diagram* (MDD) representations [13] for encoding sets of reachable states and *matrix diagram representations* (MxD) for transition relations of the models.

However, a significant complication arises when the variables of the system have unknown bounds. For such systems, *on-the-fly techniques* [7,19] for transition relation construction are used, that require expansion of transition relations every time new bounds for variables are discovered which amounts to additional changes to compute-table entries. While the information in the transition relation is crucial to the application of symbolic techniques, its repeated construction in the form of MxD affects the overall efficiency of symbolic methods involved in reachability set analysis. When the events of a system are such that the enabling of each event is co-dependent on multiple variables, and the firing of the event affecting each variable is independent of other variables, the re-building of transition relations becomes an overhead because the growth of variable bound is a function of the variable itself. For the aforementioned systems, a more efficient technique would be to use a static representation of transition relations in order to restrict the modification of compute-table entries to the development of the reachability set alone. The focus and contribution of this paper is to devise a data structure, called an *implicit relation forest*, that encodes the events of such systems and uses the saturation algorithm for reachability set generation with reduced time and memory expense. To provide strong evidence of improvement in reachability set generation process, the paper proposes a modified saturation algorithm that uses implicit relations to conduct experiments and compares performance results with that of the on-the-fly saturation algorithm.

The rest of the paper is organized as follows. Section 2 defines the class of models we consider, and briefly recalls decision diagrams and on-the-fly saturation. Section 3 introduces implicit relation forests, details their construction from a model, and presents the saturation algorithm modified to use the alternate representation for model events. Section 4 discusses the related work and compares them with implicit relations. Section 5 describes the experimental evaluation of the devised method on an extensive set of Petri net models collected from the annual Model Checking Competition (MCC) [1]. Finally, Sect. 6 draws conclusions and discusses future research directions.

2   Background

This section describes the class of models that we consider in the paper, recalls the basics of decision diagrams, MDDs, and MxDs, and the saturation algorithm.

2.1 Model Definition

Rather than restricting our discussion to a particular formalism, we consider a class of generic high-level discrete-state models that includes many existing formalisms. A discrete-state model $\mathcal{M}$ is defined by a tuple $(\mathcal{V}, \mathcal{E}, \mathbf{i}_0, \Delta)$ where:

- $\mathcal{V} = \{v_1, v_2, \ldots, v_L\}$ is a finite set of *state variables* of the model. Each state variable v_k can assume a value from the set of natural numbers. A (*global*) state $\mathbf{i}$ of $\mathcal{M}$ is then an L-tuple $(i_1, i_2, \ldots, i_L) \in \mathbb{N}^L$.
- $\mathcal{E} = \{e_1, e_2, \ldots, e_{|\mathcal{E}|}\}$ is a finite set of *events* of the model.
- $\mathbf{i}_0 \in \mathbb{N}^L$ is the initial state of the model.
- $\Delta : \mathbb{N}^L \times \mathcal{E} \nrightarrow \mathbb{N}^L$ is the next state (partial) function. If $\Delta(\mathbf{i}, e)$ is defined, we say event e is *enabled* in state $\mathbf{i}$, and if event e *occurs* then the model changes from state $\mathbf{i}$ into state $\Delta(\mathbf{i}, e)$. Otherwise, if $\Delta(\mathbf{i}, e)$ is undefined, we say event e is *disabled* in state $\mathbf{i}$.

We require that each event e can be expressed using L *local* next state (partial) functions, $\Delta_{e,1}, \ldots, \Delta_{e,L}$, such that

$$\Delta((i_1, \ldots, i_L), e) = (\Delta_{e,1}(i_1), \ldots, \Delta_{e,L}(i_L)).$$

In other words, the value of a single local variable v_k is enough to disable an event, and the change in state variable v_k when an event occurs may depend only on state variable v_k. Furthermore, each event is deterministic: for a given input state, an event can produce at most one output state. However, the model may be nondeterministic, as several events may be enabled in a given state.

For example, an ordinary Petri net [16] can be expressed using our model: the set $\mathcal{V}$ can correspond to the set of Petri net places, the set $\mathcal{E}$ can correspond to the set of Petri net transitions that are unit-weighted by definition, the initial state $\mathbf{i}_0$ will correspond to the initial marking of the Petri net, and Δ will correspond to the Petri net firing rules. Specifically, for a place p_k and a transition t, $\Delta_{t,k}(i_k)$ is defined if i_k is greater or equal to the number of edges from p_k to t, and $\Delta_{t,k}(i_k) = j_k$ if $j_k - i_k$ equals the number of edges from t to p_k minus the number of edges from p_k to t. Petri nets with inhibitor arcs can also be expressed. An example "fork-join" Petri net model is shown in Fig. 1, where the circles correspond to places and the squares correspond to transitions. Transition t_1 performs a fork operation and transition t_6 performs a join operation.

Petri nets with marking-dependent arc cardinalities can sometimes be expressed using our model. Effectively, if the cardinality on an edge from p_i to t or from t to p_i depends on the number of tokens in a place p_j, then places p_i and p_j could be grouped together in a single model variable $v_l \in \mathcal{V}$. Alternatively, transition t can be split into several model events, each representing a portion of t. For example, we might use events $e_1, e_2, e_3, \ldots$ where event e_n simulates transition t but only when there are exactly n tokens present in place p_i. These modifications are necessary to express $\Delta(\mathbf{i}, t)$ in terms of local functions $\Delta_{e,l}(i_l)$

that depend only on the local model state variable. Transition guards can be handled in a similar manner. This is the *Kronecker consistency* requirement discussed in [6], and in practice it limits the applicability of our approach to models that can be obtained by merging only a few places together and whose number of events (including those obtained from splitting transitions) is small.

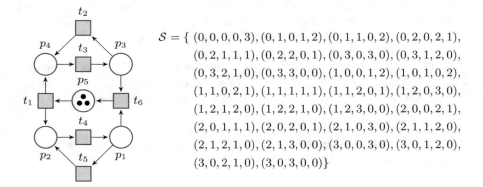

$$S = \{\, (0,0,0,0,3), (0,1,0,1,2), (0,1,1,0,2), (0,2,0,2,1),$$
$$(0,2,1,1,1), (0,2,2,0,1), (0,3,0,3,0), (0,3,1,2,0),$$
$$(0,3,2,1,0), (0,3,3,0,0), (1,0,0,1,2), (1,0,1,0,2),$$
$$(1,1,0,2,1), (1,1,1,1,1), (1,1,2,0,1), (1,2,0,3,0),$$
$$(1,2,1,2,0), (1,2,2,1,0), (1,2,3,0,0), (2,0,0,2,1),$$
$$(2,0,1,1,1), (2,0,2,0,1), (2,1,0,3,0), (2,1,1,2,0),$$
$$(2,1,2,1,0), (2,1,3,0,0), (3,0,0,3,0), (3,0,1,2,0),$$
$$(3,0,2,1,0), (3,0,3,0,0)\}$$

Fig. 1. A fork-join Petri net model (left) and its reachability set (right).

For a given model $\mathcal{M} = (\mathcal{V}, \mathcal{E}, \mathbf{i}_0, \Delta)$, we can define the following.

- The next state function for event e, $\mathcal{N}_e : \mathbb{N}^L \to 2^{\mathbb{N}^L}$, is defined as $\mathcal{N}_e(\mathbf{i}) = \{\mathbf{j} : \Delta(\mathbf{i}, e) = \mathbf{j}\}$. We then define the overall next state function as $\mathcal{N}(\mathbf{i}) = \bigcup_{e \in \mathcal{E}} \mathcal{N}_e(\mathbf{i})$, which gives the set of states reachable via the occurrence of one event from a single starting state, and further extend this to sets of starting states: $\mathcal{N}(\mathcal{I}) = \bigcup_{\mathbf{i} \in \mathcal{I}} \mathcal{N}(\mathbf{i})$.
- The *reachability set* $\mathcal{S} \subseteq \mathbb{N}^L$ is the set of states reachable via the occurrence of zero or more events from the initial state $\mathbf{i}_0$, and is the least fixed point satisfying $\mathcal{S} = \{\mathbf{i}_0\} \cup \mathcal{S} \cup \mathcal{N}(\mathcal{S})$.

As an example, the reachability set $\mathcal{S}$ is shown in Fig. 1 for the fork-join Petri net model, where a state is shown as $(p_1, p_2, p_3, p_4, p_5)$.

The focus of this paper is on algorithms to generate the set $\mathcal{S}$, using decision diagrams. This is a necessary first step for many types of analysis, including verification of safety properties or model checking of more complex properties specified in a temporal logic. For this work, we assume that $\mathcal{S}$ is finite (in general there is no guarantee of this). Note that the set $\mathcal{S}$ is finite if and only if every state variable is bounded. We do not require knowledge of these bounds *a priori*; instead, our reachability set generation algorithm will discover these bounds.

2.2 Multi-valued Decision Diagrams and Matrix Diagrams

An ordered multi-valued decision diagram (MDD) [13] defined over the sequence of L domain variables $(u_L, \ldots, u_1)$, with a given variable order such that $u_l \succ u_k$

iff $l > k$ and a specified domain for each variable $\mathcal{D}(u_k) = \{0, 1, 2, \ldots, n_k - 1\}$, is a directed acyclic edge-labelled graph where:

- Each node m is associated with some variable, denoted as $m.var$.
- There are two *terminal* nodes, **0** and **1**. These are associated with a special variable u_0, satisfying $u_k \succ u_0$ for any domain variable u_k.
- Each *non-terminal* node m is associated with a domain variable u_k and $\forall i_k \in \mathcal{D}(u_k)$, there is an edge labelled with i_k pointing to a child $m[i_k]$.
- The variable associated with any child $m[i_k]$ of a non-terminal node m is guided by the variable order such that $m.var \succ m[i_k].var$.

A node m in an MDD encodes a function $f_m : \mathcal{D}(u_L) \times \cdots \times \mathcal{D}(u_1) \to \{0, 1\}$, defined recursively by

$$f_m(i_L, \ldots, i_1) = \begin{cases} m, & \text{if } m.var = u_0 \\ f_{m[i_k]}(i_L, \ldots, i_1), & \text{if } m.var = u_k \succ u_0 \end{cases}$$

Non-terminal node n is a *duplicate* of node m if $n.var = m.var$, and if $n[i] = m[i], \forall i \in \mathcal{D}(n.var)$. Note that duplicate nodes n and m encode the same function: $f_m = f_n$. Non-terminal node m is *redundant* if $m[i] = m[0], \forall i \in \mathcal{D}(m.var)$; note that f_m is independent of variable $m.var$ in this case and $f_m = f_{m[0]}$. An MDD is *fully reduced* if it contains no duplicate nodes and no redundant nodes. It can be shown that fully reduced MDDs are a canonical form: any function can be represented uniquely ($f_m = f_n$ if and only if $m = n$). For our work, we instead use *zero reduced* MDDs, which contain no duplicate nodes, and *require* redundant nodes except for terminal node **0**. More formally, for any non-terminal node m with $m.var = u_k$, we require, for all i, that either $m[i].var = u_{k-1}$ or $m[i] = \mathbf{0}$. This is done because we allow the MDD domain variables to grow, or equivalently, we assume the MDD domain variables are unbounded but each MDD node contains only finitely many non-zero children.

Given a model $\mathcal{M} = (\mathcal{V}, \mathcal{E}, \mathbf{i}_0, \Delta)$, a finite set of states $\mathcal{X} \subset \mathbb{N}^L$ can be encoded as an MDD as follows.

- The MDD domain variables $(u_L, \ldots, u_1)$ correspond to the model state variables $\mathcal{V}$. For simplicity of presentation, we assume that $\forall i, u_i = v_i$; in practice, the variables can be ordered differently and there can be non-trivial mappings from state variables to domain variables (for example, several state variables could be collected into a single domain variable).
- The set $\mathcal{X}$ can be encoded by an MDD node m such that f_m is the characteristic function for the set $\mathcal{X}$: $f_m(i_L, \ldots, i_i) = 1$ iff $(i_1, \ldots, i_L) \in \mathcal{X}$.

The MDD encoding of $\mathcal{S}$, for the fork-join Petri net, is shown in Fig. 2. To increase readability, only paths that lead to terminal node **1** are shown.

An ordered matrix diagram (MxD) [14] is defined similarly to an MDD, except that each non-terminal edge is labelled with a *pair*:

- Each *non-terminal* node m is associated with a domain variable u_k and $\forall (i_k, j_k) \in \mathcal{D}(u_k) \times \mathcal{D}(u_k)$, there is an edge labelled with (i_k, j_k) pointing to a child $m[i_k, j_k]$ such that $m.var \succ m[i_k, j_k].var$.

A node m in an MxD encodes a function $f_m : \mathcal{D}^2(u_L) \times \cdots \times \mathcal{D}^2(u_1) \to \{0, 1\}$, given by $f_m = f_{m,L}$, where $f_{m,L}$ is defined recursively as

$$f_{m,L}(i_L, j_L, \ldots, i_1, j_1) = \begin{cases} m, & \text{if } L = 0 \\ f_{m[i_L, j_L], L-1}(i_L, j_L, \ldots, i_1, j_1), & \text{if } m.var = u_L \\ f_{m,L-1}(i_L, j_L, \ldots, i_1, j_1), & \text{if } u_L \succ m.var \ \wedge \ i_L = j_L \\ 0, & \text{otherwise.} \end{cases}$$

The definition of *duplicates* is similar to MDDs. A non-terminal node m is an *identity* node if (1) $m[i,i] = m[0,0], \forall i \in \mathcal{D}(m.var)$, and (2) $m[i,j] = \mathbf{0}, \forall i \neq j$. An MxD is *reduced* if it contains no duplicate nodes and no identity nodes. In practice, MxDs can be implemented as MDDs on twice as many domain variables. If the MxD domain variables are $(u_L, \ldots, u_1)$, then the MDD domain variables are $(u_L, u'_L, \ldots, u_1, u'_1)$, with the variable ordering defined such that u'_i immediately follows u_i. However, the MDDs are not fully reduced, but instead use a special *identity reduction* for primed variables [10].

Given a model $\mathcal{M} = (\mathcal{V}, \mathcal{E}, \mathbf{i}_0, \Delta)$ and bounds for each state variable, function $\mathcal{N}_e$ can be encoded as an MxD as follows. Again, the domain variables correspond to the state variables $\mathcal{V}$. Then we use an MxD node m such that $f_m(i_L, j_L, \ldots, i_1, j_1) = 1$ iff $(j_L, \ldots, j_1) \in \mathcal{N}_e((i_L, \ldots, i_1))$. The MxD encoding for $\mathcal{N}_{t_1}$, for the fork-join Petri net example, is shown in Fig. 2. Note that levels p_3, p'_3, p_1, p'_1 are skipped because places p_3 and p_1 are completely unaffected by transition t_1, and thus its occurrence does not change state variables p_3 and p_1. Also note that, because our model requires that an event occurrence changes a state variable in isolation (without considering the values of the other state

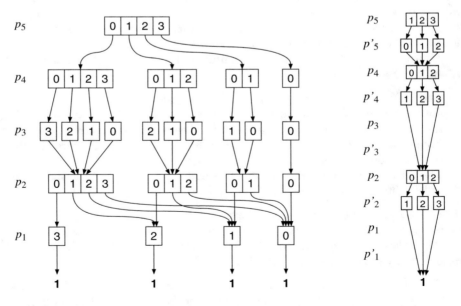

Fig. 2. MDD for $\mathcal{S}$ of fork-join Petri net (left) and MxD/MDD for t_1 (right)

variables), all MxD encodings of events will have a similar linear shape, where there can be a fanout only from u_k to u'_k and all non-zero pointers from u'_k must point to a single node.

2.3 On-the-fly Saturation Using Extensible Decision Diagrams

Given two MDD nodes m and n, encoding functions f_m and f_n, the MDD node p encoding function $f_p = f_m \oplus f_n$ for a binary operation $\oplus$ is constructed in a recursive "Apply" operation [4]. An example of this is algorithm Union, shown in Fig. 5, which constructs a new MDD encoding the union of two sets, passed as arguments and encoded as MDDs. Like most "apply" operations, Union simultaneously traverses the graphs rooted at nodes m and n, constructing a new graph rooted at p containing the result. Procedure UniqueInsert, called in line 13, eliminates duplicate nodes during construction: if the newly created node p is a duplicate of some other node q, then node p is discarded and node q is returned; otherwise, node p is added to the unique table and is returned. Duplicate computation arising from repeated recursive calls with the same arguments is avoided using a compute-table $\mathcal{C}$ (c.f. lines 4 and 14). This bounds the computational cost and the size of the resulting graph to be at worst the product of the sizes of the input graphs.

For efficiency, specialized relational product operations can be implemented (e.g., [20]) to construct the MDD for $\mathcal{N}(\mathcal{X})$ or $\mathcal{N}_e(\mathcal{X})$, when set $\mathcal{X}$ is encoded as an MDD and $\mathcal{N}$ is encoded as an MxD or MDD. A straightforward breadth-first iteration based on the fixed point equation $\mathcal{S} = \{\mathbf{i}_0\} \cup \mathcal{S} \cup \mathcal{N}(\mathcal{S})$ can then be used to generate $\mathcal{S}$. However, the iteration strategy of *node saturation* [7] can be orders of magnitude more efficient in practice and is known to terminate whenever $\mathcal{S}$ is finite.

Difficulty arises in practice for models whose state variable bounds are difficult or impossible to obtain, or are conservative. This can be alleviated using an "on-the-fly" variant of saturation [8,10,15,19], which allows variable bounds to be discovered while generating $\mathcal{S}$. This is done by distinguishing between *confirmed* local states that are known to appear in at least one reachable global state, and *unconfirmed* local states. The encoding of $\mathcal{N}$ contains all transitions out of confirmed local states, leading to confirmed or unconfirmed local states. During the saturation procedure, when an unconfirmed local state is discovered to be part of a reachable global state, it is confirmed and the encoding of $\mathcal{N}$ must be expanded to include any transitions out of the newly confirmed local state. For a Kronecker-based encoding of $\mathcal{N}$ [8], this expansion is straightforward; for more general encodings of $\mathcal{N}$ [10,15], this expansion requires rebuilding the MxD/MDD encoding of $\mathcal{N}$, and even worse, often discards compute-table information that could eliminate duplicate computations during the saturation procedure. This happens because, as the sizes of the state variable domains grow, the nodes in the encoding of $\mathcal{N}$ must also grow, and it is possible for the MxD/MDD for a next state function to change shape when the state variable domains increase. *Extensible* MDD nodes were introduced [19] to address exactly this issue, but unfortunately only certain repeating patterns can be exploited by extensible nodes.

3   Implicit Relations

Motivated by the requirement of current on-the-fly saturation methods to update (with varying degrees of computational overhead) their encodings of $\mathcal{N}$ as state variable domains increase in size, in this section we introduce *implicit relations* to encode $\mathcal{N}$ independently of the variable domain size. This new representation retains useful properties of MxDs, namely the ability to exploit identity structures and allowing nodes to be shared (i.e., have several incoming pointers). We also modify the on-the-fly saturation algorithm to work with implicit relations.

3.1   Definition

An ordered *implicit relation forest* defined over the sequence of L domain variables $(u_L, \ldots, u_1)$ is a directed acyclic graph where:

- Each node r is associated with some variable, denoted as $r.var$.
- There is a single terminal node, $\mathbf{1}$, associated with a special variable u_0 satisfying $u_k \succ u_0$ for any domain variable u_k.
- Each non-terminal *relation node* r is associated with a domain variable u_k, and contains a partial function $r.\delta : \mathcal{D}(u_k) \nrightarrow \mathcal{D}(u_k)$.
- Relation node r contains a single outgoing edge, $r.ptr$, consistent with the variable order such that $r.var \succ r.ptr.var$.

An implicit relation forest contains $|\mathcal{E}|$ implicit relations, where each relation corresponds to an event $e \in \mathcal{E}$ and is uniquely identified by the top-most relation node. A node r in an implicit relation forest encodes a function $f_r : \mathcal{D}^2(u_L) \times \cdots \times \mathcal{D}^2(u_1) \to \{0,1\}$, given by $f_r = f_{r,L}$, where $f_{r,L}$ is defined recursively as

$$
f_{r,L}(i_L, j_L, \ldots, i_1, j_1) = \begin{cases} 1, & \text{if } L = 0 \\ f_{r.ptr,L-1}(i_L, j_L, \ldots, i_1, j_1), & \text{if } r.var = u_L \ \wedge \ j_L = r.\delta(i_L) \\ f_{r,L-1}(i_L, j_L, \ldots, i_1, j_1), & \text{if } u_L \succ r.var \ \wedge \ i_L = j_L \\ 0, & \text{otherwise.} \end{cases}
$$

Let $\mathcal{U}_k$ represent the sequence of variables $u_k, \ldots, u_1$. Then $f_{r,k}$ encodes the effect of an event on $\mathcal{U}_k$. Either u_k, a variable associated with node r, participates in the event and the post-event value of u_k is $j_k = r.\delta(i_k)$ with values of $\mathcal{U}_{k-1}$ given by $f_{r.ptr,k-1}$, or u_k does not participate in the event and the value of u_k remains unchanged, $j_k = i_k$. For the latter case, the variable u_k is not associated with r and hence, $f_{r,k}$ defines the effect of the event on the variables $\mathcal{U}_{k-1}$, recursively by $f_{r,k-1}$.

A relation node p is a *duplicate* of relation node r if $p.var = r.var$, $p.ptr = r.ptr$, and $p.\delta = r.\delta$. From now on, we assume that the implicit relation forest contains no duplicate nodes.

Algorithm BuildImplicit, shown in Fig. 5, constructs a set of relation nodes, $\mathcal{R}$, for a given model $\mathcal{M}$. It builds a relation for each event $e \in \mathcal{E}$, from the bottom up. In the algorithm, we loop over variables u_i where $\Delta_{i,e}$ is not the identity function (c.f. line 4); this corresponds to variables that either affect the enabling of e or are changed when e occurs. For each such variable u_i, we create a new relation node with function $\Delta_{i,e}$, pointing to the node below it. After eliminating duplicates (c.f. line 11), the top-level node encoding $\mathcal{N}_e$ is added to the set $\mathcal{R}$.

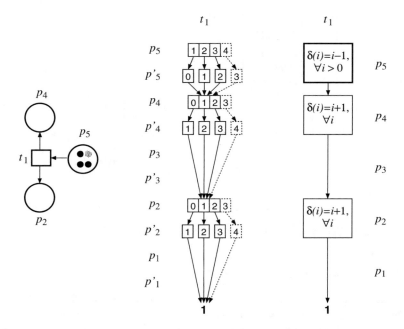

Fig. 3. Representation of transition t_1, from the fork-join Petri net model, in MxD and implicit relation. Dashed entities in the MxD represent the modification in the decision diagram due to additional token in place p_5, represented by dotted-circle.

Figure 3 compares the structure of implicit relation and MxD for transition t_1 of the fork-join Petri net model from Fig. 1. Note that for every additional token in place p_5, the MxD for this transition undergoes modification in terms of expansion of variable bound for each place. On the contrary, the implicit relation remains unchanged. The figure showcases the benefit of representing transitions of a model using implicit relation over MxD.

Figure 4 shows the implicit relation forest for all events of the fork-join Petri net model. The set $\mathcal{R}$ contains the top-most node for each event. Each node r is annotated with $r.\delta$ along with the values for which $r.\delta$ is defined. Note that the terminal node **1** is repeated for clarity purpose only. The figure also demonstrates merging of implicit relations for transitions t_6 and t_5 which is possible due to equal effects of each transition, t_6 and t_5, on the low lying variables, namely p_1. This merging of implicit relations allow f_r to encode the effect of more than one event, where r is a relation node common to multiple transitions.

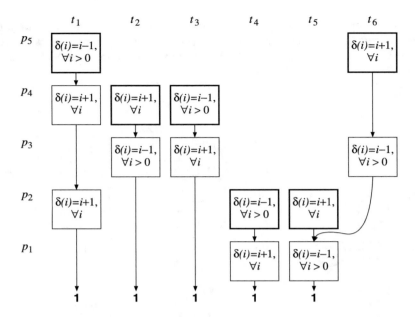

Fig. 4. The implicit relation forest for the fork-join Petri net model.

3.2   Saturation Using Relation Nodes

To use saturation, we must partition the set of relation nodes $\mathcal{R}$ into $\mathcal{R}_L, \ldots, \mathcal{R}_1$ with $\mathcal{R}_k = \{r \in \mathcal{R} : r.var = u_k\}$. Note that any event whose relation belongs to $\mathcal{R}_k$ will not be disabled by, and will not modify if it occurs, any variable $u_l \succ u_k$. The idea of saturation [7] is that, every time a node for variable u_k is created, it is *saturated* by applying the relations in set $\mathcal{R}_k$ repeatedly until a fixed point is reached. The saturation algorithm, modified to use relation nodes, is shown in Fig. 5 as procedure Saturate. It operates "in place" on a node n that has been created, but not yet added to the unique table, by repeatedly firing (c.f. line 10) the events in $\mathcal{A}[j]$ (c.f. the loop in line 8), a subset of events from $\mathcal{R}_k$ that produce same local index j on firing (c.f. the loop in line 5), over all possible values of j and adding those states to the current node (c.f. line 11). Procedure MultiRecFire, in line 10 is used to invoke RecFire for such subset of events and union the result (c.f. the lines 2, 3 of MultiRecFire) before saturating the node. It has been observed [5] that the order in which local states are explored (c.f. line 3 of Saturate) can significantly affect the efficiency of the iteration. The differences with respect to saturation using MxDs for relations are in lines 6 and 7, which obtain the local index j produced when an event fires on i using the relation node r, and the use of the common downward pointer $r.ptr$ in line 10.

Procedure RecFire, also shown in Fig. 5, is used to "fire" a relation r on node n (this determines the relational product of the set encoded by MDD node n and the relation encoded by relation node r), except that any created nodes are saturated immediately (c.f. line 15). Note that lines 8–9 handle the case where the relation graph skips a level (corresponding to an identity function), while lines 11–14 handle the case where the relation node and MDD node are at the same level. We do not give the case where the MDD graph skips a level, as this can only happen with edges that point directly to terminal node **0**. Again, the differences with respect to saturation using MxDs for relations are in lines 11, 12, and 13 which use the relation node r.

For fixed variable bounds, our modified saturation algorithm has the same complexity as the on-the fly saturation using extended decision diagrams. However, as bounds expand during saturation, relations stored using Kronecker representations, MxDs, or extended decision diagrams must all be updated to some extent for the increased bounds, with various costs for this reconstruction based on the type of storage. Even worse, some methods require discarding entries of the compute table, which can require significant duplication of computation. In contrast, no adjustments are needed to the implicit relation forest when bounds expand, and no compute table entries ever need to be discarded.

3.3   Implementation Notes

We briefly discuss some ideas for efficient implementation of implicit relation forest nodes. For a node r, its partial function $r.\delta$ can be implemented using a function pointer, an abstract class with a virtual function, or with a parse tree or similar representation for expressions.

The elimination of duplicate nodes can be done similarly to MDDs, which utilize a *unique table*. A hash signature for $r.\delta$, with the property that equal functions should produce equal signatures, can be used to reduce the number of (potentially expensive) comparisons between functions. We stress that a stray duplicate node will not affect correctness, but only the efficiency, by potentially requiring duplication in computation.

To reduce the number of calls to $r.\delta$ (another potentially expensive operation), each node can maintain an array that memorizes $r.\delta$, so that $r.\delta(i)$ must be computed at most once for each i. Clearly this is a time/memory tradeoff, and note that the memory cost for this array is similar to and does not exceed other relation representations such as MxDs. However, simple functions, like update by a constant, can be handled directly without using a function pointer or a memorization array.

Finally, we note that the flexibility of defining $r.\delta$ as a (partial) *function* allows us to easily handle the case where MDD variable u_i is not necessarily equal to the number of tokens in place p_i. This is required if u_i corresponds to more than one Petri net place. In our implementation, the value of u_i is the index of a submarking, stored in a collection of submarkings, where the submarking is over the places corresponding to MDD variable u_i.

```
mdd Union(mdd m, mdd n)
 1 if n = 1 ∨ m = 1 then return 1;
 2 if n = 0 then return m;
 3 if m = 0 then return n;
 4 if ∃p s.t. (∪, m, n, p) ∈ C or
 5            (∪, n, m, p) ∈ C then
 6    return p;
 7 k ← max(m.var, n.var);
 8 p ← new MDD node for variable u_k;
 9 for each i ∈ D(u_k) do
10    md ← (u_k ≻ m.var) ? m : m[i];
11    nd ← (u_k ≻ n.var) ? n : n[i];
12    p[i] ← Union(md, nd);
13 p ← UniqueInsert(p);
14 C ← C ∪ {(∪, m, n, p)};
15 return p;
```

```
nodeset BuildImplicit(model M)
 • Build set of relation nodes for model M
 1 R ← ∅;
 2 for each e ∈ E do
 3    Split Δ into Δ_{e,1}, ..., Δ_{e,L};
 4    V_e ← {i : Δ_{e,i} ≠ identity func.};
 5    r ← 1;
 6    for i ∈ V_e do
 7       p ← new relation node;
 8       p.var ← u_i;
 9       p.ptr ← r;
10       p.δ ← Δ_{i,e};
11       r ← UniqueInsert(p);
12    R ← R ∪ {r};
13 return R
```

```
MultiRecFire(mdd n, nodeset rSet)
 • RecFire all events in rSet on node n
 1 for each r ∈ rSet do
 2    f_r ← RecFire(n, r);
 3    f ← Union(f, f_r);
 4 return f;
```

```
Saturate(level k, mdd n)
 • Saturate node n in place using R_k.
 1 Q ← {i : n[i] ≠ 0};
 2 while Q ≠ ∅ do
 3    i ← SelectElement(Q);
 4    Q ← Q \ {i};
 5    for each r ∈ R_k do
 6       if r.δ(i) is defined then
 7          A[r.δ(i)] ← A[r.δ(i)] ∪ r.ptr
 8    for each a ∈ A do
 9       j ← a.index;
10       f ← MultiRecFire(n[i], a.rSet);
11       u ← Union(f, n[j])
12       if u ≠ n[j] then
13          n[j] ← u;
14          Q ← Q ∪ {j};
```

```
mdd RecFire(mdd n, relation r)
 • Fire r on node n and then saturate it.
 1 if n = 0 then return 0;
 2 if r = 1 then return n;
 3 if ∃m s.t. (RecFire, n, r, m) ∈ C then
 4    return m;
 5 k ← max(n.var, r.var);
 6 m ← new MDD node for variable u_k;
 7 if n.var ≻ r.var then
 8    for each i ∈ D(u_k) do
 9       m[i] ← RecFire(n[i], r);
10 else
11    for each i s.t. r.δ(i) is defined do
12       j ← r.δ(i);
13       f ← RecFire(n[i], r.ptr);
14       m[j] ← Union(m[j], f);
15 Saturate(k, m);
16 m ← UniqueInsert(m);
17 C ← C ∪ {(RecFire, n, r, m)};
18 return m;
```

Fig. 5. MDD and relation node algorithms.

4   Related Work

This section discusses our approach of saturation with implicit relations in light of related work. We examine the alternative approaches of encoding transitions and compare them against the idea of this paper.

4.1 Kronecker Representations

The saturation algorithm originally used a Kronecker representation to encode transitions [7]. Conceptually, such a scheme requires, for each model event e and each state variable v_k, a boolean matrix $\mathbf{N}_{e,k}$ used to encode function $\Delta_{e,k}$, where $\mathbf{N}_{e,k}[i_k, j_k]$ is one if and only if $\Delta_{e,k}(i_k) = j_k$. Note that $\mathbf{N}_{e,k}$ will be the identity matrix if event e does not change or depend on state variable v_k. In practice, identity matrices need not be stored explicitly, and other sophisticated schemes [6] could be used to store each $\mathbf{N}_{e,k}$. The dimension of $\mathbf{N}_{e,k}$ is the bound for state variable v_k, and it was originally assumed that this bound was known. On-the-fly saturation [8] eliminated this requirement, allowing the bounds of state variables to expand during saturation. As bounds expand, the matrices $\mathbf{N}_{e,k}$ also expand in size.

Implicit relation forests have two main advantages as compared to on-the-fly saturation using Kronecker representations. First, implicit relation forests allow for "sharing", i.e., a node can have more than one parent node, which occurs whenever two transitions have the same effect on the bottom-most k state variables. In some models, especially if transitions are split to maintain Kronecker consistency [6], this sharing can be significant, and the primary benefit is reduction of computation time, as this duplication in computation is avoided via the compute table. Second, the saturation algorithm is simpler with implicit nodes, as there is no longer a need to distinguish between "confirmed" and "unconfirmed" local states, nor is it necessary to expand matrices as local states are confirmed.

4.2 MDDs and Extensible MDDs

A 2L-variable MDD, also called MxD, suffers from deletion of relevant yet incomplete compute-table entries when the *operand* MxDs remain unchanged but the resultant MxD undergoes modification in case of discovery of new bounds for a variable. Such deletions lead to reduced efficiency, to address which *extensible* MxDs were introduced in [19]. However, implicit relations provide a more efficient format, as demonstrated in Sect. 5, for encoding transitions to tackle the overhead cost of rebuilding *extensible* MxDs for the subclass of Petri nets.

4.3 Interval Mapping Diagrams

Strehl's work [17] on *interval mapping diagrams* (IMD) provides a generalized encoding of transitions wherein the *state distance* between pre- and post-transition state variable values are stored. The *state distance* is defined by an *action operator* and *action interval*, which together formulate the net-effect of the transition, on a *predicate interval* of the state variable, which refers to the enabling condition of the transition.

Implicit relation forests have the advantage of encoding any (partial) function as an effect of a transition on a variable, in contrast to IMDs, where the *action operator* is restricted to use increment, decrement, and equality operators only.

4.4   Homomorphisms

Couvreur et. al's work [11] offers an efficient way of encoding transitions using the concept of inductive homomorphisms. The encoding is defined to work with Data Decision Diagrams (DDD) [11] and Hierarchical Set Decision Diagrams (SDD) [12]. The approach offers freedom to the user in defining transitions and is more efficient compared to prior works [8,10,15,19].

Implicit relations are an adaptation of inductive homomorphisms that work with MDD and are restrictive in terms of the nature of transitions that can be encoded. Only transitions with "firing" conditions defined as partial functions of the participant variables are compatible with implicit relations. A comparative study between tools implementing homomorphism on DDD and implicit relations on MDD is discussed in Sect. 5 to get a general overview of their performance on a set of benchmark models.

5   Experimental Evaluation

Intra-Tool Comparative Performance Analysis

We implemented the modified saturation algorithm based on implicit relations (SATIMP) in SMART [9] using Meddly [2,3] as the underlying decision diagram library. We conducted experiments to compare the performance of SATIMP with the existing "on-the-fly saturation with matrix diagrams" approach (OTFSAT) for reachability set generation on a suite of 70 Petri net models that is available as *known-models* in MCC 2018 [1]. An experimental run involves execution of SATIMP and OTFSAT on a model instance with a timeout of one hour for each approach. All experiments are run on a server of Intel Xeon CPU 2.13 GHz with 48G RAM under Linux Kernel 4.9.9. For a given experimental run, the decision diagrams used in SATIMP and OTFSAT approach have identical variable ordering to ensure fair comparison.

Every Petri net model in the suite has multiple instances characterized by scaling parameters that affect the size of model ($|\mathcal{V}| + |\mathcal{E}|$) or the initial state of the model ($\mathbf{i}_0$). In order to demonstrate the effect of size and complexity of the models on the performance of SATIMP and OTFSAT, the set of benchmark models are classified into two categories namely, Type-1 models with scaling parameters affecting model size, and Type-2 models with scaling parameters affecting the initial state. Safe petri nets form a subset of Type-1 models. Table 1 summarizes the experiments run on a subset (due to space constraint) of Type-1 and Type-2 models with key metrics of comparison as runtime, measured in seconds, and total number of pings and hits to the compute-table for saturation operation. Since, OTFSAT uses MxD, the additional computation is summarized in the column for the total number of pings and hits to the compute-table for MxD operations. The pings and hits to the compute-table provide a respective estimate of the number of decision diagram computations needed and reused for each saturation approach.

It is also important to note that the computation time spent in calculating the next-state of a variable is additional to the time spent for executing the saturation algorithm. SATIMP generates the next-state of a variable using the information stored in the implicit relations and OTFSAT spends time modifying MxD when new bounds of the variable are discovered.

Observations from Table 1, for Type-1 models, confirm that the computation time for reachability set remains fairly equal in both implementations. For these models, since the scaling parameter does not affect the bound of the variables in the model, there is comparatively less time spent on modification of MxDs. Hence, the computation time spent by SATIMP to calculate the next-state of every variable is close to the time spent by OTFSAT in construction of matrix diagrams.

On the contrary, for Type-2 models, a significant improvement in performance of SATIMP is observed. For these models, the maximum local bound of any state variable discovered during the reachability set generation is a number greater than or equal to 1 (except that maximum local bound of every state variable is not 1) as determined by the scaling parameter(s) in the model definition. This requires frequent expansion of MxD nodes to encode the effect of transitions on the local state space of each state variable. Supported by this fact and experimental results in Table 1, a few observations can be noted. First, while SATIMP is able to complete models with high scaling parameters quite early, OTFSAT either takes long time, generally increased many-fold as compared to implicit relation, or does not finish the task before timeout. In such models, a significant amount of computational time is spent in modification of the matrix diagrams as shown by the number of pings and hits to compute-table for MxDs, which is otherwise absent in implicit relations.

Second, since MxDs expand and contract during manipulation, it may require compute-table entries to be discarded. Hence, the number of pings to the saturation compute-table would be relatively higher in OTFSAT as compared to SATIMP. The dashes in the table correspond to cases in which the runtime to construct S exceeded one hour. However, no claims can be made about the models that did not complete within the timeout.

For comparing the maximum memory usage of SATIMP with that of OTFSAT, we have chosen the largest completed instances of each model from the results in Table 1. When matrix diagrams consume memory on megabytes scale, implicit relations manage to store the exact information in much lesser space. The figures in Table 2 provide substantial proof of improvement in memory usage.

Our results include metrics that are typical for efficiency comparisons between two approaches, and illustrate the efficiency of using SATIMP in terms of both computational and storage requirements. In practice, the use of implicit relations allow for reachability analysis of much larger systems as compared to that with MxDs.

Table 1. Computational requirements of saturation for reachability set generation in Type-1 models and Type-2 models.

Model	$|S|$	OffSat			SatImp			Additional MxD CT in OffSat	
		Time (sec)	Pings $\times 10^5$	Hits $\times 10^5$	Time (sec)	Pings $\times 10^5$	Hits $\times 10^5$	Pings $\times 10^3$	Hits $\times 10^3$
Type-1:									
DES 30a	1.92×10^{13}	22.60	175	34	23.91	176	34	24	16
DES 30b	1.97×10^{22}	103.97	719	201	102.74	745	203	18	11
DES 40a	3.52×10^{13}	49.27	341	62	49.54	344	63	28	18
DES 40b	3.60×10^{22}	149.64	995	256	152.69	1041	258	20	12
FlexibleBarrier 10a	6.91×10^{10}	3.01	43	26	2.96	29	43	27	12
FlexibleBarrier 12a	8.92×10^{12}	15.75	219	144	15.74	219	143	15	10
Raft 5	5.94×10^{18}	23.58	136	77	23.91	137	77	11	7
Raft 6	2.91×10^{26}	189.464	1065	621	193.92	1070	630	18	12
RWmutex r10w100	1.12×10^{3}	2.29	26	0	2.48	25	0	102	69
RWmutex r10w500	1.52×10^{3}	51.18	432	30	40.15	431	30	586	422
Type 2 :									
Angiogenesis 15	1.12×10^{15}	324.79	4873	4340	140.33	2973	2671	320	310
CircadianClock 1000	4.02×10^{15}	3334.53	6834	6753	78.02	186	136	88016	88014
FMS 100	2.70×10^{21}	8.90	340	332	4.12	220	214	341	50
FMS 200	1.95×10^{25}	78.28	3331	3294	33.50	1737	1714	50	50
GPPP C1000N10	1.42×10^{10}	1.19	5	3	0.21	3	2	43	43
GPPP C1000N100	1.14×10^{15}	440.35	1560	1413	114.88	654	543	18072	18071
Kanban 500	$7.09 \times 10^{2}6$	458.66	2558	2542	12.14	756	752	12704	12703
Kanban 1000	1.42×10^{30}	2347.18	20231	20170	72.50	5909	5891	50677	50436
Robot Manipulation 20	4.11×10^{9}	6.91	172	153	1.22	34	31	18	17
Robot Manipulation 50	8.53×10^{12}	176.05	3941	3659	33.15	871	833	253	253
SmallIOS MT1024DC256	3.27×10^{12}	971.77	7506	7475	66.13	3198	3184	24522	24521
SmallIOS MT2048DC0512	1.04×10^{14}	–	–	–	620.61	25184	25118	–	–
SmallIOS MT2048DC1024	2.46×10^{14}	–	–	–	1105.18	38082	37945	–	–
SwimmingPool 9	1.81×10^{10}	11.73	116	97	7.28	116	97	70	70
SwimmingPool 10	3.36×10^{10}	15.81	163	137	10.98	161	135	95	95

Table 2. Storage requirements by saturation algorithm for transition encodings.

Model	Memory for OTFSAT (KB)	Memory for SATIMP (KB)
DES 40b	524.00	7.55
FlexibleBarrier12a	380.81	41.33
Raft 6	452.27	41.94
RWmutex r10w500	11864.42	403.26
Angiogenesis 15	1133.90	9.28
CircadianClock 1000	959318.00	12.00
FMS 200	10175.00	26.59
GPPP 100 100	1470591.00	48.56
Kanban 1000	464342.00	16.00
Robot Manipulation 50	18020.00	14.12
Small OS 1024 256	242091.00	106.50
Swimming Pool 10	1823.60	10.88

Inter-Tool Comparative Performance Analysis

This section presents the performance comparison between the state-space generation algorithms of SMART and ITSTools [18], where the former tool uses implicit relations and MDD and the latter is based on homomorphisms, DDD and SDD. The goal of this comparative analysis is to only gauge the efficiency of SATIMP by using the well-established technique of homomorphism-based saturation as a benchmark.

A suite of 70 Petri net models from *known models* section of MCC 2018 is used in the experiments to compare the tools based on the runtime of state-space generation process. The largest instance of each model that could complete state-space generation with both SMART and ITSTools in MCC 2018, is chosen for this experiment. Table 3 shows only a subset of these experiments due to space constraint. Since the tools are based on decision diagrams and variable order is critical for efficiency of the state-space generation, identical static variable orders are used in both tools for each experimental run. SMART is tuned to run on implementation settings similar to that of MCC 2018 settings of ITS-Tools. For example, with reference to Sect. 3.3, the MDD variable u_i is adapted to be equal to the number of tokens in place p_i. However, the use of SDD in ITS-tools is omitted from the experiments to ensure fair comparison, because the construction of an SDD in ITS-tools requires auxiliary information about hierarchy of model variables and is not inferred directly from the model. All experiments are run in the same environment as described in the previous section.

In Table 3, it is observed that SMART is faster than ITSTools for 41 out of 67 models by an average of 3.125 times. These models include Kanban, Flexible Manufacturing System, House Construction and Philosophers where SMART is 235, 30, 10 and 16 times faster respectively. For 10 of the models, where

Angiogenesis, SharedMemory etc are few of them, SMART is about 0.56 times faster than ITSTools. For the remaining 16 models, ITSTools is 4.69 times faster than SMART on an average. The experimental results allow us to surmise that the performance of SMART when using implicit relations on MDD-based storage complements the performance of ITSTools that use inductive homomorphisms with DDD-based storage for saturation.

Table 3. Performance comparison between SMART and ITS-tools.

Model	Instance	Reachable States (#)	Runtime (sec)	
			ITS Tools	SMART
Kanban	100	1.7263E+19	3.37E+03	1.42E+01
FMS	200	1.9536E+25	4.64E+02	1.42E+01
SwimmingPool	6	1.6974E+09	5.26E+01	2.89E+00
Philosophers	500	3.6300E+238	4.03E+00	2.24E−01
HouseConstruction	10	1.6636E+09	6.11E+00	5.99E−01
ClientsAndServers	5	1.2551E+11	7.11E+01	8.55E+00
CircadianClock	100	4.2040E+10	1.74E+00	4.60E−01
IBMB2S565S3960	none	1.5511E+16	1.60E+01	8.65E+00
Ring	none	9.0265E+11	1.72E−01	9.34E−02
TokenRing	15	3.5358E+07	1.57E+01	1.26E+01
Referendum	100	5.1537E+47	1.07E+00	8.96E−01
SharedMemory	20	4.4515E+11	5.04E+00	7.76E+00
EnergyBus	none	2.1318E+12	5.34E+01	9.00E+01
Angiogenesis	5	4.2735E+07	5.20E−01	9.58E−01
FlexibleBarrier	4a	2.0737E+04	6.11E−02	1.51E−01
Railroad	10	2.0382E+06	2.34E+00	5.94E+00
Peterson	3	3.4079E+06	2.70E+01	7.45E+01
CSRepetitions	3	1.3407E+08	6.01E−01	2.47E+00
UtahNoC	none	4.7599E+09	5.29E+00	3.44E+01
PaceMaker	none	3.6803E+17	2.47E−01	3.07E+00

6   Conclusions and Future Work

Reachability set generation using on-the-fly saturation with MxDs is quite an improvement over explicit techniques, as it is often able to handle extremely large sets with less time. However, the computational cost for building such transition relations repeatedly during reachability set generation, creates additional focus towards handling the relations along with the state space. The transition relations undergo manipulations for the construction of next-state functions necessary for state space generation, while the underlying functions to generate the next state is available in the model itself. Implicit relations manage to encapsulate and exploit these properties. Hence, we adapted the saturation algorithm to work with implicit relations and showed how additional computations can be saved during the reachability set generation.

Saturation algorithm using implicit relations for the reachability set generation provides promising results for the defined class of models. Experimental results indicate that the costs are improved for a large set of models across different sizes, though the approach is not adapted to handle marking dependent events as discussed in Sect. 2.1. The improvement is mainly due to the essence of implicit relations to encase the properties of the system in a simple straightforward approach.

Future work should investigate modification of the implicit relations to represent Petri nets with marking-dependent arcs by disclosing value of each variable to the underlying variables in the implicit relation. We intend to create a merger between implicit relations and MxDs that will exploit their respective static and dynamic ingredients in the fusion. Allowing implicit nodes inside an MxD forest would allow us to handle models that are not "Kronecker consistent", but still get benefits for events that affect the participant variables independent of each other.

Acknowledgment. This work was supported in part by the National Science Foundation under grant ACI-1642397.

References

1. MCC: Model Checking Competition @ Petri Nets. https://mcc.lip6.fr
2. MEDDLY webpage. https://sourceforge.net/projects/meddly/
3. Babar, J., Miner, A.S.: Meddly: multi-terminal and Edge-valued Decision Diagram LibrarY. In: Proceedings of QEST, pp. 195–196. IEEE Computer Society (2010)
4. Bryant, R.E.: Symbolic boolean manipulation with ordered binary-decision diagrams. ACM Comput. Surv. **24**(3), 293–318 (1992)
5. Chung, M.-Y., Ciardo, G., Yu, A.J.: A fine-grained fullness-guided chaining heuristic for symbolic reachability analysis. In: Graf, S., Zhang, W. (eds.) ATVA 2006. LNCS, vol. 4218, pp. 51–66. Springer, Heidelberg (2006). https://doi.org/10.1007/11901914_7
6. Ciardo, G., Lüttgen, G., Miner, A.S.: Exploiting interleaving semantics in symbolic state-space generation. Form. Methods Syst. Des. **31**, 63–100 (2007)
7. Ciardo, G., Lüttgen, G., Siminiceanu, R.: Saturation: an efficient iteration strategy for symbolic state—space generation. In: Margaria, T., Yi, W. (eds.) TACAS 2001. LNCS, vol. 2031, pp. 328–342. Springer, Heidelberg (2001). https://doi.org/10.1007/3-540-45319-9_23
8. Ciardo, G., Marmorstein, R., Siminiceanu, R.: Saturation unbound. In: Garavel, H., Hatcliff, J. (eds.) TACAS 2003. LNCS, vol. 2619, pp. 379–393. Springer, Heidelberg (2003). https://doi.org/10.1007/3-540-36577-X_27
9. Ciardo, G., Miner, A.S.: SMART: simulation and Markovian analyzer for reliability and timing. In: Proceedings of IEEE International Computer Performance and Dependability Symposium (IPDS 1996), p. 60. IEEE Computer Society Press (1996)

10. Ciardo, G., Yu, A.J.: Saturation-based symbolic reachability analysis using conjunctive and disjunctive partitioning. In: Borrione, D., Paul, W. (eds.) CHARME 2005. LNCS, vol. 3725, pp. 146–161. Springer, Heidelberg (2005). https://doi.org/10.1007/11560548_13
11. Couvreur, J.-M., Encrenaz, E., Paviot-Adet, E., Poitrenaud, D., Wacrenier, P.-A.: Data decision diagrams for Petri net analysis. In: Esparza, J., Lakos, C. (eds.) ICATPN 2002. LNCS, vol. 2360, pp. 101–120. Springer, Heidelberg (2002). https://doi.org/10.1007/3-540-48068-4_8
12. Couvreur, J.-M., Thierry-Mieg, Y.: Hierarchical decision diagrams to exploit model structure. In: Wang, F. (ed.) FORTE 2005. LNCS, vol. 3731, pp. 443–457. Springer, Heidelberg (2005). https://doi.org/10.1007/11562436_32
13. Kam, T., Villa, T., Brayton, R.K., Sangiovanni-Vincentelli, A.: Multi-valued decision diagrams: theory and applications. Mult.-Valued Log. 4(1–2), 9–62 (1998)
14. Miner, A.S.: Implicit GSPN reachability set generation using decision diagrams. Perform. Eval. 56(1), 145–165 (2004). Dependable Systems and Networks - Performance and Dependability Symposium (DSN-PDS) 2002: Selected Papers
15. Miner, A.S.: Saturation for a general class of models. In: Proceedings of QEST, pp. 282–291, September 2004
16. Murata, T.: Petri nets: properties, analysis and applications. Proc. IEEE 77(4), 541–579 (1989)
17. Strehl, K., Thiele, L.: Interval diagram techniques for symbolic model checking of Petri nets. In: Proceedings of Design, Automation and Test in Europe (DATE 1999), pp. 756–757, March 1999
18. Thierry-Mieg, Y.: Symbolic model-checking using ITS-tools. In: Baier, C., Tinelli, C. (eds.) TACAS 2015. LNCS, vol. 9035, pp. 231–237. Springer, Heidelberg (2015). https://doi.org/10.1007/978-3-662-46681-0_20
19. Wan, M., Ciardo, G.: Symbolic state-space generation of asynchronous systems using extensible decision diagrams. In: Nielsen, M., Kučera, A., Miltersen, P.B., Palamidessi, C., Tůma, P., Valencia, F. (eds.) SOFSEM 2009. LNCS, vol. 5404, pp. 582–594. Springer, Heidelberg (2009). https://doi.org/10.1007/978-3-540-95891-8_52
20. Yoneda, T., Hatori, H., Takahara, A., Minato, S.: BDDs vs. Zero-suppressed BDDs: for CTL symbolic model checking of Petri nets. In: Srivas, M., Camilleri, A. (eds.) FMCAD 1996. LNCS, vol. 1166, pp. 435–449. Springer, Heidelberg (1996). https://doi.org/10.1007/BFb0031826

Taking Some Burden Off an Explicit CTL Model Checker

Torsten Liebke and Karsten Wolf[✉]

Universität Rostock, Institut für Informatik, Rostock, Germany
{torsten.liebke,karsten.wolf}@uni-rostock.de

Abstract. In the CTL category of recent model checking contests, less problems have been solved than in the Reachability and LTL categories. Hence, improving CTL model checking technology deserves particular attention. We propose to relieve a generic explicit CTL model checker. This is done by designing specialised routines that cover a large set of simple (and frequently occurring) formula types. The CTL model checker is then only applied to formulas that do not fall into any special case. For the simple queries, we may apply simple depth-first search instead of recursive search, we may use much more powerful dialects of the stubborn set reduction, and we may add additional tools for verification, such as the state equation. Our approach covers about half of the CTL category of a recent model checking contest and significantly increases the power of CTL model checking.

Keywords: CTL model checking · Partial order reduction

1 Introduction

In recent years, Computational Tree Logic (CTL, [2]) has been the category where most queries were left unsolved in the yearly Petri net model checking contest (MCC, [9]). Consequently, CTL model checking deserves particular attention with the aim of keeping pace with LTL and reachability checking. At present, leading Petri net CTL model checkers such as TAPAAL [7] and LoLA [29] use explicit model checking algorithms. Their main tool for alleviating state explosion is the stubborn set method [22] or, more general, the class of partial order reduction methods [12,18]. In essence, partial order reduction explores, in any given marking, only a subset of the enabled transitions. CTL preserving partial order reduction [11] has severe restrictions: we either find, in a given marking, a *singleton* set consisting of an invisible transition that satisfies all other conditions for a stubborn set, or we have to fire all transitions enabled in this marking. This condition is necessary for CTL preservation since otherwise the position of visible transitions with respect to branching points may not be preserved which in turn would jeopardise preservation of the branching time logic CTL.

Many CTL queries have a rather simple structure in the sense that they contain only few temporal operators. In the MCC, this might be an artifact of the

© Springer Nature Switzerland AG 2019
S. Donatelli and S. Haar (Eds.): PETRI NETS 2019, LNCS 11522, pp. 321–341, 2019.
https://doi.org/10.1007/978-3-030-21571-2_18

formula generation mechanism. However, we share the same experience with the users of our tool LoLA. Even if complicated CTL formulas occasionally occur, they are subject to several simplification approaches. Firstly, there exist many tautologies in temporal logic. Not all of them are commonly known. This way, an originally complicated formula may automatically be rewritten to a much simpler query [1]. The formula rewriting system of LoLA currently contains more than 100 rewrite rules that are based on CTL* tautologies. For Petri nets, secondly, linear programming techniques employing the Petri net state equation can be applied to the atomic propositions in the formula [1], sometimes proving them to be invariantly true or false. This way, whole subformulas of a query may collapse, enabling further rewriting based on tautology. Boolean combinations of queries can be simplified by checking the subformulas separately (thus having queries with less visible transitions in each run which propels partial order reduction). Thirdly, complicated queries may be replaced by simpler queries through modifications in the system under investigation. A simple example is the verification of relaxed soundness [8] for workflow nets. For every transition t, we have to show that there is a path to a given final place f that includes the occurrence of t. In CTL, this reads as $\mathrm{EF}(t \text{ occurs} \wedge \mathrm{EF}(f \geq 0))$. Inserting a fresh post-place p to t, the query can be simplified to $\mathrm{EF}\ (p \geq 0 \wedge f \geq 0)$. The most systematic approach of this kind is LTL model checking as a whole. The explicit verification of an LTL formula ϕ is done by modifying the system under investigation (we refer to the construction of the product system with the Büchi automaton for $\neg\phi$, [26]). In the modified system, we only need to verify $\neg\ \mathrm{GF}$ *accepting-state* instead of the arbitrarily complicated ϕ.

We conclude that explicit CTL model checking can be substantially improved through a special treatment of as many as possible of the most simple queries. Special treatment means that we apply a specific verification procedure to such queries thus avoiding the application of the generic CTL model checking routines. This approach has two obvious advantages. Firstly, some of the queries may permit the use of completely different verification technology. For example, for properties like $EF\ \phi$ or $AG\ \phi$ (with ϕ assumed not to contain additional temporal operators), we may employ the Petri net state equation for verification [28]. Secondly, a verification technique dedicated to just one class C of simple CTL queries may use a better partial order reduction: we only need to preserve C rather than whole CTL.

In this paper, we focus on the second item. We identify several classes of simple CTL queries for which specific search routines enable the use of partial order reduction methods better than CTL preserving ones. These partial order reduction methods are already known in most cases. So the actual contribution of this paper is to show that the systematic separation of simple queries from general CTL routines can indeed improve CTL model checking. In 2018, almost 70% of the CTL queries in the MCC were transferred to specific routines for simple queries in our tool LoLA. Employing these methods, LoLA could solve more than 50% of the queries that could not be solved with the generic CTL model checking algorithm.

We start with a brief introduction of the terminology of Petri nets and the temporal logic CTL. We then provide the necessary facts on the stubborn set method. In the main part of the paper, we discuss our list of simple CTL queries. We conclude with experimental results.

2   Terminology

Definition 1 (Place/transition net). *A place/transition net consists of a finite set P of places, a finite and disjoint set T of transitions, a set $F \subseteq (P \times T) \cup (T \times P)$ of arcs, a weight function $W : (P \times T) \cup (T \times P) \rightarrow \mathbb{N}$ where $[x, y] \notin F$ if and only if $W(x, y) = 0$, and a marking m_0, the initial marking. A marking is a mapping $m : P \rightarrow \mathbb{N}$.*

Definition 2 (Behaviour of a place/transition net). *Transition t is enabled in marking m if, for all $p \in P$, $W(p, t) \leq m(p)$. If t is enabled in m, t can fire, producing a new marking m' where, for all $p \in P$, $m'(p) = m(p) - W(p, t) + W(t, p)$. This firing relation is denoted as $m \xrightarrow{t} m'$. It can be extended to firing sequences by the following inductive scheme: $m \xrightarrow{\varepsilon} m$ (for the empty sequence ε), and $m \xrightarrow{w} m' \wedge m' \xrightarrow{t} m'' \implies m \xrightarrow{wt} m''$ (for a sequence w and a transition t). The reachability graph of a place/transition net N has a set of vertices that comprises of all markings that are reachable by any sequence from the initial marking of m. Every element $m \xrightarrow{t} m'$ of the firing relation ($t \in T$) defines an edge from m to m' annotated with t.*

With a matrix C where, for all $p \in P$ and $t \in T$, $C(p, t) = W(t, p) - W(p, t)$, and marking m, equation $m_0 + Cx = m$ is called the *Petri net state equation*. It has a nonnegative integer solution for every m reachable from m_0: fix a firing sequence w from m_0 to m and let, for every t, $x[t]$ be the number of occurrences of t in w. For unreachable m, the state equation may or not have nonnegative integer solutions.

In the sequel, we consider only Petri nets with finite reachability graph (i.e. bounded Petri nets).

Definition 3 (Syntax of CTL). TRUE, FALSE, FIREABLE (t) *(for $t \in T$),* DEADLOCK, *and $k_1 p_1 + \cdots + k_n p_n \leq k$ ($k_i, k \in \mathbb{Z}, p_i \in P$) are atomic propositions. $PQ = \{A, E\}$ is called the set of path quantifiers, $UT = \{X, F, G\}$ the set of unary temporal operators, and $BT = \{U, R\}$ the set of binary temporal operators.*

Every atomic proposition is a CTL formula. If ϕ and ψ are CTL formulas, so are $\neg\phi$, $(\phi \wedge \psi)$, $(\phi \vee \psi)$, $QY\phi$ (with $Q \in QP$ and $Y \in UT$), and $Q(\phi B \psi)$ (with $Q \in QP$ and $B \in BT$).

The logic LTL is defined similarly. The only difference is that the path quantifiers are not used in LTL.

Definition 4 (Semantics of CTL). *Marking m satisfies CTL formula ϕ ($m \models \phi$) according to the following inductive scheme:*

- $m \models$ TRUE, $m \not\models$ FALSE;
- $m \models$ FIREABLE (t) *if t is enabled in m;*
- $m \models$ DEADLOCK *if there is no enabled transition in m;*
- $m \models k_1 p_1 + \cdots + k_n p_n \leq k$ *if $k_1 m(p_1) + \cdots + k_n m(p_n) \leq k$;*
- $m \models \neg\phi$ *if $m \not\models \phi$;*
- $m \models (\phi \wedge \psi)$ *if $m \models \phi$ and $m \models \psi$;*
- $m \models EX\phi$ *if there is a t and an m' with $m \xrightarrow{t} m'$ and $m' \models \phi$;*
- $m \models E(\phi U \psi)$ *if there is a path $m_1 m_2 \ldots m_k$ ($m_1 = m, k \geq 1$) in the reachability graph where $m_k \models \psi$ and, for all i with $1 \leq i < k$, $m_i \models \phi$;*
- $m \models A(\phi U \psi)$ *if, for all maximal paths (i.e. infinite or ending in a deadlock) $m_1 m_2 \ldots$ in the reachability graph with $m_1 = m$, there is a k ($k \geq 1$) where $m_k \models \psi$ and, for all i with $1 \leq i < k$, $m_i \models \phi$.*

The semantics of the remaining CTL operators is defined using the tautologies $(\phi \vee \psi) \iff \neg(\neg\phi \wedge \neg\psi)$, $AX\phi \iff \neg EX\neg\phi$, $EF\phi \iff E(\text{TRUE } U\phi)$, $AF\phi \iff A(\text{TRUE } U\phi)$, $AG\phi \iff \neg EF\neg\phi$, $EG\phi \iff \neg AF\neg\phi$, $E(\phi R\psi) \iff \neg A(\neg\phi U \neg\psi)$, and $A(\phi R\psi) \iff \neg E(\neg\phi U \neg\psi)$.

A place/transition net N satisfies a CTL formula if its initial marking m_0 does.

For LTL, all temporal modalities concern the same single path. Moreover, only infinite paths are considered. A maximal finite path is transformed into an infinite path by infinitely repeating the last (deadlock) marking. Otherwise, evaluation accords with CTL. A place/transition net N satisfies an LTL formula if all paths starting from m_0 do.

3   CTL Model Checking

We consider *local* model checking, that is, we want to evaluate a given CTL formula just for the initial marking m_0. Other markings are only considered as far as necessary for determining the value at m_0. In global model checking, one would be interested in the value of the given formula in all reachable markings. As a reference for our work, we use the algorithm of [27]. In the sequel, we briefly sketch this algorithm.

We assume that, attached to every marking, there is a vector that has an entry for every subformula of the given CTL query. The value of a single entry can be true, false, or unknown. Whenever we want to access the value of a subformula ϕ in a marking m, we inspect the corresponding value. If it is unknown, we recursively launch a procedure to evaluate ϕ in m.

If ϕ is an atomic proposition or a Boolean combination of subformulas, evaluation is trivial. For evaluating a formula of shape EX ϕ' or AX ϕ', we proceed to the immediate successor states and evaluate ϕ' in those states. If the successor marking has not been visited yet, we add it to the set of visited markings and initialise its vector of values.

If ϕ has the shape $A(\psi U \chi)$, we launch a depth-first search from m, aiming at the detection of a counterexample. The search proceeds through markings that satisfy ψ, violate χ, and for which $A(\psi U \chi)$ is recorded as unknown. Whenever we leave the space of state satisfying these assumption, there is a reaction that does not require continuation of the search beyond that marking, as follows.

If χ is satisfied, or $A(\psi U \chi)$ is recorded as true in any marking m', we backtrack since there cannot be a counterexample path containing m'. If χ and ψ are violated, or $A(\psi U \chi)$ is recorded as false, we exit the search since the search stack forms a counterexample for $A(\psi U \chi)$ in m. If we hit a marking m' on the search stack, we have found a counterexample, too (a path where ψ and not χ hold forever). The depth-first search assigns a value different from unknown to all states visited during the search: For markings on the search stack (i.e. participating in the counterexample), $A(\psi U \chi)$ is false, while for states that have been visited but already removed from the search stack, $A(\psi U \chi)$ is actually true.

If ϕ has the shape $E(\psi U \chi)$, we launch a similar depth-first search, aiming at the detection of a witness path. This time, we integrate Tarjan's algorithm [21] for detecting the strongly connected components (SCC) during the search. It proceeds through markings that satisfy ψ, violate χ, and for which $E(\psi U \chi)$ is recorded as unknown. If we hit a marking m' where χ is satisfied, or $E(\psi U \chi)$ is true, we have found our witness. In states where ψ and χ are violated, or $E(\psi U \chi)$ is known to be false, we backtrack since there cannot be witness path containing such a marking. Again, we assign a value different from unknown to every marking visited during the search. Markings that are on the search stack as well as markings that are not on the search stack but appear in SCC that have not yet been completely explored, get value true. An SCC is not yet fully explored if it contains elements that are still on the search stack. Then, however, a path to the search stack extended by the remaining portion of the search stack forms a witness. For markings in SCC that have been completely explored, $E(\psi U \chi)$ is false.

We can see that existential and universal until operators are not fully symmetric. This is due to the fact that a cycle of markings that satisfy ψ and violate χ, form a counterexample for universal until but no witness for existential until. Any SCC with more than one member would contain such a cycle. Consequently, universal until will never remove markings from the search stack without closing a whole (singleton) SCC.

The remaining CTL operators can be traced back to the two until operators using tautologies. Since every search assigns values to all visited markings, the overall run time of the algorithm is $O(|\phi||R|)$ where $|\phi|$ is the length of ϕ (the number of subformulas), and $|R|$ is the number of markings reachable from m_0.

4   Partial Order Reduction—the Stubborn Set Method

Given a Petri net N and a property ϕ, the stubborn set method aims at producing a subgraph G' of the reachability graph G of N such that the evaluation of

ϕ using G' yields the same value as the evaluation on G. To this end, a set stubborn(m) of transitions is assigned to every marking m, and only enabled transitions in stubborn(m) are explored for the construction of G'.

Over the years, a consistent systematic has emerged for presenting stubborn set methods. There is a list of *principles* that should govern the selection of stubborn sets. Each principle comes with an *algorithmic approach* for computing a stubborn set that obeys that principle. Finally, there is a list of *results* stating that, if G' is computed using stubborn sets that meet some selection of principles, all properties of a certain class of properties are preserved. In the sequel, we shall list principles and results that we need for our considerations below. For some principles, there exist several variations that push results further to the limit. However, our focus here is not stubborn set theory as such but CTL model checking technology. For this reason, we selected principles such that presentation is as understandable as possible. For stronger results on stubborn sets, the reader is referred to [11,15,23,25]. We will further completely skip the algorithmic approaches as they are not necessary for understanding our argument.

In the sequel, let $N = [P, T, F, W, m_0]$ be an arbitrary fixed place/transition net.

Definition 5 (COM: The commutativity principle). stubborn(m) $\subseteq T$ *satisfies the commutativity principle (COM for short) if, for all* $w \in (T \backslash$ stubborn(m))* *and all* $t \in$ stubborn(m), $m \xrightarrow{wt} m'$ *implies* $m \xrightarrow{tw} m'$.

Definition 6 (KEY: The key transition principle). stubborn(m) $\subseteq T$ *satisfies the key transition principle (KEY for short) if* m *does not enable any transition or it contains a transition* t^* *(a* key *transition) such that, for all* $w \in (T \setminus$ stubborn(m))*, $m \xrightarrow{w} m'$ *implies that* t^* *is enabled in* m'.

Definition 7 (VIS: The visibility principle). *Transition* t *is invisible w.r.t. an LTL or CTL formula* ϕ *if, for all atomic propositions* ψ *occurring in* ϕ *and all markings* m, m', $m \xrightarrow{t} m'$ *implies that* $\psi(m)$ *holds if and only if* $\psi(m')$ *holds.* stubborn(m) $\subseteq T$ *satisfies the visibility principle for a property* ϕ *(VIS(ϕ) for short) if* stubborn(m) *contains only invisible transitions w.r.t.* ϕ, *or all transitions.*

Definition 8 (IGN: The non-ignoring principle). stubborn *satisfies the non-ignoring principle (IGN for short) if every cycle in the reduced reachability graph contains a marking where all enabled transitions are explored.*

Definition 9 (UPS: The up-set principle). *For a marking* m *and a CTL property* ϕ *such that* $m \not\models \phi$, U *is an up-set if every path from* m *to a marking that satisfies* ϕ *contains an element of* U. stubborn(m) *satisfies the up-set principle w.r.t.* ϕ *if* $m \models \phi$ *or* $U \subseteq$ stubborn(m), *for some up-set* U.

Definition 10 (BRA: The branching principle). stubborn(m) *satisfies the branching principle (BRA for short) if* stubborn(m) *contains a single enabled transition, or all enabled transitions.*

In the following propositions, let G' be a reduced reachability graph using stubborn sets that meet the principles mentioned in the assumption.

Each principle has a specific purpose for proving property preservation. In most cases, we assume that there is a path π in the full reachability graph (e.g. a witness or counterexample for the property under investigation) and show that the reduced system contains a path π that is equally fit w.r.t. the studied property. With COM, π' may execute transitions in another order than π. With KEY (in connection with COM), π' may contain transitions that are not occurring in π. With UPS, the stubborn set at m will always contain a transition of π. With VIS, visible transitions in π' appear in the same order as in π, if they appear in π'. IGN is used for making sure that all transitions of π are eventually occurring in π', and BRA is used for making sure that visible transitions are not swapped with branches in the state space other than branches that are introduced by concurrency. Again, [11,15,23,25] provide more details concerning these issues.

Proposition 1 (Preservation of deadlocks, [22]). *If the principles COM and KEY are satisfied then G' contains all deadlocks and at least one infinite path of the original reachability graph.*

Proposition 2 (Preservation of terminal SCC, [24]). *If the principles COM, KEY, and IGN are satisfied then G' contains at least one marking of every terminal SCC of the original reachability graph.*

Proposition 3 (Preservation of reachability, [15,19]). *Let ϕ be a CTL formula without temporal operators. If the principles COM and UPS(ϕ) are satisfied then $EF\phi$ is preserved.*

Proposition 4 (Preservation of LTL-X, [18,23]). *Let ϕ be an LTL property not using the X operator. If the principles COM, KEY, VIS(ϕ), and IGN are satisfied than ϕ is preserved.*

Proposition 5 (Preservation of CTL-X, [11]). *Let ϕ be a CTL formula not using the X operator. If the principles COM, KEY, VIS(ϕ), IGN, and BRA are satisfied then ϕ is preserved.*

5   Simple CTL Queries

We are now ready to discuss the advantages of separating simple CTL queries. In most cases, one of the advantages shall be the ability of using a more powerful stubborn set method. In all reported cases, we will be able to drop the very limiting BRA principle that enables reduction only in markings where just one (invisible) enabled transition is sufficient to meet all the other principles. In addition, less restrictive conditions (i.e. a smaller set of principles to be met), leads to potentially smaller stubborn sets and thus to better reduction.

The simple problems discussed below appear as pairs of an existentially and a universally quantified formula. These two formulas can be reduced to each other by negation. Hence, they permit the application of the same verification techniques.

In the sequel, let ϕ and ψ be CTL formulas without temporal operators. Experimental data refers to the tool LoLA 2 [29], applied to the benchmark of the model checking contest (MCC) 2018. We give 300 seconds for every individual query. More details on experiments can be found in Sect. 7.

5.1   AG ϕ, EF ϕ

For the reachability problem $EF\phi$, we may use stubborn set as suggested by Proposition 3 [19], or a relaxed version [15]. Both techniques have specific advantages. The first method works much better if $EF\phi$ is true while the second method has advantages if $EF\phi$ is false. Any of the methods, however, is much more powerful than the CTL-X preserving method.

For reachability, Petri net structure theory can be applied. If the Commoner's theorem [4,13] applies, EF DEADLOCK evaluates to false. The conditions of the theorem can be checked as a satisfiability problem in propositional logic (SAT) [17]. The Petri net state equation, enhanced with the refinement method proposed in [28] provides a powerful tool for verifying other reachability queries. Since the structural methods can be traced back to NP-complete problems (SAT resp. Integer Linear Programming) and therefore use only polynomial space, they can be applied in parallel to state space exploration.

The ability of LoLA to solve far beyond 90% of the queries in the reachability category of the MCC, compared to less than 70% if only a CTL model checker is applied to the CTL category, clearly confirms the conclusion to separate reachability queries from CTL model checking.

5.2   AF ϕ, EG ϕ

The CTL formula $AF\phi$ is equivalent to the LTL formula $F\phi$. The universal path quantifier is implicitly present in LTL, too, since a system satisfies an LTL formula if *all* its paths do. That is, we may apply LTL-X preserving stubborn sets instead of CTL-X preserving ones. Without the BRA principle, LTL-X preserving stubborn sets are more powerful (more than 90% success in the LTL category, compared to less than 70% success if CTL-X preserving stubborn sets are applied to all of the CTL category).

Additionally, we may completely drop the IGN principle for visible transitions. We sketch a proof for $EG\phi$. If there is no witness path (an infinite path where ϕ permanently holds) in the original reachability graph, there cannot be one in the reduced reachability graph which is a subgraph. If there is an (infinite) witness path π, then by COM, KEY, and VIS, there is an infinite path π' in the reduced system such that visible transitions of π' occur in the same order as in π. Invisible transitions in π' do not alter the value of ϕ. That is, π' witnesses $EG\phi$ as well since otherwise there would be a prefix of π where ϕ is violated, contradicting the assumption that π is a witness path.

When only COM, KEY, and VIS need to be established in stubborn set computation, we can often find much smaller stubborn sets and achieve much better state space reduction.

5.3 E (ϕ U ψ), A(ϕ R ψ)

To satisfy E (ϕ U ψ), we need to use stubborn sets that preserve two properties: first, the reachability of ψ, and second, the non-violation of ϕ. Combining the discussion for reachability (EF) and non-violation (EG), we propose the following combination of principles for the stubborn sets to be used: COM, UPS(ψ), and VIS(ϕ). We sketch the arguments for correctness of this setting. Assume the original reachability graph contains a witness path π. By UPS(ψ), this path contains a transition that is in the stubborn set used in the initial marking. By COM, we can shift the first such transition to the front of the path. By VIS(ϕ), this modification does not change the order of transitions visible for ϕ. At least the first transition of the modified path can be replayed in the reduced reachability graph. By induction, a witness path in the reduced system is established.

We obtain a combination of principles where the harmful BRA principle is absent and VIS can disregard ψ. In addition, the UPS principle preserves a shortest witness path. This accelerates the positive effect of on-the-fly model checking in all situations where E(ϕ U ψ) turns out to be true.

We can employ linear programming for checking a necessary and a sufficient condition for E(ϕ U ψ). A necessary criterion is obviously EF ψ, and the approach in [28] can be used for checking this condition. A sufficient condition is the reachability of ψ using only transitions that are invisible to ϕ, in addition to checking ϕ in the initial marking. This can be checked by removing all transitions visible for ϕ from N and applying the approach of [28] to the resulting net. With the moderate memory footprint of linear programming, the necessary and sufficient conditions can be checked in parallel to the actual depth-first search ("portfolio approach").

5.4 EGEF ϕ, AFAG ϕ

For this pair of formulas, we do not have a dedicated version of stubborn sets, so we apply CTL preserving stubborn sets for state space reduction. However, the check for the pair of temporal operators can be folded into a single depth-first search. We present the approach for EGEF ϕ. The witness path π for the EG operator is a maximal path (i.e. infinite or ending in a deadlock).

If the path ends in a deadlock, the deadlock marking has to satisfy ϕ since this is the only way for ϕ to be reachable from that marking. If the deadlock satisfies ϕ, all markings on the path satisfy EF ϕ automatically, so this case can be easily implemented. An infinite path appears in a model checker as a cycle that is reachable from m_0. For satisfying EGEF ϕ, it is necessary and sufficient that, from one of the markings m on the cycle, a marking m' is reachable that satisfies ϕ. Necessity follows immediately from the definition of the semantics of CTL. Sufficiency follows from the fact that m is reachable from all markings in π, so m' is reachable as well from all markings in π.

We record, for every marking visited in depth-first search, whether a marking satisfying ϕ can be reached. To this end, every marking that satisfies ϕ itself is marked as "can reach ϕ". In addition, whenever depth-first search backtracks from a marking that can reach ϕ, the predecessor marking is marked as well as "can reach ϕ". For detecting cycles, we use the well-known fact [14] that every cycle in a state space contains an edge from some marking m to a marking m' such that, at some stage of depth-first search, m is the top element of the search stack and m' is on the search stack as well (such an edge is called *backward edge*). During the search, we maintain information whether or not the search stack contains such m'. If this is the case while the marking on top of the stack can reach ϕ, we return *true*. If we reach a deadlock satisfying ϕ, we return *true* as well. If the search is completed without having returned true, we return *false*.

Lemma 1. *The procedure sketched above correctly evaluates EGEF ϕ.*

Proof. If we reach a deadlock satisfying ϕ, EGEF ϕ is trivially true. If we return *true* in any other situation, we have a marking m on the search stack that is member of some cycle reachable from m_0. From m, the top element m' of the stack is reachable and, from m', a marking satisfying ϕ can be reached. Hence, EGEF ϕ is true. For the other direction, assume that EGEF ϕ is true and consider a witness path π for the EG operator. If this is a finite path, the final marking must be a deadlock satisfying ϕ. Otherwise, π is infinite. The set of markings that are visited infinitely often in π is strongly connected, hence contained in an SCC C of the reachability graph. The root m^* of C (i.e. the marking of C entered first by the search) is member of some cycle (by strong connectivity). As m^* is the first marking of C entered by the search, it is target of a backward edge. This is recognised before m^* is finally left by depth-first search. Depth-first search explores all markings reachable from m^* before finally leaving m^*. That is, in the moment we are about to finally leave m^*, we know that m^* is target of a backward edge and can reach ϕ. Hence, we return true (if we have not returned true much earlier). $\qquad\square$

In addition to the combined depth-first search, we can add a check for EF ϕ as a necessary condition and AG ϕ as a sufficient criterion to a portfolio for EGEF ϕ. Again, the state equation approach can be used in order not to take too much memory away from the main search procedure.

5.5   EFEG ϕ, AGAF ϕ

We present the approach for EFEG ϕ. We check the property by nested depth-first search. The approach uses ideas from [5,6,10,14] that are concerned with the similar problem of finding accepting cycles in Büchi automata. Outer search proceeds through markings that have already proven not to be part of a ϕ-cycle (or a ϕ-deadlock). This includes markings that do not satisfy ϕ and markings where inner search has already been run. Inner search proceeds only through ϕ-markings and tries to find a cycle or a deadlock. By definition, EFEG ϕ holds if and only if a ϕ-cycle or a ϕ-deadlock is reachable from m_0. We start with outer search. Whenever we encounter a fresh ϕ-marking m, we switch to inner search. If inner search terminates without having found a cycle or deadlock, we resume outer search in m.

This procedure is very similar to the general CTL model checking algorithm. However, we may apply dedicated stubborn sets. In outer search, we distinguish markings that satisfy ϕ from markings that do not satisfy ϕ. If m does not satisfy ϕ, we use stubborn sets that satisfy COM and UPS(ϕ). If m satisfies ϕ, we have two correct combinations of principles. We can use stubborn sets that satisfy COM and UPS($\neg\phi$), or stubborn sets that satisfy COM, KEY, and VIS(ϕ). In inner search, we use stubborn sets satisfying COM, KEY, and VIS(ϕ).

Lemma 2. *A reduced reachability graph obeying the principles stated above preserves EFEG ϕ.*

Proof. Let $m_1^* \ldots m_n^*$ be a ϕ-cycle or a ϕ-deadlock (then: $n = 1$). Let $m_1 m_2 \ldots m_k$ be a path such that m_1 has been visited in outer search in the reduced reachability graph, and $m_k = m_i^*$, for some i ($1 \leq i \leq n$). Consider first the case where all m_j ($1 \leq j \leq k$) satisfy ϕ. Then inner search from m_1 will find a ϕ-cycle or ϕ-deadlock since the path

$$\pi = m_1 \ldots \left(m_k = m_i^* m_{i+1}^* \ldots m_n^* m_1^* \ldots m_{i-1}^* \right)^*$$

witnesses EG ϕ and EG ϕ is preserved by stubborn sets with COM, KEY, and VIS (see Subsect. 5.2).

Second, consider the case where m_1 does not satisfy ϕ. Since $m_k = m_i^*$ satisfies ϕ, the path from m_1 to m_k contains a transition of the up-set used in m_1, and, by the UPS principle, elements of the stubborn set used in m_1. Applying COM, we obtain an alternative path where the first transition is in the stubborn set used in m_1. Its successor meets the same properties in m_1 but with a smaller value for k.

It remains to consider the case where m_1 satisfies ϕ and the first case is not applicable. Then, for at least one q ($2 \leq q \leq k$), m_q violates ϕ. If we apply stubborn sets satisfying COM and UPS($\neg\phi$), we argue as in the second case. This yields a continuation for the witness path in the reduced reachability graph. If we obey COM, KEY, and VIS instead, we argue as follows. If a transition of the stubborn set used in m_1 occurs in π, COM yields a continuation of the path in the reduced reachability graph. Otherwise, the stubborn set in m_1 contains only invisible transitions (by VIS). Choose a key transition t^* in the stubborn set for m_1 (available via KEY). By KEY, t^* is never disabled in π. By COM, all transitions in π can still be executed after having fired t^*. The t^*-successor m' of m_1 occurs in the reduced reachability graph. The third case is applicable only a finite number of times since m' satisfies ϕ but there is no ϕ-cycle reachable in inner search from m_1. □

As in previous cases, AG ϕ is a sufficient condition for EFEG ϕ while EFϕ is necessary. The state equation approaches to these properties may be added to the portfolio for EFEG ϕ.

5.6 AGEF ϕ, EFAG ϕ, EFAGEF ϕ, AGEFAG ϕ

These properties are tightly related to terminal SCC of the reachability graph. For AGEF ϕ, every terminal SCC must contain a marking satisfying ϕ. For EFAG ϕ, there must exist a terminal SCC where all markings satisfy ϕ. For EFAGEF ϕ, a terminal SCC must exist where at least one marking satisfies ϕ, and for AGEFAG ϕ, all markings in all terminal SCC must satisfy ϕ.

By Proposition 2, stubborn sets obeying COM, KEY, and IGN preserve access to all terminal SCC of the reachability graph. Adding UPS(ϕ) for AGEF ϕ and EFAGEF ϕ (or UPS($\neg\phi$) for the other two cases) at least inside the terminal SCC preserves the properties under investigation. There are several strategies for implementing UPS in the terminal SCC. We can either require it for all markings (then KEY may be dropped) [20], or enforce a relaxed version of UPS in all markings (see [15] for details), or we may launch a depth first search using stubborn sets with COM and UPS whenever we encounter a terminal SCC in the reduced graph w.r.t. COM, KEY, and IGN.

The proposed procedure has two advantages. First, we proceed in a single depth-first search compared to the recursive approach of a CTL model checker. Second, we can drop the very problematic BRA principle. Being able to drop the VIS principle as well, the stubborn set method can achieve substantial reduction even in cases where ϕ is a property that refers to a large number of places, and causes many transitions to be visible.

For all properties considered in this subsection, AG ϕ is a sufficient condition and EF ϕ is necessary. Using the state equation approach mentioned above, we can add these checks to our portfolio. This way, we have an additional opportunity to answer the query early while using only a moderate amount of additional memory.

5.7 Formulas Starting with EX and AX

This section is concerned with formulas of the shape EXEF ϕ, EXEG ϕ, EXE(ϕ R ψ), EXE(ϕ U ψ), EXEGEF ϕ, EXEFEG ϕ, AXAG ϕ, AXAF ϕ, AXA(ϕ R ψ), AXA(ϕ U ψ), AXAGAF ϕ and AXAFAG ϕ. We explicitly discuss the existentially quantified ones. Verification of these properties can be traced back to the respective formula without the leading EX operator. All we need to do is to explore all enabled transitions of m_0, and not to store m_0. That is, whenever m_0 is visited during the search, it is treated as fresh marking and a stubborn set can be used. Other than this, the same stubborn set approaches as discussed earlier are applicable.

5.8   Single-Path Formulas

In this section, we discuss a larger class of CTL formulas. We aim at applying
LTL model checking instead of CTL model checking. This way, the BRA principle
may be skipped. Switching to an LTL model checker is actually a good idea,
given the better success rate of tools like LoLA in the LTL category of the
MCC. According to [3], removing the path quantifiers of a CTL formula yields
the only candidate to be an equivalent LTL formula. But this candidate may or
may not turn out to be indeed equivalent. The ACTL formulas where equivalence
can be achieved can be characterised [16]. We chose to apply the approach to a
collection of CTL formulas that can be more easily be recognised by a rewriting
system.

LTL is a linear time temporal logic. That is, a counterexample for an LTL
formula is always a single maximal path of the system. In contrast, CTL is a
branching time temporal logic. This means that the counterexample is a subtree
of the computation tree (the unrolling of the reachability graph). For instance,
a witness for EGEF ϕ consists of a maximal path where, for each marking
a finite path to a state satisfying ϕ branches off. Even with the observations
made in Subsect. 5.4, the structure remains more complicated than a single path.
However, in several cases, the branching structure collapses into a single path.
Consider EFEG ϕ. Here, we only need a finite path to the first state of a ϕ-cycle
(or deadlock), extended with the cycle itself. It is precisely a counterexample
for the LTL formula GF $\neg\phi$ that is obtained by negating EFEG ϕ to AGAF
$\neg\phi$ and then dropping the universal path quantifiers. In the sequel, we shall
exhibit a class of CTL formulas where this approach is applicable. We call them
single-path formulas. They may contain only existential path quantifiers or only
universal path quantifiers. In the next definition, let a *state predicate* be a CTL
formula without any temporal operator.

Definition 11 (Existential single-path formula). *If ϕ and ψ are existential
single-path formulas and ω is a state predicate, then the following formulas are
existential single-path formulas:*

– ω *(the base of the inductive definition);*
– *EG ω;*
– *EF ϕ;*
– *E(ω U ϕ);*
– *E(ϕ R ω);*
– *$\phi \lor \psi$;*
– *$\phi \land \omega$;*

Universal single-path formulas are defined accordingly:

Definition 12 (Universal single-path formula). *If ϕ and ψ are universal
single-path formulas and ω is a state predicate, then the following formulas are
universal single-path formulas:*

– ω *(the base of the inductive definition);*
– *AF ω;*
– *AG ϕ;*

- $A(\omega\ R\ \phi);$
- $A(\phi\ U\ \omega);$
- $\phi \wedge \psi;$
- $\phi \vee \omega;$

The class of single-path formulas covers several cases discussed earlier in this paper. However, the results above are stronger then the results we shall obtain now, so the separate treatment is indeed justified. It is easy to see that the negation of an existential single-path formula is indeed a universal single-path formula and vice versa. That is, we may restrict subsequent considerations to universal single-path formulas.

For a universal single-path formula ϕ, let LTL(ϕ) be the formula obtained from ϕ by removing all path quantifiers. We claim:

Lemma 3. *Let ϕ be a universal single-path formula and N a Petri net. Then N satisfies ϕ if and only if N satisfies LTL(ϕ).*

Proof. We show that violation of ϕ implies violation of LTL(ϕ) and violation of LTL(ϕ) implies violation of ϕ. We proceed by induction, according to Definition 12.

Case ω (state predicate): In both CTL and LTL, a state predicate is violated if it does not hold in the initial marking.

Case AF ω: In both CTL and LTL, a counterexample is a maximal path where all markings violate ω. Since ω is a state predicate, it directly refers to the markings on the path.

Case $A(\omega\ R\ \phi)$: A counterexample for A(ω R ϕ) is a finite path to a marking where all but the last marking violate ω and the last marking violates ϕ. As ω is a state predicate, the intermediate markings as such violate ω. Hence, the path, extended by a counterexample path for ϕ at the final marking (which exists by induction hypothesis) yields a path that is a counterexample for LTL(A(ω R ϕ)). For the other direction, consider a counterexample for LTL(A(ω R ϕ)). It must have a suffix serving as a counterexample for LTL(ϕ). Hence the first marking of that path violates ϕ (using once more the induction hypothesis). The markings that are not part of the considered suffix violate ω, so the full path is a counterexample for A(ω R ϕ).

Case $A(\phi\ U\ \omega)$: A counterexample can either be a maximal path where ω is violated in every marking (then apply the argument of Case AF ω) or a path where ω is violated until both ω and ϕ are violated (then apply the argument of case A(ω R ϕ)).

Case AG ϕ: This case can be traced back to Case A(ω R ϕ) using the tautology AG $\phi \iff$ A(false R ϕ).

Case $\phi \wedge \psi$: If ϕ is violated, there is a counterexample for ϕ for which the induction hypothesis may be applied. Otherwise, there is a counterexample for ψ for which again the induction hypothesis applies.

Case $\phi \vee \omega$: In this case, ϕ and ω are violated. Since ω is a state predicate, only the initial marking of the path is concerned. Hence, the induction hypothesis applied to ϕ yields the desired result.   □

Using Lemma 3 the considered fragment of CTL can be verified using an LTL model checker. As another option, we may use a CTL model checker but apply LTL preserving stubborn sets. Existential single-path formulas can be verified by checking their negation.

5.9   Boolean Combinations

If a CTL formula is a Boolean combination of subformulas, we may check the subformulas individually. Doing that, the subformulas often have a smaller set of visible transitions, so some of the stubborn set principles are stronger for a subformula than for the whole formula. Some subformulas may contain the X operator, so the stubborn set method can be applied at least to the subformulas not containing the X operator. Some subformulas may fall into any of the classes considered above, so their verification may be accelerated.

In a setting with distributed memory, the subformulas can be verified in parallel. With shared memory, a parallel execution is not necessarily recommendable since the individual verification procedures compete for memory which may lead to memory exhaustion in all procedures while verification could have been successful if the whole memory were available for either of the procedures.

To get the most out of our accelerated procedures in a shared memory setting, we rate subformulas according to their simplicity. Then, the simplest formulas are checked first. This way, we get an increased probability that the result of the Boolean combination can already be determined (by a true subformula of a disjunction or a false subformula of a conjunction) before the procedures for the most complicated formulas have been launched.

Our rating works as follows. The simplest category consists of subformulas that do not contain temporal operators. They are true, false, or can be evaluated by just inspecting the initial marking. Second category consists of formulas that contain only X operators. They can be verified by exploring the state space to a very limited depth. Then follow categories for the simple cases studied above. The simplicity of these categories is mainly influenced by our experience concerning their performance in the MCC. Then follow the categories LTL-X, CTL-X, LTL, and CTL (in this order). For the last categories, applicability of stubborn sets is the distinguishing feature.

6   Preprocessing

It has already been recognised [1] that formulas should be carefully preprocessed before running a model checking procedure. Atomic propositions may turn out to be always true or always false, proven by the infeasibility of a linear program that can be derived from the proposition and the Petri net state equation. Once some of the propositions have been identified as true or false, whole subformulas may turn out to be true or false as well. This way, a significant number of formulas can be evaluated without running a model checker at all. For other formulas, the remaining model checking problem is simpler than the original one.

In the remainder of this section, we add a few observations to the findings of [1]. We have two objectives. First, we want to increase the number of situations where one of the special routines discussed in the previous section can be applied. Second, we want to increase the power of the stubborn set method.

Boolean Operators. Some tautologies, such as AG ($\phi \wedge \psi$) $\iff$ (AG $\phi \wedge$ AG ψ) can be applied in both directions. Applying it from right to left decreases the number of temporal operators. However, the operator in general applies to a more complicated subformula, with more transitions being visible. Applying the formula from left to right leads to a formula with more temporal operators that, however, work on a smaller subformula. Stubborn sets potentially work better. Moreover, we increase the likelihood that the Boolean operator then becomes the root of the formula tree and the results of Sect. 5.9 are applicable. Hence, our tool LoLA uses an orientation of tautologies that prefers pushing Boolean operators towards the root of the formula tree.

X Operators. We also try to push X operators towards the root to the formula tree. To this end, we apply tautologies such as EFEX ϕ $\iff$ EXEF ϕ from left to right. This way, we increase the likelihood that we finally obtain one of the formulas considered in Sect. 5.7. Moreover, we get larger subformulas that do not contain an X operator. Since CTL preserving stubborn sets work only on X-free formulas, stubborn reduction is not applicable to a formula containing an X operator as a whole. However, when an X-free subformula of a CTL formula is evaluated on some level of recursion in the procedure sketched in Sect. 3, there is no reason not to apply stubborn sets. Hence, the rewriting strategy improves the applicability of stubborn set reduction.

Traps. When investigating atomic propositions, [1] mainly employs the Petri net state equation. In quite some situations where the state equation is not able to prove a proposition to be invariantly true or false, a trap can actually help. A trap is a set Q of places that, once containing a token, always keeps at least one token. This is formally established by requiring that every transition that consumes tokens from any place in Q, also produces a token on some place in Q. Consider an atomic proposition $k_1 p_1 + \cdots + k_n p_n \geq 1$ with all k_i being positive. If $\{p_1, \ldots, p_n\}$ includes a trap that has at least one token in the initial marking, the proposition is invariantly true. Existence of a trap is easily checked. We start with $\{p_1, \ldots, p_n\}$ and remove places where some transition consumes tokens while not producing tokens on any of the places. This way, we obtain the maximal trap included in $\{p_1, \ldots, p_n\}$.

Embedded Place Invariants. A place invariant i assigns a weight $i(p)$ to every place p such that the weighted sum of tokens remains constant for all reachable markings. Place invariants can be found by solving the system of equations $C^T i = 0$, where C is the incidence matrix of N. With a place invariant i, the equation $i(p_1)m(p_1) + \cdots + i(p_n)m(p_n) = im_0$ holds for all reachable markings in a net with set of places $\{p_1, \ldots, p_n\}$. Sometimes, such invariants can be used for simplifying an atomic proposition. Consider as an example the proposition $p_1 + 2p_2 + p_3 \geq 2$ and assume that there is a place invariant that yields the equation $p_1 + p_2 = 1$. Then the atomic proposition can be simplified to $p_2 + p_3 \geq 1$.

It is not constant but does no longer mention p_1. Consequently, the set of visible transitions may become smaller since the environment of p_1 does no longer need to be considered as visible (unless transitions still appear in the environment of p_2 or p_3). With a smaller set of visible transitions, better stubborn set reduction may be expected (in particular regarding the VIS principle). In general, a helpful invariant can be systematically computed as the solution of a linear program. Consider an atomic proposition of the shape $k_1 p_1 + \cdots + k_m p_m + l_1 q_1 \cdots + l_n q_n$ op k, where all p_j and q_j are places, all k_j are positive integers, all l_j are negative integers, $op \in \{=, \neq, <, >, \leq, \geq\}$, and k is an integer. The linear program looks for the largest possible invariant where the coefficients are between 0 and k_i (resp. l_i): Maximise $i(p_1) + \cdots + i(p_m) - i(q_1) - \cdots - i(q_n)$ where $C^T i = 0$, and $0 \leq i(p_j) \leq k_j$ (for $1 \leq j \leq m$), and $l_j \leq i(q_j) \leq 0$ (for $1 \leq j \leq n$). If the linear program is feasible, subtracting the resulting solution from the atomic proposition may or may not lead to less mentioned places but is guaranteed not to add places to the formal sum of the proposition.

7 Experimental Validation

We implemented the methods discussed in the paper in our tool LoLA (implementation does not yet cover EXEFEG, EXEGEF, and the universal counterparts AXAGAF and AXAFAG). For evaluating the methods, we use the benchmark provided by the MCC 2018 [9]. We used the formulas provided in the CTL category. While the nets of the MCC are contributed by the community, the formulas are actually generated automatically, and are to a certain degree random.

In 2018, a total of 767 place/transition Petri nets were used in the MCC. For every net, 32 CTL formulas are provided. This makes 24544 individual verification problems. For 3704 problems (15.1%), the initial rewriting process yielded a formula that does not contain any temporal operator. Here, sufficiently many atomic propositions have been found to be invariantly true or false. Resulting formulas can be evaluated by just inspecting the initial marking, so no actual run of a model checker is necessary. 13366 problems (54.5%), after rewriting, fall into some of the categories mentioned in Sect. 5. That is, we need to run the generic CTL model checker only for 30.4% of the CTL problems in the MCC!

For the 13366 problems where application of a special routine is possible, we compared the proposed routine with a run of the generic CTL model checking procedure. To this end, we used 300 seconds of execution time and unlimited memory for every problem instance. Experiments were executed on our machine Ebro. This machine has been used for executing parts of the actual MCC in recent years. It has 32 physical cores running at 2.7 GHz and 1 TB of RAM. Memory overflow was no issue within the 300 seconds given to each instance.

Table 1 lists the results of our experiments. It shows that specialised routines in total are more successful than the CTL model checking procedure. For the formulas where specialised routines have been found, we increased the success rate from 69.2% to 85.7%. In other words, specialised routines are able to solve more

Table 1. Comparison between CTL model checking procedure (CTL) and special routine (as proposed in this paper).

Formula type	Number	CTL		Special		Improvement	
	(#)	#	%	#	%	#	%
EF ϕ, AG ϕ	2471	1438	58.2	2300	93.1	862	34.9
EG ϕ, AF ϕ	1767	1625	92.0	1670	94.5	45	2.5
E(ϕ R ψ), A(ϕ U ψ)	168	157	93.5	160	95.2	3	1.8
E(ϕ U ψ), A(ϕ R ψ)	318	187	58.8	198	62.3	11	3.5
EFEG ϕ, AGAF ϕ	515	340	66.0	431	83.7	91	17.7
EGEF ϕ, AFAG ϕ	385	276	71.7	277	71.9	1	0.3
EXEF ϕ, AXAG ϕ	353	193	54.7	319	90.4	126	35.7
EXEG ϕ, AXAF ϕ	197	177	89.8	178	90.4	1	0.5
EXE(ϕ R ψ), AXA(ϕ U ψ)	19	17	89.5	18	94.7	1	5.3
EXE(ϕ U ψ), AXA(ϕ R ψ)	33	20	60.6	24	72.7	4	12.1
EFAG ϕ, AGEF ϕ	884	286	32.4	343	38.8	57	6.4
EFAGEF ϕ, AGEFAG ϕ	13	3	23.1	6	46.2	3	23.1
Single-Path	421	275	65.3	295	70.1	20	4.8
Boolean	5822	4250	73.0	5239	90.0	989	17.0
All	13366	9244	69.2	11458	85.7	2214	16.5

than half of the cases where a generic CTL model checker was not successful. This means that the proposed approach proved to be effective.

The table also shows that success is very unevenly distributed over the various formula types. The big success of reachability (EF ϕ) is of course to be expected and can be quoted to the large portfolio that included search with very powerful stubborn sets, the state equation approach, and the use of the siphon/trap property.

On the other edge of the spectrum, the little success for EGEF ϕ is well explained by the fact that we still need to apply the CTL-X preserving stubborn set method, so we have a more efficient exploration of the state space but the state space as such remains the same. In case of EXEG ϕ, the CTL model checker left only 20 problems open. That is, there is not much room for improvement. Problems in the MCC can be separated into the categories "easy enough for everybody", "too hard for everybody", and "battleground". The first category refers to nets with rather small state space. Here, every approach is able to get a result in time. In the second category, we have nets with very large state spaces and dense dependencies between transitions. At least explicit model checkers that depend on the reduction power of the stubborn set method, have no chance to verify such systems. This means that progress in model checking mainly refers to the battleground category. We should aim at covering the problems in this category as much as possible. Returning to the EXEG ϕ category, the little success may very well be due to the fact that only one of the 20 formulas left open by the CTL model checker actually fell into the battleground category. Consequently, we do *not* conclude that the special routine for EXEG ϕ is ineffective as such. Given the fact that we may apply more powerful stubborn sets, we have reason to believe that the procedure would be more effective on a different benchmark, with more EXEG formulas in the battleground category. In consequence, the large bandwidth of success rates in the different formula types does not jeopardise the general conclusion in favour of using specialised routines.

The checks for EF ϕ as a necessary condition and AG ϕ as a sufficient condition, which we run in parallel to the depth-first search, provided solutions to 305 problems (1.2% of the whole CTL category).

8   Conclusion

We proposed to relieve the CTL model checker by providing specialised support for a large set of simple CTL queries. Special treatment permits the use of much more powerful stubborn set dialects. In addition, the Petri net state equation may be employed for solving the problem, or for checking necessary or sufficient conditions in a portfolio approach. In the MCC, specialised routines are applicable to more than half of the problems. In the introduction, we argued that a significant percentage of simple queries has to be expected in practice, too.

With our approach we increased the success rate for simple formulas by 16.5% in the MCC benchmark. This is a remarkable achievement since none of the additionally solved problems falls into the "easy enough for everybody" category. Over half of the simple problems left unsolved by the CTL model checker can now be solved. The performance demonstrated here with the MCC benchmark (that uses randomly generated formulas) can be repeated in other situations with meaningful formulas. Unfortunately, there is not enough space to report details.

Offering the new methods, LoLA unfortunately does not yet reach the performance of TAPAAL [7], the 2018 winner of the MCC CTL category. TAPAAL offers some techniques that have not (yet) been implemented in LoLA. For instance, TAPAAL uses sophisticated net reduction as another form of preprocessing [7].

Future work could include finding more formula types that permit any improvement in verification. In addition, some of the ideas of this paper could be integrated into a CTL model checker itself. For instance, treating AGEF, or even AGEFAG, in a single depth-first search should be possible even if that pair of operators occurs in the middle of a more complex CTL formula. In addition, the proposed stubborn set dialects may not necessarily be the optimal ones for the respective formula type. Finding alternative stubborn set methods for larger classes of formulas, we may ultimately be able to have a dedicated dialect of stubborn sets for every subformula of a CTL query. Finally, the initially proposed idea of modifying the net for the sake of simplifying a CTL query has not been systematically explored yet.

References

1. Bønneland, F., Dyhr, J., Jensen, P.G., Johannsen, M., Srba, J.: Simplification of CTL formulae for efficient model checking of petri nets. In: Khomenko, V., Roux, O.H. (eds.) PETRI NETS 2018. LNCS, vol. 10877, pp. 143–163. Springer, Cham (2018). https://doi.org/10.1007/978-3-319-91268-4_8

2. Clarke, E.M., Emerson, E.A.: Design and synthesis of synchronization skeletons using branching time temporal logic. In: Kozen, D. (ed.) Logic of Programs 1981. LNCS, vol. 131, pp. 52–71. Springer, Heidelberg (1982). https://doi.org/10.1007/BFb0025774

3. Clarke, E.M., Draghicescu, I.A.: Expressibility results for linear-time and branching-time logics. In: de Bakker, J.W., de Roever, W.-P., Rozenberg, G. (eds.) REX 1988. LNCS, vol. 354, pp. 428–437. Springer, Heidelberg (1989). https://doi.org/10.1007/BFb0013029

4. Commoner, F.: Deadlocks in Petri Nets. Applied Data Research Inc., Wakefield, Massachusetts, Report CA-7206-2311 (1972)

5. Courcoubetis, C., Vardi, M.Y., Wolper, P., Yannakakis, M.: Memory-efficient algorithms for the verification of temporal properties. Formal Methods Syst. Des. **1**(2/3), 275–288 (1992)

6. Couvreur, J.-M., Duret-Lutz, A., Poitrenaud, D.: On-the-fly emptiness checks for generalized Büchi automata. In: Godefroid, P. (ed.) SPIN 2005. LNCS, vol. 3639, pp. 169–184. Springer, Heidelberg (2005). https://doi.org/10.1007/11537328_15

7. David, A., Jacobsen, L., Jacobsen, M., Jørgensen, K.Y., Møller, M.H., Srba, J.: TAPAAL 2.0: integrated development environment for timed-arc petri nets. In: Flanagan, C., König, B. (eds.) TACAS 2012. LNCS, vol. 7214, pp. 492–497. Springer, Heidelberg (2012). https://doi.org/10.1007/978-3-642-28756-5_36

8. Dehnert, J., Rittgen, P.: Relaxed soundness of business processes. In: Dittrich, K.R., Geppert, A., Norrie, M.C. (eds.) CAiSE 2001. LNCS, vol. 2068, pp. 157–170. Springer, Heidelberg (2001). https://doi.org/10.1007/3-540-45341-5_11

9. Kordon, F., et al.: Homepage of the Model Checking Contest, June 2018. http://mcc.lip6.fr/

10. Geldenhuys, J., Valmari, A.: More efficient on-the-fly LTL verification with tarjan's algorithm. Theor. Comput. Sci. **345**(1), 60–82 (2005)

11. Gerth, R., Kuiper, R., Peled, D.A., Penczek, W.: A partial order approach to branching time logic model checking. Inf. Comput. **150**(2), 132–152 (1999)

12. Godefroid, P., Wolper, P.: A partial approach to model checking. Inf. Comput. **110**(2), 305–326 (1994)

13. Hack, M.H.T.: Analysis of Production Schemata by Petri Nets. Master's thesis, MIT, Dept. Electrical Engineering, Cambridge (1972)

14. Holzmann, G.J., Peled, D.A., Yannakakis, M.: On nested depth first search. In: Proceedings 2nd SPIN Workshop, pp. 23–32 (1996)

15. Kristensen, L.M., Schmidt, K., Valmari, A.: Question-guided stubborn set methods for state properties. Formal Methods Syst. Des. **29**(3), 215–251 (2006)

16. Maidl, M.: The common fragment of CTL and LTL. In: Proceedings of FOCS, pp. 643–652. IEEE Computer Society (2000)

17. Oanea, O., Wimmel, H., Wolf, K.: New algorithms for deciding the siphon-trap property. In: Lilius, J., Penczek, W. (eds.) PETRI NETS 2010. LNCS, vol. 6128, pp. 267–286. Springer, Heidelberg (2010). https://doi.org/10.1007/978-3-642-13675-7_16

18. Peled, D.: All from one, one for all: on model checking using representatives. In: Courcoubetis, C. (ed.) CAV 1993. LNCS, vol. 697, pp. 409–423. Springer, Heidelberg (1993). https://doi.org/10.1007/3-540-56922-7_34
19. Schmidt, K.: Stubborn sets for standard properties. In: Donatelli, S., Kleijn, J. (eds.) ICATPN 1999. LNCS, vol. 1639, pp. 46–65. Springer, Heidelberg (1999). https://doi.org/10.1007/3-540-48745-X_4
20. Schmidt, K.: Stubborn sets for model checking the EF/AG fragment of CTL. Fundam. Inform. **43**(1–4), 331–341 (2000)
21. Tarjan, R.E.: Depth-first search and linear graph algorithms. SIAM J. Comput. **1**(2), 146–160 (1972)
22. Valmari, A.: Stubborn sets for reduced state space generation. In: Rozenberg, G. (ed.) ICATPN 1989. LNCS, vol. 483, pp. 491–515. Springer, Heidelberg (1991). https://doi.org/10.1007/3-540-53863-1_36
23. Valmari, A.: The state explosion problem. In: Reisig, W., Rozenberg, G. (eds.) ACPN 1996. LNCS, vol. 1491, pp. 429–528. Springer, Heidelberg (1998). https://doi.org/10.1007/3-540-65306-6_21
24. Valmari, A.: Stubborn set methods for process algebras. In: Proceedings of DIMACS Workshop on Partial Order Methods in Verification, vol. 29, pp. 213–231 (1997)
25. Valmari, A., Hansen, H.: Stubborn set intuition explained. In: Koutny, M., Kleijn, J., Penczek, W. (eds.) Transactions on Petri Nets and Other Models of Concurrency XII. LNCS, vol. 10470, pp. 140–165. Springer, Heidelberg (2017). https://doi.org/10.1007/978-3-662-55862-1_7
26. Vardi, M.Y.: An automata-theoretic approach to linear temporal logic. In: Moller, F., Birtwistle, G. (eds.) Logics for Concurrency. LNCS, vol. 1043, pp. 238–266. Springer, Heidelberg (1996). https://doi.org/10.1007/3-540-60915-6_6
27. Vergauwen, B., Lewi, J.: A linear local model checking algorithm for CTL. In: Best, E. (ed.) CONCUR 1993. LNCS, vol. 715, pp. 447–461. Springer, Heidelberg (1993). https://doi.org/10.1007/3-540-57208-2_31
28. Wimmel, H., Wolf, K.: Applying CEGAR to the Petri net state equation. Logical Methods Comput. Sci. **8**(3) (2012)
29. Wolf, K.: Petri net model checking with LoLA 2. In: Khomenko, V., Roux, O.H. (eds.) PETRI NETS 2018. LNCS, vol. 10877, pp. 351–362. Springer, Cham (2018). https://doi.org/10.1007/978-3-319-91268-4_18

Saturation Enhanced with Conditional Locality: Application to Petri Nets

Vince Molnár[1,2(✉)] and István Majzik[1]

[1] Fault Tolerant Systems Research Group,
Department of Measurement and Information Systems,
Budapest University of Technology and Economics, Budapest, Hungary
[2] MTA-BME Lendület Cyber-Physical Systems Research Group,
Budapest, Hungary
molnarv@mit.bme.hu

Abstract. The saturation algorithm for symbolic state space generation has proved to be an efficient way to tackle the state space explosion problem in the verification of concurrent, asynchronous systems. Since its original publication in 2001, several variants and extensions have been introduced. The reason for altering the algorithm in these variants is often specific to how it handles transitions. Saturation heavily relies on the notion of locality: transitions tend to affect only some of the state variables. The saturation effect, however, can be achieved and even enhanced with a weaker notion of locality, which we call conditional locality. In this paper, we define a generalized version of the saturation algorithm (GSA) for multi-valued decision diagrams that works with conditional locality and show that it enables the direct usage of transition relations that previously required a specialized algorithm such as variants of constrained saturation. Focusing on Petri nets, we also empirically demonstrate on models of the Model Checking Contest that the GSA often outperforms the original saturation algorithm whenever conditional locality can be exploited and has virtually no overhead for other models.

Keywords: Generalized saturation · Symbolic model checking ·
Formal verification · Conditional locality

1 Introduction

Model checking is a formal verification technique that looks for specified behavioral patterns in a discrete-state system by exploring its state space. Even though we can sometimes avoid the full exploration of the state space, the huge number of reachable states in non-trivial systems tend to limit the applicability of model checking. Concurrent, asynchronous systems are especially problematic for approaches based on a total ordering of events, because the interleaving of the behavior of independent components can easily cause a combinatoric explosion.

S. Donatelli and S. Haar (Eds.): PETRI NETS 2019, LNCS 11522, pp. 342–361, 2019.
https://doi.org/10.1007/978-3-030-21571-2_19

This problem has been tackled in many ways, one of which is symbolic model checking with decision diagrams. Decision diagrams can efficiently encode large state spaces by exploiting the regularities between states. Furthermore, the symbolic encoding of states and transitions enables the efficient computation of next states by working with sets of states and relations. Even though this technique was a great step forward [2], simpler exploration strategies like breadth-first search suffered from the large size of intermediate decision diagrams.

Decision diagrams have an interesting property: their size is not proportional to the number of encoded states. In fact, after some point, adding more states will *reduce* the size of the diagram because more and more regularities will be introduced. This is what the *saturation algorithm*, first introduced in [3], exploits.

To saturate means *to fill completely*. The main idea of the algorithm is to saturate smaller parts of a decision diagram before moving on to larger parts. Specifically, saturation processes decision diagrams in a bottom-up fashion, exploring the state space starting with transitions that do not require any component whose state variable is higher in a variable ordering than the processed level – transitions that are *local* on the currently processed variables. This is often possible because in concurrent systems, transitions usually affect only a small number of the components. With this strategy, saturation can keep all subdiagrams empirically small.

The saturation algorithm initially required the Kronecker condition of the transition relation, but this restriction was removed when [10] introduced *matrix-diagrams* and [5] did the same with decision diagrams with 2 levels per variable.

Efficient saturation-based model checking for computational-tree logic (CTL) has been introduced in [12]. The main novelty was the introduction of *constrained saturation*, which provides an efficient way to handle constraints on the state space. Constrained saturation takes a set of states – the constraint – and performs state space exploration without leaving the constraint. With the modified algorithm, it is possible to avoid intersecting the transition relations with the constraint, which preserves the locality of transitions and the beneficial properties of saturation.

Building on constrained saturation, [11] introduced further extensions to support the verification of linear temporal logic (LTL) and [8] proposed a new approach for the model checking of prioritized Petri nets. Both of them proposed ways to preserve locality for a transition relation that is composed of simple transitions and additional constraints (such as synchronization between the system and the property automaton or enabledness based on priorities).

In this paper, we propose a new algorithm for saturation that generalizes the attempts of preserving locality in the approaches above. We introduce *conditional locality* to relax the original notion of locality and automatically handle transition relations that previously required a form of constrained saturation to process efficiently (such as [8,11]). In addition to generalizing a family of algorithms, using conditional locality can increase the saturation effect, which is intuitively associated with better performance. We investigate this effect in the context of Petri nets, where we empirically show that the *generalized saturation algorithm* (GSA) can be orders of magnitude faster than the original saturation algorithm (presented in detail in [4]) and is virtually never slower.

The main motivation of conditional locality is to compute fixed points even more locally. Saturation ignores variables that are independent of the processed events to avoid computing the fixed point for each of their valuations. With conditional locality, we can ignore even those variables that are not written but read by an event (because they will not change), but compute the fixed point as many times as the value of those variables would cause a different result. The intuition is that the resulting nodes will be part of the final decision diagram more often than those created by the original saturation algorithm, leading to less intermediate nodes and therefore improved performance.

The most important related work is [10], where the authors propose a method to *split* a transition relation such that the resulting relations are as local as possible. The key idea is to extract relations which do not depend on the variables higher in the variable ordering and therefore the method works well when the transition relation is a "sum" of such a relation and another one (i.e. $\mathcal{R} = \mathcal{R}_1 \cup \mathcal{R}_2$). Our approach also handles the cases when the relation is the result of "removing" certain cases from a transition that normally does not depend on a variable (i.e. $\mathcal{R} = \mathcal{R}_1 \setminus \mathcal{R}_2$). Another work that is similar in spirit is [9], where the dependencies of high-level transitions on state variables are more fine-grained than *dependent* and *independent*, which enables a more compact encoding and more efficient update of the transition relation. Our approach also refines this dependency relation to relax the notion of locality.

The key novelties introduced in this paper are the following: *(1)* the introduction of *conditional locality* to relax the original notion of locality; *(2)* the *generalization* of a family of saturation-based algorithms using conditional locality; and *(3)* an *empirical demonstration* of the efficiency of the proposed approach on Petri nets. The paper is structured as follows. Section 2 presents the formalisms and notations used in the rest of the paper. Section 3 introduces conditional locality and the generalized saturation algorithm. The empirical evaluation on Petri nets is in Sect. 4. Finally, Sect. 5 concludes the paper.

2   Background

In this section we summarize the theoretical background of our work and introduce the necessary notations. First, we briefly present Petri nets in Sect. 2.1, then we introduce partitioned transition systems in Sect. 2.2. Building on the latter, we define locality in Sect. 2.3, then formalize multi-valued decision diagrams for encoding states (Sect. 2.4) and abstract next-state diagrams for encoding transition relations (Sect. 2.5). Finally, we present the saturation and constrained saturation algorithms in Sect. 2.6.

2.1   Petri Nets

Petri nets are a widely used formalism to model concurrent, asynchronous systems. The formal definition of a Petri net (including inhibitor arcs) is as follows (see Fig. 1 for an illustration of the notations).

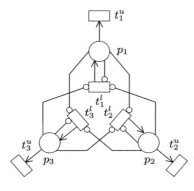

$P = \{p_1, p_2, p_3\} = V;\ K = 3;\ \text{ordering: } x_k = p_k;$
$M_0 = (p_1 \leftarrow 0, p_2 \leftarrow 0, p_3 \leftarrow 0);\ D(p_i) = \{0, 1\};$
$S^0 = \{(0, 0, 0)\};\ T = \{t_1^l, t_1^u, t_2^l, t_2^u, t_3^l, t_3^u\} = \mathcal{E};$
$W^-(t_i^u, p_i) = W^+(t_i^u, p_i) = W^\circ(t_i^l, p_j) = 1$ (other weights are 0); $\mathcal{N}_{t_1^l} = \{\big((0, 0, 0), (1, 0, 0)\big)\};$

Examples: t_1^l is locally read-only and t_1^u is locally invariant on p_2 and p_3; $Supp(t_1^u) = \{p_1\},\ Top(t_1^u) = p_1,\ Supp(t_1^l) = \{p_1, p_2, p_3\},\ Top(t_1^l) = p_3.$

Examples: t_1^l is conditionally local over $\{p_1\}$ with respect to $(0, 0)_{(\{p_2, p_2\})};\ Supp_c(t_1^l) = \{p_1\},\ Guard(t_1^l) = \{p_2, p_3\},\ Top_c(t_1^l) = p_1.$

Fig. 1. Petri net model of 3 concurrent processes locking (t_i^l) and unlocking (t_i^u) a mutually exclusive resource. Examples for the interpretations of the various notations introduced in Sects. 2.1–2.3 and 3.1 are given on the right.

Definition 1 (Petri net). *A Petri net is a tuple $PN = (P, T, W, M_0)$ where:*

- *P is the set of* places *(defining state variables);*
- *T is the set of* transitions *(defining behavior) such that $P \cap T = \emptyset$;*
- *$W = W^- \sqcup W^+ \sqcup W^\circ$ is a multiset of three types of* arcs *(the weight function), where $W^-, W^\circ : P \times T \to \mathbb{N}$ and $W^+ : T \times P \to \mathbb{N}$ are the set of input arcs, inhibitor arcs and output arcs, respectively;*
- *$M_0 : P \to \mathbb{N}$ is the* initial marking, *i.e. the number of* tokens *on each place.*

The three types of weight functions describe the structure of the Petri net: there is an input or output arc between a place p and a transition t iff $W^-(p, t) > 0$ or $W^+(t, p) > 0$, respectively, and there is an inhibitor arc iff $W^\circ(p, t) < \infty$.

The state of a Petri net is defined by the current marking $M : P \to \mathbb{N}$. The dynamic behavior of a Petri net is described as follows. A transition t is *enabled* iff $\forall p \in P : M(p) \in \big[W^-(p, t), W^\circ(p, t)\big)$. Any enabled transition t may fire non-deterministically, creating the new marking M' of the Petri as follows: $\forall p \in P : M'(p) = M(p) - W^-(p, t) + W^+(t, p)$. We denote the firing of transition t in marking M resulting in M' with $M \xrightarrow{t} M'$. A marking M_i is *reachable* from the initial marking if there exists a sequence of markings such that $M_0 \xrightarrow{t_1} M_1 \xrightarrow{t_2} \cdots \xrightarrow{t_i} M_i$. The set of reachable markings (i.e. the *state space* of the Petri net) is denoted by $\mathcal{S}_r$. This work assumes $\mathcal{S}_r$ to be finite.

2.2 Partitioned Transition Systems

A generic model for saturation is usually called a *partitioned transition system* (PTS), where high-level events (causing transitions) and their dependencies on state variables are preserved to partition the low-level next-state relations and localize the effect of transitions [4]. In decision diagram-based model checking, such models usually come with a user-specified variable ordering.

Definition 2 (Variable ordering). *A variable ordering over variables V ($|V| = K$) is a total ordering of elements of V that defines a sequence. The variable in position k of the sequence is denoted by x_k. We will say that x_1 is the* lowest *and x_K is the* highest *in the ordering. We will use the notations $V_{\leq k} = \{x_1, \ldots, x_k\}$ and $V_{>k} = \{x_{k+1}, \ldots, x_K\}$ for sets of variables constituting a prefix or suffix (respectively) of the sequence.*

With a specified variable ordering, the formal definition of a PTS is as follows (again see Fig. 1 for an illustration on the example model).

Definition 3 (Partitioned Transition System). *A partitioned transition system is a tuple $M = (V, D, S^0, \mathcal{E}, \mathcal{N})$ where:*

- *$V = \{x_1, \ldots, x_K\}$ is the finite set of variables with an arbitrary but well-defined variable ordering;*
- *D is the domain function such that $D(x_k) \subseteq \mathbb{N}$ for all $x_k \in V$;*
- *$S^0 \subseteq \hat{S}$ is the set of initial states, where $\hat{S} = \prod_{x \in V} D(x)$ is the potential state space (the shape of which is unaffected by the chosen variable ordering);*
- *$\mathcal{E}$ is the set of high-level events, specifying groups of individual transitions;*
- *$\mathcal{N} \subseteq \hat{S} \times \hat{S}$ is the transition relation partitioned by $\mathcal{E}$ such that $\mathcal{N} = \bigcup_{\alpha \in \mathcal{E}} \mathcal{N}_\alpha$. We often use $\mathcal{N}$ as a function returning the "next states": $\mathcal{N}(\mathbf{s}) = \{\mathbf{s}' | (\mathbf{s}, \mathbf{s}') \in \mathcal{N}\}$ and $\mathcal{N}(S) = \bigcup_{\mathbf{s} \in S} \mathcal{N}(\mathbf{s})$.*

A (concrete) state of the system is a vector $\mathbf{s} \in \hat{S}$, where each variable x_k has a value from the corresponding domain: $\mathbf{s}[k] \in D(x_k)$. A partial state over variables X is a vector assigning a specific value to variables in X and $\top$ (undefined) to those in $V \setminus X$. Sets of partial states are denoted by $S_{(X)}$ and when significant, a single partial state is denoted by $\mathbf{s}_{(X)}$. A partial state $\mathbf{s}_{(X)}$ matches a concrete state $\mathbf{s}$ if $\mathbf{s}[k] = \mathbf{s}_{(X)}[k]$ for every $x_k \in X$, denoted by $\mathbf{s} \in \mathcal{M}(\mathbf{s}_{(X)})$.

2.3   Locality

Exploiting the information preserved in a PTS, we can define different relationships between an event and a variable (illustrated in Fig. 1).

Definition 4 (Locally read-only). *An event α is* locally read-only *on variable x_k if for any $(\mathbf{s}, \mathbf{s}') \in \mathcal{N}_\alpha$ we have that $\mathbf{s}[k] = \mathbf{s}'[k]$. Informally, the value of x is never modified by the transitions of event α.*

While the locally read-only property guarantees that the value of the variable will not change, the event can still depend on the information stored in the variable. The following property forbids this as well.

Definition 5 (Locally invariant). *An event α is* locally invariant *on variable x_k if it is locally read-only and for any $(\mathbf{s}, \mathbf{s}') \in \mathcal{N}_\alpha$ and $v \in D(x_k)$ we also have $(\mathbf{s}_{[x_k \leftarrow v]}, \mathbf{s}'_{[x_k \leftarrow v]}) \in \mathcal{N}_\alpha$, where $\mathbf{s}_{[x_k \leftarrow v]}$ is a state where the value of variable x_k is v, but all other variables have the same value as in $\mathbf{s}$. Informally, the value of x does not affect the outcome of event α.*

With the help of local invariance, we can now define locality, the central notion of the saturation algorithm.

Definition 6 (Locality). *An event $\alpha \in \mathcal{E}$ is said to be* local *over variables $X \subseteq V$ if it is locally invariant on variables in $V \setminus X$. If X is minimal (i.e. the event is dependent on variables in X) then we say that X is the set of* supporting *variables of α: $Supp(\alpha) = X$. The variable with the highest index among the supporting variables (according to a variable order) is the* top *variable $(Top(\alpha))$ of α. We use $\mathcal{E}_k = \{\alpha \mid Top(\alpha) = x_k\}$ and $\mathcal{N}_k = \bigcup_{\alpha \in \mathcal{E}_k} \mathcal{N}_\alpha$ to denote events and their next-state relations whose top variable is the x_k.*

The next-state relation of an event α local on variables $Supp(\alpha) = X$ can be defined over partial states $S_{(X)}$, because no other information is required to compute its image. This enables a compact representation and clever iteration strategies like saturation.

2.4 State Space Encoded in Multi-valued Decision Diagrams

Saturation works with different types of decision diagrams. This paper addresses the version that uses multi-valued decision diagrams to encode the state space.[1]

Definition 7 (Multi-valued decision diagram). *An ordered quasi-reduced multi-valued decision diagram (MDD) over a set of variables V $(|V| = K)$, a variable ordering and domains D is a tuple $(\mathcal{V}, lvl, children)$ where:*

- $\mathcal{V} = \bigcup_{k=0}^{K} \mathcal{V}_k$ *is the set of* nodes, *where items of $\mathcal{V}_0$ are the terminal nodes $\mathbf{1}$ and $\mathbf{0}$, the rest $(\mathcal{V}_{>0} = \mathcal{V} \setminus \mathcal{V}_0)$ are internal nodes $(\mathcal{V}_i \cap \mathcal{V}_j = \emptyset$ if $i \neq j)$;*
- $lvl : \mathcal{V} \to \{0, 1, \dots, K\}$ *assigns non-negative level numbers to each node, associating them with variables according to the variable ordering (nodes in $\mathcal{V}_k = \{n \in \mathcal{V} \mid lvl(n) = k\}$ belong to variable x_k for $1 \leq k \leq K$ and are terminal nodes for $k = 0)$;*
- $children : \mathcal{V}_{>0} \times \mathbb{N} \to \mathcal{V}$ *defines edges between nodes labeled with elements of $\mathbb{N}$ (denoted by $n[i] = children(n, i)$, $n[i]$ is left-associative), such that for each node $n \in \mathcal{V}_k$ $(k > 0)$ and value $i \in D(x_k) : lvl(n[i]) = lvl(n) - 1$ or $n[i] = \mathbf{0}$; as well as $n[i] = \mathbf{0}$ if $i \notin D(x_k)$;*
- *for every pair of nodes $n, m \in \mathcal{V}_{>0}$, if for all $i \in \mathbb{N} : n[i] = m[i]$, then $n = m$.*

An MDD node $n \in \mathcal{V}_k$ encodes a set of partial states $S(n) = S_{(V_{\leq k})}$ over variables $V_{\leq k}$ such that for each $\mathbf{s} \in S_{(V_{\leq k})}$ the value of $n[\mathbf{s}[k]] \cdots [\mathbf{s}[k]]$ (recursively indexing n with components of $\mathbf{s}$) is $\mathbf{1}$ and for all $\mathbf{s} \notin S_{(V_{\leq k})}$ it is $\mathbf{0}$.

There are efficient recursive algorithms that compute the result of set operations directly on MDDs (e.g. union is described in [4]).

An interesting property of MDDs is that the number of nodes does not grow proportionally with the size of the encoded set. In fact, the size of an MDD can decrease when adding new states because of the exploited regularities. This phenomenon can be observed on Fig. 4, where each MDD from left to right encodes one more state, but has either 3 internal nodes or 5. Also note that the right-most MDD encodes the state space of the Petri net from Fig. 1.

---

[1] See [6] for saturation with hierarchical set decision diagrams.

2.5   Next-State Representations

We have introduced a generalization of next-state representations compatible with saturation in [8] – we will build on this notion heavily in the generalization of saturation variants.

Definition 8 (Abstract next-state diagram). *An* abstract next-state diagram *over a set of variables* V *(*$|V| = K$*) and corresponding domains* D *is a tuple* $(\mathcal{D}, lvl, next)$

- $\mathcal{D} = \sqcup_{i=0}^{K} \mathcal{D}_i$ *is the set of* next-state descriptors *(NS descriptor or descriptor for short), where items of* $\mathcal{D}_0$ *are the terminal identity* **1** *and the terminal empty* **0** *descriptors, the rest (*$\mathcal{D}_{>0} = \mathcal{D} \setminus \mathcal{D}_0$*) are* non-terminal *descriptors;*
- $lvl : \mathcal{D} \to \{0, 1, \dots, K\}$ *assigns non-negative* level numbers *to each descriptor, associating them with variables (descriptors in* $\mathcal{D}_k = \{d \in \mathcal{D} \mid lvl(d) = k\}$ *belong to variable* x_k *for* $1 \le k \le K$ *and are terminal nodes for* $k = 0$*);*
- $next : \mathcal{D} \times \mathbb{N} \times \mathbb{N} \to \mathcal{D}$ *is the indexing function that given a descriptor* d *and a pair of "before" and "after" variable values returns another descriptor* d' *such that* $lvl(d') = lvl(d) - 1$ *or* $d' = \mathbf{0}$*. Also denoted by* $d[v, v'] = d' \Leftrightarrow (d, v, v', d') \in next$ *(with* $d, d' \in \mathcal{D}$*,* $v, v' \in \mathbb{N}$*,* $d[v, v']$ *is left-associative) and* $d[\mathbf{s}, \mathbf{s}'] = d[v_K, v'_K] \cdots [v_1, v'_1]$*. We require for any* $v, v', v'' \in \mathbb{N}$ *and* $v \ne v'$ *that* $\mathbf{1}[v, v] = \mathbf{1}$*,* $\mathbf{1}[v, v'] = \mathbf{0}$*, and* $\mathbf{0}[v, v''] = \mathbf{0}$*.*

The abstract NS descriptor $d \in \mathcal{D}_k$ *encodes the relation* $\mathcal{N}(d) \subseteq \mathbb{N}^K \times \mathbb{N}^K$ *iff for all* $\mathbf{s}, \mathbf{s}' \in \mathbb{N}^K$ *the following holds:*

$$\big((\mathbf{s}, \mathbf{s}') \in \mathcal{N}(d) \Leftrightarrow d[\mathbf{s}, \mathbf{s}'] = \mathbf{1}\big) \wedge \big((\mathbf{s}, \mathbf{s}') \notin \mathcal{N}(d) \Leftrightarrow d[\mathbf{s}, \mathbf{s}'] = \mathbf{0}\big)$$

Decision diagram-based representations such as MDDs with $2K$ levels or matrix decision diagrams naturally implement abstract next-state diagrams – descriptors are nodes of the diagram, the identity descriptor is the terminal one node (**1**), the empty descriptor is the terminal zero node (**0**) and the indexing is the same (in case of MDDs with $2K$ levels $d[x, x']$ is implemented by $d[x][x']$). The main difference between these representations and abstract next-state diagrams is that the latter are *abstract* – they can have any representation as long as it can be mapped to the definition and they can be compared for equality.

In case of Petri nets, the simplest representation is the weight function of the net. Given a Petri net with $K = |P|$ places each constituting a separate state variable (p_k denoting the kth variable in the ordering encoding the number of tokens on place $p \in P$), a mapping to an abstract next-state diagram for every transition $t \in T$ is as follows.

- The set of descriptors is $\mathcal{D}_i = \mathbb{N} \times \mathbb{N} \times \mathbb{N} \times \mathcal{D}_{i-1}$ for $1 \le i \le K$, i.e. tuples of the input weight, inhibitor weight and output weight for p_i and the descriptor of for the next place if the transition is enabled with respect to p_i ($\mathcal{D}_0 = \{\mathbf{1}, \mathbf{0}\}$).
- For the *next* function, if $d = (v^-, v^\circ, v^+, d')$, the result of indexing $d[i, j]$ is d' if $v^- \le i < v^\circ$ and $j = i - v^- + v^+$ and $\mathbf{0}$ otherwise.

Input: MDD node n, c
Output: saturated MDD node n'

1 **if** $n = 0$ or $n = 1$ or $c = 0$ **then**
　return n
2 **if** $\neg\text{SATCACHEGET}(n, c, n')$ **then**
3 　　$n' \leftarrow new\text{MDDNODE}(lvl(n))$
4 　　**for each** i **where** $n[i] \neq 0$ **do**
5 　　　$\text{CONFIRM}(lvl(n), i)$
6 　　　$n'[i] \leftarrow \text{CONSSATURATE}(n[i], c[i])$
7 　　$n' \leftarrow \text{CHECKIN}(n')$
8 　　$D \leftarrow$ descriptors for $\mathcal{E}_{lvl(n)}$
9 　　**repeat**
10 　　　$changed \leftarrow$ **false**
11 　　　**for each** $d \in D$ **do**
12 　　　　$n'' \leftarrow \text{CONSFIRE}(n', c, d)$
13 　　　　**if** $n' \neq n''$ **then**
14 　　　　　$n' \leftarrow n''$, $changed \leftarrow$ **true**
15 　　**until** $\neg changed$
16 　　$\text{SATCACHEPUT}(n, d, n')$
17 **return** n'

(a) Procedure CONSSATURATE.

Input: MDD node n, c, NS descriptor d
Output: states in c in or reachable from
　　n through d + node saturated

1 **if** $n = 0$ or $d = 0$ **then return** 0
2 **if** $d = 1$ **then return** n
3 $n' \leftarrow new\text{MDDNODE}(lvl(n))$
4 **for each** i, j **where** $n[i] \neq 0$ and
　$c[j] \neq 0$ and $d[i, j] \neq 0$ **do**
5 　$s \leftarrow \text{CONSRECFIRE}(n[i], c[j], d[i, j])$
6 　**if** $s \neq 0$ **then** $\text{CONFIRM}(lvl(n), j)$
7 　$n'[j] \leftarrow n'[j] \cup s$
8 $n' \leftarrow \text{CHECKIN}(n')$
9 **return** $n' \cup n$

(b) Procedure CONSFIRE.

Input: MDD node n, c, NS descriptor d
Output: states in c reachable from n
　　through d + node saturated

1 **if** $n = 0$ or $d = 0$ **then return** 0
2 **if** $d = 1$ **then return** $n \cap c$
3 **if** $\neg\text{RECFIRECACHEGET}(n, c, d, n'')$
　then
4 　$n' \leftarrow new\text{MDDNODE}(lvl(n))$
5 　**for each** i, j **where** $n[i] \neq 0$ and
　　$c[j] \neq 0$ and $d[i, j] \neq 0$ **do**
6 　　$s \leftarrow \text{CONSRECFIRE}(n[i], c[j], d[i, j])$
7 　　**if** $s \neq 0$ **then** $\text{CONFIRM}(lvl(n), j)$
8 　　$n'[j] \leftarrow n'[j] \cup s$
9 　$n' \leftarrow \text{CHECKIN}(n')$
10 　$n'' \leftarrow \text{CONSSATURATE}(n', d)$
11 　$\text{RECFIRECACHEPUT}(n, c, d, n'')$
12 **return** n''

(c) Procedure CONSRECFIRE.

(d) NS descriptors of t_i^l.

Fig. 2. Pseudocode of constrained saturation and NS descriptors.

- For each transition t the corresponding descriptor is defined recursively: $d_0 = 1$ and $d_i = \big(W^-(p_i, t), W^\circ(p_i, t), W^+(p_i, t), d_{i-1}\big)$.

Figure 2d illustrates the NS descriptors of transitions t_i^l of the example Petri net. A descriptor $d = (v^-, v^\circ, v^+, d')$ is denoted by a node with (v^-, v°, v^+) and d' is denoted by an arrow from d. Descriptors can be shared between transitions.

2.6 Saturation

Saturation is a symbolic state space generation algorithm working on decision diagrams [4]. Formally, given a PTS M, its goal is to compute the set of states

S that are reachable from the initial states S^0 through transitions in $\mathcal{N}$: $S = S^0 \cup \mathcal{N}(S^0) \cup \mathcal{N}(\mathcal{N}(S^0)) \cdots = \mathcal{N}^*(S^0)$, where $\mathcal{N}^*$ is the reflexive and transitive closure of $\mathcal{N}$. This is equivalent to computing the least fixed point $S = S \cup \mathcal{N}(S)$ that contains S^0. The main strength of saturation is a recursive computation of this fixed point, which is based on the following definitions.

Definition 9 (Saturated state space). *Given a partitioned transition system* M, *a set of (partial) states* $S_{(X)}$ *over variables* X *is saturated iff* $S_{(X)} = S_{(X)} \cup \mathcal{N}_X(S_{(X)})$, *where* $\mathcal{N}_X = \bigcup_{\alpha \mid Supp(\alpha) \subseteq X} \mathcal{N}_\alpha$.

A saturated state space is a fixed point of $\mathcal{N}_X$. In model checking, the goal of state-space exploration is to find a least fixed point $S = S \cup \mathcal{N}(S)$ that contains the initial states S^0. Saturation computes this fixed point recursively based on the structure of a decision diagram.

Definition 10 (Saturated node). *Given a partitioned transition system* M, *an MDD node* n *on level* $lvl(n) = k$ *is saturated iff it encodes a set of (partial) states* $S(n)$ *that is saturated. Equivalently, the node is saturated iff all of its children* $n[i]$ *are saturated and* $S(n) = S(n) \cup \mathcal{N}_k(S(n))$, *where* $\mathcal{N}_k = \bigcup_{\alpha \mid Top(\alpha) = x_k} \mathcal{N}_\alpha$ *for* $1 \leq k \leq K$ *and* $\mathcal{N}_0 = \emptyset$.

As suggested by the definition, locality is mainly used to compute a *Top* value for each event, which is the lowest level on which fixed point computation involving the event can happen. By definition, the terminal nodes **1** and **0** are saturated because they do not have child nodes and $\mathcal{N}_0$ is empty. The saturation algorithm is then easily defined as a recursive algorithm that given a node n computes the *least* fixed point $S(n_s) = S(n_s) \cup \mathcal{N}_k(S(n_s))$ that contains $S(n)$, making sure that child nodes are always saturated by recursion. When applied on a node encoding the set of initial states, the result will be a node encoding the states reachable through transitions in $\mathcal{N}$.

The motivation of this decision diagram-driven strategy comes from the observation that larger sets may often be encoded in smaller MDDs. By exploring as many variations in the lower variables as possible, intermediate diagrams may be much smaller than in traditional BFS and chaining BFS strategies (also described in [4]). Another intuition is that in an MDD encoding the set of reachable states, all nodes are by definition saturated – therefore it is impractical to create nodes which have unsaturated child nodes. In other words, a saturated node has a chance of being in the final MDD, while an unsaturated one has not.

The Constrained Saturation Algorithm. The *constrained saturation algorithm* has been introduced in [12] to limit the exploration inside the boundaries of a predefined set of states (the constraint). Even though this is possible with the original algorithm by removing transitions in $\mathcal{N}$ that end in states not inside the constraint, it would damage the locality of events by making them dependent on additional variables (the event has to decide whether it is leaving the constraint or not). Constrained saturation avoids this by traversing an MDD representation of the constraint along with the MDD of the state space, and deciding the enabledness of events when firing them.

Formally, given a constraint set C, the goal of constrained saturation is to compute the least fixed point $S = S \cup (\mathcal{N}(S) \cap C)$ that contains the initial states inside the constraint $S^0 \cap C$. Definitions 9 and 10 are modified as follows.

Definition 11 (Saturated state space with constraint). *Given a partitioned transition system M and a constraint C, a set of (partial) states $S_{(X)}$ over variables X is saturated iff $S_{(X)} = S_{(X)} \cup (\mathcal{N}_X(S_{(X)}) \cap C)$, where $\mathcal{N}_X = \bigcup_{\alpha | Supp(\alpha) \subseteq X} \mathcal{N}_\alpha$.*

Definition 12 (Saturated node with constraint). *Given a partitioned transition system M and a constraint node n_c $(S(n_c) = C)$, an MDD node n on level $lvl(n) = k$ is saturated iff it encodes a set of (partial) states $S(n)$ that is saturated with respect to constraint C. Equivalently, the node is saturated iff all of its children $n[i]$ are saturated with respect to constraint node $n_c[i]$, and $S(n) = S(n) \cup (\mathcal{N}_k(S(n)) \cap C)$, where $\mathcal{N}_k = \bigcup_{\alpha | Top(\alpha) = x_k} \mathcal{N}_\alpha$ for $1 \leq k \leq K$ and $\mathcal{N}_0 = \emptyset$.*

The recursive computation of $\mathcal{N}_k(S(n)) \cap C$ is done by simultaneously traversing n with the source states, the descriptor d of $\mathcal{N}_k$ with source and target states, and n_c with target states. Note that n_c does not encode the partial state determined by the path through which recursion reached the current node, but "remembers" just enough to decide if the transition is allowed based only on the rest of the state.

Figures 2a–c present the pseudocode of the constrained saturation algorithm. To retrieve the pseudocode of the original saturation algorithm, one should assume that at any point $c \neq \mathbf{0}$ and $c[i] \neq \mathbf{0}$ for any i. The pseudocode also contains a stub for the CONFIRM procedure that serves for the on-thy-fly update of the transition relations whenever new states are found (as described in [4] and enhanced in [9]).

The CONSSATURATE procedure starts by checking the terminal cases. Line 2 checks if the same problem has already been solved. Caching – as in all operations on decision diagrams – is crucial to have optimal performance. If there is no matching entry in the cache, the algorithm recursively saturates the children of the input node n, calling CONFIRM for every encountered local state. The resulting node is checked for uniqueness in line 7 and is replaced by an already existing node if necessary (to preserve MDD canonicity). In line 8, we get the NS descriptors for each event belonging to the current level, then iteratively apply them again and again in lines 9–14 until no more states are discovered – a fixed point is reached. This version of the iteration is called *chaining* and is discussed in [4].

The result of firing an event on a set of states is computed by CONSFIRE and CONSRECFIRE. The only differences between them are that CONSRECFIRE also saturates the resulting node before returning it and also caches it – CONSFIRE is called as part of a saturation process so this is not necessary. The common parts (3–7 in CONSFIRE and 4–8 in CONSRECFIRE) compute the resulting node by recursively processing their child nodes. It is important to note that the arguments of the recursive call are $n[i]$, $c[j]$ and $d[i, j]$, that is, n is traversed along the source state and c is traversed along the target state. The recursive saturation of the result node in CONSRECFIRE in line 10 ensures that child nodes of the currently saturated node always stay saturated during the fixed point computation in accordance with Definitions 10 and 12.

3 The Generalized Saturation Algorithm

As we could see, the motivation of the constrained saturation algorithm (and all of its variants like those in [8,11]) is to handle a modified transition relation without losing locality. This paper generalizes these attempts by introducing the notion of conditional locality, a concept that expresses the most important consideration of all kinds of saturation: computing fixed points as locally (i.e. low in the decision diagram) as possible. This intuition has been discussed in Sect. 2.6, and the conclusion – that saturated nodes have a chance of being in the final MDD – can be used to improve the definitions to enhance this effect even further, which we do in the *generalized saturation algorithm* (GSA).

3.1 Conditional Locality

The concept of locality enables the saturation algorithm to ignore the value of variables outside the support of the currently processed event because it does not depend on them in any way. The result is that a fixed point can be calculated over partial states, which has to be computed *only once* regardless of the number of matching concrete states. The main motivation of conditional locality is to ignore even those variables that are not written but read by an event and compute the fixed point over even shorter partial states, but as many times as the value of those variables would cause a different result. The intuition is that the resulting nodes will be part of the final MDD more often than those created by the original saturation algorithm, leading to less intermediate nodes and therefore improved performance. Concepts are again illustrated in Fig. 1.

Definition 13 (Conditional locality). *An event $\alpha \in \mathcal{E}$ is said to be conditionally local over variables X and with respect to condition variables Y ($X \cap Y = \emptyset$) iff it is local over $X \cup Y$ and locally read-only on variables in Y. If Y is maximal and $X \cup Y = Supp(\alpha)$, then we call $Y = Guard(\alpha)$ the guard variables and $X = Supp_c(\alpha)$ the conditional support of α. The variable with the highest index among the conditionally supporting variables (according to a variable order) is the conditional top variable ($Top_c(\alpha)$) of α.*

The (full) next-state relation of a PTS can be automatically repartitioned based on conditional locality. The resulting partitions (events) will either be locally read-only on a variable or will always change its value (behaviors like "test-and-set" may combine these and be read-only sometimes but change the value other times – in this case, we can split the next-state relation). A special case of this repartitioning is built into the GSA as described in Sect. 3.2.

The following definition of conditionally saturated state spaces and MDD nodes can be considered as relaxations of Definitions 9 and 10 based on conditional locality.

Definition 14 (Conditionally saturated state space). *Given a partitioned transition system M, a set of (partial) states $S_{(X)}$ over variables X is conditionally saturated with respect to the partial state $\mathbf{s}_{(Y)}$ $(Y \subseteq V \setminus X)$ iff $S' = S' \cup \mathcal{N}_X(S')$, where $S' = \{\mathbf{s}_{(Y)}\} \times S_{(X)}$ and $\mathcal{N}_X = \bigcup_{\alpha | Supp_c(\alpha) \subseteq X, Guard(\alpha) \subseteq X \cup Y} \mathcal{N}_\alpha$.*

Note that a set of (partial) states $S_{(X)}$ over variables X that is conditionally saturated with respect to a zero-length state $\mathbf{s}_{(\emptyset)}$ is also saturated over X, therefore the goal is the same as before: generate a minimal, conditionally saturated set of states S with respect to $\mathbf{s}_{(\emptyset)}$ that contains the initial states S^0.

Definition 15 (Conditionally saturated node). *Given a partitioned transition system M, an MDD node n on level $lvl(n) = k$ is conditionally saturated with respect to the partial state $\mathbf{s}_{(V_{>k})}$ iff it encodes a set of (partial) states $S(n)$ that is conditionally saturated with respect to $\mathbf{s}_{(V_{>k})}$.*

The equivalent definition in terms of child nodes is now phrased as a theorem.

Theorem 1 (Conditionally saturated node – recursive definition). *Given a partitioned transition system M, an MDD node n on level $lvl(n) = k$ is conditionally saturated with respect to the partial state $\mathbf{s}_{(V_{>k})}$ iff (1) all of its children $n[i]$ are conditionally saturated with respect to $\mathbf{s}_{(V_{>k-1})}$, $\mathbf{s}_{(V_{>k-1})} \in \mathcal{M}(\mathbf{s}_{(V_{>k})})$ and $\mathbf{s}_{(V_{>k-1})}[k] = i$; and (2) $S' = S' \cup \mathcal{N}_k(S')$, where $S' = \{\mathbf{s}_{(V_{>k})}\} \times S(n)$ and $\mathcal{N}_k = \bigcup_{\alpha | Top_c(\alpha) = x_k} \mathcal{N}_\alpha$ for $1 \leq k \leq K$ and $\mathcal{N}_0 = \emptyset$.*

Proof. We prove only that a node described in the theorem encodes a conditionally saturated set of states. To prove the fixed point, we have to show that for any state $\mathbf{s} \in \{\mathbf{s}_{(V_{>k})}\} \times S(n)$ we have $\mathcal{N}_{V_{\leq k}}(\mathbf{s}) \subseteq \{\mathbf{s}_{(V_{>k})}\} \times S(n)$. Note that $\mathcal{N}_{V_{\leq k}} = \bigcup_{i=0}^{k} \mathcal{N}_k$ because if $Supp_c(\alpha) \subseteq V_{\leq k}$ then $Top_c(\alpha) \leq k$ and $Guard(\alpha) \subseteq V_{\leq k} \cup V_{>k} = V$ always holds. Assume there is a state $\mathbf{s}' \in \mathcal{N}_{V_{\leq k}}(\mathbf{s})$ that is not in $\{\mathbf{s}_{(V_{>k})}\} \times S(n)$. We know that $(\mathbf{s}, \mathbf{s}') \in \mathcal{N}_l$ for some $l \leq k$. If $l = k$ then we have a direct contradiction with the second requirement of the theorem. If $l < k$, then $\mathbf{s}'[k] = \mathbf{s}[k] = i$, because the transition cannot change the value of x_k. Because the first requirement of the theorem says that $n[i]$ is conditionally saturated with respect to $\mathbf{s}_{(V_{>k-1})}$ as defined above, and $\mathcal{N}_l \subseteq \mathcal{N}_{V_{\leq k-1}}$, it follows that $\mathbf{s}'$ must be in $\{\mathbf{s}_{(V_{>k-1})}\} \times S(n[i]) \subseteq \{\mathbf{s}_{(V_{>k})}\} \times S(n)$.

Based on Theorem 1 and the observation after Definition 14, the set of reachable states is encoded as a conditionally saturated MDD node on level K.

The key difference compared to Definitions 9 and 10 is the inclusion of a partial state with respect to which we can define a fixed point. Because we consider the repartitioned events that are now conditionally local, the partial state can be used to bind their guard variables, which will specify their effect on the variables in their conditional support. Since the guard variables are not changed when executing the transitions, we can compute a fixed point on only those variables that are in the conditional support of the event.

Even though the definition uses a partial state to define the fixed point, it is generally enough to traverse the NS descriptors just like the constraint in constrained saturation: whenever we navigate to $n[i]$, we should also navigate through $d[i, i]$. The resulting descriptor will characterize all the partial states that cause the same behavior in the rest of the transitions.

3.2 Detailed Description of the GSA

The pseudocode for the GSA is presented in Fig. 3. The inputs are an MDD node n encoding the initial states S^0 of a PTS, and a NS descriptor d representing the whole next-state relation $\mathcal{N}$. Since the algorithm will automatically partition the next-state relation based on conditional locality, d can be an union of all d_α (descriptors for events).

Sometimes, computing the full next-state relation is not practical, either because of its cost (e.g. we have to change representation) or because we want to use chaining in the fixed point computation. An advantage of abstract next-state diagrams is the ability to represent operations in a lazy manner. For example, the union of two descriptors may be represented by extending the set of descriptors $\mathcal{D}$ with elements of $\mathcal{D} \times \mathcal{D} \times \{union\}$ $(lvl((d_1, d_2, union)) = lvl(d_1) = lvl(d_2))$ and extending $next$ such that $(d_1, d_2, union)[i, j]$ is: $\mathbf{1}$ if d_1 or d_2 is $\mathbf{1}$; d_1 if d_2 is $\mathbf{0}$; d_2 if d_1 is $\mathbf{0}$; and $(d_1[i, j], d_2[i, j], union)$ otherwise. The *lazy descriptor* $(d_1, d_2, union)$ will not be equivalent to any non-lazy descriptor (even if they encode the same relation), but will be equivalent to $(d_1, d_2, union)$ or $(d_2, d_1, union)$, which is not optimal cache-wise but is often better than pre-computing the union. This approach can be generalized to more than two operands.

Compared to (constrained) saturation in Figs. 2a–c, the main differences and points of interest are listed below. In SATURATE:

- Next-state descriptors are not retrieved for each level, but are a parameter.
- Recursive saturation of child nodes in line 7 passes $d[i, i]$ as the NS descriptor to use on the lower level $k - 1$, which encodes a set of transitions that do not modify the variable associated to this level (and any above), therefore they are conditionally local over $V_{\leq k-1}$ with respect to the partial state specified by the SATURATE procedures currently on the call stack.
- Cache lookup in line 3 considers d instead of the partial state specified by the call stack because every partial state leading to d would produce the same result.
- In the fixed-point iteration in line 10 the SPLIT procedure is used to retrieve the operands of a lazy union descriptor to support chaining. It may be implemented in any other way as long as the returned set of descriptors cover the relation encoded by the descriptor passed as argument.
- In lines 6 and 9, the UPDATE procedure supports on-the-fly next-state relation building by providing a hook for replacing parts of d.

In SATFIRE:

- There are two descriptors: d_s for recursive saturation and d_f to fire.
- In the loop computing local successors in line 4 we omit locally read-only transitions $(i \neq j)$, because they will be processed by recursive saturation.
- In the recursive firing in line 5, d_s is indexed by $[y, y]$ because (like in constrained saturation) the resulting node will be $n'[y]$ (and therefore $d_s[y, y]$ describes the conditionally local transitions), while d_f is indexed as usual.

In SATRECFIRE:

- Cache lookup in line 3 considers both next-state descriptors.
- In the loop computing local successors in line 5 we now *consider* every transition even if they are read-only, (on some level above they changed a variable).
- Recursive saturation in line 9 will use d_s (which is still conditionally local).

3.3   Constrained Saturation as an Instance of the GSA

With the automatic partitioning offered by the GSA, next-state relations that motivated the introduction of constrained saturation and its variants can now be directly encoded into the transition relation without any cost. This is because a constraint is a *guard*, therefore it can cause an event only to become read-only on a variable instead of independent, but will still never write it. Adding a constraint will never raise the conditional top variable of events, but it can raise their unconditional top variable in many cases, which is associated with degraded performance.

Indeed, the handling of d_s in the GSA is very similar to the handling of the constraint node – we could say that our algorithm uses the next-state relation itself as a constraint. Combining this with the flexibility of abstract NS descriptors (lazy descriptors in particular), we get the properties of constrained saturation enhanced with every difference between the original saturation algorithm and the GSA (see Sect. 3.4).

We illustrate the usage of abstract NS descriptors for variants of constrained saturation with the kind of constraint used in the original constrained saturation algorithm [12].

Definition 16. (Constrained next-state descriptor). *Given a NS descriptor d and a constraint node c, the* constrained next-state descriptor d_c *describing* $\mathcal{N}(d_c) = \mathcal{N}(d) \setminus (\mathbb{N}^K \times S(c))$ *is a tuple $d_c = (d, c)$ with $lvl(d_c) = lvl(d) = lvl(c)$, and $d_c[i, j]$ is: $\mathbf{0}$ if $d[i, j] = \mathbf{0}$ or $c[j] = \mathbf{0}$; and $(d[i, j], c[j])$ otherwise.*

3.4   Discussion

To estimate the efficiency of the algorithm, we will consider the advantages and disadvantages of the different modifications. First and foremost it is important to note that if $Top_c(\alpha) = Top(\alpha)$ for every event α, then the GSA degrades to

Input: MDD node n,
　　　　 NS descriptor d
Output: saturated MDD node n'

```
1  if n ∈ {0, 1} or d ∈ {0, 1} then
2  |  return n

3  if ¬SATCACHEGET(n, d, n') then
4  |  n' ← new MDDNODE(lvl(n))
5  |  for each i where n[i] ≠ 0 do
6  |  |  CONFIRM(lvl(n), i), UPDATE(d)
7  |  |  n'[i] ← SATURATE(n[i], d[i, i])
8  |  repeat
9  |  |  changed ← false, UPDATE(d)
10 |  |  for each d_f ∈ Split(d) do
11 |  |  |  n'' ← SATFIRE(n', d, d_f)
12 |  |  |  if n' ≠ n'' then
13 |  |  |  |  n' ← n'', changed ← true
14 |  |  until ¬changed
15 |  SATCACHEPUT(n, d, n')
16 return n'
```

(a) Procedure SATURATE.

Input: MDD node n,
　　　　 NS descriptor for saturate d_s,
　　　　 NS descriptor for fire d_f
Output: the result of firing d from the
　　　　 states n with the children
　　　　 saturated

```
1  if n = 0 or d = 0 then return 0
2  if d = 1 then return n
3  n' ← new MDDNODE(lvl(n))
4  for each x, y where x ≠ y and
   d[x, y] ≠ 0 and n[x] ≠ 0  do
5  |  m ← SATRECFIRE(d[y, y], d[x, y], n[x])
6  |  if m ≠ 0 then CONFIRM(lvl(n), y)
7  |  n'[y] ← n'[y] ∪ m
8  CHECKIN(n')
9  return n'
```

(b) Procedure SATFIRE.

Input: NS descriptor for saturate d_s, NS descriptor for fire d_f, MDD node n
Output: saturated MDD node n'' (after firing d_f from n saturated with d_s)

```
1  if n = 0 or d = 0 then return 0
2  if d = 1 then return n
3  if ¬RECFIRECACHEGET(d_s, d_f, n, n'') then
4  |  n' ← new MDDNODE(lvl(n))
5  |  for each x, y where d[x, y] ≠ 0 and n[x] ≠ 0  do
6  |  |  m ← SATRECFIRE(d_s[y, y], d_f[x, y], n[x])
7  |  |  if m ≠ 0 then CONFIRM(lvl(n), y)
8  |  |  n'[y] ← n'[y] ∪ m
9  |  CHECKIN(n'), n'' ← SATURATE(n', d_s), RECFIRECACHEPUT(d_s, d_f, n, n'')
10 return n''
```

(c) Procedure SATRECFIRE.

Fig. 3. Pseudocode of the GSA.

the original saturation algorithm from [4] or the corresponding constrained saturation algorithm from [8,11,12] with no difference in the iteration strategy and the virtually zero overhead of handling the next-state relation as a parameter. In every other case, there may be a complex interplay between the advantages and disadvantages discussed below.

An advantage of using conditional locality is that $Top_c(\alpha) \leq Top(\alpha)$, i.e. we can potentially use event α when saturating a node on a lower level, which is intuitively better because it raises the chance that the resulting node will be

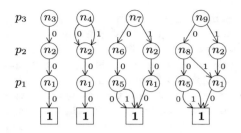 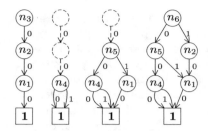

Fig. 4. A (degraded) run of saturation on the example model: n_3 encodes S^0, $n_4 = n_3 \cup \mathcal{N}_{t_3^l}(n_3)$, $n_7 = n_4 \cup \mathcal{N}_{t_1^l}(n_4)$, the state space is $n_9 = n_7 \cup \mathcal{N}_{t_2^l}(n_7)$. Note that t_i^u does not reach new states.

Fig. 5. A run of the GSA on the example model: n_3 encodes S^0, n_4 is the saturated n_1 (after firing t_1^l), n_5 is the saturated n_2 (after firing t_2^l) and n_6 is the state space.

part of the final diagram. Figures 4 and 5 illustrate the MDDs that are created while exploring the state space of the example Petri net model from Fig. 1 with saturation and the GSA. Saturation is degraded to a chaining version of BFS because every transition that can yield a new state is dependent on all variables. In the unfortunate case of firing t_3^l, t_1^l and t_2^l in this order, the number of created nodes will be 9 compared to the 6 nodes created by the GSA, which can still exploit the read-only dependencies.

A direct price of this is the diversification of cache entries. By repartitioning the events, we may introduce a lot more next-state relations to process, and it is not evident if their smaller size and the enhanced saturation effect can compensate this. Furthermore, by keeping track of d_s (the descriptor to saturate with), we spoil the cache of saturation due to the following.

Whenever we navigate through $d[i, i]$, we remember something from i in the context of the next-state relation, yielding a potentially large number of different descriptors to saturate with. The original saturation algorithm saturates each MDD node only once, because it uses the same next-state relation every time. In the GSA, we saturate every pair of different MDD node and NS descriptor, so the diversity of descriptors can be a crucial factor. In the extreme case, when at least one event remembers every value along the path (for example because it copies them to other variables below), caching can degrade to the point where everything will need to be computed from scratch.

The other extreme is when each event remembers only one thing from the values bound above: whether it is enabled or not (e.g. when variables are compared to constants in guard expressions). Fortunately, this is the case with Petri nets: each transition will check variables locally and decide whether it is still enabled or not. This means that given a descriptor d representing transitions in T, the number of possible successors for $d[i, i]$ will be $O(|T|)$ (n values can partition $\mathbb{N}$ into $n + 1$ partitions, each transition may contribute 2 values – one for an input arc and one for an inhibitor arc), but this number will also be limited by the number of non-zero child nodes of the saturated MDD node.

Given the facts that transitions of Petri nets are inherently conditionally local without repartitioning, and many nets are bounded (often safe), model checking of Petri net models with the GSA can be expected to yield favorable results. In fact, the experiment presented in Sect. 4 shows that the GSA is superior to the original saturation algorithm on every model that we analyzed.

For other types of models, we have yet to investigate the efficiency of the algorithm and the balance of benefits and overhead. It might be the case that we have to refine the read-only dependency into "local" and "global" evaluation (depending on whether we have to remember the value of the variable or can evaluate it immediately) and use conditional locality only with the "local" case. We also have to note that the efficient update of the next-state descriptors is not trivial and subject to future work.

4   Evaluation

In this section, we present the results of our experiments performed on a large set of Petri net models.

4.1   Research Questions

We have two main research questions about the GSA, both comparing it to the original saturation algorithm (SA) from [4] (results should apply to constrained saturation as well). Both questions will be answered by measuring the relevant metrics for each algorithm and comparing the results for each benchmark model.

We expect that (1) the GSA will be identical to the SA when conditional locality cannot be exploited; as well as in other cases (2) the GSA will create less MDD nodes than the SA and (3) in these cases it will be faster than the SA.

4.2   The Benchmark

We have implemented both the original saturation algorithm and the GSA in Java. Both variants used the same libraries for MDDs and next-state descriptors, and their source code differs only in the points discussed in Sect. 3.2.

We used the latest set of 743 available models from the Model Checking Contest 2018 [7], excluding only the Glycolytic and Pentose Phosphate Pathways (G-PPP) model with a parameter of 10–1000000000 (because the initial marking cannot be represented on 32-bit signed integers). We generated a variable ordering for each model using the *sloan* algorithm recommended by [1], and a *modified sloan* algorithm where we omitted read-only dependencies when building the dependency graph (motivated by the notion of conditional locality). We ran state space exploration 3 times on each model with each ordering, measuring several metrics of the algorithms. We will report the median of the running time of the algorithms (excluding the time of loading the model) and the total number of MDD nodes created during each run, as well as the size of the state space and the final MDD for each model and each ordering.

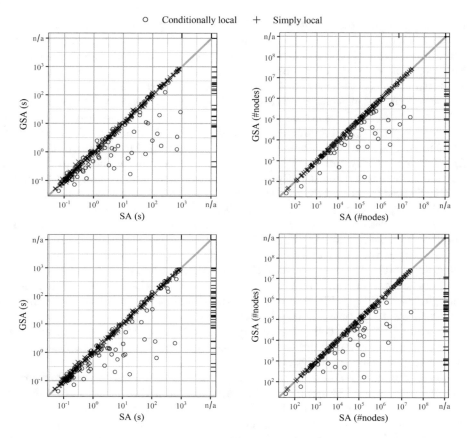

Fig. 6. Main results of the experiment: running times and total number of created nodes with sloan ordering (top row) and modified sloan ordering (bottom row).

Measurements were conducted on a bare-metal server machine rented from the Oracle Cloud (BM.Standard.E2.64), with 64 cores and 512 GB of RAM, running Ubuntu 18.04 and Java 11. Three processes were run simultaneously, each with a maximum Java heap size of 100 GB and stack size of 512 MB. No process has run out of memory and the combined CPU utilization never exceeded 70%. Timeout was 20 min (including loading the model and writing results).

4.3   Results

The main results of the experiments can be seen in Fig. 6. Every point represents a model (dashes on the side means a timeout), classified into two groups: "simply local" if none of the events had a read-only top variable and "conditionally local" otherwise. In the "simply local" group we expected no difference because the GSA should degrade into the original saturation algorithm, which was supported by the results. In the other group we were optimistic about the balance of advantages and disadvantages as discussed in Sect. 3.4, but the results were even better than what we expected. As the plots show, a significant part of the "conditionally local" models are below the reference diagonal, meaning that the GSA were often orders of magnitudes faster.

With the sloan ordering, 274 models were in the "conditionally local" group and the GSA was at least twice as fast as the SA in 53 cases. With the modified ordering, these numbers are 69 out of 298. In one case (*SmallOperatingSystem-MT0256DC0128*), the SA managed to finish just in time while the GSA exceeded the timeout (scaling was similar for smaller instances). Models where the GSA finished successfully but the SA exceeded the timeout with the sloan ordering include instances of *CloudDeployment*, *DiscoveryGPU*, *DLCround*, *DLCshifumi*, *EGFr*, *Eratosthenes*, *MAPKbis*, *Peterson* and *Raft*; and with the modified ordering also *AirplaneLD*, *BART*, *Dekker*, *FlexibleBarrier*, *NeoElection*, *ParamProductionCell*, *Philosopher*, *Ring* and *SharedMemory*. Analyzing these models in detail may provide insights about when the GSA is especially efficient.

Looking at the plots about the number of created MDD nodes (i.e. the size of the unique table) reveals that our expectations about less intermediate diagrams were correct and this probably has direct influence on the execution time. Even though not visible in Fig. 6, interactive data analysis revealed that the model instances are more-or-less located at the same point on the execution time and node count plots. The collected data also suggests a linear relationship between the number of created nodes and the execution time, but this is rather a lower bound than a general prediction.

As an auxiliary result and without any illustration, we also report that out of the 117 cases when the sloan ordering and the modified ordering were different and we have data about the final MDD size, the modified sloan ordering produced smaller final MDDs 69 times and larger MDDs 39 times. This motivates further work on variable orderings like in [1]. We have also compared the SA with sloan ordering and the GSA with the modified sloan ordering to find that the GSA with the modified sloan ordering was better in 78 cases and worse in 16 cases (considering only at least a factor of 2 in both cases).

5   Conclusions

In this paper, we have formally introduced the *generalized saturation algorithm* (GSA), a new saturation algorithm enhanced with the notion of *conditional locality*. We have shown that the GSA generalizes a family of constrained saturation variants and discussed the effects of using conditional locality. We have empirically evaluated our approach on Petri nets from the Model Checking Contest to find that the GSA has virtually no overhead compared to the original saturation algorithm, but can outperform it by orders of magnitude when conditional locality can be exploited.

We have made theoretical considerations and prepared the algorithm to be compatible with a wide range of next-state representations as well as the on-the-fly update approach described in [4]. The GSA seems to be superior to the original saturation algorithm on Petri net models, but its efficiency over more general classes of models is yet to be explored.

Acknowledgments. This work has been partially supported by Nemzeti Tehetség Program, Nemzet Fiatal Tehetségeiért Ösztöndíj 2018 (NTP-NFTÖ-18).

References

1. Amparore, E.G., Donatelli, S., Beccuti, M., Garbi, G., Miner, A.S.: Decision diagrams for Petri nets: a comparison of variable ordering algorithms. Trans. Petri Nets Other Models Concurr. **13**, 73–92 (2018)
2. Burch, J.R., Clarke, E.M., McMillan, K.L., Dill, D.L., Hwang, L.J.: Symbolic model checking: 10^{20} states and beyond. Inf. Comput. **98**(2), 142–170 (1992)
3. Ciardo, G., Lüttgen, G., Siminiceanu, R.: Saturation: an efficient iteration strategy for symbolic state-space generation. In: Margaria, T., Yi, W. (eds.) TACAS 2001. LNCS, vol. 2031, pp. 328–342. Springer, Heidelberg (2001). https://doi.org/10.1007/3-540-45319-9_23
4. Ciardo, G., Marmorstein, R., Siminiceanu, R.: The saturation algorithm for symbolic state-space exploration. Int. J. Softw. Tools Technol. Transf. **8**(1), 4–25 (2006)
5. Ciardo, G., Yu, A.J.: Saturation-based symbolic reachability analysis using conjunctive and disjunctive partitioning. In: Borrione, D., Paul, W. (eds.) CHARME 2005. LNCS, vol. 3725, pp. 146–161. Springer, Heidelberg (2005). https://doi.org/10.1007/11560548_13
6. Hamez, A., Thierry-Mieg, Y., Kordon, F.: Hierarchical set decision diagrams and automatic saturation. In: van Hee, K.M., Valk, R. (eds.) PETRI NETS 2008. LNCS, vol. 5062, pp. 211–230. Springer, Heidelberg (2008). https://doi.org/10.1007/978-3-540-68746-7_16
7. Kordon, F., et al.: Complete Results for the 2018 Edition of the Model Checking Contest, June 2018. http://mcc.lip6.fr/2018/results.php
8. Marussy, K., Molnár, V., Vörös, A., Majzik, I.: Getting the priorities right: saturation for prioritised Petri nets. In: van der Aalst, W., Best, E. (eds.) PETRI NETS 2017. LNCS, vol. 10258, pp. 223–242. Springer, Cham (2017). https://doi.org/10.1007/978-3-319-57861-3_14
9. Meijer, J., Kant, G., Blom, S., van de Pol, J.: Read, write and copy dependencies for symbolic model checking. In: Yahav, E. (ed.) HVC 2014. LNCS, vol. 8855, pp. 204–219. Springer, Cham (2014). https://doi.org/10.1007/978-3-319-13338-6_16
10. Miner, A.S.: Implicit GSPN reachability set generation using decision diagrams. Perform. Eval. **56**(1–4), 145–165 (2004)
11. Molnár, V., Vörös, A., Darvas, D., Bartha, T., Majzik, I.: Component-wise incremental LTL model checking. Form. Asp. Comput. **28**(3), 345–379 (2016)
12. Zhao, Y., Ciardo, G.: Symbolic CTL model checking of asynchronous systems using constrained saturation. In: Liu, Z., Ravn, A.P. (eds.) ATVA 2009. LNCS, vol. 5799, pp. 368–381. Springer, Heidelberg (2009). https://doi.org/10.1007/978-3-642-04761-9_27

Parametrics and Combinatorics

Parameterized Analysis of Immediate Observation Petri Nets

Javier Esparza⬥, Mikhail Raskin⬥, and Chana Weil-Kennedy(✉)⬥

Technical University of Munich, Munich, Germany
{esparza,raskin,chana.weilkennedy}@in.tum.de

Abstract. We introduce immediate observation Petri nets, a class of interest in the study of population protocols (a model of distributed computation), and enzymatic chemical networks. In these areas, relevant analysis questions translate into *parameterized* Petri net problems: whether an infinite set of Petri nets with the same underlying net, but different initial markings, satisfy a given property. We study the parameterized reachability, coverability, and liveness problems for immediate observation Petri nets. We show that all three problems are in PSPACE for infinite sets of initial markings defined by counting constraints, a class sufficiently rich for the intended application. This is remarkable, since the problems are already PSPACE-hard when the set of markings is a singleton, i.e., in the non-parameterized case. We use these results to prove that the correctness problem for immediate observation population protocols is PSPACE-complete, answering a question left open in a previous paper.

Keywords: Petri nets · Reachability analysis ·
Parameterized verification · Population protocols

1 Introduction

We study the theory of *immediate observation Petri nets*, a class of Petri nets with applications to the study of population protocols and chemical reaction networks, two models of distributed computation.

Population protocols are a formalism for the study of ad hoc networks of tiny computing devices without any infrastructure. They were introduced by Angluin *et al.* [5], and have been very intensely studied, in particular in recent years (see e.g. [1–3,10]). The model postulates a "soup" of finite-state, indistinguishable agents interacting in pairs. Formally, a population protocol has a finite set of states Q and a set of transitions of the form $(q_1, q_2) \mapsto (q_3, q_4)$, which allow two agents in states q_1 and q_2 to interact and simultaneously move to q_3 and q_4. A global state of the protocol, called a *configuration*, is a mapping C that

---

This project has received funding from the European Research Council (ERC) under the European Union's Horizon 2020 research and innovation programme under grant agreement No. 787367 (PaVeS).

S. Donatelli and S. Haar (Eds.): PETRI NETS 2019, LNCS 11522, pp. 365–385, 2019.
https://doi.org/10.1007/978-3-030-21571-2_20

assigns to each state q the current number $C(q)$ of agents in q. A protocol has a set of initial configurations. Intuitively, each initial configuration corresponds to an input, and the purpose of a protocol is to compute a boolean output, 0 or 1, for each input. A protocol outputs b for a given initial configuration C if in all fair runs starting at C (with respect to a certain fairness condition), all agents eventually agree to output b. So, loosely speaking, population protocols compute by reaching a stable consensus. The *predicate computed by a protocol* is the function that assigns to each initial configuration C the boolean output computed by the protocol when started at C.

Even this very abstract description shows that a population protocol is "nothing but" a (place/transition) Petri net: a state corresponds to a place, a transition of the protocol to a net transition with two input and two output places, an agent to a token, and a configuration to a marking. In the last years, this connection was exploited to address the problem of proving population protocols correct. The fundamental correctness problem for population protocols asks, given a protocol and a predicate, whether the protocol computes the predicate. This question was proved decidable in [12,13], but, unfortunately, the same papers also showed that the correctness problem is at least as hard as Petri net reachability, and so of non-elementary complexity [9].

In their seminal paper on the expressive power of population protocols [6], Angluin *et al.* defined subclasses corresponding to different communication primitives between agents. In the standard model, agents communicate through *rendez-vous*: transitions $(q_1, q_2) \mapsto (q_3, q_4)$ formalize that both partners exchange full information about their current states, and update them based on it. Angluin *et al.* introduced *immediate observation* protocols, called IO protocols for short, whose transitions have the form $(q_1, q_2) \mapsto (q_1, q_3)$. Intuitively, in an IO protocol an agent can change its state from q_2 to q_3 by *observing* that another agent is in state q_1; the agent in state q_1 may not even know that it is being observed. A characterization of the predicates computable by IO protocols was given in [6], and in [14] Esparza *et al.* studied the complexity of the correctness problem. They showed that it was PSPACE-hard and solvable in EXPSPACE, and left the problem of closing this gap for future research.

In this paper we study the theory of *immediate observation Petri nets* (IO nets), the Petri nets underlying immediate observation protocols. Our initial motivation is their application to population protocol problems, especially the gap just mentioned. However, IO nets also model *networks of enzymatic chemical reactions*, in which an enzyme E catalyzes the formation of product P from substrate S [7,16]. An example of application of Petri net techniques to such a network is presented in [4].[1]

Analysis problems for population protocols or chemical networks are *parametric* in the number of agents or the number of molecules. In other words, they ask whether the system satisfies a property *for any number of agents* or *for*

---

[1] The Petri nets of [4] are in fact slightly more general than IO nets, but equivalent to them for properties that depend only on the reachability graph, as are the net properties studied in [4].

any number of molecules. When formalized as Petri nets problems, they become questions of the form "does an *infinite set* of Petri nets differing only in their initial markings satisfy a given property?" We investigate parameterized versions of the standard reachability, coverability, and liveness problems for IO nets in which the set of initial markings is a *cube*, i.e., a set of markings obtained by attaching to each place a lower bound and an upper bound (possibly infinite) for the number of tokens. We prove that, remarkably, while the standard problems are PSPACE-hard even in the non-parameterized case, they remain in PSPACE in the parameterized case. This is in strong contrast with the situation for more general classes of nets. For example, while the non-parametric problems are in PSPACE for conservative nets or 1-safe nets, their "cube-versions" become EXPSPACE-hard or even non-elementary. As an application of our results, we close the gap left open in [14], and prove that the correctness problem for IO protocols is PSPACE-complete.

For space reasons, all missing proofs and some technical details are relegated to the full version of this article [15].

2   Preliminaries

Multisets. A *multiset* on a finite set E is a mapping $C \colon E \to \mathbb{N}$, i.e. for any $e \in E$, $C(e)$ denotes the number of occurrences of element e in C. Let $\wrangle e_1, \ldots, e_n \wrangle$ denote the multiset C such that $C(e) = |\{j \mid e_j = e\}|$. Operations on $\mathbb{N}$ like addition or comparison are extended to multisets by defining them component wise on each element of E. Subtraction is allowed as long as each component stays non-negative. We define $|C| \stackrel{\text{def}}{=} \sum_{e \in E} C(e)$ the sum of the occurrences of each element in C. Given a total order $e_1 \prec e_2 \prec \cdots \prec e_n$ on E, a multiset C can be equivalently represented by the vector $(C(e_1), \ldots, C(e_n)) \in \mathbb{N}^n$.

Place/Transition Petri Nets with Weighted Arcs. A *Petri net* N is a triple (P, T, F) consisting of a finite set of *places* P, a finite set of *transitions* T and a *flow function* $F \colon (P \times T) \cup (T \times P) \to \mathbb{N}$.

A *marking* M is a multiset on P, and we say that a marking M puts $M(p)$ *tokens* in place p of P. The *size* of M, denoted by $|M|$, is the total number of tokens in M. The *preset* $^\bullet t$ and *postset* $t^\bullet$ of a transition t are the multisets on P given by $^\bullet t(p) = F(p, t)$ and $t^\bullet(p) = F(t, p)$. A transition t is *enabled* at a marking M if $^\bullet t \leq M$, i.e. $^\bullet t$ is component-wise smaller or equal to M. If t is enabled then it can be *fired*, leading to a new marking $M' = M - {}^\bullet t + t^\bullet$. We note this $M \xrightarrow{t} M'$.

Reachability and Coverability. Given $\sigma = t_1 \ldots t_n$ we write $M \xrightarrow{\sigma} M_n$ when $M \xrightarrow{t_1} M_1 \xrightarrow{t_2} M_2 \ldots \xrightarrow{t_n} M_n$, and call σ a *firing sequence*. We write $M' \xrightarrow{*} M''$ if $M' \xrightarrow{\sigma} M''$ for some $\sigma \in T^*$, and say that M'' is *reachable* from M'. A marking M *covers* another marking M', written $M \geq M'$ if $M(p) \geq M'(p)$ for all places p. A marking M is *coverable* from M' if there exists a marking M'' such that $M' \xrightarrow{*} M'' \geq M$.

Conservative Petri Nets. A *Petri net* $N = (P, T, F)$ is *conservative* if there is a mapping $I \colon P \to \mathbb{Q}_{>0}$ such that $\sum_{p \in P} I(p) \cdot {}^\bullet t(p) = \sum_{p \in P} I(p) \cdot t^\bullet(p)$ for all t. Further, N is **1**-*conservative* if it is conservative with I equal to 1 over all P (see [17]). It follows immediately from the definitions that if N is conservative and $M \xrightarrow{*} M'$, then $\sum_{p \in P} I(p) \cdot M(p) = \sum_{p \in P} I(p) \cdot M'(p)$.

3   A Primer on Population Protocols

As mentioned in the introduction, a population protocol consists of a set of states Q and a set of transitions $T \subseteq Q^2 \times Q^2$. A transition $((q_1, q_2), (q_3, q_4)) \in T$ is denoted $(q_1, q_2) \mapsto (q_3, q_4)$. A *configuration* is a multiset of states. A configuration, say C, such that $C(q_1) = 2$ and $C(q_2) = 1$, indicates that currently there are two agents in state q_1 and one agent in state q_2. The connection to Petri nets is immediate: The Petri net modeling a protocol has one place for each state, and one transition for every transition of the protocol. If transition t of the Petri net models $(q_1, q_2) \mapsto (q_3, q_4)$, then ${}^\bullet t = \{q_1, q_2\}$, and $t^\bullet = \{q_3, q_4\}$. An agent in state q is modeled by a token in place q. A configuration C with $C(q)$ agents in state q is modeled by the marking putting $C(q)$ tokens in place q for every $q \in Q$. Observe that the transitions of the net do not change the total number of tokens, and so we have:

Fact 1. *Petri nets obtained from population protocols are **1**-conservative.*

Population protocols are designed to compute predicates $\varphi \colon \mathbb{N}^k \to \{0, 1\}$. We first give an informal explanation of how a protocol computes a predicate, and then a formal definition using Petri net terminology. A protocol for φ has a distinguished set of input states $\{q_1, q_2, \ldots, q_k\} \subseteq Q$. Further, each state of Q, initial or not, is labeled with an *output*, either 0 or 1. Assume for example $k = 2$. In order to compute $\varphi(n_1, n_2)$, we first place n_i agents in q_i for $i = 1, 2$, and 0 agents in other states. This is the initial configuration of the protocol for the input (n_1, n_2). Then we let the protocol run. The protocol satisfies that in every *fair* run starting at the initial configuration (fair runs are defined formally below), eventually all agents reach states labeled with 1, and stay in such states forever, or they reach states of labeled with 0, and stay in such states forever. So, intuitively, in all fair runs all agents eventually "agree" on a boolean value. By definition, this value is the result of the computation, i.e, the value of $\varphi(n_1, n_2)$.

Formally, and in Petri net terms, fix a Petri net $N = (P, T, F)$ with $|{}^\bullet t| = 2 = |t^\bullet|$ for every transition t. Further, fix a set $I = \{p_1, \ldots, p_k\}$ of *input places*, and a *function* $O \colon P \to \{0, 1\}$. A marking M of N is a b-*consensus* if $M(p) > 0$ implies $O(p) = b$. A b-consensus M is *stable* if every marking reachable from M is also a b-consensus. A firing sequence $M_0 \xrightarrow{t_1} M_1 \xrightarrow{t_2} M_2 \cdots$ of N is *fair* if it is finite and ends at a deadlock marking, or if it is infinite and the following condition holds for all markings M, M' and $t \in T$: if $M \xrightarrow{t} M'$ and $M = M_i$ for infinitely many $i \geq 0$, then $M_j \xrightarrow{t_{j+1}} M_{j+1} = M \xrightarrow{t} M'$ for infinitely many $j \geq 0$. In other words, if a fair sequence reaches a marking infinitely often, then all the transitions enabled at that marking will be fired infinitely often from that marking. A fair firing sequence *converges to* b if there is $i \geq 0$ such that M_j is a b-consensus for every marking $j \geq i$ of the sequence. For every $\boldsymbol{v} \in \mathbb{N}^k$ with $|\boldsymbol{v}| \geq 2$ let $M_{\boldsymbol{v}}$ be the marking given by $M_{\boldsymbol{v}}(p_i) = \boldsymbol{v}_i$ for every $p_i \in I$, and $M_{\boldsymbol{v}}(p) = 0$ for every $p \in P \setminus I$. We call $M_{\boldsymbol{v}}$ the *initial marking for input* $\boldsymbol{v}$. The net N *computes the predicate* $\varphi \colon \mathbb{N}^k \to \{0, 1\}$ if for every $\boldsymbol{v} \in \mathbb{N}^k$, every fair firing sequence starting at $M_{\boldsymbol{v}}$ converges to b.

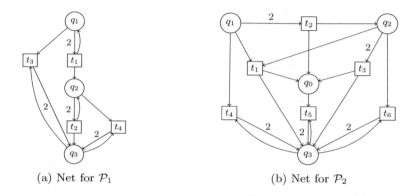

(a) Net for $\mathcal{P}_1$ \hspace{4em} (b) Net for $\mathcal{P}_2$

Fig. 1. Petri nets underlying population protocols.

Example 1. We exhibit two population protocols that compute the predicate $\varphi(x) \stackrel{\text{def}}{=} [x \geq 3]$, and their corresponding Petri nets.

The first protocol $\mathcal{P}_1$ has states $Q_1 = \{q_1, q_2, q_3\}$ and transitions $(q_a, q_a) \mapsto (q_{a+1}, q_a)$ and $(q_a, q_3) \mapsto (q_3, q_3)$ for $a = 1, 2$. The only input state is q_1. States q_1 and q_2 are labeled with 0, and state q_3 with 1. The Petri net for $\mathcal{P}_1$ is shown in Fig. 1a. The initial marking for input x puts x tokens on q_1, and no token elsewhere. If $x \geq 3$, then every fair firing sequence eventually reaches the deadlock marking with x tokens in q_3 and no tokens elsewhere (indeed, transitions t_3 and t_4 ensure that after a token reaches q_3, eventually all other tokens move to q_3 as well). So the agents eventually reach consensus 1. If $x < 3$, then no firing sequence ever puts a token in q_3 and so, since both q_1 and q_2 have output 0, the agents reach consensus 0.

The second protocol $\mathcal{P}_2$ has place set $Q_2 = \{q_0, q_1, q_2, q_3\}$, and transitions $(q_a, q_b) \mapsto (q_0, q_{\min(a+b,3)})$ for $0 < a, b < 3$, and $(q_a, q_3) \mapsto (q_3, q_3)$ for $0 \leq a < 3$. The Petri net for $\mathcal{P}_2$ is shown in Fig. 1b. Again, the only input state is q_1. States q_0, q_1, q_2 are labeled with 0, and state q_3 is labeled with 1. The reader can check that, as in the first protocol, the agents eventually reach consensus 1 from an input x iff $x \geq 3$.

Both these protocols could be generalized to calculate $[x \geq n]$ for any natural $n \geq 1$.

Immediate Observation Protocols. When two agents of a population protocol communicate, they can both simultaneously change their states. This corresponds to communication by *rendez-vous*. In [6], Angluin *et al.* introduced *immediate observation* protocols, corresponding to a more restricted communication mechanisms. One of the agents observes the state of the other agent, and updates its own state accordingly; the observed agent does not change its state, since it may not even know that it is being observed. Transitions are of the form $(q_s, q_o) \mapsto (q_d, q_o)$, where q_o is the state of the observed agent. In the paper, they showed that the predicates computable by immediate observation protocols are exactly those described by counting constraints, a formalism introduced in Sect. 7.

Example 2. Protocol $\mathcal{P}_1$ of Example 1 is immediate observation, but $\mathcal{P}_2$ is not.

Verifying Population Protocols. Not every population protocol is well designed. For some inputs $(n_1, \ldots, n_k)$ the protocol can have fair runs that never converge, or fair runs converging to the wrong value $1 - \varphi(n_1, \ldots, n_k)$. This raises the question of how to automatically verify that a protocol correctly computes a predicate. The main difficulty is to prove convergence to the right value *for each of the infinitely many possible inputs*. In Petri net terms, we have to show that the net derived from the protocol satisfies a property *for infinitely many initial markings*. So, strictly speaking, we have to show that an infinite collection of Petri nets satisfies a given property. We call problems of this kind *parameterized*.

4   Parameterized Analysis Problems

Standard analysis problems for Petri nets concern one initial marking. For example, the *reachability problem* (*coverability problem*) consists of, given a net N and two markings M, M' of N, deciding if M is reachable (coverable) from M'. Parameterized problems, like the correctness problem for population protocols, involve an *infinite* set of initial markings. In order to study their complexity, it is necessary to specify the shape of the set. For the applications to population protocols and chemical networks the following definition is adequate:

Definition 1. *A set $\mathcal{M}$ of markings of a net N is a* cube *if there are mappings $L: P \to \mathbb{N}$ and $U: P \to \mathbb{N} \cup \{\infty\}$ such that $M \in \mathcal{M}$ iff $L \leq M \leq U$. We call L and U the* lower *and* upper *bound of $\mathcal{M}$, respectively, and use the notation $(L, U) \overset{def}{=} \mathcal{M}$. The* cube-reachability (cube-coverability) *problem consists of, given a net N and cubes $\mathcal{M}, \mathcal{M}'$ of N, deciding if there are markings $M \in \mathcal{M}, M' \in \mathcal{M}'$, such that M is reachable (coverable) from M'.*

Observe that, if the set of places of the Petri net corresponding to a population protocol is $\{p_1, \ldots, p_n, p_{n+1}, \ldots, p_{n+m}\}$, where $p_1, \ldots, p_n$ are the initial places, then the set of input configurations corresponds to the cube (L, U) where $L(p_i) = 0$ for every $1 \leq i \leq n + m$, and $U(p_i) = \infty$ for $1 \leq i \leq n$ and $U(p_i) = 0$ for $n + 1 \leq i \leq n + m$.

In general, parameterized problems are much harder than non-parameterized ones. Consider for example the class of conservative Petri nets which, by Fact 1 contains all nets derived from population protocols. We have:

Theorem 2. *For* **1**-*conservative Petri nets:*

- *Reachability, coverability, and liveness are in* PSPACE.
- *Cube-reachability and cube-coverability are as hard as for general Petri nets, and so non-elementary and* EXPSPACE-*hard, respectively.*

In the rest of the paper we introduce immediate observation Petri nets, the class of Petri nets corresponding to immediate observation protocols and enzymatic reaction networks, and study the cube-reachability, coverability, and liveness problems. We prove that, while the problems are PSPACE-hard even for single markings, their cube versions *remain* PSPACE. This pinpoints the essential property of the class: loosely speaking, deciding standard problems for infinitely many markings is not harder than deciding them for one marking.

5   Immediate Observation Petri Nets

We introduce the class of immediate observation Petri nets (IO nets) and then show that the reachability, coverability, and liveness problems are PSPACE-hard for this class.

Definition 2. *A transition t of a Petri net is an* immediate observation tran- *sition if there are three places p_s, p_d, p_o, not necessarily distinct, such that $^\bullet t = \{p_s, p_o\}$ and $t^\bullet = \{p_d, p_o\}$. We call p_s, p_d, p_o the* source, destination, *and* observed *places of t, respectively. A Petri net is an* immediate observation *net if and only if all its transitions are immediate observation transitions.*

Following the useful convention of population protocols, we write $t = (p_s, p_o) \mapsto (p_d, p_o)$.

Example 3. The Petri net illustrated in Fig. 1a is an immediate observation Petri net.

We show that the standard simulation of bounded-tape Turing machines by 1-safe Petri nets, as described for example in [8, 11], can be modified to produce an IO net (actually, a 1-safe IO net). Using this result, we can then easily prove that the reachability, coverability, and liveness problems are PSPACE-hard. Since a set consisting of a single marking is a special case of a cube, the result carries over to the cube-versions of the problems.

We fix a deterministic Turing machine M with set of control states Q, alphabet Σ containing the empty symbol $\lrcorner$, and partial transition function $\delta \colon Q \times \Sigma \to Q \times \Sigma \times D$ $(D = \{-1, +1\})$. We let K denote an upper bound on the number of tape cells visited by the computation of M on empty tape. The *implementation* of M is the IO Petri net N_M described below.

Places of N_M. The net N_M contains two sets of *cell places* and *head places* modelling the state of the tape cells and the head, respectively. The cell places are:

- $off[\sigma, n]$ for each $\sigma \in \Sigma$ and $1 \le n \le K$. A token on $off[\sigma, n]$ denotes that cell n contains symbol σ, and the cell is "off", i.e., the head is not on it.
- $on[\sigma, n]$ for each $\sigma \in \Sigma$ and $1 \le n \le K$, with analogous intended meaning.

The head places are:

- $at[q, n]$ for each $q \in Q$ and $1 \le n \le K$. A token on $at[q, n]$ denotes that the head is in control state q and at cell n.
- $move[q, \sigma, n, d]$ for each $q \in Q$, $\sigma \in \Sigma$, $1 \le n \le K$ and every $d \in D$ such that $1 \le n + d \le K$. A token on $move[q, \sigma, n, d]$ denotes that head is in control state q, has left cell n after writing symbol σ on it, and is currently moving in the direction given by d.

Transitions of N_M. Intuitively, the implementation of M contains a set of *cell transitions* in which a cell observes the head and changes its state, and a set of *head transitions* in which the head observes a cell. Further, each of these sets contains transitions of two types. The set of cell transitions contains:

- Type **1a**: $(off[\sigma, n] \ , \ at[q, n]) \mapsto (on[\sigma, n] \ , \ at[q, n])$ for every state $q \in Q$, symbol $\sigma \in \Sigma$, and cell $1 \le n \le K$.
  The n-th cell, currently *off*, observes that the head is on it, and switches itself *on*.
- Type **1b**: $(on[\sigma, n] \ , \ move[q, \sigma', n, d]) \mapsto (off[\sigma', n] \ , \ move[q, \sigma', n, d])$ for every $q \in Q$, $\sigma \in \Sigma$, and $1 \le n \le K$ such that $1 \le n + d \le K$.
  The n-th cell, currently *on*, observes that the head has left after writing σ', and switches itself *off* (accepting the character the head intended to write).

The set of head transitions contains:

- Type **2a**: $(at[q, n], \ on[\sigma, n]) \mapsto (move[\delta_Q(q, \sigma), \delta_\Sigma(q, \sigma), n, \delta_D(q, \sigma)], \ on[\sigma, n])$ for every $q \in Q$, $\sigma \in \Sigma$, and $1 \le n \le K$ such that $1 \le n + \delta_D(q, \sigma) \le K$.
  The head, currently on cell n, observes that the cell is *on*, writes the new symbol on it, and leaves.
- Type **2b**: $(move[q, \sigma, n, d] \ , \ off[\sigma, n]) \mapsto (at[q, n+d] \ , \ off[\sigma, n])$ for every $q \in Q$, $\sigma \in \Sigma$, and $1 \le n \le K$ such that $1 \le n + d \le K$.
  The head, currently moving, observes that the old cell has turned *off*, and places itself on the new cell.

This concludes the definition of N_M. In Theorem 3 below we formalize the relation between the Turing machine M and its implementation N_M, using the following definition.

Definition 3. *Given a configuration c of M with control state q, tape content $\sigma_1 \sigma_2 \cdots \sigma_K$, and head on cell $n \le K$, we denote M_c the marking that puts a token in $off[\sigma_i, i]$ for each $1 \le i \le K$, a token in $at[q, n]$, and no tokens elsewhere.*

Now we state our simulation theorem and hardness result.

Theorem 3. *For every two configurations c, c' of M that write at most K cells: $c \to c'$ iff $M_c \xrightarrow{t_1 t_2 t_3 t_4} M_{c'}$ in N_M for some transitions t_1, t_2, t_3, t_4 of types $\mathbf{1a}$, $\mathbf{2a}$, $\mathbf{1b}$, $\mathbf{2b}$, respectively.*

Theorem 4. *The reachability, coverability and liveness problems for IO nets are* PSPACE-*hard.*

6 The Pruning Theorem

In this section, we present the fundamental property of immediate observation nets that entails most of the results in this paper: the Pruning Theorem.

The Pruning Theorem intuitively states that if M is coverable from a marking M'', then it is also coverable from a "small" marking $S'' \leq M''$, where "small" means $|S''| \leq |M| + |P|^3$. We state the theorem below, and then build up to its proof which is presented in Sect. 6.3.

Theorem 5 (Pruning Theorem). *Let $N = (P, T, F)$ be an IO net, let M be a marking of N, and let $M'' \xrightarrow{*} M'$ be a firing sequence of N such that $M' \geq M$. There exist markings S'' and S' such that*

$$
\begin{array}{ccc}
M'' & \xrightarrow{\;\;*\;\;} & M' \geq M \\
\geq & & \geq \\
S'' & \xrightarrow{\;\;*\;\;} & S' \geq M
\end{array}
$$

and $|S''| \leq |M| + |P|^3$.

It is easy to see that for $M' = M$ the Pruning Theorem holds, because since N is conservative, $|M''| = |M|$ and we can choose $S'' = M''$. It is also not difficult to find a non-IO net for which the theorem does not hold.

Example 4. Consider our IO net represented in Fig. 1a. There is a firing sequence $(30, 0, 1) \xrightarrow{*} (0, 0, 31)$ where $(0, 0, 31)$ covers $(0, 0, 2)$. By application of the Pruning Theorem, we obtain a firing sequence $(3, 0, 1) \xrightarrow{*} (0, 0, 4) \geq (0, 0, 2)$ where $|(3, 0, 1)| = 4$ is smaller than $|(0, 0, 2)| + 3^3 = 29$.

Example 5 (A non-IO net for which the theorem does not hold.). To see that the IO condition cannot be replaced with conservativeness of the network, consider the net with 4 places q_1, q_2, q_3, q_4 and a single transition $(q_1, q_2) \mapsto (q_3, q_4)$. There is a firing sequence $(1000, 1000, 0, 0) \xrightarrow{*} (0, 0, 1000, 1000) \geq (0, 0, 100, 0)$. But to cover $(0, 0, 100, 0)$ from a marking below $(1000, 1000, 0, 0)$ we need to fire the transition at least 100 times. This requires a marking with $200 > 100 + 4^3$ tokens.

6.1   Trajectories, Histories, Realizability

Since the transitions of IO nets do not create or destroy tokens, we can give tokens identities. Given a firing sequence, each token of the initial marking follows a *trajectory*, or sequence of *steps*, through the places of the net until it reaches the final marking of the sequence.

Definition 4. *A* trajectory *is a sequence* $\tau = p_1 \ldots p_n$ *of places. We denote* $\tau(i)$ *the i-th place of* τ. *The i-th step of* τ *is the pair* $\tau(i)\tau(i+1)$ *of adjacent places.*

   A history *is a multiset of trajectories of the same length. The length of a history is the common length of its trajectories. Given a history H of length n and index $1 \leq i \leq n$, the i-th marking of H, denoted M_H^i, is defined as follows: for every place p, $M_H^i(p)$ is the number of trajectories $\tau \in H$ such that $\tau(i) = p$. The markings M_H^1 and M_H^n are called the* initial *and* final *markings of H.*

   A history H of length $n \geq 1$ is realizable *in an IO net N if there exist transitions of N $t_1, \ldots, t_{n-1}$ and numbers $k_1, \ldots, k_{n-1} \geq 0$ such that $M_H^1 \xrightarrow{t_1^{k_1}} M_H^2 \cdots M_H^{n-1} \xrightarrow{t_{n-1}^{k_{n-1}}} M_H^n$. Notice that a history of length 1 is always realizable.*

Remark 1. Notice that there may be more than one realizable history corresponding to a firing sequence in an IO net, because the firing sequence does not keep track of which token goes where, while the history does.

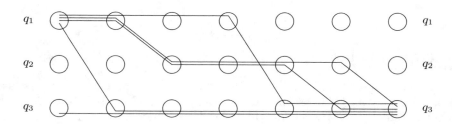

Fig. 2. Realizable history in our IO net with three states.

Example 6. Histories can be graphically represented. Consider Fig. 2 which illustrates a history H of length 7. It consists of five trajectories: one trajectory from q_3 to q_3 passing only through q_3, and four trajectories from q_1 to q_3 which follow different place sequences. H's first marking is $M_H^1 = (4, 0, 1)$ and H's seventh and last marking is $M_H^7 = (0, 0, 5)$. History H is realizable in the IO net N of Fig. 1a which has place set $\{q_1, q_2, q_3\}$ and transitions $t_1 = (q_1, q_1) \mapsto (q_2, q_1), t_2 = (q_2, q_2) \mapsto (q_3, q_2), t_3 = (q_1, q_3) \mapsto (q_3, q_3)$ and $t_4 = (q_2, q_3) \mapsto (q_3, q_3)$. Indeed $M_H^1 \xrightarrow{t_3 t_1^2 t_3 t_2 t_4} M_H^7$.

   We define a class of histories sufficient for describing all the firing sequences for IO nets.

Definition 5. *A step $\tau(i)\tau(i+1)$ of a trajectory τ is* horizontal *if $\tau(i) = \tau(i+1)$, and* non-horizontal *otherwise.*

A history H *of length n is* well-structured *if for every $1 \le i \le n-1$ one of the two following conditions hold:*

- *For every trajectory $\tau \in H$, the i-th step of τ is horizontal.*
- *For every two trajectories $\tau_1, \tau_2 \in H$, if the i-th steps of τ_1 and τ_2 are non-horizontal, then they are equal.*

We then have the following result, whose proof can be found in the full version of this paper.

Lemma 1. *Let N be an IO net. Then $M \xrightarrow{*} M'$ iff there exists a well-structured history realizable in N with M and M' as initial and final markings.*

We now proceed to give a syntactic characterization of the well-structured realizable histories.

Definition 6. *H is* compatible *with N if for every trajectory τ of H and for every non-horizontal step $\tau(i)\tau(i + 1)$ of τ, the net N contains a transition $(\tau(i), p_o) \mapsto (\tau(i+1), p_o)$ for some place p_o and H contains a trajectory τ' with $\tau'(i) = \tau'(i + 1) = p_o$.*

Lemma 2. *Let N be an IO net. A well-structured history is realizable in N iff it is compatible with N.*

Example 7. In the realizable history H of Fig. 2, all the trajectories are such that the third step is horizontal. For every step except the third, all the non-horizontal steps are equal, so H is well-structured. For N the IO net of Fig. 1a, H is indeed compatible with N.

6.2   Pruning Histories

We start by introducing bunches of trajectories.

Definition 7. *A* bunch *is a multiset of trajectories with the same length and the same initial and final place.*

Example 8. Figure 2's realizable history is constituted of a trajectory from q_3 to q_3 and a bunch B with initial place q_1 and final place q_3 made up of four different trajectories.

We show that every well-structured realizable history containing a bunch of size larger than $|P|$ can be "pruned", meaning that the bunch can be replaced by a smaller one, while keeping the history well-structured and realizable.

Lemma 3. *Let N be an IO net. Let H be a well-structured history realizable in N containing a bunch $B \subseteq H$ of size larger than $|P|$. There exists a nonempty bunch B' of size at most $|P|$ with the same initial and final places as B, such that the history $H' \overset{def}{=} H - B + B'$ (where $+, -$ denote multiset addition and subtraction) is also well-structured and realizable in N.*

Proof. Let P_B be a set of all places visited by at least one trajectory in the bunch B. For every $p \in P_B$ let $f(p)$ and $l(p)$ be the earliest and the latest moment in time when this place has been used by any of the trajectories (the first and the last occurrence can be in different trajectories).

Let $\tau_p, p \in P_B$ be a trajectory that first goes to p by the moment $f(p)$, then waits there until $l(p)$, then goes from p to the final place. To go to and from p it uses fragments of trajectories of B.

We will take $B' = \{\tau_p \mid p \in P_B\}$ and prove that replacing B with B' in H does not violate the requirements for being a well-structured history realizable in N. Note that we can copy the same fragment of a trajectory multiple times.

First let us check the well-structuring condition. Note that we build τ_p by taking fragments of existing trajectories and using them at the exact same moments in time, and by adding some horizontal fragments. Therefore, the set of non-horizontal steps in B' is a subset (if we ignore multiplicity) of the set of non-horizontal steps in B, and the replacement operation cannot increase the set of non-horizontal steps occurring in H.

Now let us check compatibility with N. Consider any non-horizontal step in H' in any trajectory at position $(i, i+1)$. By construction, the same step at the same position is also present in H. History H is realizable in N and thus by Lemma 2 it is compatible with N, so H contains an enabling horizontal step $p_o p_o$ in some trajectory at that position $(i, i+1)$. There are two cases: either that step $p_o p_o$ was provided by a bunch being pruned, or by a bunch not affected by pruning. In the first case, note that the place p_o of this horizontal step must be first observed no later than i, and last observed not earlier than $i + 1$. This implies $f(p_o) \leq i < i+1 \leq l(p_o)$. As H' contains a horizontal step $p_o p_o$ for all positions between $f(p_o)$ and $l(p_o)$, in particular it contains it at position $(i, i+1)$. In the second case the same horizontal step is present in H' as a part of the same trajectory.

So H' is well-structured and compatible with N, and thus by Lemma 2 realizable in N. □

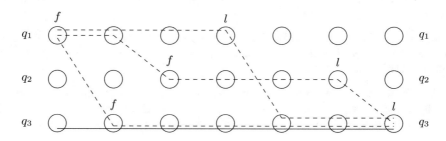

Fig. 3. History H of Fig. 2 after pruning.

Example 9. Consider the well-structured realizable history of Fig. 2, leading from $(4, 0, 1)$ to $(0, 0, 5)$, which covers marking $(0, 0, 2)$. Bunch B from q_1 to q_3 is of size four which is bigger than $|P| = 3$. The set P_B of places visited by trajectories of B is equal to P. Figure 3 is annotated with the first and last moments $f(p)$ and $l(p)$ for $p \in P_B$. Lemma 3 applied to H and B "prunes" bunch B into B' made up of trajectories $\tau_{q_1}, \tau_{q_2}, \tau_{q_3}$, drawn in dashed lines in Fig. 3. Notice that in this example, the non-horizontal 5-th step in H does not appear in the new well-structured and realizable history $H' = H - B + B'$. History H' is such that $M_{H'}^1 = (3, 0, 1) \xrightarrow{t_3 t_1 t_3 t_4} (0, 0, 4) = M_{H'}^7$ and $M_{H'}^7 \geq (0, 0, 2)$.

6.3   Proof of the Pruning Theorem

Using Lemma 3 we can now finally prove the Pruning Theorem:

Proof. (of Theorem 5). Let $M'' \xrightarrow{*} M' \geq M$. By Lemma 1, there is a well-structured realizable history H with M'' and M' as initial and final markings, respectively. Let $H_M \subset H$ be an arbitrary sub(multi)set of H with final marking M, and define $H' = H - H_M$. Further, for every $p, p' \in P$, let $H'_{p,p'}$ be the bunch of all trajectories of H' with p and p' as initial and final places, respectively. We have

$$H' = \sum_{p,p' \in P} H'_{p,p'}$$

So H' is the union of $|P|^2$ (possibly empty) bunches. Apply Lemma 3 to each bunch of H' with more than $|P|$ trajectories yields a new history

$$H'' = \sum_{p,p' \in p} H''_{p,p'}$$

such that $|H''_{p,p'}| \leq |P|$ for every $p, p' \in P$, and such that the history $H'' + H_M$ is well-structured and realizable.

Let S'' and S' be the initial and final markings of $H'' + H_M$. We show that S'' and S' satisfy the required properties:

- $S'' \xrightarrow{*} S'$, because $H'' + H_M$ is well-structured and realizable.
- $S' \geq M$, because $H_M \subseteq H'' + H_M$.
- $|S''| \leq |M| + |P|^3$ because $|H'' + H_M| = \sum_{p',p} |H''_{p,p'}| + |H_M| \leq |P|^2 \cdot |P| + |M| = |M| + |P|^3$.

This concludes the proof.   □

Remark 2. A slight modification of our construction allows one to prove Theorem 5 (but not Lemma 3) with $2|P|^2$ overhead instead of $|P|^3$. We provide more details in the full version [15]. However, since some results of Sect. 7 explicitly rely on Lemma 3, we prove Theorem 5 as a consequence of Lemma 3 for simplicity.

7   Counting Constraints and Counting Sets

In this section we first briefly recall *counting constraints* [14][2], a class of constraints that allow us to finitely represent (possibly infinite) sets of markings, called *counting sets*. We prove Theorem 6, a powerful result stating that counting sets of IO nets are closed under reachability, and giving a very tight relation between the sizes of the constraints representing a counting set, and the set of markings reachable from it. Theorem 6 strongly improves on Theorem 18 of [14].

Counting Constraints and Counting Sets. Recall Definition 1 which defines a *cube* of a net N as a set of markings given by a lower bound $L: P \to \mathbb{N}$ and an upper bound $U: P \to \mathbb{N} \cup \{\infty\}$, written (L, U), and such that $M \in (L, U)$ iff $L \leq M \leq U$. In the rest of the paper, the term cube will refer both to the set of markings and to the description by upper and lower bound (L, U). A *counting constraint* is a formal finite union of cubes, i.e. a formal finite union of upper and lower bound pairs of the form (L, U). The semantics of a counting constraint is called a *counting set* and it is the union of the cubes defining the counting constraint. The counting set for a counting constraint Γ is denoted $[\![\Gamma]\!]$. Notice that one counting set can be the semantics of different counting constraints. For example, consider a net with just one place p_1. Let $(L, U) = (1, 3)$, $(L', U') = (2, 4)$, $(L'', U'') = (1, 4)$. The counting constraints $(L, U) \cup (L', U')$ and (L'', U'') define the same counting set. It is easy to show (see also [14]) that counting sets are closed under Boolean operations.

Measures of Counting Constraints. Let $C = (L, U)$ be a cube, and let $\Gamma = \bigcup_{i=1}^{m} C_i$ be a counting constraint. We use the following notations:

$$\|C\|_l \stackrel{\text{def}}{=} \sum_{p \in P} L(p) \qquad \|C\|_u \stackrel{\text{def}}{=} \sum_{\substack{p \in P \\ U(p) < \infty}} U(p) \text{ (and 0 if } U(p) = \infty \text{ for all } p).$$

$$\|\Gamma\|_l \stackrel{\text{def}}{=} \max_{i \in [1,m]} \{\|C_i\|_l\} \quad \|\Gamma\|_u \stackrel{\text{def}}{=} \max_{i \in [1,m]} \{\|C_i\|_u\}$$

We call $\|C\|_l$ the *L-norm* and $\|C\|_u$ the *U-norm* of C. Similarly for Γ. We recall Proposition 5 of [14] for the norms of the union, intersection and complement.

Proposition 1. *Let Γ_1, Γ_2 be counting constraints.*

- *There exists a counting constraint Γ with $[\![\Gamma]\!] = [\![\Gamma_1]\!] \cup [\![\Gamma_2]\!]$ such that $\|\Gamma\|_u \leq \max\{\|\Gamma_1\|_u, \|\Gamma_2\|_u\}$ and $\|\Gamma\|_l \leq \max\{\|\Gamma_1\|_l, \|\Gamma_2\|_l\}$.*
- *There exists a counting constraint Γ with $[\![\Gamma]\!] = [\![\Gamma_1]\!] \cap [\![\Gamma_2]\!]$ such that $\|\Gamma\|_u \leq \|\Gamma_1\|_u + \|\Gamma_2\|_u$ and $\|\Gamma\|_l \leq \|\Gamma_1\|_l + \|\Gamma_2\|_l$.*
- *There exists a counting constraint Γ with $[\![\Gamma]\!] = \mathbb{N}^n \setminus [\![\Gamma_1]\!]$ such that $\|\Gamma\|_u \leq n\|\Gamma_1\|_l$ and $\|\Gamma\|_l \leq n\|\Gamma_1\|_u + n$.*

---

[2] Actually, our counting constraints correspond to the "counting constraints in normal form" of [14]. We shorten the name, because we never need counting constraints not in normal form.

Predecessors and Successors of Counting Sets. Fix an IO net $N = (P, T, F)$. The sets of predecessors and successors of a set $\mathcal{M}$ of markings of N are defined as follows: $pre^*(\mathcal{M}) \stackrel{\text{def}}{=} \{M' | \exists M \in \mathcal{M} . M' \stackrel{*}{\rightarrow} M\}$, and $post^*(\mathcal{M}) \stackrel{\text{def}}{=} \{M | \exists M' \in \mathcal{M} . M' \stackrel{*}{\rightarrow} M\}$.

Lemma 4. *Let (L, U) be a cube of an IO net N of place set P. For all $M' \in pre^*(L, U)$, there exists a cube (L', U') such that*

1. *$M' \in (L', U') \subseteq pre^*(L, U)$, and*
2. *$\|(L', U')\|_l \leq \|(L, U)\|_l + |P|^3$ and $\|(L', U')\|_u \leq \|(L, U)\|_u$.*

Proof. Let M' be a marking of $pre^*(L, U)$. There exists a marking $M \in (L, U)$ such that $M' \longrightarrow M$, and $M \geq L$. The construction from the Pruning Theorem applied to this firing sequence yields markings S', S such that

$$
\begin{array}{ccc}
M' & \stackrel{*}{\longrightarrow} M & \geq L \\
\geq & \geq & \\
S' & \stackrel{*}{\longrightarrow} S & \geq L
\end{array}
$$

and $|S'| \leq |L| + |P|^3$. Since M is in (L, U), we have $U \geq M \geq S \geq L$ and so marking S is in (L, U) and S' is in $pre^*(L, U)$.

We want to find L', U' satisfying the conditions of the Lemma, i.e. such that $M' \in (L', U')$ and $(L', U') \subseteq pre^*(L, U)$. We define L' as equal to marking S' over each place of P. The following part of the proof plays out in the setting of the Pruning Theorem section, in which the tokens are de-anonymized. Let H_M be a well-structured realizable history from M' to M. Let p be a place of P. We want to define $U'(p)$. Consider $\mathcal{B}_p^M$ the set of bunches in history H_M that have p as an initial place. For every bunch B, let f_B be the final place of the bunch. We define $U'(p)$ depending on the final places of bunches in $\mathcal{B}_p^M$.

Case 1. There exists a bunch B in $\mathcal{B}_p^M$ whose final place f_B is such that $U(f_B) = \infty$. In this case we define $U'(p)$ to be ∞.

Case 2. For all bunches B in $\mathcal{B}_p^M$, the final place f_B of B is such that $U(f_B) < \infty$. In this case we define $U'(p)$ to be $\sum_{B \in \mathcal{B}_p^M} size(B)$, where $size(B)$ is the number of trajectories with multiplicity in B, and 0 if $\mathcal{B}_p^M$ is empty.

Let us show that (L', U') has the properties we want. The number of tokens in marking M' at place $p \in P$ is the sum of the sizes of the bunches that start from p in history H_M. That is, $M'(p) = \sum_{B \in \mathcal{B}_p^M} size(B)$ which is exactly $U'(p)$ when $U'(p)$ is finite. Thus for all $p \in P$, $M'(p) \leq U'(p)$ and $M'(p) \geq S'(p) = L'(p)$, so M' is in (L', U').

The construction from the Pruning Theorem "prunes" history H_M from M' to M into a well-structured realizable history H_S from S' to S with the same set of non-empty bunches. We are going to show that $(L', U') \subseteq pre^*(L, U)$ by "boosting" the bunches of history H_S to create histories H_R which will start in any marking R' of (L', U') and end at some marking R in (L, U). For any constant $k \in \mathbb{N}$, a bunch B of history H_S is *boosted by k* into a bunch B' by

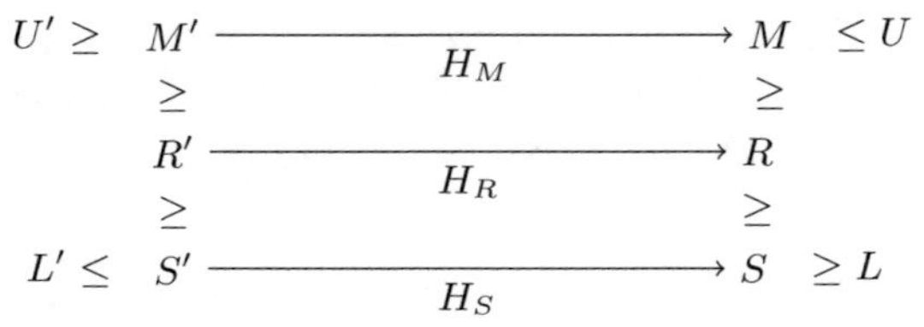

$$U' \geq \quad M' \xrightarrow{ H_M } M \quad \leq U$$
$$\geq \qquad\qquad\qquad\qquad \geq$$
$$R' \xrightarrow{ H_R } R$$
$$\geq \qquad\qquad\qquad\qquad \geq$$
$$L' \leq \quad S' \xrightarrow{ H_S } S \quad \geq L$$

Fig. 4. Boosting H_S into H_R.

selecting any trajectory τ in B and augmenting its multiplicity by k to create a new bunch B' of size $size(B) + k$.

Let R' be a marking in (L', U'). We construct a new history H_R starting in R', and we prove that its final place is in (L, U). What we aim to build is illustrated in Fig. 4. We initialize H_R as the bunches of history H_S. We call $\mathcal{B}_p^S$ the set of the bunches of H_S starting in p.

For p such that there is a bunch $B_S \in \mathcal{B}_p^S$ with infinite $U(f_{B_S})$, i.e. such that $\mathcal{B}_p^S$ is in *Case 1* defined above, we take this bunch B_S and boost it by $R'(p) - S'(p)$ into a new bunch B_R. Informally, we need not worry about exceeding the bound U on the final place of the trajectories of B_R, because this place is f_{B_S} and its upper bound is infinite. The number of trajectories starting in p in history H_R is now $R'(p)$.

Otherwise, p is such that $\mathcal{B}_p^S$ is in *Case 2*, so we know that $R'(p) \leq M'(p)$ because $U'(p)$ was defined to be $M'(p)$. Each bunch in $\mathcal{B}_p^S$ in history H_S has a corresponding bunch in history H_M because the pruning operation never erases a bunch completely, it only diminishes its size. We can boost all bunches in $\mathcal{B}_p^S$ to the size of the corresponding bunches in H_M and not exceed the finite bounds of U on the final places of these bunches. We arbitrarily select bunches in $\mathcal{B}_p^S$ which we boost so that the sum of the size of bunches in $\mathcal{B}_p^S$ is equal to $R'(p)$.

Now by construction, history H_R starts in marking R', and it ends in a marking R such that $S \leq R \leq U$, as every bunch is either boosted to a size no greater than it had in H_M, or leads to a place p with $U(p) = \infty$. Since $S \geq L$, this implies that $R \geq L$ and so $R \in (L, U)$ and $R' \in pre^*(L, U)$.

Finally, we show that the norms of (L', U') are bounded. For the L-norm, we simply add up the tokens in $S = L'$. Thus by the Pruning theorem

$$\|(L', U')\|_l \leq |L| + |P|^3 \leq \|(L, U)\|_l + |P|^3.$$

By definition of the U-norm, $\|(L', U')\|_u = \sum_{p \in P | U'(p) < \infty} U'(p)$. If $U'(p) < \infty$ then $\mathcal{B}_p^M$ of history H_M is in Case 2 and there is no bunch $B \in \mathcal{B}_p^M$ going from p to a final place f_B such that $U(f_B) = \infty$. So the set of bunches B starting in a place p such that $U'(p) < \infty$ is included in the set of bunches B' such that $U(f_{B'}) < \infty$, and thus

$$\sum_{p \in P | U'(p) < \infty} U'(p) = \sum_{p \in P | U'(p) < \infty} \left(\sum_{B \in \mathcal{B}_p^M} size(B) \right) \leq \sum_{B | U(f_B) < \infty} size(B).$$

Now $\sum_{B|U(f_B)<\infty} size(B)$ in history H_M is exactly $\sum_{p\in P|U(p)<\infty} M(p)$. Since $M \in (L, U)$, for all places we have $M(p) \le U(p)$ and so

$$\sum_{p\in P|U'(p)<\infty} U'(p) \le \sum_{p\in P|U(p)<\infty} M(p) \le \sum_{p\in P|U(p)<\infty} U(p).$$

So by definition of the norm, $\|(L', U')\|_u \le \|(L, U)\|_u$.   □

This result entails the main theorem of the section.

Theorem 6. *Let N be an IO net with a set P of places, and let S be a counting set. Then $pre^*(S)$ is a counting set and there exist counting constraints Γ and Γ' satisfying $[\![\Gamma]\!] = S$, $[\![\Gamma']\!] = pre^*(S)$ and we can bound the norm of Γ' by*

$$\|\Gamma'\|_u \le \|\Gamma\|_u \ and \ \|\Gamma'\|_l \le \|\Gamma\|_l + |P|^3$$

The same holds for $post^*$ by using the net with reversed transitions.

Proof (Sketch). Lemma 4 gives "small" cubes such that $pre^*(S)$ is the union of these cubes. Since there are only a finite number of such "small" cubes, this union is finite and $pre^*(S)$ is a counting set. The bounds on the norms of $pre^*(S)$ are derived from the bounds on the norms of these cubes.

Remark 3. Theorem 6 is a dramatic improvement on Theorem 18 of [14], which could only give a much higher bound for $\|\Gamma'\|_l$:

$$\|\Gamma'\|_l \le (\|\Gamma\|_l + \|\Gamma\|_u)^{2^{\mathcal{O}(|P|^2 \log |P|)}} \text{ instead of } \|\Gamma'\|_l \le \|\Gamma\|_l + |P|^3.$$

8   Cube Problems for IO Nets Are in PSPACE

We prove that the cube-reachability, cube-coverability, and cube-liveness problems for IO nets are in PSPACE.

Theorem 7. *The cube-reachability and cube-coverability problems for IO nets are in PSPACE.*

Proof. Let us first consider cube-reachability. Let N be an IO net with set of places P, and let S_0 and S be cubes. Some marking of S is reachable from some marking of S_0 iff $post^*(S_0) \cap S \ne \emptyset$. Let Γ_0 and Γ be two counting constraints for S_0 and S respectively. By Theorem 6 and Proposition 1, there exists a counting constraint Γ' such that $[\![\Gamma']\!] = post^*(S_0) \cap S$, and such that $\|\Gamma'\|_u \le \|\Gamma_0\|_u + \|\Gamma\|_u$ and $\|\Gamma'\|_l \le \|\Gamma_0\|_l + |P|^3 + \|\Gamma\|_l$. Therefore, $post^*(S_0) \cap S \ne \emptyset$ holds iff $post^*(S_0) \cap S$ contains a "small" marking M satisfying $|M| \le \|\Gamma_0\|_l + |P|^3 + \|\Gamma\|_l$. The PSPACE decision procedure takes the following steps: **(1)** Guess a "small" marking $M \in S$. **(2)** Check that M belongs to $post^*(S_0)$.

The algorithm for **(2)** is to guess a marking $M_0 \in S_0$ such that $|M_0| = |M|$, and then guess a firing sequence (step by step), leading from M_0 to M. This can be performed in polynomial space because each marking along the path is of size $|M|$, and we only need to store the current marking to check if it is equal to M.

Now for cube-coverability. Again let N be an IO net with set of places P, and let S_0 and S be cubes. In particular let $S = (L, U)$ for some upper and lower bounds L, U. Some marking of S is coverable from some marking of S_0 iff $post^*(S_0) \cap S_\infty \neq \emptyset$, where S_∞ is the cube defined by lower bound L and upper bound ∞ on all places. From here we proceed with the same **PSPACE** decision procedure as above.   □

Notice that cube-reachability and coverability can be extended to counting set-reachability and coverability simply by virtue of a counting set being a finite union of cubes.

Recall that a marking M_0 of an IO net N is *live* if for every marking M reachable from M_0 and for every transition t of N, some marking reachable from M enables t. The cube-liveness problem consists of deciding if, given a net N and a cube $\mathcal{M}$ of markings of N, every marking of $\mathcal{M}$ is live.

Theorem 8. *The cube-liveness problem for IO nets is in* **PSPACE.**

Proof. Let N be an IO net with set of places P, and $\mathcal{M}$ a cube. Let $t = (p_s, p_o) \mapsto (p_d, p_o)$ be a transition of N. The set $En(t)$ of markings that enable t contains the markings that put at least one token in p_s and at least one token in p_o (unless $p_s = p_o$ in which case there should be at least two tokens in that place). Clearly, $En(t)$ is a cube. Then $\overline{pre^*(En(t))}$ is the set of markings M from which one cannot execute transition t anymore by any firing sequence starting in M. So the set $\mathcal{L}$ of live markings of N is given by

$$\mathcal{L} = pre^* \overline{\left(\bigcup_{t \in T} \overline{pre^*(En(t))} \right)}$$

Deciding whether $\mathcal{M} \subseteq \mathcal{L}$ is equivalent to deciding whether $\mathcal{M} \cap \overline{\mathcal{L}} = \emptyset$ holds, or, equivalently, whether $\bigcup_{t \in T} \overline{pre^*(En(t))}$ is reachable from $\mathcal{M}$. By definition, the cube describing $En(t)$ has an L-norm equal to 2 and U-norm equal to 0. By Theorem 6 and Proposition 1, there exists a counting constraint Γ' such that $[\![\Gamma']\!] = \bigcup_{t \in T} \overline{pre^*(En(t))}$ and its norms are of size polynomial in $|P|$. So by Theorem 7 this reachability problem can be solved in **PSPACE** in the size of the input, i.e. net N and set $\mathcal{M}$.   □

9   Application: Correctness of IO Protocols Is PSPACE-complete

In [14], Esparza *et al.* studied the correctness problem for immediate observation protocols. The problem asks, given a protocol and a predicate, whether the protocol computes the predicate. In order to study the complexity of the problem we need to restrict ourselves to a class of predicates representable by finite means. Fortunately, Angluin *et al.* have shown in [6] that IO protocols compute exactly the predicates representable by counting constraints, i.e., the predicates $\varphi \colon \mathbb{N}^k \to \{0,1\}$ for which there is a counting constraint Γ such that $\varphi(\boldsymbol{v}) = 1$ iff $\boldsymbol{v}$ satisfies Γ. So we can formulate the problem as follows: given a counting constraint Γ and an IO protocols with a suitable set of input states, does it compute the predicate described by Γ? It is shown in [14] that the problem is PSACE-hard and in EXPSPACE, and closing this gap was left for future research.

In Petri net terms, the correctness problem for IO nets asks, given an IO net N and a counting constraint Γ, whether N computes Γ (formally defined in Sect. 3). We use the Pruning Theorem and the results of this paper to show that the correctness problem for IO nets, and so for IO protocols, is PSPACE-complete.

We present a proposition that characterizes the nets N that compute a given predicate $\varphi \colon \mathbb{N}^k \to \{0,1\}$. On top of the definitions of Sect. 3, we need some notations. For $b \in \{0,1\}$:

- $\mathcal{I}_b = \{M_{\boldsymbol{v}} \mid \varphi(\boldsymbol{v}) = b\}$, i.e., $\mathcal{I}_1$ ($\mathcal{I}_0$) denotes the initial markings of N for the input vectors satisfying (not satisfying) φ.
- $\mathcal{C}_b$ denotes the set of b-consensuses of N.
- $\mathcal{ST}_b \stackrel{\text{def}}{=} \overline{pre^*\left(\overline{\mathcal{C}_b}\right)}$ denotes the set of stable consensuses of N (the complement of the markings from which one can reach a non-b-consensus).

Proposition 2. *Let N be an IO net, let I be a set of input places, and let $\varphi \colon \mathbb{N}^k \to \{0,1\}$ be a predicate where $k = |I|$. Net N computes φ iff $post^*(\mathcal{I}_b) \subseteq pre^*(\mathcal{ST}_b)$ holds for $b \in \{0,1\}$.*

We can now show:

Theorem 9. *The correctness problem for IO nets is PSPACE-complete.*

Proof. Let N be an IO net with P its set of places, I a set of input places of size k, and $\varphi \colon \mathbb{N}^k \to \{0,1\}$ a predicate described by some counting constraint Γ_φ. Recall that $\mathcal{ST}_b$ is given by $\overline{pre^*(\overline{\mathcal{C}_b})}$ where $\mathcal{C}_b$, for $b \in \{0,1\}$, can be represented by the cube defined by upper bound equal to 0 on all places $p_i \in O^{-1}(1-b)$ and ∞ otherwise, and the lower bound equal to 0 everywhere. The condition for correctness of Proposition 2 can be rewritten as

$$post^*(\mathcal{I}_b) \cap \overline{pre^*(\mathcal{ST}_b)} = \emptyset. \tag{1}$$

Deciding (1) is equivalent to deciding whether $\overline{pre^*(\mathcal{ST}_b)}$ is reachable from $\mathcal{I}_b$. The cube describing $\mathcal{C}_b$ has upper and lower norm equal to 0. By Theorem 6 and

Proposition 1, there exists a counting constraint Γ_b such that $[\![\Gamma_b]\!] = \overline{pre^*(\mathcal{ST}_b)}$ and its norms are of size polynomial in $|P|$. Set $\mathcal{I}_b$ is a counting set described by either Γ_φ or its complement. So by Theorem 7 this reachability problem can be solved in **PSPACE**.

The proof for **PSPACE**-hardness reduces from the acceptance problem for deterministic Turing machines running in linear space, and is in the full version [15]. $\square$

10 Conclusion

Many modern distributed systems are parameterized, and they have to be modeled as an infinite set of Petri nets differing only in their initial markings. This leads to a new class of *parameterized* analysis problems, which typically are much harder to solve that standard ones. We have shown that, remarkably, this is not the case for immediate observation Petri nets, a subclass of **1**-conservative nets able to model immediate observation protocols and enzymatic chemical reaction networks. We have proved that the parameterized reachability, coverability, and liveness problems are **PSPACE**-complete, which is also the complexity of their non-parameterized versions. Current research on population protocols or networks considers quantitative properties like, in the case of population protocols, the computation of the expected time to stabilization. In future research we plan to study algorithms for these questions.

Acknowledgments. We thank three anonymous reviewers for numerous suggestions to improve readability, and Pierre Ganty for many helpful discussions.

References

1. Alistarh, D., Aspnes, J., Eisenstat, D., Gelashvili, R., Rivest, R.L.: Time-space trade-offs in population protocols. In: Proceedings of the Twenty-Eighth Annual ACM-SIAM Symposium on Discrete Algorithms (SODA), pp. 2560–2579 (2017)
2. Alistarh, D., Aspnes, J., Gelashvili, R.: Space-optimal majority in population protocols. In: Proceedings of the Twenty-Ninth Annual ACM-SIAM Symposium on Discrete Algorithms (SODA), pp. 2221–2239 (2018)
3. Alistarh, D., Gelashvili, R.: Recent algorithmic advances in population protocols. SIGACT News **49**(3), 63–73 (2018)
4. Angeli, D., De Leenheer, P., Sontag, E.D.: A Petri net approach to the study of persistence in chemical reaction networks. Math. Biosci. **210**(2), 598–618 (2007)
5. Angluin, D., Aspnes, J., Diamadi, Z., Fischer, M.J., Peralta, R.: Computation in networks of passively mobile finite-state sensors. In: Proceedings of the 23rd Annual ACM Symposium on Principles of Distributed Computing (PODC), pp. 290–299 (2004)
6. Angluin, D., Aspnes, J., Eisenstat, D., Ruppert, E.: The computational power of population protocols. Distrib. Comput. **20**(4), 279–304 (2007)
7. Baldan, P., Cocco, N., Marin, A., Simeoni, M.: Petri nets for modelling metabolic pathways: a survey. Nat. Comput. **9**(4), 955–989 (2010)
8. Cheng, A., Esparza, J., Palsberg, J.: Complexity results for 1-safe nets. Theor. Comput. Sci. **147**(1&2), 117–136 (1995)

9. Czerwinski, W., Lasota, S., Lazic, R., Leroux, J., Mazowiecki, F.: The reachability problem for Petri nets is not elementary (extended abstract). CoRR, abs/1809.07115 (2018)
10. Elsässer, R., Radzik, T.: Recent results in population protocols for exact majority and leader election. Bull. EATCS **126** (2018)
11. Esparza, J.: Decidability and complexity of Petri net problems — an introduction. In: Reisig, W., Rozenberg, G. (eds.) ACPN 1996. LNCS, vol. 1491, pp. 374–428. Springer, Heidelberg (1998). https://doi.org/10.1007/3-540-65306-6_20
12. Esparza, J., Ganty, P., Leroux, J., Majumdar, R.: Verification of population protocols. In: CONCUR. LIPIcs, vol. 42, pp. 470–482. Schloss Dagstuhl - Leibniz-Zentrum fuer Informatik (2015)
13. Esparza, J., Ganty, P., Leroux, J., Majumdar, R.: Verification of population protocols. Acta Informatica **54**(2), 191–215 (2017)
14. Esparza, J., Ganty, P., Majumdar, R., Weil-Kennedy, C.: Verification of immediate observation population protocols. In: CONCUR. LIPIcs, vol. 118, pp. 31:1–31:16. Schloss Dagstuhl - Leibniz-Zentrum fuer Informatik (2018)
15. Esparza, J., Raskin, M., Weil-Kennedy, C.: Parameterized analysis of immediate observation petri nets. CoRR, abs/1902.03025 (2019)
16. Marwan, W., Wagler, A., Weismantel, R.: Petri nets as a framework for the reconstruction and analysis of signal transduction pathways and regulatory networks. Nat. Comput. **10**(2), 639–654 (2011)
17. Mayr, E.W., Weihmann, J.: A framework for classical Petri net problems: conservative petri nets as an application. In: Ciardo, G., Kindler, E. (eds.) PETRI NETS 2014. LNCS, vol. 8489, pp. 314–333. Springer, Cham (2014). https://doi.org/10.1007/978-3-319-07734-5_17

The Combinatorics of Barrier Synchronization

Olivier Bodini[1], Matthieu Dien[2], Antoine Genitrini[3],
and Frédéric Peschanski[3(✉)]

[1] Université Paris-Nord – LIPN – CNRS UMR 7030, Villetaneuse, France
`Olivier.Bodini@lipn.univ-paris13.fr`
[2] Université de Caen – GREYC – CNRS UMR 6072, Caen, France
`Matthieu.Dien@unicaen.fr`
[3] Sorbonne University – LIP6 – CNRS UMR 7607, Paris, France
`{Antoine.Genitrini,Frederic.Peschanski}@lip6.fr`

Abstract. In this paper we study the notion of synchronization from the point of view of combinatorics. As a first step, we address the quantitative problem of counting the number of executions of simple processes interacting with synchronization barriers. We elaborate a systematic decomposition of processes that produces a symbolic integral formula to solve the problem. Based on this procedure, we develop a generic algorithm to generate process executions uniformly at random. For some interesting sub-classes of processes we propose very efficient counting and random sampling algorithms. All these algorithms have one important characteristic in common: they work on the control graph of processes and thus do not require the explicit construction of the state-space.

Keywords: Barrier synchronization · Combinatorics ·
Uniform random generation

1 Introduction

The objective of our (rather long-term) research project is to study the *combinatorics* of concurrent processes. Because the mathematical toolbox of combinatorics imposes strong constraints on what can be modeled, we study *process calculi* with a very restricted focus. For example in [5] the processes we study can only perform atomic actions and fork child processes, and in [4] we enrich this primitive language with *non-determinism*. In the present paper, our objective is to isolate another fundamental "feature" of concurrent processes: *synchronization*. For this, we introduce a simple process calculus whose only non-trivial concurrency feature is a principle of *barrier synchronization*. This is here understood intuitively as the single point of control where multiple processes have to "meet" before continuing. This is one of the important building blocks for concurrent and parallel systems [13].

This research was partially supported by the ANR MetACOnc project ANR-15-CE40-0014.

S. Donatelli and S. Haar (Eds.): PETRI NETS 2019, LNCS 11522, pp. 386–405, 2019.
https://doi.org/10.1007/978-3-030-21571-2_21

Combinatorics is about "counting things", and what we propose to count in our study is the number of executions of processes wrt. their "syntactic size". This is a symptom of the so-called "combinatorial explosion", a defining characteristic of concurrency. As a first step, we show that counting executions of concurrent processes is a difficult problem, even in the case of our calculus with limited expressivity. Thus, one important goal of our study is to investigate interesting sub-classes for which the problem becomes "less difficult". To that end, we elaborate in this paper a systematic decomposition of arbitrary processes, based on only four rules: (B)ottom, (I)ntermediate, (T)op and (S)plit. Each rule explains how to "remove" one node from the control graph of a process while taking into account its contribution in the number of possible executions. Indeed, one main feature of this BITS-decomposition is that it produces a symbolic integral formula to solve the counting problem. Based on this procedure, we develop a generic algorithm to generate process executions uniformly at random. Since the algorithm is working on the control graph of processes, it provides a way to statistically analyze processes without constructing their state-space explicitly. In the worst case, the algorithm cannot of course overcome the hardness of the problem it solves. However, depending on the rules allowed during the decomposition, and also on the strategy adopted, one can isolate interesting sub-classes wrt. the counting and random sampling problem. We identify well-known "structural" sub-classes such as *fork-join parallelism* [11] and *asynchronous processes with promises* [15]. For some of these sub-classes we develop dedicated and efficient counting and random sampling algorithms. A large sub-class that we find particularly interesting is what we call the "BIT-decomposable" processes, i.e. only allowing the three rules (B), (I) and (T) in the decomposition. The counting formula we obtain for such processes is of a linear size (in the number of atomic actions in the processes, or equivalently in the number of vertices in their control graph). We also discuss informally the typical shape of "BIT-free" processes.

The outline of the paper is as follows. In Sect. 2 we introduce a minimalist calculus of barrier synchronization. We show that the control graphs of processes expressed in this language are isomorphic to arbitrary partially ordered sets (Posets) of atomic actions. From this we deduce our rather "negative" starting point: counting executions in this simple language is intractable in the general case. In Sect. 3 we define the BITS-decomposition, and we use it in Sect. 4 to design a generic uniform random sampler. In Sect. 5 we discuss various sub-classes of processes related to the proposed decomposition, and for some of them we explain how the counting and random sampling problem can be solved efficiently. In Sect. 6 we propose an experimental study of the algorithm toolbox discussed in the paper.

Note that some technical complement and proof details are deferred to an external "companion" document. Moreover we provide the full source code developed in the realm of this work, as well as the benchmark scripts. All these complement informations are available online[1].

---

[1] cf. https://gitlab.com/ParComb/combinatorics-barrier-synchro.git.

Related Work

Our study intermixes viewpoints from concurrency theory, order-theory as well as combinatorics (especially enumerative combinatorics and random sampling). The *heaps combinatorics* (studied in e.g. [1]) provides a complementary interpretation of concurrent systems. One major difference is that this concerns "true concurrent" processes based on the trace monoid, while we rely on the alternative *interleaving semantics*. A related uniform random sampler for networks of automata is presented in [3]. Synchronization is interpreted on words using a notion of "shared letters". This is very different from the "structural" interpretation as joins in the control graph of processes. For the generation procedure [1] requires the construction of a "product automaton", whose size grows exponentially in the number of "parallel" automata. By comparison, all the algorithms we develop are based on the control graph, i.e. the space requirement remains polynomial (unlike, of course, the time complexity in some cases). Thus, we can interpret this as a space-time trade-of between the two approaches. A related approach is that of investigating the combinatorics of *lassos*, which is connected to the observation of state spaces through linear temporal properties. A uniform random sampler for lassos is proposed in [16]. The generation procedure takes place *within* the constructed state-space, whereas the techniques we develop do not require this explicit construction. However lassos represent infinite runs whereas for now we only handle finite (or finite prefixes) of executions.

A coupling from the past (CFTP) procedure for the uniform random generation of linear extensions is described, with relatively sparse details, in [14]. The approach we propose, based on the continuous embedding of Posets into the hypercube, is quite complementary. A similar idea is used in [2] for the enumeration of Young tableaux using what is there called the *density method*. The paper [12] advocates the uniform random generation of executions as an important building block for *statistical model-checking*. A similar discussion is proposed in [18] for *random testing*. The *leitmotiv* in both cases is that generating execution paths *without* any bias is difficult. Hence a uniform random sampler is very likely to produce interesting and complementary tests, if comparing to other test generation strategies.

Our work can also be seen as a continuation of the *algorithm and order* studies [17] orchestrated by Ivan Rival in late 1980's only with powerful new tools available in the modern combinatorics toolbox.

2   Barrier Synchronization Processes

The starting point of our study is the small process calculus described below.

Definition 1 (Syntax of barrier synchronization processes). *We consider countably infinite sets $\mathcal{A}$ of (abstract) atomic actions, and $\mathcal{B}$ of barrier names. The set $\mathcal{P}$ of processes is defined by the following grammar:*

$$
\begin{aligned}
P, Q ::= {} & 0 && \textit{(termination)} \\
& | \; \alpha.P && \textit{(atomic action and prefixing)} \\
& | \; \langle B \rangle P && \textit{(synchronization)} \\
& | \; \nu(B)P && \textit{(barrier and scope)} \\
& | \; P \parallel Q && \textit{(parallel)}
\end{aligned}
$$

The language has very few constructors and is purposely of limited expressivity. Processes in this language can only perform atomic actions, fork child processes and interact using a basic principle of *synchronization barrier*. A very basic process is the following one:

$$\nu(B) \; [\mathsf{a}_1.\langle B \rangle \; \mathsf{a}_2.0 \parallel \langle B \rangle \mathsf{b}_1.0 \parallel \mathsf{c}_1.\langle B \rangle \; 0]$$

This process can initially perform the actions a_1 and c_1 in an arbitrary order. We then reach the state in which all the processes agrees to synchronize on B:

$$\nu(B) \; [\langle B \rangle \; \mathsf{a}_2.0 \parallel \langle B \rangle \mathsf{b}_1.0 \parallel \langle B \rangle \; 0]$$

The possible next transitions are: $\xrightarrow{\mathsf{a}_2} \mathsf{b}_1.0 \xrightarrow{\mathsf{b}_1} 0$, alternatively $\xrightarrow{\mathsf{b}_1} \mathsf{a}_2.0 \xrightarrow{\mathsf{a}_2} 0$

In the resulting states, the barrier B has been "consumed".

The operational semantics below characterize processes transitions of the form $P \xrightarrow{\alpha} P'$ in which P can perform action α to reach its (direct) derivative P'.

Definition 2 (Operational semantics).

$$
\frac{}{\alpha.P \xrightarrow{\alpha} P} \; \text{(act)} \qquad
\frac{P \xrightarrow{\alpha} P'}{P \parallel Q \xrightarrow{\alpha} P' \parallel Q} \; \text{(lpar)} \qquad
\frac{Q \xrightarrow{\alpha} Q'}{P \parallel Q \xrightarrow{\alpha} P \parallel Q'} \; \text{(rpar)}
$$

$$
\frac{\mathsf{sync}_B(P)=Q \quad \mathsf{wait}_B(Q) \quad P \xrightarrow{\alpha} P'}{\nu(B)P \xrightarrow{\alpha} \nu(B)P'} \; \text{(lift)} \qquad
\frac{\mathsf{sync}_B(P)=Q \quad \neg\mathsf{wait}_B(Q) \quad Q \xrightarrow{\alpha} Q'}{\nu(B)P \xrightarrow{\alpha} Q'} \; \text{(sync)}
$$

$$
\textit{with:} \quad
\begin{bmatrix}
\mathsf{sync}_B(0)=0 \\
\mathsf{sync}_B(\alpha.P)=\alpha.P \\
\mathsf{sync}_B(P\|Q)=\mathsf{sync}_B(P)\|\mathsf{sync}_B(Q) \\
\mathsf{sync}_B(\nu(B)P)=\nu(B)P \\
\forall C \neq B, \; \mathsf{sync}_B(\nu(C)P)=\nu(C) \; \mathsf{sync}_B(P) \\
\mathsf{sync}_B(\langle B \rangle P)=P \\
\forall C \neq B, \; \mathsf{sync}_B(\langle C \rangle P)=\langle C \rangle P
\end{bmatrix}
\quad
\begin{bmatrix}
\mathsf{wait}_B(0)=\textsf{false} \\
\mathsf{wait}_B(\alpha.P)=\mathsf{wait}_B(P) \\
\mathsf{wait}_B(P\|Q)=\mathsf{wait}_B(P)\vee\mathsf{wait}_B(Q) \\
\mathsf{wait}_B(\nu(B)P)=\textsf{false} \\
\forall C \neq B, \; \mathsf{wait}_B(\nu(C)P)=\mathsf{wait}_B(P) \\
\mathsf{wait}_B(\langle B \rangle P)=\textsf{true} \\
\forall C \neq B, \; \mathsf{wait}_B(\langle C \rangle P)=\mathsf{wait}_B(P)
\end{bmatrix}
$$

The rule (sync) above explains the synchronization semantics for a given barrier B. The rule is non-trivial given the broadcast semantics of barrier synchronization. The definition is based on two auxiliary functions. First, the function $\mathsf{sync}_B(P)$ produces a derivative process Q in which all the possible synchronizations on barrier B in P have been effected. If Q has a sub-process that cannot yet synchronize on B, then the predicate $\mathsf{wait}_B(Q)$ is true and the synchronization on B is said incomplete. In this case the rule (sync) does not apply, however the transitions *within* P can still happen through (lift).

2.1   The Control Graph of a Process

We now define the notion of a (finite) execution of a process.

Definition 3 (execution). *An execution σ of P is a finite sequence $\langle \alpha_1, \ldots, \alpha_n \rangle$ such that there exist a set of processes $P'_{\alpha_1}, \ldots, P'_{\alpha_n}$ and a path $P \xrightarrow{\alpha_1} P'_{\alpha_1} \cdots \xrightarrow{\alpha_n} P'_{\alpha_n}$ with $P'_{\alpha_n} \not\rightarrow$ (no transition is possible from P'_{α_n}).*

We assume that the occurrences of the atomic actions in a process expression have all distinct labels, $\alpha_1, \ldots, \alpha_n$. This is allowed since the actions are uninterpreted in the semantics (cf. Definition 2). Thus, each action α in an execution σ can be associated to a unique *position*, which we denote by $\sigma(\alpha)$. For example if $\sigma = \langle \alpha_1, \ldots, \alpha_k, \ldots, \alpha_n \rangle$, then $\sigma(\alpha_k) = k$.

The behavior of a process can be abstracted by considering the *causal ordering relation* wrt. its atomic actions.

Definition 4 (cause, direct cause). *Let P be a process. An action α of P is said a cause of another action β, denoted by $\alpha < \beta$, iff for any execution σ of P we have $\sigma(\alpha) < \sigma(\beta)$. Moreover, α is a direct cause of β, denoted by $\alpha \prec \beta$ iff $\alpha < \beta$ and there is no γ such that $\alpha < \gamma < \beta$. The relation $<$ obtained from P is denoted by $\mathscr{PO}(P)$.*

Obviously $\mathscr{PO}(P)$ is a *partially ordered set* (poset) with *covering* $\prec$, capturing the *causal ordering* of the actions of P. The covering of a partial order is by construction an *intransitive directed acyclic graph* (DAG), hence the description of $\mathscr{PO}(P)$ itself is simply the transitive closure of the covering, yielding $O(n^2)$ edges over n elements. The worst case (maximizing the number of edges) is a complete bipartite graph with two sets of $2n$ vertices connected by n^2 edges (cf. Fig. 1).

$$\nu(B) \begin{bmatrix} \alpha_1.\langle B \rangle \parallel \alpha_2.\langle B \rangle \parallel \ldots \parallel \alpha_n.\langle B \rangle \\ \langle B \rangle.\beta_1 \parallel \langle B \rangle.\beta_2 \parallel \ldots \parallel \langle B \rangle.\beta_n \end{bmatrix}$$

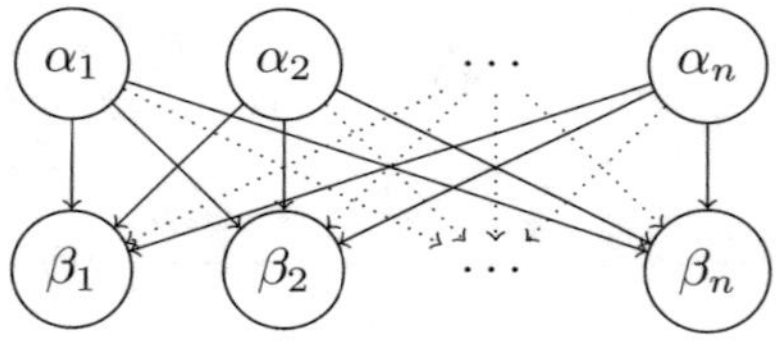

Fig. 1. A process of size $2n$ and its control graph with $2n$ nodes and n^2 edges.

For most practical concerns we will only consider the covering, i.e. the intransitive DAG obtained by the *transitive reduction* of the order. It is possible to direclty construct this control graph, according to the following definition.

Definition 5 (Construction of control graphs). *Let P be a process term. Its control graph is $ctg(P) = \langle V, E \rangle$, constructed inductively as follows:*

$$\begin{bmatrix} ctg(0) = \langle \emptyset, \emptyset \rangle & ctg(\nu(B)P) = \bigotimes_{\langle B \rangle} ctg(P) \\ ctg(\alpha.P) = \alpha \rightsquigarrow ctg(P) & ctg(P \parallel Q) = ctg(P) \cup ctg(Q) \\ ctg(\langle B \rangle P) = \langle B \rangle \rightsquigarrow ctg(P) & with \; \langle V_1, E_1 \rangle \cup \langle V_2, E_2 \rangle = \langle V_1 \cup V_2, E_1 \cup E_2 \rangle \end{bmatrix}$$

$$with \begin{cases} x \rightsquigarrow \langle V, E \rangle = \langle V \cup \{x\}, \{(x,y) \mid y \in srcs(E) \vee (E = \emptyset \wedge y \in V)\} \rangle \\ srcs(E) = \{y \mid (y,z) \in E \wedge \nexists x, \; (x,y) \in E\} \\ \bigotimes_{\langle B \rangle} \langle V, E \rangle = \langle V \setminus \{\langle B \rangle\}, E \setminus \begin{array}{l} \{(x,y) \mid x \neq y \wedge (x = \langle B \rangle \vee y = \langle B \rangle)\} \\ \cup \{(\alpha, \beta) \mid \{(\alpha, \langle B \rangle), (\langle B \rangle, \beta)\} \subseteq E\} \end{array} \rangle \end{cases}$$

Given a control graph Γ, the notation $x \leadsto \Gamma$ corresponds to prefixing the graph by a single atomic action. The set $\mathsf{srcs}(E)$ corresponds to the *sources* of the edges in E, i.e. the vertices without an incoming edge. And $\bigotimes_{\langle B \rangle} \Gamma$ removes an explicit barrier node and connect all the processes ending in B to the processes starting from it. In effect, this realizes the synchronization described by the barrier B. We illustrate the construction on a simple process below:

$$\mathsf{ctg}(\nu(B)\nu(C)[\langle B \rangle \langle C \rangle a.0 || \langle B \rangle \langle C \rangle b.0])$$
$$= \bigotimes_{\langle B \rangle} \bigotimes_{\langle C \rangle} (\mathsf{ctg}(\langle B \rangle \langle C \rangle a.0) \cup \mathsf{ctg}(\langle B \rangle \langle C \rangle b.0))$$
$$= \bigotimes_{\langle B \rangle} \bigotimes_{\langle C \rangle} \langle \{\{\langle B \rangle, \langle C \rangle, a\}, \{(\langle B \rangle, \langle C \rangle), (\langle C \rangle, a)\}\rangle\}$$
$$\cup \langle \{\{\langle B \rangle, \langle C \rangle, b\}, \{(\langle B \rangle, \langle C \rangle), (\langle C \rangle, b)\}\rangle\}$$
$$= \bigotimes_{\langle B \rangle} \bigotimes_{\langle C \rangle} \langle \{\{\langle B \rangle, \langle C \rangle, a, b\}, \{(\langle B \rangle, \langle C \rangle), (\langle C \rangle, a), (\langle C \rangle, b)\}\rangle\}$$
$$= \bigotimes_{\langle B \rangle} \langle \{\{\langle B \rangle, a, b\}, \{(\langle B \rangle, a), (\langle B \rangle, b)\}\rangle\}$$
$$= \langle \{a, b\}, \emptyset \rangle$$

The graph with only two unrelated vertices and no edge is the correct construction. Now, slightly changing the process we see how the construction fails for deadlocked processes.

$$\mathsf{ctg}(P) = \bigotimes_{\langle B \rangle} \bigotimes_{\langle C \rangle} (\mathsf{ctg}(\langle B \rangle \langle C \rangle a.0) \cup \mathsf{ctg}(\langle C \rangle \langle B \rangle b.0))$$
$$= \bigotimes_{\langle B \rangle} \bigotimes_{\langle C \rangle} \langle \{\{\langle B \rangle, \langle C \rangle, a\}, \{(\langle B \rangle, \langle C \rangle), (\langle C \rangle, a)\}\rangle\}$$
$$\cup \langle \{\{\langle C \rangle, \langle B \rangle, b\}, \{(\langle C \rangle, \langle B \rangle), (\langle B \rangle, b)\}\rangle\}$$
$$= \bigotimes_{\langle B \rangle} \bigotimes_{\langle C \rangle} \langle \{\{\langle B \rangle, \langle C \rangle, a, b\}, \{(\langle B \rangle, \langle C \rangle), (\langle C \rangle, a), (\langle C \rangle, \langle B \rangle), (\langle B \rangle, b)\}\rangle\}$$
$$= \bigotimes_{\langle B \rangle} \langle \{\{\langle B \rangle, a, b\}, \{(\langle B \rangle, \langle B \rangle), (\langle B \rangle, a), (\langle B \rangle, b)\}\rangle\}$$
$$= \langle \{a, b\}, \{(\langle B \rangle, \langle B \rangle), (\langle B \rangle, a), (\langle B \rangle, b)\}\rangle$$

In the final step, the barrier $\langle B \rangle$ cannot be removed because of the self-loop. So there are two witnesses of the fact that the construction failed: there is still a barrier name in the process, and there is a cycle in the resulting graph.

Theorem 1. *Let P be a process, then P has a deadlock iff $\mathsf{ctg}(P)$ has a cycle. Moreover, if P is deadlock-free (hence it is a DAG) then $(\alpha, \beta) \in \mathsf{ctg}(P)$ iff $\alpha \prec \beta$ (hence the DAG is intransitive).*

Proof (idea). The proof is not difficult but slightly technical. The idea is to extend the notion of execution to go "past" deadlocks, thus detecting cycles in the causal relation. The details are given in companion document. □

In Fig. 2 (left) we describe a system Sys written in the proposed language, together with the covering of $\mathscr{PO}(Sys)$, i.e. its control graph (right). We also indicate the number of its possible executions, a question we address next.

2.2 The Counting Problem

One may think that in such a simple setting, any behavioral property, such as the counting problem that interests us, could be analyzed efficiently e.g. by a simple induction on the syntax. However, the devil is well hidden inside the box because of the following fact.

Theorem 2. *Let U be a partially ordered set. Then there exists a barrier synchronization process P such that $\mathscr{PO}(P)$ is isomorphic to U.*

$Sys = \text{init}.\nu(G_1, G_2, J_1).$

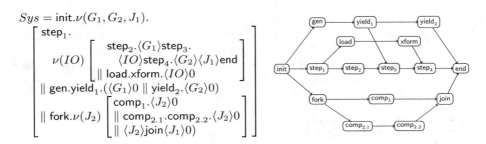

Fig. 2. An example process with barrier synchronizations (left) and its control graph (right). The process is of size 16 and it has exactly 1975974 possible executions.

Proof. (sketch). Consider G the (intransitive) covering DAG of a poset U. We suppose each vertex of G to be uniquely identified by a label ranging over $\alpha_1, \alpha_2, \ldots, \alpha_n$. The objective is to associate to each such vertex labeled α a process expression P_α. The construction is done *backwards*, starting from the *sinks* (vertices without outgoing edges) of G and bubbling-up until its *sources* (vertices without incoming edges).

There is a single rule to apply, considering a vertex labeled α whose children have already been processed, i.e. in a situation depicted as follows:

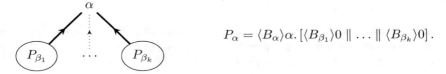
$$P_\alpha = \langle B_\alpha \rangle \alpha. \left[\langle B_{\beta_1} \rangle 0 \parallel \ldots \parallel \langle B_{\beta_k} \rangle 0 \right].$$

In the special case α is a sink we simply define $P_\alpha = \langle B_\alpha \rangle \alpha.0$. In this construction it is quite obvious that $\alpha \prec \beta_i$ for each of the β_i's, provided the barriers $B_\alpha, B_{\beta_1}, \ldots, B_{\beta_k}$ are defined somewhere in the outer scope.

At the end we have a set of processes $P_{\alpha_1}, \ldots, P_{\alpha_n}$ associated to the vertices of G and we finally define $P = \nu(B_{\alpha_1}) \ldots \nu(B_{\alpha_n}) [P_{\alpha_1} \parallel \ldots \parallel P_{\alpha_n}]$.

That $\mathscr{PO}(P)$ has the same covering as U is a simple consequence of the construction.   □

Corollary 1. *Let P be a non-deadlocked process. Then $\langle \alpha_1, \ldots, \alpha_n \rangle$ is an execution of P if it is a linear extension of $\mathscr{PO}(P)$. Consequently, the number of executions of P is equal to the number of linear extensions of $\mathscr{PO}(P)$.*

We now reach our "negative" result that is the starting point of the rest of the paper: there is no efficient algorithm to count the number of executions, even for such simplistic barrier processes.

Corollary 2. *Counting the number of executions of a (non-deadlocked) barrier synchronization process is $\sharp P$-complete[2].*

---

[2] A function f is in $\sharp P$ if there is a polynomial-time non-deterministic Turing machine M such that for any instance x, $f(x)$ is the number of executions of M that accept x as input. See https://en.wikipedia.org/wiki/%E2%99%AFP-complete.

This is a direct consequence of [8] since counting executions of processes boils down to counting linear extensions in (arbitrary) posets.

3   A Generic Decomposition Scheme and Its (Symbolic) Counting Algorithm

We describe in this section a generic (and symbolic) solution to the counting problem, based on a systematic decomposition of finite Posets (thus, by Theorem 1, of process expressions) through their covering DAG (i.e. control graphs).

3.1   Decomposition Scheme

In Fig. 3 we introduce the four decomposition rules that define the BITS-decomposition. The first three rules are somehow straightforward. The (B)-rule (resp. (T)-rule) allows to consume a node with no outgoing (resp. incoming) edge and one incoming (resp. outgoing) edge. In a way, these two rules consume the "pending" parts of the DAG. The (I)-rule allows to consume a node with exactly one incoming and outgoing edge. The final (S)-rule takes two incomparable nodes x, y and decomposes the DAG in two variants: the one for $x \prec y$ and the one for the converse $y \prec x$.

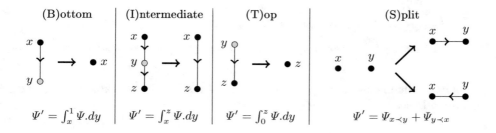

$$\Psi' = \int_x^1 \Psi.dy \qquad \Psi' = \int_x^z \Psi.dy \qquad \Psi' = \int_0^z \Psi.dy \qquad \Psi' = \Psi_{x \prec y} + \Psi_{y \prec x}$$

Fig. 3. The BITS-decomposition and the construction of the counting formula.

We now discuss the main interest of the decomposition: the incremental construction of an integral formula that solves the counting problem. The calculation is governed by the equations specified below the rules in Fig. 3, in which the current formula Ψ is updated according to the definition of Ψ' in the equations.

Theorem 3. *The numerical evaluation of the integral formula built by the BITS-decomposition yields the number of linear extensions of the corresponding Poset. Moreover, the applications of the BITS-rules are confluent, in the sense that all the sequences of (valid) rules reduce the DAG to an empty graph[3].*

---

[3] At the end of the decomposition, the DAG is in fact reduced to a single node, which is removed by an integration between 0 and 1.

The precise justification of the integral computation and the proof for the theorem above are postponed to Sect. 3.2 below. We first consider an example.

Example 1. Illustrating the BITS-decomposition scheme.

$$\Psi = 1 \qquad \Psi' = \int_0^{x_2} \Psi \mathrm{d}x_1 \qquad \Psi'' = \frac{\Psi'_{x_3 \prec x_4}}{+ \Psi'_{x_4 \prec x_3}} \qquad \Psi''' = \int_{x_4}^{x_8} \Psi''_{x_4 \prec x_3} \mathrm{d}x_7$$

The DAG to decompose (on the left) is of size 8 with nodes $x_1, \ldots, x_8$. The decomposition is non-deterministic, multiple rules apply, e.g. we could "consume" the node x_7 with the (I) rule. Also, the (S)plit rule is always enabled. In the example, we decide to first remove the node x_1 by an application of the (T) rule. We then show an application of the (S)plit rule for the incomparable nodes x_3 and x_4. The decomposition should then be performed on two distinct DAGs: one for $x_3 \prec x_4$ and the other one for $x_4 \prec x_3$. We illustrate the second choice, and we further eliminate the nodes x_7 then x_5 using the (I) rule, etc. Ultimately all the DAGs are decomposed and we obtain the following integral computation:

$$\Psi = \int_{x_2=0}^{1} \int_{x_4=x_2}^{1} \int_{x_3=x_4}^{1} \int_{x_6=x_3}^{1} \int_{x_8=x_6}^{1} \int_{x_5=x_3}^{x_8} \int_{x_7=x_4}^{x_8}$$
$$\left(\mathbb{1}_{|x_4 \prec x_3} \cdot \int_{x_1=0}^{x_2} 1 \cdot \mathrm{d}x_1 + \mathbb{1}_{|x_3 \prec x_4} \cdot \int_{x_1=0}^{x_2} 1 \cdot \mathrm{d}x_1 \right) \mathrm{d}x_7 \mathrm{d}x_5 \mathrm{d}x_8 \mathrm{d}x_6 \mathrm{d}x_3 \mathrm{d}x_4 \mathrm{d}x_2 = \frac{8+6}{8!}.$$

The result means that there are exactly 14 distinct linear extensions in the example Poset.

3.2   Embedding in the Hypercube: The Order Polytope

The justification of our decomposition scheme is based on the continuous embedding of posets into the hypercube, as investigated in [19].

Definition 6 (order polytope). *Let $P = (E, \prec)$ be a poset of size n. Let C be the unit hypercube defined by $C = \{(x_1, \ldots, x_n) \in \mathbb{R}^n \mid \forall i, \ 0 \le x_i \le 1\}$. For each constraint $x_i \prec x_j \in P$ we define the convex subset $S_{i,j} = \{(x_1, \ldots, x_n) \in \mathbb{R}^n \mid x_i \le x_j\}$, i.e. one of the half spaces obtained by cutting $\mathbb{R}^n$ with the hyperplane $\{(x_1, \ldots, x_n) \in \mathbb{R}^n \mid x_i - x_j = 0\}$. Thus, the order polytope C_P of P is:*

$$C_p = \bigcap_{x_i \prec x_j \in P} S_{i,j} \cap C$$

Each linear extension, seen as total orders, can similarly be embedded in the unit hypercube. Then, the order polytopes of the linear extensions of a poset P form a partition of the Poset embedding C_p as illustrated in Fig. 4.

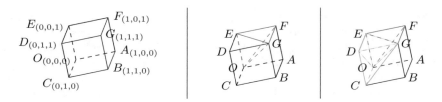

Fig. 4. From left to right: the unit hypercube, the embedding of the total order $1 \prec 2 \prec 3$ and the embedding of the poset $P = (\{1, 2, 3\}, \{1 \prec 2\})$ divided in its three linear extensions.

The number of linear extensions of a poset P, written $|\mathscr{LE}(P)|$, is then characterized as a volume in the embedding.

Theorem 4. *([19, Corollary 4.2]) Let P be a Poset of size n then its number of linear extensions $|\mathscr{LE}(P)| = n! \cdot Vol(C_P)$ where $Vol(C_P)$ is the volume, defined by the Lebesgue measure, of the order polytope C_P.*

The integral formula introduced in the BITS-decomposition corresponds to the computation of $Vol(C_p)$, hence we may now give the key-ideas of Theorem 3.

Proof (Theorem 3, sketch). We begin with the (S)-rule. Applied on two incomparable elements x and y, the rule partitions the polytope in two regions: one for $x \prec y$ and the other for $y \prec x$. Obviously, the respective volume of the two disjoint regions must be added.

We focus now on the (I)-rule. In the context of Lebesgue integration, the classic Fubini's theorem allows to compute the volume V of a polytope P as an iteration on integrals along each dimension, and this in all possible orders, which gives the confluence property. Thus,

$$V = \int_{[0,1]^n} \mathbb{1}_P(\mathbf{x})\mathrm{d}\mathbf{x} = \int_{[0,1]} \cdots \int_{[0,1]} \mathbb{1}_P((x, y, z, \dots))\mathrm{d}x\mathrm{d}y\mathrm{d}z \dots,$$

$\mathbb{1}_P$ being the indicator function of P such that $\mathbb{1}_P((x, y, z, \dots)) = \prod_{\alpha \text{ actions}} \mathbb{1}_{P_\alpha}(\alpha)$, with P_α the projection of P on the dimension associated to α. By convexity of P, the function $\mathbb{1}_{P_y}$ is the indicator function of a segment $[x, z]$. So the following identity holds: $\int_P \mathbb{1}_{P_y}(y)\mathrm{d}y = \int_x^z \mathrm{d}y$. Finally, the two other rules (T) and (B) are just special cases (taking $x = 0$, alternatively $z = 1$). □

Corollary 3. *(Stanley [19]) The order polytope of a linear extension is a simplex and the simplices of the linear extensions are isometric, thus of the same volume.*

4 Uniform Random Generation of Process Executions

In this section we describe a generic algorithm for the uniform random generation of executions of barrier synchronization processes. The algorithm is based on the

BITS-decomposition and its embedding in the unit hypercube. It has two essential properties. First, it is directly working on the control graphs (equivalently on the corresponding poset), and thus does not require the explicit construction of the state-space of processes. Second, it generates possible executions of processes at random according to the uniform distribution. This is a guarantee that the sampling is not biased and reflects the actual behavior of the processes[4].

Algorithm 1. Uniform sampling of a simplex of the order polytope

---

    function SAMPLEPOINT($\mathcal{I} = \int_a^b f(y_i) \, dy_i$)
      $C \leftarrow$ eval($\mathcal{I}$) ; $U \leftarrow$ UNIFORM(a, b)
      $Y_i \leftarrow$ the solution t of $\int_a^t \frac{1}{C} f(y_i) \, dy_i = U$
      if f is not a symbolic constant **then**
        SAMPLEPOINT($f\{y_i \leftarrow Y_i\}$)
      else return the Y_i's

The starting point of Algorithm 1 (cf. previous page) is a Poset over a set of points $\{x_1, \ldots, x_n\}$ (or equivalently its covering DAG). The decomposition scheme of Sect. 3 produces an integral formula $\mathcal{I}$ of the form $\int_0^1 F(y_n, \ldots, y_1) \, dy_n \cdots dy_1$. with F a symbolic integral formula over the points $x_1, \ldots, x_n$. The y variables represent a permutation of the poset points giving the order followed along the decomposition. Thus, the variable y_i corresponds to the i-th removed point during the decomposition. We remind the reader that the evaluation of the formula $\mathcal{I}$ gives the number of linear extensions of the partial order. Now, starting with the complete formula, the variables y_1, y_2, ... will be eliminated, in turn, in an "outside-in" way. Algorithm 1 takes place at the i-th step of the process. At this step, the considered formula is of the following form:

$$\int_a^b \underbrace{\left(\int \cdots \int 1 \, dy_n \cdots dy_{i+1} \right)}_{f(y_i)} dy_i.$$

Note that in the subformula $f(y_i)$ the variable y_i may only occur (possibly multiple times) as an integral bound.

    In the algorithm, the variable C gets the result of the numerical computation of the integral $\mathcal{I}$ at the given step. Next we draw (with UNIFORM) a real number U uniformly at random between the integration bounds a and b. Based on these two intermediate values, we perform a numerical solving of variable t in the integral formula corresponding to the *slice* of the polytope along the hyperplan $y_i = U$. The result, a real number between a and b, is stored in variable Y_i. The justification of this step is further discussed in the proof sketch of Theorem 5 below.

---

[4] The Python/Sage implementation of the random sampler is available at the following location: https://gitlab.com/ParComb/combinatorics-barrier-synchro/blob/master/code/RandLinExtSage.py.

If there remains integrals in $\mathcal{I}$, the algorithm is applied recursively by substituting the variable y_i in the integral bounds of $\mathcal{I}$ by the numerical value Y_i. If no integral remains, all the computed values Y_i's are returned. As illustrated in Example 2 below, this allows to select a specific linear extension in the initial partial ordering. The justification of the algorithm is given by the following theorem.

Theorem 5. *Algorithm 1 uniformly samples a point of the order polytope with a $\mathcal{O}(n)$ complexity in the number of integrations.*

Proof (sketch). The problem is reduced to the uniform random sampling of a point p in the order polytope. This is a classical problem about marginal densities that can be solved by slicing the polytope and evaluating incrementally the n continuous random variables associated to the coordinates of p. More precisely, during the calculation of the volume of the polytope P, the last integration (of a monovariate polynomial $p(y)$) done from 0 to 1 corresponds to integrate the slices of P according the last variable y. So, the polynomial $p(y)/\int_0^1 p(y)dy$ is nothing but the density function of the random variable Y corresponding to the value of y. Thus, we can generate Y according to this density and fix it. When this is done, we can inductively continue with the previous integrations to draw all the random variables associated to the coordinates of p. The linear complexity of Algorithm 1 follows from the fact that each partial integration deletes exactly one variable (which corresponds to one node). Of course at each step a possibly costly computation of the counting formula is required. $\qquad\square$

We now illustrate the sampling process based on Example 1 (page 9).

Example 2. First we assume that the whole integral formula has already been computed. To simplify the presentation we only consider (S)plit-free DAGs *i.e.* decomposable without the (S) rule. Note that it would be easy to deal with the (S)plit rule: it is sufficient to uniformly choose one of the DAG processed by the (S)-rule w.r.t. their number of linear extensions.

Thus we will run the example on the DAG of Example 1 where the DAG corresponding to "$x_4 \prec x_3$" as been randomly chosen (with probability $\frac{8}{14}$) *i.e.* the following formula holds:

$$\int_0^1 \left(\int_{x_2}^1 \int_{x_4}^1 \int_{x_3}^1 \int_{x_6}^1 \int_{x_4}^{x_8} \int_{x_3}^{x_8} \int_0^{x_2} dx_1 dx_5 dx_7 dx_8 dx_6 dx_3 dx_4 \right) dx_2 = \frac{8}{8!}.$$

In the equation above, the sub-formula between parentheses would be denoted by $f(x_2)$ in the explanation of the algorithm. Now, let us apply the Algorithm 1 to that formula in order to sample a point of the order polytope. In the first step the normalizing constant C is equal to $\frac{8!}{8}$, we draw U uniformly in $[0,1]$ and so we compute a solution of $\frac{8!}{8} \int_0^t \ldots dx_2 = U$. That solution corresponds to the second coordinate of a the point we are sampling. And so on, we obtain values for each of the coordinates:

$$\begin{cases} X_1 = 0.064\ldots, & X_2 = 0.081\ldots, & X_3 = 0.541\ldots, & X_4 = 0.323\ldots, \\ X_5 = 0.770\ldots, & X_6 = 0.625\ldots, & X_7 = 0.582\ldots, & X_8 = 0.892\ldots \end{cases}$$

These points belong to a simplex of the order polytope. To find the corresponding linear extension we compute the rank of that vector *i.e.* the order induced by the values of the coordinates correspond to a linear extension of the original DAG:

$$(x_1, x_2, x_4, x_3, x_7, x_6, x_5, x_8).$$

This is ultimately the linear extension returned by the algorithm.

5   Classes of Processes that are BIT-Decomposable (or Not)

Thanks to the BITS decomposition scheme, we can generate a counting formula for any (deadlock-free) process expressed in the barrier synchronization calculus, and derive from it a dedicated uniform random sampler. However the (S)plit rule generates two summands, thus if we cannot find common calculations between the summands the resulting formula can grow exponentially in the size of the concerned process. If we avoid splits in the decomposition, then the counting formula remains of linear size. This is, we think, a good indicator that the sub-class of so-called "BIT-decomposable" processes is worth investigating for its own sake. In this Section, we first give some illustrations of the expressivity of this subclass, and we then study the question of what it is to be *not* BIT-decomposable. By lack of space, the discussion in this Section remains rather informal with very rough proof sketches, and more formal developments are left for a future work. Also, the first two subsections are extended results based on previously published papers (respectively [6] and [7]).

5.1   From Tree Posets to Fork-Join Parallelism

If the control-graph of a process is decomposed with only the B(ottom) rule (or equivalently the T(op) rule), then it is rather easy to show that its shape is that of a *tree*. These are processes that cannot do much beyond forking sub-processes. For example, based on our language of barrier synchronization it is very easy to encode e.g. the (rooted) binary trees:

$$T ::= 0 \mid \alpha.(T \parallel T) \quad \text{or e.g.} \quad T ::= 0 \mid \nu B \ (\alpha.\langle B\rangle 0 \parallel \langle B\rangle T \parallel \langle B\rangle T)$$

The good news is that the combinatorics on trees is well-studied. In the paper [4] we provide a thorough study of such processes, and in particular we describe very efficient counting and uniform random generation algorithms. Of course, this is not a very interesting sub-class in terms of concurrency.

Thankfully, many results on trees generalize rather straightforwardly to *fork-join parallelism*, a sub-class we characterize inductively in Table 1. Informally, this proof system imposes that processes use their synchronization barriers according to a *stack discipline*. When synchronizing, only the last created barrier is available, which exactly corresponds to the traditional notion of a *join* in concurrency. Combinatorially, there is a correspondence between these processes

Table 1. A proof system for fork-join processes.

$$\frac{}{\sigma \vdash_{FJ} 0} \quad \frac{\sigma \vdash_{FJ} P}{\sigma \vdash_{FJ} \alpha.P} \quad \frac{\sigma \vdash_{FJ} P \quad \sigma \vdash_{FJ} Q}{\sigma \vdash_{FJ} P \parallel Q} \quad \frac{B{::}\sigma \vdash_{FJ} P}{\sigma \vdash_{FJ} \nu(B)\, P} \quad \frac{\sigma \vdash_{FJ} P}{B{::}\sigma \vdash_{FJ} \langle B \rangle.P}$$

and the class of *series-parallel Posets*. In the decomposition both the (B) and the (I) rule are needed, but following a tree-structured strategy. Most (if not all) the interesting questions about such partial orders can be answered in (low) polynomial time.

Theorem 6 (cf. [6]). *For a fork join process of size n the counting problem is of time complexity $O(n)$ and we developed a bit-optimal uniform random sampler with time complexity $O(n\sqrt{n})$ on average.*

5.2   Asynchronism with Promises

We now discuss another interesting sub-class of processes that can also be characterized inductively on the syntax of our process calculus, but this time using the three BIT-decomposition rules (in a controlled manner). The strict stack discipline of fork-join processes imposes a form of *synchronous* behavior: all the forked processes must terminate before a join may be performed. To support a limited form of *asynchronism*, a basic principle is to introduce *promise* processes.

Table 2. A proof system for promises.

$$\frac{}{\emptyset \vdash_{ctrl} 0} \quad \frac{\pi \vdash_{ctrl} P}{\pi \vdash_{ctrl} \alpha.P} \quad \frac{\pi \vdash_{ctrl} P}{\pi \cup \{B\} \vdash_{ctrl} \langle B \rangle.P} \quad \frac{B \notin \pi \quad \pi \cup \{B\} \vdash_{ctrl} P \quad Q \uparrow_B}{\pi \vdash_{ctrl} \nu(B)\,(P \parallel Q)}$$

$$\text{with } Q \uparrow_B \text{ iff } Q \equiv \alpha.R \text{ and } R \uparrow_B \text{ or } Q \equiv \langle B \rangle.0$$

In Table 2 we define a simple inductive process structure composed as follows. A *main control thread* can perform atomic actions (at any time), and also fork a sub-process of the form $\nu(B)\,(P \parallel Q)$ but with a strong restriction:

- a single barrier B is created for the sub-processes to interact.
- the left sub-process P must be the continuation of the main control thread,
- the right sub-process Q must be a promise, which can only perform a sequence of atomic actions and ultimately synchronize with the control thread.

We are currently investigating this class as a whole, but we already obtained interesting results for the *arch-processes* in [7]. An arch-process follows the constraint of Table 2 but adds further restrictions. The main control thread can still spawn an arbitrary number of promises, however there must be two separate phases for the synchronization. After the first promise synchronizes, the main control thread cannot spawn any new promise. In [7] a supplementary constraint is added (for the sake of algorithmic efficiency): each promise must perform exactly one atomic action, and the control thread can only perform actions when all the promises are running. In this paper, we remove this rather artificial constraint considering a larger, and more useful process sub-class.

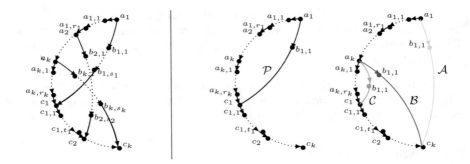

Fig. 5. The structure of an arch-process (left) and the inclusion-exclusion counting principle (right).

In Fig. 5 (left) is represented the general structure of a generalized arch-process. The a_i's actions are the promise forks, and the synchronization points are the c_j's. The constraint is thus that all the a_i's occur before the c_j's.

Theorem 7. *The number of executions of an arch-process can be calculated in $O(n^2)$ arithmetic operations, using a dynamic programming algorithm based on memoization.*

Proof (idea). A complete proof is provided in [7] for "simple" arch-processes, and the generalization is detailed in the companion document. We only describe the *inclusion-exclusion* principle on which our counting algorithm is based. Figure 5 (right) describes this principles (we omit the representation of the other promises to obtain a clear picture of our approach). Our objective is to count the number of execution contributed by a single promise with atomic action $b_{1,1}$. If we denote by $\ell_{\mathcal{P}}$ this contribution, we reformulate it as a combination $\ell_{\mathcal{P}} = \ell_{\mathcal{A}} - \ell_{\mathcal{B}} + \ell_{\mathcal{C}}$ as depicted on the rightmost part of Fig. 5. First, we take the "virtual" promise $\mathcal{A}$ going from the starting point a_1 of $\ell_{\mathcal{P}}$ until the end point c_k of the main thread. Of course there are two many possibilities if we only keep $\mathcal{A}$. An over-approximation of what it is to remove is the promise $\mathcal{B}$ going from the start of the last promise (at point a_k) until the end. But this time we removed too many possibilities, which corresponds to promise $\mathcal{C}$. The latter is thus reinserted in the count. Each of these three "virtual" promises have a simpler counting procedure. To guarantee the quadratic worst-time complexity (in the number of arithmetic operations), we have to memoize the intermediate results. We refer to the companion document for further details.   □

From this counting procedure we developed a uniform random sampler following the principles of the *recursive method*, as described in [10].

Theorem 8. *Let $\mathcal{P}$ be a promise-process of size n with $k \geq n$ promises. A random sampler of $O(n^4)$ time-complexity (in the number of arithmetic operations) builds uniform executions.*

The algorithm and the complete proof are detailed in the companion document. One notable aspect is that in order to get rid of the forbidden case of executions associated to the "virtual" promise $\mathcal{B}$ we cannot only do rejection (because the induced complexity would be exponential). In the generalization of arch-processes, we proceed by case analysis: for each possibility for the insertion of $b_{1,1}$ in the main control thread we compute the relative probability for the associated process $\mathcal{P}$. This explains the increase of complexity (from $O(n^2)$ to $O(n^4)$) if compared to [7].

5.3 BIT-Free Processes

The class of BIT-decomposable processes is rather large, and we in fact only uncovered two interesting sub-classes that can be easily captured inductively on the process syntax. The relatively non-trivial process *Sys* of Fig. 2 is also interestingly BIT-decomposable. We now adopt the complementary view of trying to understand the combinatorial structure of a so called "BIT-free" process, which is *not* decomposable using only the (B), (I) and (T) rules.

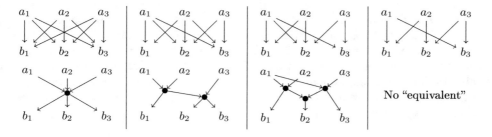

Fig. 6. Typical BIT-free substructures, and their BIT "equivalent" (when possible).

The BIT-free condition implies the occurrence of structures similar to the ones depicted on Fig. 6. These structures are composed of a set of "bottom" processes (the b_i's) waiting for "top" processes (the a_j') according to some synchronization pattern. We represent the whole possibilities of size 3 (up-to order-isomorphism) in the upper-part of the figure. The upper-left process is a complete (directed) bipartite graph, which can in fact be "translated" to a BIT-decomposable process as seen on the lower-part of the figure. This requires the introduction of a single "synchronization point" between the two process groups. This transformation preserves the number of executions and is Poset-wise equivalent. At each step "to the right" of Fig. 6, we remove a directed edge. In the second and third processes (in the middle), we also have an equivalent with respectively two and three synchronization points. In these cases, the number of linear extensions is not preserved but the "nature" of the order is respected: the interleavings of the initial atomic actions are the same. The only non-transformable structure, let's say the one "truly" BIT-free is the rightmost process. Even if we introduce synchronization points (we need at least three of them), the structure would not become BIT-decomposable. In terms of order theory such a structure is called a *Crown* poset. In [9] it is shown that the counting problem is already ♯-P complete for partial orders of height 2, hence directed bipartite digraphs similar to the structures of Fig. 6. One might wonder if this is still the case when these structures cannot occur, especially in the case of BIT-decomposable processes. This is for us a very interesting (and open) problem.

6　Experimental Study

In this section, we put into use the various algorithms for counting and generating process executions uniformly at random. Table 3 summarizes these algorithms and the associated worst-case time complexity bounds (when known). We implemented all the algorithms in Python 3, and we did not optimize for efficiency, hence the numbers we obtain only give a rough idea of their performances. For the sake of reproducibility, the whole experimental setting is available in the companion repository, with explanations about the required dependencies and usage. The computer we used to perform the benchmark is a standard laptop PC with an I7-8550U CPU, 8Gb RAM running Manjaro Linux. As an initial experiment, the example of Fig. 2 is BIT-decomposable, so we can apply the BIT and CFTP algorithms. The counting (of its 1975974 possible executions) takes about 0.3 s and it takes about 9 ms to uniformly generate an execution with the BIT sampler, and about 0.2 s with CFTP. For "small" state spaces, we observe that BIT is always faster than CFTP (Table 4).

Table 3. Summary of counting and uniform random sampling algorithms (time complexity figures with n: number of atomic actions).

Algorithm	Class	Count.	Unif. Rand. Gen.	Reference
FJ	Fork-join	$O(n)$	$O(n \cdot \sqrt{n})$ on average	[6]
ARCH	Arch-processes	$O(n^2)$	$O(n^4)$ worst case	[7]/Theorem 8
BIT	BIT-decomposable	?	?	Theorem 3
CFTP[a]	All processes	–	$O(n^3 \cdot log\ n)$ expected	[14]

[a]The CFTP algorithm is the only one we did not design, but only implement. Its complexity is $O(n^3 \cdot log\ n)$ (randomized) expected time.

Table 4. Benchmark results for BIT-decomposable classes: FJ and Arch.

FJ size	$\sharp\mathscr{LE}$	FJ gen	(count)	BIT gen	(count)	CFTP gen
10	19	0.00001 s	(0.0002 s)	0.0006 s	(0.03 s)	0.04 s
30	10^9	0.00002 s	(0.0002 s)	0.02 s	(0.03 s)	1.8 s
40	$6\cdot10^6$	0.00004 s	(0.0003 s)	3.5 s	(5.2 s)	5.6 s
63	$4\cdot10^{29}$	0.0005 s	(0.03 s)	Mem. crash	(Crash)	55 s
217028	$2\cdot10^{292431}$	8.11 s	(3.34 s)	Mem. crash	(Crash)	Timeout

Arch size	$\sharp\mathscr{LE}$	ARCH gen	(count)	BIT gen	(count)	CFTP gen
10:2	43	0.00002 s	(0.00004 s)	0.002 s	(0.000006 s)	0.04 s
30:2	$9.8\cdot10^8$	0.003 s	(0.0009 s)	0.000007 s	(0.0004 s)	1.5 s
30:4	$6.9\cdot10^{10}$	0.001 s	(0.005 s)	0.000007 s	(0.004 s)	2.5 s
100:2	$1.3\cdot10^{32}$	0.75 s	(0.16 s)	Mem. crash	(Crash)	[6] 5.6 s
100:32	$1\cdot10^{53}$	2.7 s	(0.17 s)	Mem. crash	(Crash)	[6] 5.9 s
200:66	10^{130}	54 s	(31 s)	Mem. crash	(Crash)	Timeout

For arch-processes of size 100 with 2 arches or 32, the cftp algorithm timeouts (30s) for almost all of the input graphs.

For a more thorough comparison of the various algorithms, we generated random processes (uniformly at random among all processes of the same size) in the classes of fork-join (FJ) and arch-processes as discussed in Sect. 5, using our own Arbogen tool[5] or an ad hoc algorithm for arch-processes (presented in the companion repository). For the fork-join structures, the size is simply the number of atomic actions in the process. It is not a surprise that the dedicated algorithms we developed in [6] outperforms the other algorithms by a large margin. In a few second it can handle extremely large state spaces, which is due to the large "branching factor" of the process "forks". The arch-processes represent a more complex structure, thus the numbers are less "impressive" than in the FJ case. To generate the arch-processes (uniformly at random), we used the number of atomic actions as well as the number of spawned promises as main parameters. Hence an arch of size '$n{:}k$' has n atomic actions and k spawned promises. Our dedicated algorithm for arch-process is also rather effective, considering the state-space sizes it can handle. In less than a minute it can generate an execution path uniformly at random for a process of size 200 with 66 spawned promises, the state-space is in the order of 10^{130}. Also, we observe that in all our tests the observable "complexity" is well below $O(n^4)$. The reason is that we perform the pre-computations (corresponding to the worst case) in a just-in-time (JIT) manner, and in practice we only actually need a small fractions of the computed values. However the random sampler is much more efficient with the separate precomputation. As an illustration, for arch-processes of size 100 with 32 arches, the sampler becomes about 500 times faster. However the memory requirement for the precomputation grows very quickly, so that the JIT variant is clearly preferable.

---

[5] Arbogen is uniform random generation for context-free grammar structures: cf. https://github.com/fredokun/arbogen.

In both the FJ and arch-process cases the current implementation of the BIT algorithms is not entirely satisfying. One reason is that the strategy we employ for the BIT-decomposition is quite "oblivious" to the actual structure of the DAG. As an example, this strategy handles fork-joins far better than arch-processes. In comparison, the CFTP algorithm is less sensitive to the structure, it performs quite uniformly on the whole benchmark. We are still confident that by handling the integral computation natively, the BIT algorithms could handle much larger state-spaces. For now, they are only usable up-to a size of about 40 nodes (already corresponding to a rather large state space).

7   Conclusion and Future Work

The process calculus presented in this paper is quite limited in terms of expressivity. In fact, as the paper makes clear it can only be used to describe (intransitive) directed acyclic graphs! However we still believe it is an interesting "core synchronization calculus", providing the minimum set of features so that processes are isomorphic to the whole combinatorial class of partially ordered sets. Of course, to become of any practical use, the barrier synchronization calculus should be complemented with e.g. non-deterministic choice (as we investigate in [4]). Moreover, the extension of our approach to iterative processes remains full of largely open questions.

Another interest of the proposed language is that it can be used to define process (hence poset) sub-classes in an inductive way. We give two illustrations in the paper with the *fork-join* processes and *promises*. This is complementary to definitions wrt. some combinatorial properties, such as the "BIT-decomposable" vs. "BIT-free" sub-classes. The class of arch-processes (that we study in [7] and generalize in the present paper) is also interesting: it is a combinatorially-defined sub-class of the inductively-defined asynchronous processes with promises. We see as quite enlightening the meeting of these two distinct points of view.

Even for the "simple" barrier synchronizations, our study is far from being finished because we are, in a way, also looking for "negative" results. The counting problem is hard, which is of course tightly related to the infamous "combinatorial explosion" phenomenon in concurrency. We in fact believe that the problem remains intractable for the class of BIT-decomposable processes, but this is still an open question that we intend to investigate furthermore. By delimiting more precisely the "hardness" frontier, we hope to find more interesting sub-classes for which we can develop efficient counting and random sampling algorithms.

Acknowledgment. We thank the anonymous reviewers as well as our "shepard" for helping us making the paper better and hopefully with less errors.

References

1. Abbes, S., Mairesse, J.: Uniform generation in trace monoids. In: Italiano, G.F., Pighizzini, G., Sannella, D.T. (eds.) MFCS 2015. LNCS, vol. 9234, pp. 63–75. Springer, Heidelberg (2015). https://doi.org/10.1007/978-3-662-48057-1_5
2. Banderier, C., Marchal, P., Wallner, M.: Rectangular young tableaux with local decreases and the density method for uniform random generation (short version). In: GASCom 2018, Athens, Greece, June 2018
3. Basset, N., Mairesse, J., Soria, M.: Uniform sampling for networks of automata. In: Concur 2017, LIPIcs, vol. 85, pp. 36:1–36:16. Schloss Dagstuhl (2017)
4. Bodini, O., Genitrini, A., Peschanski, F.: The combinatorics of non-determinism. In: FSTTCS 2013, LIPIcs, vol. 24, pp. 425–436. Schloss Dagstuhl (2013)
5. Bodini, O., Genitrini, A., Peschanski, F.: A quantitative study of pure parallel processes. Electron. J. Comb. 23(1), pp. P1.11, 39 (2016)
6. Bodini, O., Dien, M., Genitrini, A., Peschanski, F.: Entropic uniform sampling of linear extensions in series-parallel posets. In: Weil, P. (ed.) CSR 2017. LNCS, vol. 10304, pp. 71–84. Springer, Cham (2017). https://doi.org/10.1007/978-3-319-58747-9_9
7. Bodini, O., Dien, M., Genitrini, A., Viola, A.: Beyond series-parallel concurrent systems: the case of arch processes. In: Analysis of Algorithms, AofA 2018, LIPIcs, vol. 110, pp. 14:1–14:14 (2018)
8. Brightwell, G., Winkler, P.: Counting linear extensions is #P-complete. In: STOC, pp. 175–181 (1991)
9. Dittmer, S., Pak, I.: Counting linear extensions of restricted posets. arXiv e-prints arXiv:1802.06312, February 2018
10. Flajolet, P., Zimmermann, P., Cutsem, B.V.: A calculus for the random generation of labelled combinatorial structures. Theor. Comput. Sci. **132**(2), 1–35 (1994)
11. Gerbessiotis, A.V., Valiant, L.G.: Direct bulk-synchronous parallel algorithms. J. Parallel Distrib. Comput. **22**(2), 251–267 (1994)
12. Grosu, R., Smolka, S.A.: Monte Carlo model checking. In: Halbwachs, N., Zuck, L.D. (eds.) TACAS 2005. LNCS, vol. 3440, pp. 271–286. Springer, Heidelberg (2005). https://doi.org/10.1007/978-3-540-31980-1_18
13. Hensgen, D., Finkel, R.A., Manber, U.: Two algorithms for barrier synchronization. Int. J. Parallel Prog. **17**(1), 1–17 (1988)
14. Huber, M.: Fast perfect sampling from linear extensions. Discrete Math. **306**(4), 420–428 (2006)
15. Liskov, B., Shrira, L.: Promises: linguistic support for efficient asynchronous procedure calls in distributed systems. In: PLDI 1988, pp. 260–267. ACM (1988)
16. Oudinet, J., Denise, A., Gaudel, M.-C., Lassaigne, R., Peyronnet, S.: Uniform Monte-Carlo model checking. In: Giannakopoulou, D., Orejas, F. (eds.) FASE 2011. LNCS, vol. 6603, pp. 127–140. Springer, Heidelberg (2011). https://doi.org/10.1007/978-3-642-19811-3_10
17. Rival, I. (ed.): Algorithms and Order. NATO Science Series, vol. 255. Springer, Dordrecht (1989). https://doi.org/10.1007/978-94-009-2639-4
18. Sen, K.: Effective random testing of concurrent programs. In: Automated Software Engineering ASE 2007, pp. 323–332. ACM (2007)
19. Stanley, R.P.: Two poset polytopes. Discrete Comput. Geom. **1**, 9–23 (1986)

Parameter Synthesis for Bounded Cost Reachability in Time Petri Nets

Didier Lime[1($\boxtimes$)], Olivier H. Roux[1], and Charlotte Seidner[2]

[1] École Centrale de Nantes, LS2N UMR CNRS 6004, Nantes, France
Didier.Lime@ec-nantes.fr
[2] Université de Nantes, LS2N UMR CNRS 6004, Nantes, France

Abstract. We investigate the problem of parameter synthesis for time Petri nets with a cost variable that evolves both continuously with time, and discretely when firing transitions. More precisely, parameters are rational symbolic constants used for time constraints on the firing of transitions and we want to synthesise all their values such that the cost variable stays within a given budget.

We first prove that the mere existence of such values for the parameters is undecidable. We nonetheless provide a symbolic semi-algorithm that is proved both sound and complete when it terminates. We also show how to modify it for the case when parameters values are integers. Finally, we prove that this modified version terminates if parameters are bounded. While this is to be expected since there are now only a finite number of possible parameter values, this is interesting because the computation is symbolic and thus avoids an explicit enumeration of all those values. Furthermore, the result is a symbolic constraint representing a finite union of convex polyhedra that is easily amenable to further analysis through linear programming.

We finally report on the implementation of the approach in Romeo, a software tool for the analysis of hybrid extensions of time Petri nets.

1 Introduction

So-called *priced* or *cost timed* models are suitable for representing real-time systems whose behaviour is constrained by some resource consuming (be it energy or CPU time, for instance) and for which we need to assess the total cost accumulated during their execution. Such models can even describe whether the evolution of the cost during the run is caused by staying in a given state (continuous cost) or by performing a given action (discrete cost). Thus, the task of finding if the model can reach some "good" states while keeping the overall cost under a given bound (or, further, finding the minimum cost) can prove of interest in many real-life applications, such as optimal scheduling or production line planning.

This work is partially supported by the ANR national research program PACS (ANR-14-CE28-0002).

S. Donatelli and S. Haar (Eds.): PETRI NETS 2019, LNCS 11522, pp. 406–425, 2019.
https://doi.org/10.1007/978-3-030-21571-2_22

Timed models, however, require a thorough knowledge of the system for their analysis and are thus difficult to build in the early design stages, when the system is not fully identified. Even when all timing constraints are known, the whole design process must often be caried out afresh, whenever the environment changes. To obtain such valuable characteristics as flexibility and robustness, the designer may want to relax constraints on some specifications by allowing them a wider range of values. To this end, parametric reasoning is particularly relevant for timed models, since it allows designers to use parameters instead of definite timing values.

We therefore propose to tackle the definition and analysis of models that support both (linear) cost functions and timed parameters.

Related Work. Parametric timed automata (PTA) [3] extend timed automata [2] to overcome the limits of checking the correctness of the systems with respect to definite timing constraints. The reachability-emptiness problem, which tests whether there exists a parameter valuation such that the automaton has an accepting run, is fundamental to any verification process but is undecidable [3]. L/U automata [13] use each parameter either as a lower bound or as an upper bound on clocks. The reachability-emptiness problem is decidable for this model, but the state-space exploration, which would allow for explicit synthesis of all the suitable parameter valuations, still might not terminate [15]. To obtain decidability results, the approach described in [15] does not rely on syntactical restrictions on guards and invariants, but rather on restricting the parameter values to bounded integers. From a practical point of view, this subclass of PTA is not that restrictive, since the time constraints of timed automata are usually expressed as natural (or perhaps rational) numbers.

In [4], the authors have proved the decidability of the optimal-cost problem for Priced Timed Automata with non-negative costs. In [7,8,16], the computation of the optimal-cost to reach a goal location is based on a forward exploration of zones extended with linear cost functions. In [12], the authors have improved this approach, so as to ensure termination of the forward exploration algorithm, even when clocks are not bounded and costs are negative, provided that the automaton has no negative cost cycles. In [1], the considered model is a timed arc Petri net, under weak firing semantics, extended with rate costs associated with places and firing costs associated with transitions. The computation of the optimal-cost for reaching a goal marking is based on similar techniques to [4]. In [11], the authors have investigated the optimal-cost reachability problem for time Petri nets where each transition has a firing cost and each marking has a rate cost (represented as a linear rate cost function over markings). To compute the optimal-cost to reach a goal marking, the authors have revisited the state class graph method to include costs.

Our Contribution. We propose in Sect. 2 an extension of time Petri nets with costs (both discrete and continuous with time) and timing parameters, i.e., rational symbolic constants used in the constraints on the firing times of transitions.

Within this formalism, we define two problems dealing with parametric reachability within a bounded cost. We prove in Sect. 3 that the existence of a parameter valuation to reach a given marking under a given bounded cost is undecidable. This proof adapts a 2-counter machine encoding first proposed in [14] for PTA. To our knowledge it is the first time a direct Petri net encoding is provided and the adaptation is not trivial. We give in Sect. 4 a symbolic semi-algorithm that computes all such parameter valuations when it terminates, and we prove its correctness. We propose in Sect. 5 a variant of this semi-algorithm that computes integer parameter valuations and prove in Sect. 6 its termination provided those parameter valuations are bounded and the cost of each run is uniformly lower-bounded for integer parameter valuations. This technique is symbolic and avoids the explicit enumeration of all possible parameter valuations. The basic underlying idea of using the integer hull operator was first investigated in [15] for PTA, but this is the first time that it is adapted and proved to work with state classes for time Petri nets, and the fact that it naturally also preserves costs for integer parameter valuations is new and very interesting. We finally describe in Sect. 7 the implementation of the approach in the tool Romeo by analysing a small scheduling case-study.

2   Parametric Cost Time Petri Nets

2.1   Preliminaries

We denote the set of natural numbers (including 0) by $\mathbb{N}$, the set of integers by $\mathbb{Z}$, the set of rational numbers by $\mathbb{Q}$ and the set of real numbers by $\mathbb{R}$. We note $\mathbb{Q}_{\geq 0}$ (resp. $\mathbb{R}_{\geq 0}$) the set of non-negative rational (resp. real) numbers. For $n \in \mathbb{N}$, we let $[\![0, n]\!]$ denote the set $\{i \in \mathbb{N} \mid i \leq n\}$. For a finite set X, we denote its size by $|X|$.

Given a set X, we denote by $\mathcal{I}(X)$, the set of non empty real intervals that have their finite end-points in X. For $I \in \mathcal{I}(X)$, $\underline{I}$ denotes its left end-point if I is left-bounded and $-\infty$ otherwise. Similarly, $\overline{I}$ denotes the right end-point if I is right-bounded and ∞ otherwise. We say that an interval I is non-negative if $I \subseteq \mathbb{R}_{\geq 0}$. Moreover, for any $d \in \mathbb{R}_{\geq 0}$ and any non-negative interval I, we let $I \ominus d$ be the interval defined by $\{\theta - d \mid \theta \in I \wedge \theta - d \geq 0\}$. Note that this is again a non-negative interval.

Given sets V and X, a V-valuation (or simply valuation when V is clear from the context) of X is a mapping from X to V. We denote by V^X the set of V-valuations of X. When X is finite, given an arbitrary fixed order on X, we often equivalently consider V-valuations as vectors of $V^{|X|}$. Given a V-valuation v of X and $Y \subseteq X$, we denote by $v_{|Y}$ the projection of v on Y, i.e., the valuation on Y such that $\forall x \in Y, v_{|Y}(x) = v(x)$.

2.2   Time Petri Nets with Costs and Parameters

Definition 1 (Parametric Cost Time Petri Net (pcTPN)). *A* Parametric Cost Time Petri Net *(pcTPN) is a tuple* $\mathcal{N} = (P, T, \mathbb{P}, {}^{\bullet}., .^{\bullet}, m_0, I_s, cost_t, cost_m)$ *where*

- P is a finite non-empty set of places,
- T is a finite set of transitions such that $T \cap P = \emptyset$,
- $\mathbb{P}$ is a finite set of parameters,
- $^\bullet. : T \to \mathbb{N}^P$ is the backward incidence mapping,
- $.^\bullet : T \to \mathbb{N}^P$ is the forward incidence mapping,
- $m_0 \in \mathbb{N}^P$ is the initial marking,
- $I_s : T \to \mathcal{I}(\mathbb{N} \cup \mathbb{P})$ is the (parametric) static firing interval function,
- $cost_t : T \to \mathbb{Z}$ is the discrete cost function, and
- $cost_m : \mathbb{N}^P \to \mathbb{Z}$ is the cost rate function.

Given a parameterized object x (be it a pcTPN, a function, an expression, etc.), and a $\mathbb{Q}$-valuation v of parameters, we denote by $v(x)$ the corresponding non-parameterized object, in which each parameter a has been replaced by the value $v(a)$.

A *marking* is an $\mathbb{N}$-valuation of P. For a marking $m \in \mathbb{N}^P$, $m(p)$ represents a number of *tokens* in place p. A transition $t \in T$ is said to be *enabled* by a given marking $m \in \mathbb{N}^P$ if for all places p, $m(p) \geq {}^\bullet t(p)$. We also write $m \geq {}^\bullet t$. We denote by $\mathsf{en}(m)$ the set of transitions that are enabled by the marking m: $\mathsf{en}(m) = \{t \in T \mid m \geq {}^\bullet t\}$.

Firing an enabled transition t from marking m leads to a new marking $m' = m - {}^\bullet t + t^\bullet$. A transition $t' \in T$ is said to be *newly enabled* by the firing of a transition t from a given marking $m \in \mathbb{N}^P$ if it is enabled by the new marking but not by $m - {}^\bullet t$ (or it is itself fired). We denote by $\mathsf{newen}(m, t)$ the set of transitions that are newly enabled by the firing of t from the marking m: $\mathsf{newen}(m, t) = \{t' \in \mathsf{en}(m - {}^\bullet t + t^\bullet) \mid t' \notin \mathsf{en}(m - {}^\bullet t) \text{ or } t = t'\}$.

A *state* of the net $\mathcal{N}$ is a tuple (m, I, c, v) in $\mathbb{N}^P \times \mathcal{I}(\mathbb{R}_{\geq 0})^T \times \mathbb{R} \times \mathbb{Q}^{\mathbb{P}}_{\geq 0}$, where: m is a marking of $\mathcal{N}$, I is called the interval function and associates a *temporal interval* to each transition enabled by m. Value c is the cost associated with that state and valuation v assigns a rational value to each parameter for the state.

Definition 2 (Semantics of a pcTPN). *The semantics of a pcTPN is a timed transition system* $(Q, Q_0, \to)$ *where:*

- $Q \subseteq \mathbb{N}^P \times \mathcal{I}(\mathbb{R}_{\geq 0})^T \times \mathbb{R} \times \mathbb{Q}^{\mathbb{P}}_{\geq 0}$
- $Q_0 = \{(m_0, I_0, 0, v) \mid v \in \mathbb{Q}^{\mathbb{P}}_{\geq 0}, \forall t \in T, v(I_s(t)) \neq \emptyset\}$ *where* $\forall t \in \mathsf{en}(m_0), I_0(t) = I_s(t)$
- $\to$ *consists of two types of transitions:*
  - *discrete transitions:* $(m, I, c, v) \xrightarrow{t \in T} (m', I', c', v)$ *iff*
    * $m \geq {}^\bullet t$, $m' = m - {}^\bullet t + t^\bullet$ *and* $v(\underline{I(t)}) = 0$,
    * $\forall t' \in \mathsf{en}(m')$
      · $I'(t') = I_s(t')$ *if* $t' \in \mathsf{newen}(m, t)$,
      · $I'(t') = I(t')$ *otherwise*
    * $c' = c + cost_t(t)$
  - *time transitions:* $(m, I, c, v) \xrightarrow{d \in \mathbb{R}_{\geq 0}} (m, I \ominus d, c', v)$, *iff* $\forall t \in \mathsf{en}(m)$, $\overline{(I \ominus d)(t)} \geq 0$ *and* $c' = c + cost_m(m) * d$.

A run of a pcTPN $\mathcal{N}$ is a (finite or infinite) sequence $q_0 a_0 q_1 a_1 q_2 a_2 \cdots$ such that $q_0 \in Q_0$, for all $i > 0$, $q_i \in Q$, $a_i \in T \cup \mathbb{R}_{\geq 0}$ and $q_i \xrightarrow{a_i} q_{i+1}$. The set of runs of $\mathcal{N}$ is denoted by $\mathsf{Runs}(\mathcal{N})$. We note $(m, I, c, v) \xrightarrow{t@d} (m', I', c', v)$ for the sequence of elapsing $d \geq 0$ followed by the firing of the transition t. We denote by $\mathsf{sequence}(\rho)$ the projection of the run ρ over T: for a run $\rho = q_0 \xrightarrow{t_0@d_0} q_1 \xrightarrow{t_1@d_1} q_2 \xrightarrow{t_2@d_2} q_3 \xrightarrow{t_3@d_3} \cdots$, we have $\mathsf{sequence}(\rho) = t_0 t_1 t_2 t_3 \cdots$. We write $q \xrightarrow{t} q'$ if there exists $d \geq 0$ such that $q \xrightarrow{t@d} q'$.

For a finite run ρ we denote by $\mathsf{last}(\rho)$ the last state of ρ and by $\mathsf{lastm}(\rho)$ its marking. A state (m, I, c, v) is said to be *reachable* if there exists a finite run ρ of the net, with $\mathsf{last}(\rho) = (m, I, c, v)$. A marking m is reachable for parameter valuation v, if there exists some I and c such that (m, I, c, v) is reachable.

For $k \in \mathbb{N}$ and parameter valuation v, the (Cost) Time Petri net $v(\mathcal{N})$ is said to be k-bounded if for all reachable markings m, and all places p, $m(p) \leq k$. We say that $v(\mathcal{N})$ is bounded if there exists k such that it is k-bounded.

The *cost* $\mathsf{cost}(\rho)$ of a finite run ρ, with last state (m, I, c, v) is c. Since we are interested in minimising the cost, the *cost* of a sequence of transitions σ is defined as $\mathsf{cost}(\sigma) = \inf_{\rho \in \mathsf{Runs}(\mathcal{N}), \mathsf{sequence}(\rho) = \sigma} \mathsf{cost}(\rho)$. For the sake of the clarity of the presentation, we consider only closed intervals (or right-open to infinity) so this infimum is actually a minimum.

2.3 Parametric Cost Problems

Given a set of target markings Goal, the problems we are interested in are:

1. the existential problem: Given a finite maximum cost value $c_{\max}$, is there a parameter valuation v such that some marking in Goal is reachable with a cost less than $c_{\max}$ in $v(\mathcal{N})$?
2. the synthesis problem: Given a finite maximum cost value $c_{\max}$, compute all the parameter valuations v such that some marking in Goal is reachable with a cost less than $c_{\max}$ in $v(\mathcal{N})$.

We prove in Sect. 3 that the existential problem is undecidable.

3 Undecidability Results

The existential parametric time bounded reachability problem for bounded parametric time Petri nets asks whether a given target marking is reachable for some valuation of the parameter(s) within $c_{\max}$ time units. This is a special case of the existential cost bounded reachability problem defined in Sect. 2, with no discrete cost and a uniform cost rate of 1. Proposition 1 therefore implies the undecidability of that more general problem.

Proposition 1. *Existential parametric time bounded reachability is undecidable for bounded parametric time Petri nets.*

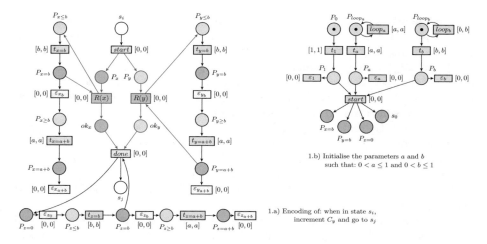

Fig. 1. Increment Gadget (left) and Initial gadget (right)

Proof. Given a bounded parametric time Petri net $\mathcal{N}$, we want to decide whether there exists some parameter valuation v such that some given marking can be reached within c_{max} time units in $v(\mathcal{N})$. The idea of this proof was first sketched in [14] for parametric timed automata. We encode the halting problem for two-counter machines, which is undecidable [18], into the existential problem for parametric timed Petri nets. Recall that a 2-counter machine $\mathcal{M}$ has two non-negative counters (here C_x and C_y), a finite number of states and a finite number of transitions, which can be of the form: (1) when in state s_i, increment a counter and go to s_j; (2) when in state s_i, decrement a counter and go to s_j; (3) when in state s_i, if a counter is null then go to s_j, otherwise block. The machine starts in state s_0 and halts when it reaches a particular state s_{halt}.

Given such a machine $\mathcal{M}$, we now provide an encoding as a parametric time Petri net $\mathcal{N}_{\mathcal{M}}$: each state s_i of the machine is encoded as place, which we also call s_i. The encoding of the 2-counter machine $\mathcal{M}$ is as follows: it uses two rational-valued parameter a and b, and three gadgets shown in Fig. 1a modelling three clocks x, y, z. Recall that, for a state (m, I, c, v), the enabling time of an enabled transition t is $v(\overline{I_s(t)} - \overline{I(t)})$. For the gadget modelling the clock x, the value of the clock x is equal to: (i) the enabling time of the transition $t_{x=b}$ when $P_{x\le b}$ is marked; (ii) b when $P_{x\le b}$ is marked; (iii) the sum of b and the enabling time of the transition $t_{x=a+b}$ when $P_{x\ge b}$ is marked (note that this value is lower than $a + b$); (iv) $a + b$ when $P_{x=a+b}$ is marked; (v) an unknown (an irrelevant) value in all other cases.

The gadget encoding the increment instruction of C_y is given in Fig. 1a. The clocks x and y store the value of each counter C_x and C_y as follows $x = b - a.C_x$ and $y = b - a.C_y$ when $z = 0$. The zero-test gadget is given in Fig. 2. We use the initial gadget in Figure 1b to initialise a and b such that $0 < a \le 1$ and $0 < b \le 1$. The system is studied over 1 time unit.

Increment: We start from some encoding configuration: $x = b - a.C_x$, $y = b - a.C_y$ and $z = 0$ in a marking such that the places $P_{z=0}$ and s_i are marked. After the firing of the transition *start*, there is an interleaving of the transitions $R(x)$ and $R(y)$ that go through the gadget. Finally, we can fire the transition *done* when $z = b$ (i.e. $b - a.C_x$ later) and we have $z = 0$, $x = b - a.C_x$ and $y = b - a(C_y + 1)$ as expected. Moreover, $v(\mathcal{N}_\mathcal{M})$ will block for all the parameter valuations v which not correctly encode the machine.

Decrement: By replacing the arc from $P_{z=b}$ to *done* by an arc from $P_{z=a+b}$ to *done*, the only difference in the previous reasoning is that the elapsing time to fire *done* is increased of a. Then we obtain $z = 0$, $x = b + a - a.C_x = b - a.(C_x - 1)$ and $y = b - a.C_y$ corresponding to the decrement of C_x.

We can obtain symmetrically (by swapping x and y) the increment of C_x and the decrement of C_y.

Both the increment gadget and the zero-test gadget require b time units, and the decrement gadget requires $(a + b)$ time units. Since the system executes over 1 time unit, for any value of $a > 0$ and $b > 0$, the number of operations that the machine can perform is finite. We consider two cases:

1. Either the machine halts, both counters C_x and C_y are bounded (let c their maximum value) and the halting and finite execution of the machine is within m steps. If $c = 0$ then the machine is a sequence of m zero-test taking $m.b$ time units and the parametric Petri net $\mathcal{N}_\mathcal{M}$ can go within 1 time unit to a marking m_{halt} if $0 < a \leq 1$ and $0 < b \leq \frac{1}{m}$. If $c > 0$, since an instruction requires at most $a + b$ time units, if $a + b \leq \frac{1}{m}$ and if $0 < a \leq \frac{b}{c}$ then there exists a run that correctly simulates the machine, and eventually reaches m_{halt} within 1 time unit.

   This set of valuations is non-empty: for example if $c = 0$, then we can choose $a = b = \frac{1}{m}$ and if $c > 0$, then, since $m \geq c$, we can choose $a = \frac{b}{m}$ and $b = \frac{1}{1+m}$ hence $a = \frac{1}{m(1+m)}$.

2. Or the machine does not halt. A step requires at least b time units then for any value v of the parameters, after a maximum number of steps (at most $\frac{1}{b}$), one whole time unit will elapse without $v(\mathcal{N}_\mathcal{M})$ reaching m_{halt}.   □

4   A Symbolic Semi-algorithm for Parameter Synthesis

4.1   State Classes

We now introduce the notion of state classes for pcTPNs. It was originally introduced for time Petri nets in [9,10], and extended for timing parameters in [21], and for costs in [11]. We show that those two extensions seamlessly blend together.

For an arbitrary sequence of transitions $\sigma = t_1 \ldots t_n \in T^*$, let C_σ be the set of all states that can be reached by the sequence σ from any initial state q_0: $C_\sigma = \{q \in Q | q_0 \xrightarrow{t_1} q_1 \cdots \xrightarrow{t_n} q\}$. All the states of C_σ share the same marking and can therefore be written as a pair (m, D) where m is the common marking

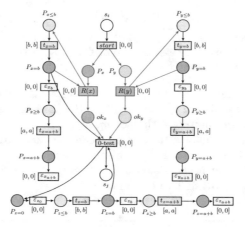

Fig. 2. Encoding 0-test over bounded-time: when in state s_i, if $C_x = 0$ then go to s_j

and, if we note $\text{en}(m) = \{t_1, \ldots, t_n\}$, then D is the set of points $(\theta_1, \ldots, \theta_n, c, v)$ such that $(m, I, c, v) \in C_\sigma$ and for all $t_i \in \text{en}(m)$, $\theta_i \in I(t_i)$. For short, we will often write $(\boldsymbol{\theta}, c, v)$ for such a point, with $\boldsymbol{\theta} = (\theta_1, \ldots, \theta_n)$ and a small abuse of notation. We denote by Θ the set of θ_i variables, of which we have one per transition of the net: for the sake of simplicity, we will usually use the same index to denote for instance that θ_i corresponds to transition t_i.

C_σ is called a *state class* and D is its *firing domain*.

Lemma 1 equivalently characterises state classes, as a straigthforward reformulation of the definition:

Lemma 1. *For all classes $C_\sigma = (m, D)$, $(\boldsymbol{\theta}, c, v) \in D$ if and only if there exists a run ρ in $v(\mathcal{N})$, and $I : \text{en}(m) \to \mathcal{I}(\mathbb{Q}_{\geq 0})$, such that $\text{sequence}(\rho) = \sigma$, $(m, I, c) = \text{last}(\rho)$, and $\boldsymbol{\theta} \in I$.*

From Lemma 1, we can then deduce a characterisation of the "next" class, obtained by firing a firable transition from some other class. This is expressed by Lemma 2.

Lemma 2. *Let $C_\sigma = (m, D)$ and $C_{\sigma.t_f} = (m', D')$, we have:*

$$(\boldsymbol{\theta}', c', v) \in D' \text{ iff } \exists (\boldsymbol{\theta}, c, v) \in D \text{ s.t. } \begin{cases} \forall t_i \in \text{en}(m), \theta_i - \theta_f \geq 0 \\ \forall t_i \in \text{en}(m - {}^\bullet t_f), \theta'_i = \theta_i - \theta_f \\ \forall t_i \in \text{newen}(m, t_f), \theta'_i \in v(I_s)(t_i) \\ c' = c + \text{cost}_m(m) * \theta_f + \text{cost}_t(t_f) \end{cases}$$

Proof. Consider $(\boldsymbol{\theta}', c', v) \in D'$. Then by Lemma 1, there exists a run ρ' in $v(\mathcal{N})$, and $I' : \text{en}(m) \to \mathcal{I}(\mathbb{Q}_{\geq 0})$, such that $\text{sequence}(\rho') = \sigma.t_f$, $(m', I', c') = \text{last}(\rho')$, and $\boldsymbol{\theta}' \in I'$. Consider the prefix ρ of ρ' such that $\text{sequence}(\rho) = \sigma$. The last state of ρ can be written (m, I, c, v) for some I and c. We know that t_f is fired from

(m, I, c, v) so there exists some delay d such that $\underline{I(t_f)} \leq d$ and for all other transitions t_i enabled by m, $\overline{I(t_i)} \geq d$. Furthermore, $c = c' - \text{cost}_m(m) * d - \text{cost}_t(t_f)$. It follows that there exists a point $\boldsymbol{\theta} \in I$ with the desired properties.
The other direction is similar.   □

Note that according to Lemma 2, D' is not empty if and only if there exists $(\boldsymbol{\theta}, c, v)$ in D such that for all $t_i \in \text{en}(m)$, $\theta_i \geq \theta_f$. In that case we say that t_f is *firable* from (m, D) and note $t_f \in \text{firable}((m, D))$.

From Lemma 2, it follows that $C_{\sigma.t_f}$ can be computed from C_σ using Algorithm 1. Note that it is formally the same algorithm as in [11].

Given a class C and a transition t firable from C, we note $\text{Next}(C, t)$ the result of applying Algorithm 1 to C and t.

Algorithm 1. Successor (m', D') of (m, D) by firing t_f

1: $m' \leftarrow m - {}^\bullet t_f + t_f^\bullet$
2: $D' \leftarrow D \wedge \bigwedge_{i \neq f, t_i \in \text{en}(m)} \theta_f \leq \theta_i$
3: for all $t_i \in \text{en}(m - {}^\bullet t_f), i \neq f$, add variable θ_i' to D', constrained by $\theta_i = \theta_i' + \theta_f$
4: add variable c' to D', constrained by $c' = c + \theta_f * \text{cost}_m(m) + \text{cost}_t(t_f)$
5: eliminate (by projection) variables c, θ_i for all i from D'
6: for all $t_j \in \text{newen}(m, t_f)$, add variable θ_j' to D', constrained by $\theta_j' \in I_s(t_j)$

Let $C_0 = (m_0, D_0)$ be the initial class. Domain D_0 is defined by the constraints $\forall t_i \in \text{en}(m_0), \theta_i \in I_s(t_i)$, $\forall t \in T, I_s(t) \neq \emptyset$, and $c = 0$. This gives a convex polyhedron of $\mathbb{R}_{\geq 0}^{|\text{en}(m_0)| + |\mathbb{P}| + 1}$; since all the operations on domains in Algorithm 1 are polyhedral, all the domains of state classes are also convex polyhedra. Note that only enabled transitions are constrained in the domain of a state class.

Naturally, we define the *cost* of state class C_σ as $\text{cost}(C_\sigma) = \text{cost}(\sigma)$.

4.2   The Synthesis Semi-algorithm

In Algorithm 2, we explore the symbolic state-space in a classic manner. Whenever a goal marking is encountered we collect the parameter valuations that allowed that marking to be reached with a cost less or equal to $c_{\max}$.

The PASSED list records the visited symbolic states. Instead of checking new symbolic states for membership, we test a weaker relation denoted by $\preccurlyeq$: does there exist a visited state allowing more behaviors with a cheaper cost?

For any state class $C = (m, D)$ and any point $(\boldsymbol{\theta}, v) \in D_{|\Theta \cup \mathbb{P}}$, the optimal cost of $(\boldsymbol{\theta}, v)$ in D is defined by $\text{cost}_D(\boldsymbol{\theta}, v) = \inf_{(\boldsymbol{\theta}, c, v) \in D} c$.

Definition 3. *Let $C = (m, D)$ and $C' = (m', D')$ be two parametric cost state classes. We say that C is subsumed by C', which we denote by $C \preccurlyeq C'$ iff $m = m'$, $D_{|\Theta \cup \mathbb{P}} \subseteq D'_{|\Theta \cup \mathbb{P}}$, and for all $(\boldsymbol{\theta}, v) \in D_{|\Theta \cup \mathbb{P}}$, $\text{cost}_{D'}(\boldsymbol{\theta}, v) \leq \text{cost}_D(\boldsymbol{\theta}, v)$.*

Algorithm 2. Symbolic semi-algorithm computing all parameter valuations such that some markings are reachable with a bounded cost.

```
1: PolyRes ← ∅
2: Passed ← ∅
3: Waiting ← {(m₀, D₀)}
4: while Waiting ≠ ∅ do
5:     select Cσ = (m, D) from Waiting
6:     if m ∈ Goal then
7:         PolyRes ← PolyRes ∪ (D ∩ (c ≤ cmax))|P
8:     end if
9:     if for all C' ∈ Passed, Cσ ⋠ C' then
10:        add Cσ to Passed
11:        for all t ∈ firable(Cσ), add Cσ.t to Waiting
12:    end if
13: end while
14: return PolyRes
```

The following result is a fairly direct consequence of Definition 3:

Lemma 3. *Let C_{σ_1} and C_{σ_2} be two state classes such that $C_{\sigma_1} \preccurlyeq C_{\sigma_2}$.*

If a transition sequence σ is firable from C_{σ_1}, it is also firable from C_{σ_2} and $\mathsf{cost}(C_{\sigma_1.\sigma}) \geq \mathsf{cost}(C_{\sigma_2.\sigma})$.

Proof. Let $C_{\sigma_1} = (m_1, D_1)$ and $C_{\sigma_2} = (m_2, D_2)$. From Definition 3, for any point $(\boldsymbol{\theta}, c_1, v) \in D_1$, there exists a point $(\boldsymbol{\theta}, c_2, v) \in D_2$ such that $c_2 \leq c_1$. This implies that: (i) $\mathsf{cost}(C_{\sigma_1}) \geq \mathsf{cost}(C_{\sigma_2})$; (ii) if transition t is firable from C_{σ_1}, then it is firable from C_{σ_2} and $\mathsf{Next}(C_{\sigma_1}, t) \preccurlyeq \mathsf{Next}(C_{\sigma_2}, t)$. And the result follows by a straightforward induction. □

While $\preccurlyeq$ can be checked using standard linear algebra techniques, we can also reduce it to standard inclusion on polyhedra by removing the upper bounds on cost (an operation called cost relaxation) [11].

Lemma 4. *The following invariant holds after each iteration of the while loop in Algorithm 2: for all $C_\sigma = (m, D) \in$ Passed,*

1. for all prefixes σ' of σ, $C_{\sigma'} \in$ Passed;
2. if $m \in$ Goal then $\left(D \cap (c \leq c_{\max})\right)_{|\mathbb{P}} \subseteq$ PolyRes;
3. if t is firable from C_σ
    – either $C_{\sigma.t} \in$ Waiting,
    – or there exists $C' \in$ Passed such that $C_{\sigma.t} \preccurlyeq C'$.

Proof. We prove this lemma by induction. Before the while loop starts, Passed is empty so the invariant is true. Let us now assume that the invariant holds for all iterations up to the n-th one, with $n \geq 0$, and that Waiting $\neq \emptyset$. Let $C_\sigma \in$ Waiting be the selected class at line 5; to check whether the invariant still holds at the end of the $(n + 1)$-th iteration, we only have to test the case where C_σ is added to Passed (which means that the condition at line 9 is true). We can then check each part of the invariant:

1. C_σ was picked from WAITING (line 5); except for the initial class (for which σ is empty, and therefore has no prefix), it means that, in a previous iteration, there was a sequence σ' and a transition $t \in$ firable$(C_{\sigma'})$ such that $\sigma = \sigma'.t$ (line 11) and $C_{\sigma'} \in$ PASSED (line 10). Since we add at most one state class to PASSED at each iteration, $C_{\sigma'}$ was added in a previous iteration and we can apply to it the induction hypothesis, which allows us to prove the first part of the invariant;

2. lines 6 and 7 obviously imply the second part of the invariant;

3. if $C_\sigma \in$ PASSED, then the condition of the if on line 9 is true and then for any transition t that is firable from C_σ, $C_{\sigma.t}$ is added to WAITING (line 11) so the third part of the invariant holds for C_σ. Nevertheless, C_σ itself is no longer in WAITING, and it is (except for the initial state class) the successor of some state class in PASSED. But then we have only two possibilities: either C_σ has been added to PASSED in line 10 if the condition on line 9 was true, and certainly $C_\sigma \preccurlyeq C_\sigma$, or there exists $C' \in$ PASSED such that $C_\sigma \preccurlyeq C'$ if that condition was false. Therefore the third part of the invariant holds.

Both the basis case and the induction step are true: the result follows by induction. □

Proposition 2. *After any iteration of the while loop in Algorithm 2:*

1. *if $v \in$ PolyRes, then there exists a run ρ in $v(\mathcal{N})$ such that $\mathsf{cost}(\rho) \le c_{\max}$ and $\mathsf{lastm}(\rho) \in$ Goal.*
2. *if WAITING $= \emptyset$ then, for all parameter valuations v such that there exists a run ρ in $v(\mathcal{N})$ such that $\mathsf{cost}(\rho) \le c_{\max}$ and $\mathsf{lastm}(\rho) \in$ Goal, we have $v \in$ PolyRes.*

Proof. 1. By induction on the while loop: initially, PolyRes is empty so the result holds trivially. Suppose it holds after some iteration n, and consider iteration $n + 1$. Let $v \in$ PolyRes after iteration $n + 1$. If v was already in PolyRes after iteration n then we can apply the induction hypothesis. Otherwise it means that if $C_\sigma = (m, D)$ is the class examined at iteration $n + 1$, then $m \in$ Goal and $v \in \left(D \cap (c \le c_{\max})\right)_{|\mathbb{P}}$. This means that there exists some point $(\boldsymbol{\theta}, c, v) \in D$ with $c \le c_{\max}$. By Lemma 1, this means that there exists a run ρ such that $(m, I, c, v) = \mathsf{last}(\rho)$, for some I such that $\boldsymbol{\theta} \in I$, and therefore $\mathsf{lastm}(\rho) \in$ Goal and $\mathsf{cost}(\rho) \le c_{\max}$.

2. Let v be a parameter valuation such that there exists a run ρ in $v(\mathcal{N})$ such that $\mathsf{cost}(\rho) \le c_{\max}$ and $\mathsf{lastm}(\rho) \in$ Goal. Let $\sigma = \mathsf{sequence}(\rho)$. We proceed by induction on the length n of the biggest suffix σ_2 of σ such that, either σ_2 is empty or, if we note $\sigma = \sigma_1\sigma_2$, with the first element of σ_2 being transition t, then $C_{\sigma_1 t} \notin$ PASSED.

If $n = 0$, then $C_\sigma = (m, D) \in$ PASSED. By Lemma 1, $v \in D_{|\mathbb{P}}$ and $m \in$ Goal. From the latter, we have $\left(D \cap (c \le c_{\max})\right)_{|\mathbb{P}} \subseteq$ PolyRes and therefore $v \in$ PolyRes because $v \in \left(D \cap (c \le c_{\max})\right)_{|\mathbb{P}}$.

Consider now $n > 0$ and assume the property holds for $n-1$. Since $n > 0$, then there exists a transition t and a sequence σ_3 such that $\sigma_2 = t.\sigma_3$. By definition of σ_2, we have $C_{\sigma_1} \in$ Passed but $C_{\sigma_1.t} \notin$ Passed. By Lemma 4, since Waiting $= \emptyset$, there must exists some class $C_{\sigma'}$ such that $C_{\sigma_1.t} \preceq C_{\sigma'}$. From Lemma 3, sequence σ_3 is also firable from $C_{\sigma'}$ and $C_{\sigma'.\sigma_3} = (m, D')$, with $\mathsf{cost}(C_{\sigma'.\sigma_3}) \leq \mathsf{cost}(C_\sigma) \leq c_{\max}$. By Lemma 1, there exists thus a run ρ' in $v(\mathcal{N})$, with $\mathsf{sequence}(\rho') = \sigma'.\sigma_3$, $\mathsf{lastm}(\rho') \in$ Goal and $\mathsf{cost}(\rho') \leq c_{\max}$. Also, from Lemma 4 (item 1), we know that for all prefixes of σ', the corresponding state class is in Passed, so the biggest suffix of $\sigma'.\sigma_3$ as defined above in the induction hypothesis has length less or equal to $n-1$, and the induction hypothesis applies to ρ', which allows to conclude.   □

In particular, if the algorithm terminates, then the waiting list is empty and PolyRes is exactly the solution to the synthesis problem.

5   Restricting to Integer Parameters

Obviously, in general, (semi-)Algorithm 2 will not terminate, since the emptiness problem for the set it computes is undecidable.

To ensure termination, we can however follow the methodology of [15]: we require that parameters are bounded integers and, instead of just enumerating the possible parameter values, we propose a modification of the symbolic state computation to compute these integer parameters symbolically. For this we rely on the notion of integer hull.

We call *integer valuation* a $\mathbb{Z}$-valuation. Note that a $\mathbb{Z}$-valuation is also an $\mathbb{R}$-valuation, and given a set D of $\mathbb{R}$-valuations, we denote by $\mathsf{Ints}(D)$ the set of integer valuations in D.

The *convex hull* of a set D of valuations, denoted by $\mathsf{Conv}(D)$, is the intersection of all the convex sets of valuations that contain D.

The *integer hull* of a set D of valuations, denoted by $\mathsf{IH}(D)$, is defined as the convex hull of the integer valuations in D: $\mathsf{IH}(D) = \mathsf{Conv}(\mathsf{Ints}(D))$.

For a state class $C = (m, D)$, we write $\mathsf{IH}(C)$ for $(m, \mathsf{IH}(D))$.

Before we see how our result can be adapted for the restriction to integer parameter valuations, and from there how we can enforce termination of the symbolic computations when parameters are assumed to be bounded, we need some results on the structure of the polyhedra representing firing domains of cost TPNs.

By the Minkowski-Weyl Theorem (see e.g. [20]), every convex polyhedron can be either described as a set of linear inequalities, as seen above, or by a set of *generators*. More precisely, for the latter: if d is the dimension of polyhedron P, there exists $v_1, \ldots, v_p, r_1, \ldots, r_s \in \mathbb{R}^d$, such that for all points $x \in P$, there exists $\lambda_1, \ldots, \lambda_p \in \mathbb{R}, \mu_1, \ldots, \mu_s \in \mathbb{R}_{\geq 0}$ such that $\sum_i \lambda_i = 1$ and $x = \sum_i \lambda_i v_i + \sum_i \mu_i r_i$. The v_i's are called the *vertices* of P and the r_i's are the *extremal rays* of P. The latter correspond to the directions in which the polyhedron is infinite. In our case, they correspond to transitions with a (right-)infinite static interval, and possibly the cost.

A classic property of vertices, which can also be used as a definition, is as follows: v is a vertex of P iff for all non-null vectors $x \in \mathbb{R}^d$, either $v + x \notin P$ or $v - x \notin P$ (or both), $+$ and $-$ being understood component-wise.

Proposition 3. *Let $\mathcal{N}$ be a (non-parametric) cost TPN and let $C = (m, D)$ be one of its state classes, then D has integer vertices.*

Proof. We have proved in [11] that the domain D of a state class of a cost TPNs, with removed upper bounds on cost (so-called relaxed classes), can be partitioned into a union of simpler polyhedra $\bigcup_{i=1}^n D_i$ that have the following key properties: (1) by projecting the cost out we obtain a convex polyhedron $D_{i|\Theta}$ with integer vertices (actually a *zone*, as in [9,17]), and (2) these simpler polyhedra all have exactly one constraint on the cost variable, i.e., of the form $c \geq \ell(\boldsymbol{\theta})$, with integer coefficients. Note that the same result can be obtained, with the same technique, if we consider non-relaxed state classes, except that, we also have an upper bound on cost that is always greater or equal to the lower bound. We prove in Lemma 5 that each of these simpler polyhedra also has integer vertices. Since D and each of the D_i's are convex and since $D = \bigcup_i D_i$, D is equal to the convex hull of the vertices of the D_i's and therefore D also has integer vertices.

Lemma 5. *Let D be a convex polyhedron on variables $\theta_1, \ldots, \theta_n, c$ such that the projection of P on the θ variables has integer vertices, and there are two constraints on c of the form $c \geq \ell(\theta_1, \ldots, \theta_n)$ and $c \leq \ell'(\theta_1, \ldots, \theta_n)$, with ℓ and ℓ' linear terms with integer coefficients, such that $\ell(\theta_1, \ldots, \theta_n) \leq \ell'(\theta_1, \ldots, \theta_n)$, for all values of the θ_i's.*

Then, the vertices of D are the points $(\theta_1, \ldots, \theta_n, \ell(\theta_1, \ldots, \theta_n))$ and $(\theta_1, \ldots, \theta_n, \ell'(\theta_1, \ldots, \theta_n))$ such that $(\theta_1, \ldots, \theta_n)$ is a vertex of $D_{|\Theta}$, and they are integer points.

Proof. Recall here that we consider all constraints in D to be non-strict so all polyhedra are topologically closed. The reasoning extends with no difficulty to non-necessarily-closed polyhedra by considering so-called *closure points* in addition to vertices [6].

Consider a non-vertex point $\boldsymbol{\theta}$ in $D_{|\Theta}$ and let $(\boldsymbol{\theta}, c)$ be a point of D. Then using the form of the unique cost constraint, we have $c \geq \ell(\boldsymbol{\theta})$. Now since $\boldsymbol{\theta}$ is not a vertex, there exists a vector x such that both $\boldsymbol{\theta} + x$ and $\boldsymbol{\theta} - x$ belong to $D_{|\Theta}$. Then, for sure, $(\boldsymbol{\theta} + x, \ell(\boldsymbol{\theta} + x)) \in D$ and $(\boldsymbol{\theta} - x, \ell(\boldsymbol{\theta} - x)) \in D$. And since ℓ is linear, $(\boldsymbol{\theta} + x, \ell(\boldsymbol{\theta}) + \ell(x)) \in D$, i.e., $(\boldsymbol{\theta}, \ell(\boldsymbol{\theta})) + (x, \ell(x)) \in D$. And similarly, $(\boldsymbol{\theta}, \ell(\boldsymbol{\theta})) - (x, \ell(x)) \in D$. Using again the form of the unique cost constraint, and the fact that $c \geq \ell(\boldsymbol{\theta})$, we finally have $(\boldsymbol{\theta}, c) + (x, \ell(x)) \in D$ and $(\boldsymbol{\theta}, c) - (x, \ell(x)) \in D$, that is, $(\boldsymbol{\theta}, c)$ is not a vertex of D.

By contraposition, any vertex of D extends a vertex of $D_{|\Theta}$, and using a last time the form of the cost constraint, any vertex of D, is of the form $(\boldsymbol{\theta}, \ell(\boldsymbol{\theta}))$, with $\boldsymbol{\theta}$ a vertex of $D_{|\Theta}$: suppose $(\boldsymbol{\theta}, c)$ is a vertex of D, with $c > \ell(\boldsymbol{\theta})$, then for x defined with $c - \ell(\boldsymbol{\theta})$ on the cost variable, and 0 on all other dimensions, we clearly have both $(\boldsymbol{\theta}, c) + x$ and $(\boldsymbol{\theta}, c) - x$ in D, which is a contradiction.

We conclude by remarking that, since $D_{|\Theta}$ has integer vertices, all the coordinates of $\boldsymbol{\theta}$ are integers, and since ℓ has integer coefficients then $\ell(\boldsymbol{\theta})$ is an integer.

We can deal with the upper bound defined by ℓ' in exactly the same way. $\square$

From Proposition 3, we can prove the following lemma that will be very useful in the subsequent proofs.

Lemma 6. *Let (m, D) be a state class of a pcTPN and let $(\boldsymbol{\theta}, c, v)$ be a point in D.*
If v is an integer valuation, then $(\boldsymbol{\theta}, c, v) \in \mathsf{IH}(D)$.

Proof. Since $(\boldsymbol{\theta}, c, v) \in D$ then $(\boldsymbol{\theta}, c) \in v(D)$. By Proposition 3, $v(D)$ being the firing domain of a state class in a (non-parametric) cost TPN, it has integer vertices, and therefore $v(D) = \mathsf{IH}(v(D))$. Point $(\boldsymbol{\theta}, c)$ is therefore a convex combination of integer points in $v(D)$. Clearly, for all integer points $(\boldsymbol{\theta}', c')$ in $v(D)$, we have that $(\boldsymbol{\theta}', c', v)$ is an integer point of D. Since D is convex, this implies that $(\boldsymbol{\theta}, c, v) \in \mathsf{IH}(D)$. $\square$

When we restrict ourselves to integer parameter but continue to work symbolically, we need to adjust the definitions of the firability of a transition from a class and of the cost of a class.

First, a transition t_f is firable for integer parameter valuations from a class (m, D), call this $\mathbb{N}^\mathbb{P}$-firable, if there exists an *integer* parameter valuation v and a point $(\boldsymbol{\theta}, c, v)$ in D such that for all transitions $t_i \in \mathsf{en}(m), \theta_i \geq \theta_f$.

Lemma 7. *Let $C = (m, D)$ be a state class. Transition $t_f \in \mathsf{en}(m)$ is $\mathbb{N}^\mathbb{P}$-firable from C if and only if it is firable (not necessarily $\mathbb{N}^\mathbb{P}$-firable) from $(m, \mathsf{IH}(D))$.*

Proof. $\Leftarrow$: trivial because $\mathsf{IH}(D) \subseteq D$.
$\Rightarrow$: since t_f is $\mathbb{N}^\mathbb{P}$-firable from C, there exists an integer parameter valuation v, and $(\boldsymbol{\theta}, c, v) \in D$ such that for all transitions $t_i \in \mathsf{en}(m), \theta_i \geq \theta_f$. And the result follows from Lemma 6 because v is an integer valuation. $\square$

Second, the cost of a class $C = (m, D)$, for integer parameters, is $\mathsf{cost}_\mathbb{N}(C) = \inf_{(\boldsymbol{\theta}, c, v) \in D, v \in \mathbb{N}^\mathbb{P}} c$.
Lemma 8 is a direct consequence of Lemma 6:

Lemma 8. *Let (m, D) be a state class. We have: $\mathsf{cost}_\mathbb{N}((m, D)) = \mathsf{cost}((m, \mathsf{IH}(D)))$.*

Lemma 9. *If v is an integer parameter valuation, then for all classes $C_\sigma = (m, D)$, $(\boldsymbol{\theta}, c, v) \in \mathsf{IH}(D)$ if and only if there exists a run ρ in $v(\mathcal{N})$, and $I : \mathsf{en}(m) \to \mathcal{I}(\mathbb{Q}_{\geq 0})$, such that $\mathsf{sequence}(\rho) = \sigma$, $(m, I, c) = \mathsf{last}(\rho)$, and $\boldsymbol{\theta} \in I$.*

Proof. $\Rightarrow$: if $(\boldsymbol{\theta}, c, v) \in \mathsf{IH}(D)$ then it is also in D and the result follows from Lemma 1.
$\Leftarrow$: by Lemma 1, we know that there exists some $(\boldsymbol{\theta}, c, v) \in D$, and since v is an integer valuation, by Lemma 6, $(\boldsymbol{\theta}, c, v) \in \mathsf{IH}(D)$. $\square$

Lemma 10. *Let C_{σ_1} and C_{σ_2} be two state classes such that $\mathsf{IH}(C_{\sigma_1}) \preccurlyeq \mathsf{IH}(C_{\sigma_2})$. If a transition sequence σ is $\mathbb{N}^{\mathbb{P}}$-firable from C_{σ_1} it is also $\mathbb{N}^{\mathbb{P}}$-firable from C_{σ_2} and $\mathit{cost}_{\mathbb{N}}(C_{\sigma_1}.\sigma) \geq \mathit{cost}_{\mathbb{N}}(C_{\sigma_2}.\sigma)$.*

Proof. Let $C_{\sigma_1} = (m_1, D_1)$ and $C_{\sigma_2} = (m_2, D_2)$. From Definition 3, for any point $(\boldsymbol{\theta}, c_1, v) \in \mathsf{IH}(D_1)$, there exists a point $(\boldsymbol{\theta}, c_2, v) \in \mathsf{IH}(D_2)$ such that $c_2 \leq c_1$. With Lemmas 7 and 8, this implies that: (i) $\mathit{cost}_{\mathbb{N}}(C_{\sigma_1}) \geq \mathit{cost}_{\mathbb{N}}(C_{\sigma_2})$; (ii) if transition t is $\mathbb{N}^{\mathbb{P}}$-firable from C_{σ_1}, then it is $\mathbb{N}^{\mathbb{P}}$-firable from C_{σ_2} and $\mathsf{Next}(C_{\sigma_1}, t) \preccurlyeq \mathsf{Next}(C_{\sigma_2}, t)$. And, as before, the result follows by a straightforward induction. $\square$

Algorithm 3. Restriction of (semi-)Algorithm 2 to integer parameter valuations.

```
1:  PolyRes ← ∅
2:  PASSED ← ∅
3:  WAITING ← {(m₀, D₀)}
4:  while WAITING ≠ ∅ do
5:      select Cσ = (m, D) from WAITING
6:      if m ∈ Goal then
7:          PolyRes ← PolyRes ∪ (IH(D) ∩ (c ≤ cmax))|ℙ
8:      end if
9:      if for all C' ∈ PASSED, IH(Cσ) ⋠ IH(C') then
10:         add Cσ to PASSED
11:         for all t ∈ firable(IH(Cσ)), add Cσ.t to WAITING
12:     end if
13: end while
14: return PolyRes
```

Using Lemma 9 instead of Lemma 1, and Lemma 10 instead of Lemma 3 in the proof of Proposition 2, we get the following proposition, stating the completeness and soundness of Algorithm 3.

Proposition 4. *After any iteration of the while loop in Algorithm 3:*

1. *if $v \in$ PolyRes and v is an integer parameter valuation then there exists a run ρ in $v(\mathcal{N})$ such that $\mathsf{cost}(\rho) \leq c_{\max}$ and $\mathsf{lastm}(\rho) \in$ Goal.*
2. *if WAITING $= \emptyset$ then for all integer parameter valuations v such that there exists a run ρ in $v(\mathcal{N})$ such that $\mathsf{cost}(\rho) \leq c_{\max}$ and $\mathsf{lastm}(\rho) \in$ Goal, we have $v \in$ PolyRes.*

In Algorithm 3, we compute state classes as usual then handle them via their integer hulls. We can actually simply integrate integer hulls at the end of Algorithm 1 and use Algorithm 2 with this updated successor computation as proved by Lemma 11.

Lemma 11. *Let (m, D) be a state class of a pcTPN $\mathcal{N}$, and t a transition firable from C. Let $(m', D') = \mathsf{Next}((m, D), t)$ and $(m'', D'') = \mathsf{Next}((m, \mathsf{IH}(D)), t)$. Then $m'' = m'$ and $\mathsf{IH}(D'') = \mathsf{IH}(D')$.*

Proof. The equality of markings is trivial so we focus on firing domains.

By definition of the integer hull, we have $\mathsf{IH}(D) \subseteq D$. Since the computation of the next class domain is non-decreasing with respect to inclusion, we then have $D'' \subseteq D'$. Taking the integer hull is also non-decreasing wrt. inclusion, so $\mathsf{IH}(D'') \subseteq \mathsf{IH}(D')$.

Consider now an integer point $(\boldsymbol{\theta}', c', v)$ in D'. Then $(\boldsymbol{\theta}', c') \in v(D')$. Consider state class computations in the (non-parametric) cost TPN $v(\mathcal{N})$: there exists some point $(\boldsymbol{\theta}, c)$ in $v(D)$ such that $(m', \boldsymbol{\theta}', c') \in \mathsf{Next}((m, \{(\boldsymbol{\theta}, c)\}), t)$. Since $(\boldsymbol{\theta}, c, v)$ thus belongs to D and since v is an integer parameter valuation, by Lemma 6, we have that $(\boldsymbol{\theta}, c, v) \in \mathsf{IH}(D)$. Thus $(\boldsymbol{\theta}', c', v) \in D''$ and since it is an integer point, it is in $\mathsf{IH}(D'')$. □

6   Termination of Algorithm 3

We now consider that parameter valuations are bounded by some value $M_1 \in \mathbb{N}$ (and that they still have integer values). We also assume that, for all integer parameter valuations, there exists $M_2 \in \mathbb{Z}$ such that for all runs ρ in $v(\mathcal{N})$, $\mathsf{cost}(\rho) \geq M_2$: this allows us, as in [11,12], to keep Algorithm 3 simple by doing away with negative cost loop-checking. Finally, we assume the net itself is bounded: there exists $M_3 \in \mathbb{N}$ such that for all reachable markings m, for all places p, $m(p) \leq M_3$.

To prove the termination of Algorithm 3 under these assumptions, we consider $\succcurlyeq$ the symmetric relation to $\preccurlyeq$, such that $x \succcurlyeq y$ iff $y \preccurlyeq x$. We prove that it is a well quasi-order (wqo), i.e., that for every infinite sequence of state classes, there exist C and C' in the sequence, with C strictly preceding C' such that $C \succcurlyeq C'$. This implies that the exploration of children in Algorithm 3 will always eventually stop.

Proposition 5. *Let $\mathcal{N}$ be a bounded pcTPN, with bounded integer parameters and such that the cost of all runs is uniformly lower-bounded for all integer parameter valuations.*

Relation $\succcurlyeq$ is well-quasiorder on the set of state classes of $\mathcal{N}$.

Proof. Consider an infinite sequence $C_0, C_1, C_2, \dots$ of state classes. Let $C_i = (m_i, D_i)$.

From [11], we know that $\succcurlyeq$ is a wqo for the state classes of bounded (non parametric) cost TPNs. So for each integer parameter valuation v, and using a classic property of wqo we can extract a subsequence of $v(C_0), v(C_1), \dots$ that is completely ordered by $\succcurlyeq$. And since, we have a finite number of such parameter valuations, we can extract an infinite subsequence $C_{i_0}, C_{i_1}, \dots$ such that for all integer parameter valuations v, $v(C_{i_0}) \succcurlyeq v(C_{i_1}) \succcurlyeq \cdots$.

Let us consider two of those: C_{i_r} and C_{i_s}, with $r < s$.

Since $\mathsf{IH}(D_{i_s})$ has integer vertices, and for any integer parameter valuation, $v(C_{i_r}) \succcurlyeq v(C_{i_s})$, which implies that $v(D_{i_s}) \subseteq v(D_{i_r})$, then all the vertices of D_{i_s} are also in D_{i_r}. Now assume that some extremal ray $\boldsymbol{r}$ of D_{i_s} is not in D_{i_r}. Then starting from some vertex $\boldsymbol{x}$ of D_{i_r}, there must be some $\lambda \leq 0$ such that $\boldsymbol{x} + \lambda r \notin D_{i_s}$ and the same holds for any $\lambda' \geq \lambda$ (by convexity). But since r has rational coordinates for some value of λ', $\lambda' r$ is an integer vector and so is $\boldsymbol{x} + \lambda' r$, which contradicts the fact that $v(D_{i_s}) \subseteq v(D_{i_r})$, for all integer parameter valuations v, and in particular $(\boldsymbol{x} + \lambda' r)_{|\mathbb{P}}$. We can therefore conclude that $D_{i_r} \subseteq D_{i_s}$ and we now proceed to proving that D_{i_s} is also "cheaper" than D_{i_r}.

We use another property of the vertices of convex polyhedra: vertices of a convex polyhedron of dimension n defined by m inequalities $\sum_{k=1}^{n} a_{kl} x_k \leq b_l$, for $j \in [1..m]$ are solutions of a system of n linearly independent equations $\sum_{k=1}^{n} a_{kl} x_k = b_l$, with l in a subset of size n of $[1..m]$.

Now consider the polyhedron D obtained from $\mathsf{IH}(D_{i_r})$, with its cost variable c, by adding one variable c' constrained by the cost inequalities of $\mathsf{IH}(D_{i_s})$. Clearly, since c and c' are not constrained together, the vertices of D are those of $\mathsf{IH}(D_{i_r})$, extended with the corresponding minimal and maximal values of c', and symmetrically those of $\mathsf{IH}(D_{i_s})$, extended with the corresponding minimal and maximal values of c'. Since the inequalities constraining c and c' have integer coefficients, and $\mathsf{IH}(D_{i_s})$ and $\mathsf{IH}(D_{i_r})$ have integer vertices, D also has integer vertices.

For the i-th lower-bound inequality on c, and the j-th lower-bound inequality on c', we define E_{ij} as D in which we transform both constraints into equalities. Clearly, from the property above, this does not add any new vertex, but it may remove some. Second, by construction, we have $\bigcup_{ij} E_{ij} = \{(\boldsymbol{\theta}, \min_{(\boldsymbol{\theta},c,v) \in \mathsf{IH}(D_{i_r})} c, \min_{(\boldsymbol{\theta},c,v) \in \mathsf{IH}(D_{i_s})} c) | \boldsymbol{\theta} \in \mathsf{IH}(D_r)_{|\Theta}\}$. If we minimize $c - c'$ over E_i, we know from the theory of linear programming that the minimum is obtained at a vertex of E_{ij}, and therefore, in particular, for an integer valuation v of the parameters, and an integer vector $\boldsymbol{\theta}$ of D_{i_r}. Since we have $v(C_{i_r}) \succcurlyeq v(C_{i_s})$, we then know that for these values of the theta variables and parameters, $c \leq c'$. This means that this holds for the whole of E_{ij}, and finally that $C_{i_r} \succcurlyeq C_{i_s}$. $\square$

7   Case Study

We now consider a scheduling problem where some tasks include *runnables*, a key concept of the AUTomotive Open System ARchitecture (AUTOSAR), the open standard for designing the architecture of vehicle software [5]. Runnables represent the functional view of the system and are executed by the runtime of the software component [19]. For their execution they are mapped to tasks and a given runnable can be split across different tasks to introduce parallelism, for instance. In industrial practice, runnables that interact a lot are mapped to the same task, in particular when they perform functions with the same period.

In this example, we consider 3 non-preemptive, periodic tasks T1, T2 and T3, on which have already been mapped some runnables that interact together; we add another independent runnable whose code must be split between tasks T1 and T2:

- the period of task T1 is 100 time units; T1 includes a "fixed part", independent from the new runnable and whose execution lasts 22 t.u.;
- the period of T2 is 200 t.u.; T2 also has a fixed part lasting 28 t.u.;
- the period of T3 is 400 t.u.; its execution lasts 11 t.u.;
- the period of the runnable is 200 t.u.; its execution lasts 76 t.u.; parameter a denotes the duration of the section that is executed in T1[1].

The processing unit consists of 2 cores C0 and C1; T3 can only execute on C0 whereas both T1 and T2 can execute on either core. When both cores are idle, the cost is null; when only one core is busy, the cost is equal to 2/t.u.; when both cores are busy, the cost is equal to 3/t.u. Any optimised strategy to divide the runnable over T1 and T2 and to allocate these tasks to C0 or C1 must therefore favour the cases where both cores are in the same state.

Figure 3 presents the model for this problem[2]. The associated cost function is: $2*(C0 \neq C1) + 3*C0*C1 + 1000*(W1*(R1C0 + R1C1) + W2*(R2C0 + R2C1) + W3*R3C0)$, where the name of a place (e.g. $R1C0$) represents its marking[3].

We limit the study of the system to the first 400 t.u., at the end of which T1 has been executed 4 times, T2 twice and T3 once. A preliminary analysis (not

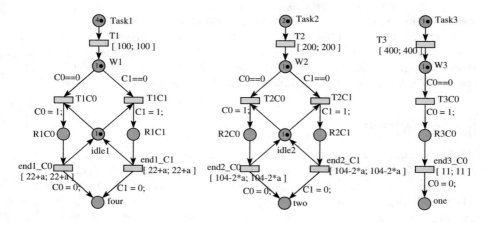

Fig. 3. Offline non preemptive scheduling problem

---

[1] Every 200 t.u., since T1 is executed twice as often as T2, T1 is running during $(22+a)*2 = 44 + 2a$ t.u. whereas T2 is running during $28 + (76 - 2a) = 104 - 2a$ t.u.

[2] To ensure a correct access to the cores, we could have added one place for each core and some arcs on each task to capture and release them but the resulting net would have been quite unreadable. Instead, we chose to add 2 integer variables C0 and C1 (both initialised to 0); a variable equal to 0 (resp. 1) obviously means the corresponding core is idle (resp. busy).

[3] The last term ensures that such cases where an instance of a task is activated while a previous one is running are heavily penalised.

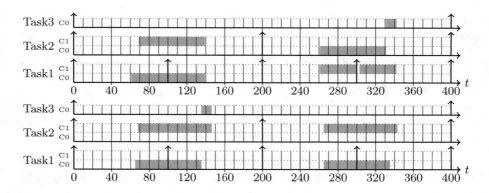

Fig. 4. Gantt charts for $a = 17$ (above) and $a = 13$ (below)

detailed here for the sake of concision) showed that the lowest cost is 466. By setting our maximal cost to this value, we then check the following property with our Romeo tool: EF four==4 and two==2 and one==1 and cost≤466. The answer provided by Romeo is that the property is true iff $a \in [13, 17]$. We then set a to 17; Romeo yields the following timed trace, in which the notation T1@t1 means that transition T1 is fired at date t1: T1C0@61, T2C1@69, T1@100, end1_C0@100, T1C0@100, end1_C0@139, end2_C1@139, T1@200, T2@200, T2C0@261, T1C1@261, T1@300, end1_C1@300, T1C1@303, end2_C0@331, T3C0@331, end3_C0@342, end1_C1@342.

From this trace, we obtain the Gantt chart in Fig. 4 (above). Setting a to 13 yields another timed trace, resulting in the Gantt chart in Fig. 4 (below). In both cases, we can see that both cores are busy during 148 t.u. (and for 11 t.u., only one is idle), which confirms our analysis on the optimised strategy above.

8   Conclusion

We have proposed a new Petri net-based formalism with parametric timing and cost features, thus merging two classic lines of work. For this formalism, we define an existential problem and a synthesis problem for parametric reachability within a bounded cost. We prove that the former is undecidable but we nonetheless give and prove a symbolic semi-algorithm for the latter. We finally propose a variant of the synthesis algorithm suitable for integer parameter valuations and prove its termination when those parameter valuations are bounded, and some other classic assumptions. This symbolic algorithm avoids the explicit enumeration of all possible parameter valuations. It is implemented in our tool Romeo and we have reported on a case-study addressing a scheduling problem, and inspired by the AUTOSAR standard.

Further work includes computing the optimal cost as a function of parameters and investigating the case of costs (discrete and rates) as parameters.

References

1. Abdulla, P.A., Mayr, R.: Priced timed Petri nets. Log. Meth. Comput. Sci. **9**(4) (2013)
2. Alur, R., Dill, D.: A theory of timed automata. Theor. Comput. Sci. **126**(2), 183–235 (1994)
3. Alur, R., Henzinger, T.A., Vardi, M.Y.: Parametric real-time reasoning. In: ACM Symposium on Theory of Computing, pp. 592–601 (1993)
4. Alur, R., Torre, S.L., Pappas, G.J.: Optimal paths in weighted timed automata. Theor. Comput. Sci. **318**(3), 297–322 (2004)
5. AUTOSAR: specification of RTE software. Technical report 4.4.0, October 2018
6. Bagnara, R., Hill, P., Zaffanella, E.: Not necessarily closed polyhedra and the double description method. Form. Asp. Comput. **17**, 222–257 (2005)
7. Behrmann, G., et al.: Minimum-cost reachability for priced time automata. In: Di Benedetto, M.D., Sangiovanni-Vincentelli, A. (eds.) HSCC 2001. LNCS, vol. 2034, pp. 147–161. Springer, Heidelberg (2001). https://doi.org/10.1007/3-540-45351-2_15
8. Behrmann, G., Larsen, K.G., Rasmussen, J.I.: Optimal scheduling using priced timed automata. SIGMETRICS Perform. Eval. Rev. **32**(4), 34–40 (2005)
9. Berthomieu, B., Diaz, M.: Modeling and verification of time dependent systems using time Petri nets. IEEE Trans. Soft. Eng. **17**(3), 259–273 (1991)
10. Berthomieu, B., Menasche, M.: An enumerative approach for analyzing time Petri nets. In: IFIP, pp. 41–46. Elsevier Science Publishers (1983)
11. Boucheneb, H., Lime, D., Parquier, B., Roux, O.H., Seidner, C.: Optimal reachability in cost time Petri nets. In: Abate, A., Geeraerts, G. (eds.) FORMATS 2017. LNCS, vol. 10419, pp. 58–73. Springer, Cham (2017). https://doi.org/10.1007/978-3-319-65765-3_4
12. Bouyer, P., Colange, M., Markey, N.: Symbolic optimal reachability in weighted timed automata. In: Chaudhuri, S., Farzan, A. (eds.) CAV 2016. LNCS, vol. 9779, pp. 513–530. Springer, Cham (2016). https://doi.org/10.1007/978-3-319-41528-4_28
13. Hune, T., Romijn, J., Stoelinga, M., Vaandrager, F.: Linear parametric model checking of timed automata. J. Log. Algebr. Program. **52–53**, 183–220 (2002)
14. Jovanović, A.: Parametric verification of timed systems. Ph.D. thesis, École Centrale Nantes, Nantes, France (2013)
15. Jovanović, A., Lime, D., Roux, O.H.: Integer parameter synthesis for real-time systems. IEEE Trans. Softw. Eng. (TSE) **41**(5), 445–461 (2015)
16. Larsen, K., et al.: As cheap as possible: effcient cost-optimal reachability for priced timed automata. In: Berry, G., Comon, H., Finkel, A. (eds.) CAV 2001. LNCS, vol. 2102, pp. 493–505. Springer, Heidelberg (2001). https://doi.org/10.1007/3-540-44585-4_47
17. Larsen, K.G., Pettersson, P., Yi, W.: Model-checking for real-time systems. In: Reichel, H. (ed.) FCT 1995. LNCS, vol. 965, pp. 62–88. Springer, Heidelberg (1995). https://doi.org/10.1007/3-540-60249-6_41
18. Minsky, M.: Computation: Finite and Infinite Machines. Prentice Hall, Englewood Cliffs (1967)
19. Naumann, N.: AUTOSAR runtime environment and virtual function bus. Technical report, Hasso-Plattner-Institut (2009)
20. Schrijver, A.: Theory of Linear and Integer Programming. Wiley, New York (1986)
21. Traonouez, L.-M., Lime, D., Roux, O.H.: Parametric model-checking of stopwatch Petri nets. J. Univers. Comput. Sci. **15**(17), 3273–3304 (2009)

Models with Extensions

Coverability and Termination in Recursive Petri Nets

Alain Finkel[1], Serge Haddad[1,2], and Igor Khmelnitsky[1,2(✉)]

[1] LSV, ENS Paris-Saclay, CNRS, Université Paris-Saclay, Cachan, France
{alain.finkel,serge.haddad,igor.khmelnitsky}@ens-paris-saclay.fr
[2] Inria, Paris, France

Abstract. In the early two-thousands, Recursive Petri nets have been introduced in order to model distributed planning of multi-agent systems for which counters and recursivity were necessary. Although Recursive Petri nets strictly extend Petri nets and stack automata, most of the usual property problems are solvable but using non primitive recursive algorithms, even for coverability and termination. For almost all other extended Petri nets models containing a stack the complexity of coverability and termination are unknown or strictly larger than EXPSPACE. In contrast, we establish here that for Recursive Petri nets, the coverability and termination problems are EXPSPACE-complete as for Petri nets. From an expressiveness point of view, we show that coverability languages of Recursive Petri nets strictly include the union of coverability languages of Petri nets and context-free languages. Thus we get for free a more powerful model than Petri net.

Keywords: Recursive Petri nets · Expressiveness · Complexity · Coverability · Termination

1 Introduction

Verification Problems for Petri Nets. Petri net is a useful formalism for analysis of concurrent programs for several reasons. From a modelling point of view (1) due to the locality of the firing rule, one easily models concurrent activities and (2) the (a priori) unbounded marking of places allows to represent a dynamic number of activities. From a verification point of view, all usual properties are decidable. However Petri nets suffer two main limitations: they cannot model recursive features and the computational cost of verification may be very high. More precisely, all the known algorithms solving reachability are non primitive recursive (see for instance [21]) and it has been proved recently that the

A. Finkel—The work of this author was carried out in the framework of ReLaX, UMI2000 and also supported by ANR-17-CE40-0028 project BRAVAS.
S. Haddad—The work of this author was partly supported by ERC project EQualIS (FP7-308087).

S. Donatelli and S. Haar (Eds.): PETRI NETS 2019, LNCS 11522, pp. 429–448, 2019.
https://doi.org/10.1007/978-3-030-21571-2_23

reachability problem is non elementary [4]. Fortunately some interesting properties like coverability, termination and boundedness are **EXPSPACE**-complete [22] and thus still manageable by a tool. So an important research direction consists of extending Petri nets to support new modelling features while still preserving decidability of properties checking and if possible with a reasonable complexity.

Extended Petri Nets. Such extensions may partitionned between those whose states are still markings and the other ones. The simplest extension consists in adding inhibitor arcs which yields undecidability of all verification problems. However adding a single inhibitor arc preserves the decidability of the reachability, coverability, and boundedness problems [2,3,23]. When adding reset arcs, the coverability problem becomes Ackermann-complete and boundedness undecidable [24].

In ν-Petri nets the tokens are coloured where colours are picked in an infinite domain: their coverability problem is double Ackermann time complete [18]. In Petri nets equipped with a stack, the reachability and coverability problems are not only **TOWER**-hard [4,16] but their decidability status is still unknown. In branching vector addition systems with states (BVASS) a state is a set of threads with associated markings. A thread either fires a transition as in Petri nets or forks, transferring a part of its marking to the new thread. For BVASS, the reachability problem is also **TOWER**-hard [17] and its decidability is still an open problem while the coverability and the boundedness problems are 2-**EXPTIME**-complete [6]. In Petri nets with a stack, the reachability problem may be reduced to the coverability problem and both are at least not elementary while their decidability status is still unknown [16]. The analysis of subclasses of Petri nets with a stack is an active field of research [1,5,20,25]. However for none of the above extensions, the coverability and termination problems belong to **EXPSPACE**.

Recursive Petri Nets (RPN). This formalism has been introduced in order to model distributed planning of multi-agent systems for which counters and recursivity were necessary for specifying resources and delegation of subtasks [7]. Roughly speaking, a state of an RPN consists of a tree of *threads* where the local state of each thread is a marking. Any thread fires an *elementary* or *abstract* transition. When the transition is elementary, the firing updates its marking as in Petri nets; when it is abstract, this only consumes the tokens specified by the input arcs of the transition and creates a child thread initialised with the *starting marking* of the transition. When a marking of a thread covers one of the *final markings*, it may perform a *cut* transition pruning its subtree and producing in its parent the tokens specified by the output arcs of the abstract transition that created it. In RPN, reachability, boundedness and termination are decidable [11,12] by reducing these properties to reachability problems of Petri nets. So the corresponding algorithms are non elementary. Model checking is undecidable for RPN but becomes decidable for the subclass of sequential RPN [13]. In [14], several modelling features are proposed while preserving the decidability of the verification problems.

Our Contribution. We first study the expressive power of RPN from the point of view of coverability languages (reachability languages were studied in [11]). We first introduce a quasi-order on states of RPN compatible with the firing rule and establish that it is a not a well quasi-order. We show that the languages of RPN are *quite close* to recursively enumerable languages since the closure under homomorphism and intersection with a regular language is the family of recursively enumerable languages. More precisely, we show that coverability (as reachability) languages of RPN strictly include the union of context-free languages and Petri net coverability languages. On an other side, we prove that coverability languages of RPN and reachability languages of Petri nets are incomparable. In addition, we establish that the family of languages of RPN is closed by union, homomorphism but not by intersection with a regular language.

From an algorithmic point of view, we show that coverability and termination are EXPSPACE-complete, as for Petri nets. Thus the increasing of expressive power does not entail a corresponding increasing in complexity. In order to solve the coverability problem, we show that if there exists a covering sequence there exists a 'short' one (i.e. with a length at most doubly exponential w.r.t. the size of the input). The core of the proof consists in turning an arbitrary covering sequence into a structured one where all threads perform their firings in one shot. In order to solve the termination problem, we consider two cases for an infinite sequence depending (informally speaking) whether the depth of the trees corresponding to states is bounded or not along the sequence.

Outline. In Sect. 2, we introduce RPNs and state ordering and establish basic results related to these notions. In Sect. 3, we study the expressiveness of coverability languages. Then in Sects. 4 and 5, we show that the coverability and termination problems are EXPSPACE-complete. In Sect. 6, we conclude and give some perspectives to this work. All missing proofs can be found in [8].

2   Recursive Petri Nets

2.1   Presentation

An RPN has a structure akin to a 'directed rooted tree' of Petri nets. Each vertex of the tree, hereafter *thread*, is an instance of the RPN and possessing some marking on it. Each of these threads can fire three types of transitions. An *elementary* transition updates its own marking according to the usual Petri net firing rule. An *abstract* transition consumes tokens from the thread firing it and creates a new child (thread) for it. The marking of the new thread is determined according to the fired abstract transition. A *cut* transition can be fired by a thread if its marking is greater or equal than some marking in a finite set of *final markings*. Firing a cut transition, the thread erases itself and all of its descendants. Moreover, it creates tokens in its parent, which are specified by the abstract transition that created it.

Definition 1 (Recursive Petri Net). *A Recursive Petri Net is a 6-tuple* $\mathcal{N} = \langle P, T, W^{+}, W^{-}, \Omega, \mathcal{F} \rangle$ *where:*

- *P is a finite set of places;*
- *$T = T_{el} \uplus T_{ab}$ is a finite set of transitions with $P \cap T = \emptyset$, and T_{el} (respectively T_{ab}) is the subset of elementary (respectively abstract) transitions;*
- *W^- and W^+ are the $\mathbb{N}^{P \times T}$ backward and forward incidence matrices;*
- *$\Omega : T_{ab} \to \mathbb{N}^P$ is a function that labels every abstract transition with a starting marking;*
- *$\mathcal{F}$ is a finite set of final markings.*

Figure 1 graphically describes an example of an RPN with:

$$P = \{p_{ini}, p_{fin}, p_{beg}, p_{end}\} \cup \{p_{b_i}, p_{a_i} : i \le 2\};$$
$$T_{el} = \{t_{b_1}, t_{b_3}, t_{a_1}, t_{a_3}, t_{sa}, t_{sb}\} \, ; \, T_{ab} = \{t_{beg}, t_{b_2}, t_{a_2}\};$$
$$\mathcal{F} = \{p_{end}, p_{beg}\}.$$

and for instance $W^-(p_{ini}, t_{beg}) = 1$ and $\Omega(t_{b_2}) = p_{beg}$ (where p_{beg} denotes the marking with one token in place p_{beg} and zero elsewhere).

For brevity reasons, we denote by $W^+(t)$ a vector in $\mathbb{N}^P$, where for all $p \in P$ $W^+(t)(p) = W^+(p, t)$, and do the same for $W^-(t)$.

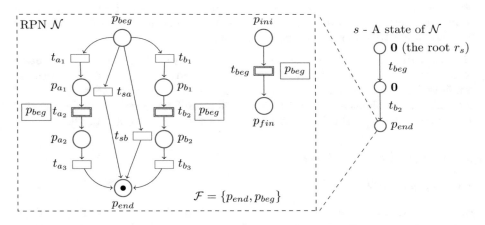

Fig. 1. An example of an RPN with the palindrome language on $\Sigma = \{a, b\}$ (see the proof of Proposition 2 in [8]).

A *state* s of an RPN is a labeled tree representing relations between threads and their associated markings. Every vertex of s is a thread and edges are labeled by abstract transitions.

Definition 2 (State of an RPN). *A state $s = \langle V, M, E, \Lambda \rangle$ of an RPN $\mathcal{N} = \langle P, T, W^+, W^-, \Omega, \mathcal{F} \rangle$ is a 4-tuple where:*

- *V is a finite set of vertices;*
- *$M : V \to \mathbb{N}^P$ is a function that labels vertices with markings;*

- $E \subseteq V \times V$ is a set of edges such that (V, E) is a Λ-labeled directed tree;
- $\Lambda : E \to T_{ab}$ is a function that labels edges with abstract transitions.

In the following, we denote by: $V_s := V$, $M_s := M$, $E_s := E$ and $\Lambda_s := \Lambda$.

For example, on the right side of Fig. 1 there is a state of the RPN $\mathcal{N}$. The state consists of three threads with markings $\mathbf{0}, \mathbf{0}$, and p_{end} (where $\mathbf{0}$ is the *null marking*) and two edges with the labels t_{beg}, t_{b_2}. Since a state consists of a directed tree, it has a *root thread* denoted by r_s.

Let s be a state of some RPN. Since a state has the structure of a directed tree every thread u has a *predecessor*, denoted by $prd(u)$, except the root. We call the vertices v for which there exists $(u, v) \in E$ a *child* of u. The *descendants* of a thread u consists of threads in the sub-tree rooted in u including u itself. We denote this set by $Des_s(u)$. Similarly the *ancestors* of u are the threads for which u is a descendant of, i.e. $Asc_s(u) = \{v \mid u \in Des_s(v)\}$. Denote by $\perp$ the empty tree. As usual two markings $m, m' \in \mathbb{N}^P$ over a set of places P are partially order as follows: $m \leq m'$ if for all places $p \in P$, $m(p) \leq m'(p)$.

The RPN moves from one state to another by one of the threads firing an elementary, abstract or cut transition. Let us present the first two kinds:

Definition 3 (operational semantics). *Let $\mathcal{N} = \langle P, T, W^+, W^-, \Omega, \mathcal{F} \rangle$ be an RPN, s be a state with some thread v and t be a transition (elementary or abstract). We say that $t \in T$ is* fireable *by v from $s = \langle V, M, E, \Lambda \rangle$ if $M(v) \geq W^-(t)$. In this case, its firing leads to the state $s' = \langle V', M', E', \Lambda' \rangle$, denoted $s \xrightarrow{(v,t)} s'$, where s' is defined below:*

- *If $t \in T_{el}$ then $s' = \langle V, M', E, \Lambda \rangle$ where $M'(u) = M(u)$ for all $u \in V \setminus \{v\}$ and $M'(v) = M(v) - W^-(t) + W^+(t)$;*
- *If $t \in T_{ab}$ then:*
  - *$V' = V \cup \{w\}$ where w is a fresh identifier ($w \notin V$);*
  - *$M'(u) = M(u)$ for all $u \in V \setminus \{v\}$, $M'(v) = M(v) - W^-(t)$ and $M'(w) = \Omega(t)$;*
  - *$E' = E \cup \{(v, w)\}$;*
  - *$\Lambda'(e) = \Lambda(e)$ for all $e \in E$ and $\Lambda((v, w)) = t$.*

Figure 2 illustrates the cases of an abstract and elementary transition firing. The first transition $t_{beg} \in T_{ab}$, is fired by the root. Its firing results in a state for which the root has a new child (denoted by v) and a new outgoing edge with label t_{beg}. The marking of the root is decreased to 0 and v gets the initial marking $\Omega(t_{beg}) = p_{beg}$. The second firing is due to an elementary transition $t_{b_1} \in T_{el}$ which is fired by v. Its firing results in a state for which the marking of v is changed to $M'_s(v) = M_s(v) + W^+(t_{b_1}) - W^-(t_{b_1}) = p_{b_1}$.

We now introduce the last type of transition: *cut transition*. Given a state $s = \langle V, M, E, \Lambda \rangle$ and a thread $v \in V$, we denote by $s \setminus v$ the state $s' = \langle V', M', E', \Lambda' \rangle$, with:

- $V' = V \setminus Des_s(v)$;

– M' is the restriction of M on V', and if $v \neq r$, $M'(prd(v)) = M(prd(v)) + W^+(\Lambda(prd(v), v))$;
– $E' = E \cap (V' \times V')$;
– Λ' is the restriction of Λ on E'.

Note that if v is the root of the tree then $s \backslash v = \perp$.

Definition 4 (τ cut transition). *Let $\mathcal{N} = \langle P, T, W^+, W^-, \Omega, \mathcal{F} \rangle$ be an RPN, $s = \langle V, M, E, \Lambda \rangle$ be a state of $\mathcal{N}$ and $v \in V$. We say that τ is* fireable *by v from s and reaches s', denoted by $s \xrightarrow{(v,\tau)} s'$, if and only if there exists $m \in \mathcal{F}$ such that $M(v) \geq m$ and $s' = s \backslash v$.*

For example, in Fig. 2 the fifth transition to be fired is the cut transition τ, fired by the thread with the marking p_{end} (denoted by w). Its firing results in a state where the thread w is erased and its parent has its marking increased by $W^+(t_{b_2}) = p_{b_2}$.

A *firing sequence* is a sequence of transition firings, written in detailed way: $s_0 \xrightarrow{(v_1,t_1)} s_1 \xrightarrow{(v_2,t_2)} \cdots \xrightarrow{(v_n,t_n)} s_n$, or when the context allows it, in a more concise way like $s_0 \xrightarrow{\sigma} s_n$ for $\sigma = (v_1, t_1)(v_2, t_2) \ldots (v_n, t_n)$. Infinite firing sequences are similarly defined. A thread is *final* (respectively *initial*) w.r.t. σ if it occurs in the final (respectively initial) state of σ. We say that $v \in Des_\sigma(u)$ if there exists $i \leq n$ such that $v \in Des_{s_i}(u)$. We call σ' a *subsequence* of σ, denoted by $\sigma' \sqsubseteq \sigma$, if there exists $i_1 < i_2 < \ldots i_k \leq n$ such that: $\sigma' = (v_{i_1}, t_{i_1})(v_{i_2}, t_{i_2}) \ldots (v_{i_k}, t_{i_k})$.

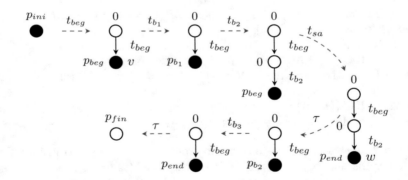

Fig. 2. Firing sequence for the RPN in Fig. 1

Remark 1. In the rest of the paper, anywhere we write "RPN $\mathcal{N}$" we will mean $\mathcal{N} = \langle P, T, W^+, W^-, \Omega, \mathcal{F} \rangle$, unless we explicitly write differently.

2.2   An Order for Recursive Petri Nets

We now define a quasi-order on the states of an RPN. Given two states s, s' of an RPN $\mathcal{N}$ we say that s is smaller or equal than s' if there is a subtree in s' which is isomorphic to s, where markings are greater or equal on each vertex, and such that the labels on the edges fulfill $W^+(t) \geq W^+(t')$.

Definition 5. *Given two states $s = \langle V, M, E, \Lambda \rangle$ and $s' = \langle V', M', E', \Lambda' \rangle$ of an RPN $\mathcal{N}$, we say that $s \preceq s'$ if and only if there exists an injective total function $f : V \rightarrow V'$ such that:*

1. *For any edge $(u, v) \in E$, we have $(f(u), f(v)) \in E'$;*
2. *For any edge $(u, v) \in E$, we have $W^+(\Lambda(u, v)) \leq W^+(\Lambda'(f(u), f(v)))$;*
3. *For any thread $v \in V$, we have $M(v) \leq M'(f(v))$.*

Figure 3 illustrates this quasi-order. The state on the right is greater than the one on the left, if $W^+(t) \geq W^+(t')$.

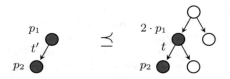

Fig. 3. Example of order between two states.

Lemma 1. *The relation $\preceq$ is a quasi-order.*

This quasi-order is not a partial order since there could be abstract transitions $t \neq t'$ with $W^+(t) = W^+(t')$.

A quasi-order $\leq$ on the states of an RPN is *strongly compatible* (as in [9]) if for all states $s \leq s'$ and firing $s \xrightarrow{(v,t)} s_1$ there exist s'_1 and a firing $s' \xrightarrow{(v',t)} s'_1$ with $s_1 \leq s'_1$.

Lemma 2. *The quasi-order $\preceq$ is strongly compatible.*

Note that, even though this quasi-order is compatible it may contain an infinite set of incomparable states (i.e. an infinite *antichain*). For example, see Fig. 4 where any two states s_i and s_j are incomparable. Indeed for any $i < j$: (1) $s_i \npreceq s_j$ since $|V_{s_j}| > |V_{s_i}|$ there cannot be any injective function from V_{s_j} to V_{s_i}, and (2) $s_j \npreceq s_i$ since for any injective function from V_{s_i} to V_{s_j} at least one of the threads with the marking p would be mapped to a thread with marking $\mathbf{0}$.

Since RPNs are not well ordered they are not well-structured transition systems [9] for which coverability is often solved in **EXPSPACE**. Therefore to solve coverability, one needs to find another way.

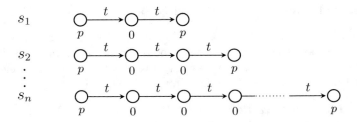

Fig. 4. An example for an antichain of states

3   Expressiveness

Expressiveness of a formalism may be defined by the family of languages that it can generate. In [11], expressiveness of RPNs was studied using reachability languages. In the next section, we are going to establish that the coverability problem for RPNs has a lower complexity than the one of the reachability problem for RPNs. Thus we want to study the coverability languages in order to determine whether this lower complexity has an impact on the expressiveness of the RPN formalism. So we equip any transition t with a *label* $\lambda(t) \in \Sigma \cup \{\varepsilon\}$ where Σ is a finite alphabet and ε is the empty word. The labelling is extended to transition sequences in the usual way with the cut transitions labelled by the empty word. Thus given a labelled marked RPN $\mathcal{N}$ and a finite subset of states S_f, the (coverability) language $\mathcal{L}(\mathcal{N}, S_f)$ is defined by:

$$\mathcal{L}(\mathcal{N}, S_f) = \{\lambda(\sigma) \mid \exists \; s_0 \xrightarrow{\sigma} s \succeq s_f \land s_f \in S_f\}$$

i.e. the set of labellings for sequences covering some state of S_f in $\mathcal{N}$.

As already announced, languages of RPNs are closed by union.

Proposition 1. *Coverability languages of RPNs are closed by union.*

The next theorem has two interesting consequences: the languages of RPNs are not closed by intersection with a regular language and this family is *quite close* to recursively enumerable languages.

Theorem 1. *Let $\mathcal{L}$ be a recursively enumerable language. Then there exists an RPN language $\mathcal{L}_1$, a regular language $\mathcal{R}_2$ and a homomorphism h such that $\mathcal{L} = h(\mathcal{L}_1 \cap \mathcal{R}_2)$.*

Proof. The result was stated in Proposition 9 of [10] for reachability languages but it also works for coverability languages since the reachability condition of the proof could easily be transformed into a coverability condition.   □

Obviously the coverability languages of RPNs include the one of PNs. In order to show its expressive power, let us introduce context-free grammars and languages. Let $G = (V, \Sigma, R, S)$ be a context-free grammar defined by V the non terminal symbols including S, the start symbol and Σ the terminal symbols.

The set of rules R is defined by $R = \{r_1, \ldots, r_n\}$ such that $r_i = (v_i, u_i)$, with $v_i \xrightarrow{r_i} u_i$, $v_i \in V$ and $u_i \in (V \cup \Sigma)^*$ a word of length n_i. W.l.o.g. one assumes that the start symbol S does not occur in the right-hand side of any production rule of G and that $n_i > 0$ except possibly for a rule $S \to \varepsilon$. Given a word $\alpha v_i \beta \in (V \cup \Sigma)^*$, an application of rule r_i yields the word $\alpha u_i \beta$, denoted $\alpha v_i \beta \xrightarrow{r_i} \alpha v_i \beta$. A *derivation* from α to β is a consecutive application of a sequence of rules, and is denoted $\alpha \xrightarrow{\sigma} \beta$. The associated language $\mathcal{L}(G)$ is defined by:

$$\mathcal{L}(G) = \{w \in \Sigma^* \mid \exists \sigma \; S \xrightarrow{\sigma} w\}$$

Proposition 2. *Context-free languages are included in coverability languages of RPNs.*

The next lemma witnesses a Petri net language interesting from an expressiveness point of view.

Proposition 3. *Let $\Sigma = \{a, b, c\}$ and $\mathcal{L}_1 = \{a^m b^n c^p \mid m \geq n \geq p\}$. Then $\mathcal{L}_1$ is the coverability language of some Petri net and is not a context-free language.*

Using the previous results, the next theorem emphasises the expressive power of coverability languages of RPNs.

Theorem 2. *Coverability languages of RPNs strictly include the union of coverability languages of PNs and context-free languages.*

Proof. The inclusion is an immediate consequence of Proposition 2. Consider the language $\mathcal{L}_2 = \mathcal{L}_1 \cup \{w\tilde{w} \mid w \in \{d, e\}^*\}$ where $\tilde{w}$ is the mirror of w. Since (1) by Proposition 1, coverability languages of RPNs are closed by union, (2) $\mathcal{L}_1$ is a PN language, and (3) the language of palindromes is a context-free language, we deduce that $\mathcal{L}_2$ is an RPN language.
PN and context-free languages are closed by homomorphism. Since the projection of $\mathcal{L}_2$ on $\{a, b, c\}$ is the language of Proposition 3, $\mathcal{L}_2$ is not a context-free language. Since the projection of $\mathcal{L}_2$ on $\{d, e\}$ is the language of palindroms, $\mathcal{L}_2$ is not a PN language (see [15]). □

The next proposition establishes that, as for Petri nets, coverability does not ensure the power of "exact counting".

Proposition 4. *Let $\Sigma = \{a, b, c\}$ and $\mathcal{L}_3 = \{a^n b^n c^n \mid n \in \mathbb{N}\}$. Then $\mathcal{L}_3$ is the reachability language of some Petri net and is not the coverability language of any RPN.*

Proof. Consider the net below and p_f be the single marking to be reached. Then its language is $\mathcal{L}_3$.

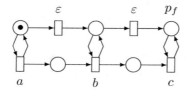

Assume that there exists a pair $(\mathcal{N}, S_f)$ such that $\mathcal{L}_3 = \mathcal{L}(\mathcal{N}, S_f)$. Define the subset of abstract transitions T_ε such that for $t \in T_\varepsilon$ there exists a firing sequence labelled by ε starting from a single thread marked by $\Omega(t)$ that reaches the empty tree. Adding a set of elementary transitions $\{t_\varepsilon \mid t \in T_\varepsilon\}$ where t and t_ε have same incidence does not modify the language of the net. For all n, let σ_n be a coverability sequence such that $\lambda(\sigma_n) = a^n b^n c^n$ and σ'_n be the prefix of σ_n whose last transition corresponds the last occurrence of a. Denote s_n the state reached by σ'_n and the decomposition by $\sigma_n = \sigma'_n \sigma''_n$. Among the possible σ_n, we select one such that s_n has a minimal number of threads.

Case 1. There exists a bound B of the depths of the trees corresponding to $\{s_n\}_{n \in \mathbb{N}}$. Let S_B be the set of states of depth at most B. Observe that $S_0 = \mathbb{N}^P$ and S_B can be identified to $\mathbb{N}^P \times \mathsf{Multiset}(T_{ab} \times S_{B-1})$. Furthermore the (component) order on $\mathbb{N}^P$ and the equality on T_{ab} are well quasi-orders. Since well quasi-ordering is preserved by the multiset operation and the cartesian product, S_B is well quasi-ordered by an quasi-order denoted $<$. By construction, $s \leq s'$ implies $s \preceq s'$. Thus there exist $n < n'$ such that $s_n \preceq s_{n'}$ which entails that $\sigma'_{n'} \sigma''_n$ is a covering sequence with trace $a^{n'} b^n c^n$ yielding a contradiction.

Case 2. The depths of the trees corresponding to s_n are unbounded. Let C be a strict upper bound of the depths of the initial state and the final states. There exists n such that the depth of s_n is greater than $(4|T_{ab}| + 1)C$. Thus in s_n, there are threads v_1, v_2 and v_3 and v_4 in the same branch at levels respectively $i_1 C$, $i_2 C$, $i_3 C$ and $i_4 C$ with $0 < i_1 < i_2 < i_3 < i_4$ created along σ_n by the firing of the same abstract transition t. Denote Tr the subtree of the final state of σ_n that matches the state to be covered and Br the branch leading to Tr in the final state. Due to the choice of C, there exists $1 \leq i \leq 4$ such that:

- Br_i, the branch from v_i to v_{i+1}, does not intersect Tr;
- either Br_i does not intersect Br or Br_i is included in Br.

For $k \in \{i, i+1\}$, consider the trace w_k of the sequence performed in the subtree rooted in v_k by the firings of σ_n.

Case $w_i = w_{i+1}$. Then one can build another covering sequence with trace $a^n b^n c^n$ by mimicking in v_i the behaviour of v_{i+1} leading to another state s_n with less threads yielding a contradiction, since s_n was supposed to have a minimal number of threads.

Case $w_i \neq w_{i+1}$. Let $w \neq \varepsilon$ the trace of the sequence performed in the subtree rooted in v_i without the trace of the sequence performed in the subtree rooted in v_{i+1}. Then one can build another covering sequence σ by mimicking in v_{i+1} the behaviour of v_i. The trace of σ is an interleaving of $a^n b^n c^n$ and w which implies that $w = a^q b^q c^q$ for some $q > 0$. Furthermore σ can be chosen in such a way that the firing subsequences in the subtrees rooted at v_i and v_{i+1} are performed in one shot which implies that its trace is $\ldots a^q a^q w_{i+1} b^q c^q b^q c^q \ldots$ yielding a contradiction. $\qquad \square$

The following corollary shows that extending coverability languages of Petri nets substituting (1) coverability by reachability or (2) Petri nets by RPNs are somewhat "orthogonal".

Corollary 1. *The families of reachability languages of Petri nets and the coverability languages of RPNs are incomparable.*

Proof. One direction is a consequence of Proposition 4 while the other direction is a consequence of Proposition 2 observing that the language of palindromes is not the reachability language of any Petri net.   □

The next corollary exhibits a particular feature of RPNs languages (e.g. not fulfilled by Petri nets or context-free languages)

Corollary 2. *Coverability languages of RPNs are not closed by intersection with a regular language.*

Proof. Due to Proposition 4, coverability languages of RPNs are strictly included in recursively enumerable languages. Since it is closed by homomorphism, Theorem 1 implies that it is not closed by intersection with a regular language.   □

4   Coverability Is EXPSPACE-Complete

Let $\mathcal{N}$ be an RPN and s_{ini}, s_{tar} be two states of $\mathcal{N}$. The *coverability problem* asks whether there exists a firing sequence $s_{ini} \xrightarrow{\sigma} s \succeq s_{tar}$. Such a sequence σ with initial and target states, is called a *covering sequence*. The section is devoted to establishing that this problem is EXPSPACE-complete. The EXPSPACE-hardness follows immediately from the EXPSPACE-hardness of the coverability problem for Petri nets [19].

In [22], Rackoff showed that the coverability problem for Petri nets belongs to EXPSPACE. More precisely, he proved that if there exists a covering sequence, then there exists a 'short' one:

Theorem 3. (Rackoff [22]). *Let $\mathcal{N}$ be a Petri net, m_{ini}, m_{tar} be markings and σ be a firing sequence such that $m_{ini} \xrightarrow{\sigma} m \geq m_{tar}$. Then there exists a sequence σ' such that $m_{ini} \xrightarrow{\sigma'} m' \geq m_{tar}$ with $|\sigma'| \leq 2^{2^{cn \log n}}$ for some constant c and n being the size of $(\mathcal{N}, m_{tar})$.*

So to solve the coverability problem on Petri nets, one guesses a sequence of length at most $2^{2^{cn \log n}}$, checking at the same time whether it is a covering sequence in exponential space. Which shows that the coverability problem belongs to NEXPSPACE = EXPSPACE by Savitch's theorem.

We follow a similar line and more specifically, we show that if there exists a covering sequence $s_{ini} \xrightarrow{\sigma} s \succeq s_{tar}$ in an RPN $\mathcal{N}$, then there exists a 'short' covering sequence σ'.

First, we establish that the final state of a covering sequence can be chosen with a limited number of threads (Proposition 5). Then we enlarge the RPN $\mathcal{N}$ with new elementary transitions getting $\widehat{\mathcal{N}}$, leaving the coverability problem unchanged. The interest of $\widehat{\mathcal{N}}$ is that a covering sequence (when it exists) can be chosen with a particular form that we call *well-sequenced* without increasing its length (Proposition 6). In order to come back to $\mathcal{N}$, we establish that the firing of an additional transition of $\widehat{\mathcal{N}}$ can be simulated by a short sequence in $\mathcal{N}$ (Proposition 7). Proposition 8 combines these intermediate results to get an upper bound for a short covering sequence.

Let σ be a firing sequence. A thread is *extremal* w.r.t. σ if it is an initial or final thread. We show that we can bound the number of extremal threads in a covering sequence. In the sequel, the size of the input of the coverability problem is denoted by η, i.e. the accumulated size of the RPN, the initial and target states. Recall that $Asc_s(v)$ is the set of ancestors of v in s.

Proposition 5. *Let $\mathcal{N}$ be an RPN and $s_{ini} \xrightarrow{\sigma} s \succeq s_{tar}$ be a covering sequence. Then there is a sequence $s_{ini} \xrightarrow{\sigma'} s' \succeq s_{tar}$ such that $|V_{s'}| \leq 3\eta$.*

Proof. If $s_{tar} = \perp$ then $\sigma' = \varepsilon$ is the appropriate sequence. Otherwise denote $f : V_{s_{tar}} \to V_s$ the injective mapping associated with $s_{tar} \preceq s$. Let $U = Asc_s(f(r_{s_{tar}})) \setminus \{f(r_{s_{tar}})\}$ be the branch in s leading to the vertex corresponding to the root of s_{tar}. Consider the set $V = V_s \setminus (U \cup f(V_{s_{tar}}))$. Then one can delete in σ all transitions fired from threads in $Des_\sigma(V)$ and those that created the threads of V and still get a covering sequence.

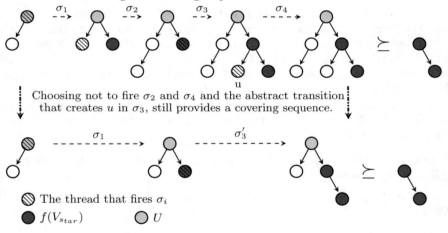

Choosing not to fire σ_2 and σ_4 and the abstract transition that creates u in σ_3, still provides a covering sequence.

◐ The thread that fires σ_i
● $f(V_{s_{tar}})$ ○ U

Now assume that on the branch U, two edges (u_1, v_1) and (u_2, v_2) are labelled by the same transition where u_2 is a descendent of u_1 and $v_1 \notin V_{s_{ini}}$. Then one can delete all transitions fired in the subbranch from v_1 to u_2 and substitute transitions (v_2, t) by transitions (v_1, t) and still get a covering sequence. So $|U \setminus V_{s_{ini}}| \leq |T_{ab}|$.

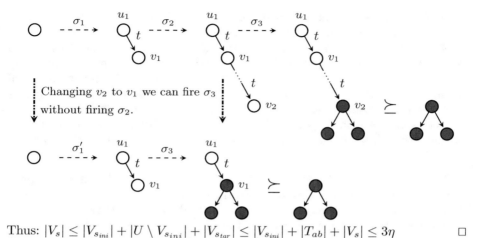

Thus: $|V_s| \leq |V_{s_{ini}}| + |U \setminus V_{s_{ini}}| + |V_{s_{tar}}| \leq |V_{s_{ini}}| + |T_{ab}| + |V_s| \leq 3\eta$ $\square$

Let $T_{ret} \subseteq T_{ab}$, the set of *returning transitions* be defined by: $t \in T_{ret}$ if there exists a firing sequence (called a *return sequence*): $s_t \xrightarrow{\sigma_t} \bot$, where $V_{s_t} = \{v_t\}$, $M_{s_t}(v_t) := \Omega(t)$, and $E_{s_t} = \Lambda_{s_t} = \emptyset$. For any $t \in T_{ret}$, we define σ_t to be some arbitrary shortest return sequence.

As mentioned before, we get $\widehat{\mathcal{N}}$ from $\mathcal{N}$ by adding elementary transitions as follows.

Definition 6. $(\widehat{\mathcal{N}})$. *Let* $\mathcal{N} = \langle P, T, W^+, W^-, \Omega, \mathcal{F} \rangle$ *be an RPN. Then* $\widehat{\mathcal{N}}$ *is an RPN where* $\widehat{\mathcal{N}} = \langle P, \widehat{T}, \widehat{W}^+, \widehat{W}^-, \Omega, \mathcal{F} \rangle$, *where* $\widehat{T}$ *is* T *with additional elementary transitions, and* $\widehat{W}^+, \widehat{W}^-$ *are updated accordingly:* $\widehat{T}_{el} = T_{el} \uplus \{t^r : t \in T_{ret}\}$, *and for any new transition* t^r, $\widehat{W}^-(t^r) = W^-(t)$ *and* $\widehat{W}^+(t^r) = W^+(t)$.

The key ingredient of the existence of a short sequence is that in $\widehat{\mathcal{N}}$ every sequence can be turned into a *well-sequenced* sequence reaching the same state. Along such a sequence, (1) there are only extremal threads, (2) firings are performed in one shot by threads, and (3) only initial threads disappear and final threads perform firings of abstract transitions.

Definition 7. *Let* $\mathcal{N}$ *be an RPN and* σ *be a firing sequence. Then* σ *is well-sequenced if* $\sigma = \sigma_1(v_1, \tau)\sigma_2(v_2, \tau)\ldots\sigma_\ell(v_\ell, \tau)\sigma_{\ell+1}\sigma_{\ell+1}^{ab}\ldots\sigma_k\sigma_k^{ab}$ *where:*

- *The threads* v_i *are initial for* $1 \leq i \leq \ell$;
- *The threads* v_i *are final for* $\ell + 1 \leq i \leq k$;
- *The firing sequence* $\sigma_i \in (\{v_i\} \times T_{el})^*$ *for* $1 \leq i \leq k$;
- *The firing sequence* $\sigma_i^{ab} \in (\{v_i\} \times T_{ab})^*$ *for* $\ell + 1 \leq i \leq k$.

Proposition 6. *Let* $\mathcal{N}$ *be an RPN and* $s \xrightarrow{\sigma} s'$ *be a firing sequence. There exists a well-sequenced firing sequence* $s \xrightarrow{\widehat{\sigma}} s'$ *in* $\widehat{\mathcal{N}}$, *with* $|\widehat{\sigma}| \leq |\sigma|$.

Proof. (sketch, full proof in [8]). By construction, σ is fireable from s to s' in $\widehat{\mathcal{N}}$. Therefore all we are left with is turning σ into a well-sequenced sequence.

Assume that an extremal thread u fires $t \in T_{ab}$ which creates a non-extremal thread v disappearing by the cut transition (v, τ) occurring in σ. For all such v's let us (1) delete from σ the step (u, t), and all the firings from $Des_\sigma(v)$ and (2) replacing the step (v, τ) by (u, t^r). After this operation, no cut transition matches the firing of an abstract transition. Assume that an initial and not final thread u fires abstract transitions. Then one deletes these firings and all firings in the descendants of u. So along σ, there are only extremal threads.

Let us establish the other requirements on σ by induction on the number of extremal threads. There are three cases when adding a new thread with maximal depth:

1. This thread is an initial thread and a leaf in the final state. Then we can push the sequence of firings it performs to the end of the sequence.
2. The thread is final and not initial. Hence the subsequence of firings it performs in σ consists of elementary transition firings that can be fired at the end of σ. Furthermore, the abstract transition that created it can be fired at the end of the firing sequence of its parent.
3. This thread is initial and not final (i.e. in $\{v_1, \ldots, v_\ell\}$). Hence the subsequence of firings it performs in σ consists of elementary transition firings possibly ended by a cut transition. If the cut transition occurs, the subsequence can be fired immediately. Otherwise, it can be omitted.

This concludes the proof.   □

In order to recover from a sequence in $\widehat{\mathcal{N}}$ a sequence in $\mathcal{N}$, for every $t \in T_{ret}$ one has to simulate the firings of a transition t^r by sequence σ_t. Therefore bounding the length of σ_t is critical.

Proposition 7. *Let $\mathcal{N}$ be an RPN and $t \in T_{ret}$. Then the returning sequence σ_t fulfills $|\sigma_t| \leq 2^{\cdot 2^{dn \log n}}$ for some constant d and $n = size(\mathcal{N})$.*

Proof. Let us enumerate $T_{ret} = \{t_1, \ldots, t_K\}$ in such a way that $i < j$ implies $|\sigma_{t_i}| \leq |\sigma_{t_j}|$. Observe first that the shortest returning sequences do not include firings of abstract transitions not followed by a matching cut transition since it could be omitted as it only deletes tokens in the thread. We argue by induction on $k \leq K$ that:

$$|\sigma_{t_k}| < 2^{k \cdot 2^{cn \log n}} \qquad \text{where } c \text{ is the Rackoff constant}$$

For $k = 1$, we know that σ_{t_1} has a minimal length over all returning sequences. Hence there are no cuts in σ_{t_1} except the last one. Due to the above observation, σ_{t_1} only includes firing of elementary transitions. Thus the Rackoff bound of Theorem 3 applies for a covering of some final marking.

Assume that the result holds for $k - 1$. Due to the requirement on lengths, σ_{t_k} only includes cuts from threads created by $t_i \in T_{ret}$ with $i < k$. Thus by Proposition 6 we get a sequence $\widehat{\sigma}_{t_k} \cdot (v_{t_k}, \tau)$ in $\widehat{\mathcal{N}}$. The sequence $\widehat{\sigma}_{t_k}$ consists of only elementary steps and does not contain any transition t_i^r with $i \geq k$. The marking reached by $\widehat{\sigma}_{t_k}$ covers some final marking, hence by Theorem 3 there exists a covering sequence $\widehat{\sigma}'_{t_k}$ such that $|\widehat{\sigma}'_{t_k}| \leq 2^{2^{cn \log n}}$. Since $\widehat{\sigma}_{t_k}$ does not contain firing of t_i^r with $i \geq k$ this also holds for $\widehat{\sigma}'_{t_k}$. Substituting any firing of t_i^r by the sequence σ_{t_i}, one gets a corresponding sequence σ'_{t_k} in $\mathcal{N}$. Using the induction hypothesis, one gets that the length of this sequence:

$$|\sigma'_{t_k}| \leq |\widehat{\sigma}_{t'_k}| 2^{(k-1)\cdot 2^{cn \log n}} \leq 2^{2^{cn \log n}} \cdot 2^{(k-1)\cdot 2^{cn \log n}} \leq 2^{k \cdot 2^{cn \log n}}$$

From minimality of σ_{t_k}, one gets $|\sigma_{t_k}| \leq |\sigma'_{t_k}| \leq 2^{k \cdot 2^{cn \log n}}$ which concludes the proof since

$$\max_{t \in T_{ret}} \{|\sigma_t|\} \leq 2^{|T_{ret}| \cdot 2^{cn \log n}} \leq 2^{n 2^{cn \log n}} \leq 2^{2^{(2c)n \log n}}.$$

$\square$

Combining all previous results, we can now bound the length of a shortest covering sequence:

Proposition 8. *Let $\mathcal{N}$ be an RPN, and $s_{ini} \xrightarrow{\sigma} s \succeq s_{tar}$. Then there exists a covering sequence of length shorter than $2^{2^{en \log \eta}}$, where e is some constant and $\eta = size(\mathcal{N}, s_{ini}, s_{tar})$.*

Proof. Using Proposition 5 we can assume that $|V_{s_{ini}} \cup V_s| \leq 3\eta$. Using Proposition 6 one gets a well-sequenced sequence $s_{ini} \xrightarrow{\widehat{\sigma}} s$ in $\widehat{\mathcal{N}}$, such that:

$$\widehat{\sigma} = \sigma_1(v_1, \tau)\sigma_2(v_2, \tau) \ldots \sigma_\ell(v_\ell, \tau)\sigma_{\ell+1}\sigma_{\ell+1}^{ab} \ldots \sigma_k \sigma_k^{ab},$$

where $\sigma_i^{ab} = (v_i, t_{i,1}) \ldots (v_i, t_{i,n_i})$. Observe that $k \leq |V_{s_{ini}} \cup V_s|$. We now show that there is a short covering sequence in $\widehat{\mathcal{N}}$. Let $f : V_{s_{tar}} \to V_s$ the function associated with $s \succeq s_{tar}$. Each of the σ_i is a sequence whose final marking of v_i covers some marking:

1. For $i \leq \ell$, a final marking of the net;
2. For $i > \ell$ and $v_i \notin f(V_{s_{tar}})$, $\sum_j W^-(t_{i,j})$;
3. For $i > \ell$ and $v_i \in f(V_{s_{tar}})$, $\sum_j W^-(t_{i,j}) + M_{s_{tar}}(f^{-1}(v_i))$.

Since all σ_i contain only elementary steps, using Theorem 3, one gets as sequence σ'_i with $|\sigma'_i| \leq 2^{2^{cn \log \eta}}$ covering the marking specified by the three cases above. Define the sequence $s \xrightarrow{\widehat{\sigma}'} s'$ where each σ_i is replaced by σ'_i. Using case 3, for all $v \in V_{s_{tar}}$ $M_{s'}(f(v)) \geq M_{s_{tar}}(v)$. Therefore $s' \succeq s_{tar}$, and the length of $\widehat{\sigma}'$ is at most:

$$|\widehat{\sigma}'| = \sum_{i=1}^{k} |\sigma'_i| + \sum_{i=1}^{\ell} |(v_i, \tau)| + \sum_{i=\ell+1}^{k} |\sigma_i^{ab}| \leq 3\eta 2^{2^{cn \log \eta}} + 3\eta + 3\eta \leq 2^{2^{2cn \log \eta}}.$$

Substituting any firing of t_i^r by σ_{t_i} in $\widehat{\sigma}'$ we get a covering sequence σ' in $\mathcal{N}$. Using Proposition 7, its length fulfills:

$$|\sigma'| \leq |\widehat{\sigma}'| \cdot 2^{2^{dn \log n}} \leq 2^{2^{en \log n}}$$

for some constant e.

$\square$

Using Proposition 8 we establish the complexity of the coverability problem.

Theorem 4. *The coverability problem for RPNs is* EXPSPACE-*complete.*

Proof. According to Proposition 8, if there is a covering sequence then there is one with length at most $2^{2^{e\eta \log \eta}}$ and no more than 4η threads. Hence one guesses a sequence of at most this length and checks simultaneously whether it is a covering sequence in exponential space. This shows that the coverability problem belongs to NEXPSPACE = EXPSPACE by Savitch theorem.   □

5   Termination Is EXPSPACE-Complete

Let $\mathcal{N}$ be an RPN and s_{ini} be an initial state of $\mathcal{N}$. The *termination problem* asks whether there exists an infinite firing sequence starting from s_{ini}. In [22] Rackoff showed that the termination problem on Petri net is solvable in exponential space:

Theorem 5 (Rackoff). *The termination problem for Petri nets is* EXPSPACE-*complete.*

We aim to show that the termination problem for RPN is EXPSPACE-complete. EXPSPACE-hardness follows immediately from EXPSPACE-hardness of the termination problem for Petri nets [19].

We first introduce and solve the *constrained termination problem* which asks whether there exists an infinite firing sequence starting from s_{ini} which does not delete any threads of s_{ini}. Accordingly, a *constrained firing sequence* is a firing sequence that does not delete any initial thread. The size of the input of the termination problem is denoted by η.

A main ingredient of the proof is a construction of an *abstract graph* related to the firing of abstract transitions. To this aim, let m be a marking then $s[m]$ is the state with a single thread whose marking is m.

Definition 8 (abstract graph). *Let $\mathcal{N}$ be an RPN and s_{ini} be an initial state. Let $G_{\mathcal{N},s_{ini}} = (V_a, E_a, M_a)$ be a labelled directed graph defined by:*

1. *$V_a = \{v_t \mid t \in T_{ab}\} \uplus V_{s_{ini}}$;*
2. *$M_a : V_a \to \mathbb{N}^P$ where for all v in $V_{s_{ini}}$, $M_a(v) = M_{s_{ini}}(v)$ and for all t in T_{ab}, $M_a(v_t) := \Omega(t)$;*
3. *$E_a \subset V_a \times V_a$ such that for all $t \in T_{ab}$ and $v \in V_a$, $(v, v_t) \in E_a$ if there exists a firing sequence σ from the state $s[M_a(v)]$ ending by a firing (v, t).*

Equivalently assertion 3 means that the edge (v, v_t) belongs to E_a if there exists a covering sequence $s[M_a(v)] \xrightarrow{\sigma} s \succeq s[W^-(t)]$. Thus building the abstract graph amounts to solving a quadratic number of coverability problems. Using Theorem 4, one can build it in EXPSPACE.

Let us illustrate the abstract graph in Fig. 5 corresponding to the RPN of Fig. 1. Here the initial state is $s[p_{ini}]$. For clarity, we have renamed the abstract

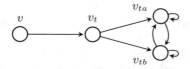

Fig. 5. An abstract graph for the RPN in Fig. 1

transitions as follows: $t := t_{beg}$, $ta := t_{a_2}$, $tb := t_{b_2}$. For instance, the existence of the edge from v_t to v_{ta} is justified by the firing sequence $(v_t, t_{a_1})(v_t, ta)$.

Let σ be an infinite firing sequence. We say that σ is *deep* if it reaches a state s whose depth is greater than $|T_{ab}| + |V_{s_{ini}}|$. Otherwise, we say that σ is *shallow*. To solve the constrained termination problem it suffices to show whether the RPN has such an infinite sequence, either shallow or deep.

The next lemma establishes that *lassos* of the abstract graph are witnesses of deep constrained infinite sequences in an RPN:

Lemma 3. *Let $\mathcal{N}$ be an RPN and s_{ini} be an initial state. Then there is a deep constrained infinite sequence starting from s_{ini} if and only if there is a path from $u \in V_{s_{ini}}$ to a cycle in $G_{\mathcal{N}, s_{ini}}$.*

We now show that for any shallow sequence σ there is a thread v which fires infinitely many times in σ.

Lemma 4. *Given $\mathcal{N}$ an RPN with an initial state s, and σ a shallow sequence. Then there is a thread v that fires infinitely many times in σ.*

Next, show that given a state consisting of one thread, one can check in **EXPSPACE** the existence of a shallow sequence which fires infinitely many times from this (root) thread.

Lemma 5. *Let $\mathcal{N}$ be an RPN and m be a marking. Then one can check whether there exists an infinite sequence starting from $s[m]$ with the root firing infinitely many times.*

Proof. There is a sequence firing infinitely many times from the root in $\mathcal{N}$ starting from $s[m]$ if and only if there is one in $\widehat{\mathcal{N}}$ starting from $s[m]$. Define $\widehat{\mathcal{N}}^e$ to be the Petri net with the same places as $\widehat{\mathcal{N}}$, and whose set of transitions is the union of the elementary transitions of $\widehat{\mathcal{N}}$ and the set $\{t^- \mid t \in T_{ab}\}$ where $W^-(t^-) = W^-(t)$ and $W^+(t^-) = 0$. We claim that there is a sequence firing infinitely many times from the root in $\widehat{\mathcal{N}}$ starting from $s[m]$ if and only if there is an infinite sequence in $\widehat{\mathcal{N}}^e$ starting from m.

For one direction, assume there exists such σ in $\widehat{\mathcal{N}}$. One eliminates in σ the cut steps by increasing occurrence order as follows. Let (v_i, τ) be a cut step and (v_j, t_j) be the step that creates v_i. Then one deletes all the steps performed by the descendants of v_i and replaces (v_j, t_j) by (v_j, t_j^r). Let σ' be the sequence obtained after this transformation. In σ', the root still fires infinitely often since no firing performed by the root has been deleted (but sometimes substituted by an elementary firing). Moreover, σ' has no more cut steps. One eliminates in σ' the abstract firings by increasing occurrence order as follows. Let (r, t_i) be an abstract firing that creates thread v. Then one deletes all the steps performed by the descendants of v and replaces (r, t_i) by (r, t_i^-). Let σ'' be the sequence obtained after this transformation. In σ'', the root still fires infinitely often since no firing performed by the root has been deleted (but sometimes substituted by an elementary firing). Moreover σ'' has only elementary steps. So it is an infinite sequence of $\widehat{\mathcal{N}}^e$.

The other direction is immediate. By Theorem 5, one can check in EXPSPACE whether there exists an infinite sequence on $\widehat{\mathcal{N}}^e$ with initial marking m.   □

Summing the results for shallow and deep sequences we get:

Proposition 9. *The constrained termination problem of RPN belongs to* EXPSPACE.

Proof. The algorithm proceeds as follows. It builds in EXPSPACE the abstract graph and it checks whether there is a constrained deep infinite sequence using the characterisation of Lemma 3. In the negative case, it looks for a constrained shallow infinite sequence. To this aim, it checks in EXPSPACE for any reachable vertex v from $V_{s_{ini}}$ in $G_{\mathcal{N}, s_{ini}}$, whether there exists an infinite sequence starting from $s[M_a(v)]$ with the root firing infinitely many times. The complexity follows from Lemma 5 while its correctness follows from Lemma 4.   □

We now prove that the termination problem is EXPSPACE by reducing it in EXPSPACE to an exponential number of instances of the constrained termination problem with similar size:

Theorem 6. *The termination problem of RPN is* EXPSPACE-*complete.*

Proof. W.l.o.g. we assume that $\mathcal{N}$ has been enlarged to $\widehat{\mathcal{N}}$ since the termination problem remains unchanged by this transformation. Let σ be an infinite firing sequence and U be the subset of initial threads that disappear along σ. Let σ' be the shortest finite prefix of σ such that all threads of U have disappeared. Using Proposition 6, σ' can be assumed to be well-sequenced. Consider σ_U, the prefix of σ' which deletes the threads of U and s_U the state reached by σ_U. State s_U is defined as follows: $V_{s_U} = V_s \setminus U$, E_{s_U} (respectively Λ_{s_U}) is the restriction of E_s (respectively Λ_s) to V_{s_U} and $M_{s_U}(v) = M_{s_{ini}}(v) + \sum_{(v,u) \in E_{s_{ini}} \wedge u \in U} W^+(\Lambda(v, u))$. Using the same proof as in Proposition 8 the length of some σ_U (when it exists) is at most doubly exponential.

An infinite firing sequence starting from s exists if and only if there exists an infinite constrained sequence from one of its reachable $s_U's'$. For any $U \subseteq V$ one checks whether s_U is reachable from s, and, in the positive case, solves the constrained termination problem for s_U. This can be done in EXPSPACE. Finally, there are only 2^η possible subsets $U \subseteq V_s$, repeating the process described above for every subset U solves the termination problem of RPN in EXPSPACE.   □

6   Conclusion

We have proven that RPN is a strict generalisation of both Petri nets and stack automata without increasing the complexity of coverability and termination problems. It remains several open problems about languages of RPN and decidability/complexity of checking properties. Here is a partial list of open problems: Is the family of covering languages of RPN included in the family of reachability languages of RPN? How to decide whether a word belongs to a coverability or reachability language of a RPN? What is the complexity of the boundedness and finiteness problems? Since the ordering posses an infinite antichain, but there exist short witnesses for coverability, does there exist an effective finite representation of the downward closure of the reachability set?

References

1. Atig, M.F., Ganty, P.: Approximating Petri net reachability along context-free traces. In: FSTTCS 2011, Mumbai, India, LIPIcs, vol. 13, pp. 152–163 (2011)
2. Bonnet, R.: The reachability problem for vector addition system with one zero-test. In: Murlak, F., Sankowski, P. (eds.) MFCS 2011. LNCS, vol. 6907, pp. 145–157. Springer, Heidelberg (2011). https://doi.org/10.1007/978-3-642-22993-0_16
3. Bonnet, R., Finkel, A., Leroux, J., Zeitoun, M.: Model checking vector addition systems with one zero-test. Logical Methods Comput. Sci. **8**(11), 1–25 (2012)
4. Czerwinski, W., Lasota, S., Lazic, R., Leroux, J., Mazowiecki, F.: The reachability problem for Petri nets is not elementary (extended abstract). CoRR, abs/1809.07115 (2018)
5. Dassow, J., Turaev, S.: Petri net controlled grammars: the case of special Petri nets. J. UCS **15**(14), 2808–2835 (2009)
6. Demri, S., Jurdziński, M., Lachish, O., Lazić, R.: The covering and boundedness problems for branching vector addition systems. J. Comput. Syst. Sci. **79**(1), 23–38 (2012)
7. Fallah Seghrouchni, A.E., Haddad, S.: A recursive model for distributed planning. In: ICMAS 1996, Kyoto, Japan, pp. 307–314 (1996)
8. Finkel, A., Haddad, S., Khmelnitsky, I.: Coverability and termination in recursive Petri nets, April 2019. https://hal.inria.fr/hal-02081019
9. Finkel, A., Schnoebelen, P.: Well-structured transition systems everywhere!. Theor. Comput. Sci. **256**(1–2), 63–92 (2001)
10. Haddad, S., Poitrenaud, D.: Decidability and undecidability results for recursive Petri nets. Technical Report 019, LIP6, Paris VI University (1999)
11. Haddad, S., Poitrenaud, D.: Theoretical aspects of recursive Petri nets. In: Donatelli, S., Kleijn, J. (eds.) ICATPN 1999. LNCS, vol. 1639, pp. 228–247. Springer, Heidelberg (1999). https://doi.org/10.1007/3-540-48745-X_14
12. Haddad, S., Poitrenaud, D.: Modelling and analyzing systems with recursive Petri nets. In: Boel, R., Stremersch, G. (eds.) Discrete Event Systems. The International Series in Engineering and Computer Science, vol. 569, pp. 449–458. Springer, Boston (2000). https://doi.org/10.1007/978-1-4615-4493-7_48
13. Haddad, S., Poitrenaud, D.: Checking linear temporal formulas on sequential recursive Petri nets. In: TIME 2001, Civdale del Friuli, Italy, pp. 198–205. IEEE Computer Society (2001)

14. Haddad, S., Poitrenaud, D.: Recursive Petri nets. Acta Inf. **44**(7–8), 463–508 (2007)
15. Lambert, J.-L.: A structure to decide reachability in Petri nets. Theor. Comput. Sci. **99**(1), 79–104 (1992)
16. Lazic, R.: The reachability problem for vector addition systems with a stack is not elementary. CoRR, abs/1310.1767 (2013)
17. Lazic,R., Schmitz, S.: Non-elementary complexities for branching vass, mell, and extensions. In: CSL-LICS 2014, Vienna, Austria, pp. 61:1–61:10. ACM (2014)
18. Lazić, R., Schmitz,S.: The complexity of coverability in ν-Petri nets. In: LICS 2016, pp. 467–476. ACM Press, New York (2016)
19. Lipton, R.J.: The reachability problem requires exponential space. Technical Report 062, Yale University, Department of Computer Science, January 1976
20. Mavlankulov, G., Othman, M., Turaev, S., Selamat, M.H., Zhumabayeva, L., Zhukabayeva, T.: Concurrently controlled grammars. Kybernetika **54**(4), 748–764 (2018)
21. Mayr, E.W.: An algorithm for the general Petri net reachability problem. SIAM J. Comput. **13**(3), 441–460 (1984)
22. Rackoff, C.: The covering and boundedness problems for vector addition systems. Theor. Comput. Sci. **6**(2), 223–231 (1978)
23. Reinhardt, K.: Reachability in Petri nets with inhibitor arcs. Electr. Notes Theor. Comput. Sci. **223**, 239–264 (2008)
24. Schnoebelen, P.: Revisiting Ackermann-hardness for lossy counter machines and reset Petri nets. In: Hliněný, P., Kučera, A. (eds.) MFCS 2010. LNCS, vol. 6281, pp. 616–628. Springer, Heidelberg (2010). https://doi.org/10.1007/978-3-642-15155-2_54
25. Zetzsche, G.: The emptiness problem for valence automata or: another decidable extension of Petri nets. In: Bojańczyk, M., Lasota, S., Potapov, I. (eds.) RP 2015. LNCS, vol. 9328, pp. 166–178. Springer, Cham (2015). https://doi.org/10.1007/978-3-319-24537-9_15

From DB-nets to Coloured Petri Nets
with Priorities

Marco Montali and Andrey Rivkin[✉]

Free University of Bozen-Bolzano, Piazza Domenicani 3, 39100 Bolzano, Italy
{montali,rivkin}@inf.unibz.it

Abstract. The recently introduced formalism of DB-nets has brought
in a new conceptual way of modelling complex dynamic systems that
equally account for the process and data dimensions, considering local
data as well as persistent, transactional data. DB-nets combine a
coloured variant of Petri nets with name creation and management
(which we call ν-CPN), with a relational database. The integration of
these two components is realized by equipping the net with special "view"
places that query the database and expose the resulting answers to the
net, with actions that allow transitions to update the content of the
database, and with special arcs capturing compensation in case of trans-
action failure. In this work, we study whether this sophisticated model
can be encoded back into ν-CPNs. In particular, we show that the meaningful fragment of DB-nets where database queries are expressed using
unions of conjunctive queries with inequalities can be faithfully encoded
into ν-CPNs with transition priorities. This allows us to directly exploit
state-of-the-art technologies such as CPN Tools to simulate and analyse
this relevant class of DB-nets.

1 Introduction

During the last decade, the Business Process Management (BPM) community
has gradually lifted its attention from process models mainly focusing on the
flow of activities to multi-perspective models that also account for the interplay
between the process and the data perspective [3,6,15]. Interestingly, the problem of modelling workflows dealing with database or document management
systems has a long tradition in the Petri net field [5,14,19]. However, the resulting frameworks do not come with a clear separation of concerns between control
flow and data-related aspects, nor with corresponding results on formal analysis.
Recent variants of high-level Petri nets have been proposed to tackle these two
challenges (see, e.g., [8,11,12,18]).

In this spectrum, the recently introduced formalism of DB-nets [12] has
brought in a new conceptual way of modelling complex dynamic systems that
equally account for the process and data dimensions, considering local data as
well as persistent, transactional data. On the one hand, a DB-net adopts a standard relational database with constraints to store persistent data. The database
can be queried through SQL/first-order queries, and updated via actions in a

© Springer Nature Switzerland AG 2019
S. Donatelli and S. Haar (Eds.): PETRI NETS 2019, LNCS 11522, pp. 449–469, 2019.
https://doi.org/10.1007/978-3-030-21571-2_24

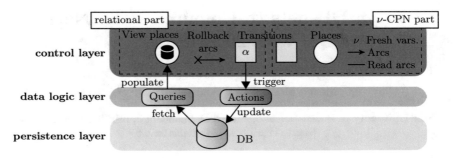

Fig. 1. The conceptual components of DB-nets

transactional way (that is, committing the update only if the resulting database satisfies all intended constraints). On the other hand, a DB-net employs a coloured variant of a Petri net with name creation and management [17] to capture the process control-flow, the injection of (possibly fresh) data such as the creation of new case identifiers [11], and tuples of typed data locally carried out by tokens. This model, which we call ν-CPN, can be seen as a fragment of standard Coloured Petri nets [7] with pattern matching on inscriptions, infinite colour domains, boolean guards, and a very limited use of SML to account for fresh data injection. This also means that ν-CPNs can be seamlessly modelled, simulated, and analysed using state-of-the-art tools such as CPN Tools.

The integration of these two components is realized in a DB-net by extending the ν-CPN with three novel constructs: *(i) view places*, special places that query the database and expose the resulting answers as coloured tokens that can be inspected but not directly consumed; *(ii) action bindings*, linking transitions to database updates by mapping inscription variables to action parameters; *(iii) rollback transition-place arcs*, capturing the emission of tokens in case a fired transition induces a failing database update, and in turn supporting the enablement of compensation transitions. All conceptual components used in the DB-net model are depicted in Fig. 1. Notably, DB-nets have been employed to formalize application integration patterns [16].

In this work, we study whether this sophisticated model can be *encoded back into ν-CPNs*, with a twofold intention. On the foundational side, we aim at understanding whether the process-data integration realized in DB-nets adds expressiveness to ν-CPNs, or it is instead conceptual, syntactic sugar. On the practical side, the existence of an encoding would allow us to directly exploit state-of-the-art tools such as CPN Tools towards simulation and analysis of DB-nets. In the case of CPN Tools, this is the only way possible when it comes to state space construction, given the fact that this feature cannot be refined through the third-party extension mechanism offered by the framework.

Specifically, we constructively show through a behavior-preserving translation mechanism that this encoding is indeed possible for a large and meaningful class of DB-nets, provided that the obtained ν-CPN is equipped with *transition priorities* [20] (a feature that is supported by virtually all CPN frameworks, including CPN Tools). Such class corresponds to DB-nets where the database is equipped with key, foreign key, and domain constraints, and where the view places query the database using unions of conjunctive queries (UCQs) with inequalities. Such query language corresponds to the widely adopted fragment of SQL consisting of select-project-join queries with filters [1].

2 The DB-net Formal Model

In this section, we briefly present the key concepts and notions used for defining DB-nets. Conceptually, a DB-net is composed of three layers (cf. Fig. 1) (1) *persistence layer*, capturing a full-fledged relational database with constraints, and used to store background data, and data that are persistent across cases; (2) *control layer*, employing a variant of CPNs to capture the process control-flow, case data, and possibly the resources involved in the process execution; (3) *data logic layer*, interconnecting in the persistence and the control layer.

Definition 1. *A db-net is a tuple $\langle \mathfrak{D}, \mathcal{P}, \mathcal{L}, \mathcal{N} \rangle$, where: (i) $\mathfrak{D}$ is a type domain; (ii) $\mathcal{P}$ is a $\mathfrak{D}$-typed persistence layer; (iii) $\mathcal{L}$ is a $\mathfrak{D}$-typed data logic layer over $\mathcal{P}$; (iv) $\mathcal{N}$ is a $\mathfrak{D}$-typed control layer over $\mathcal{L}$.*

We next formalize the framework layer by layer.

Persistence Layer. A *type domain* $\mathfrak{D}$ is a finite set of pairwise disjoint *data types* $\mathcal{D} = \langle \Delta_{\mathcal{D}}, \Gamma_{\mathcal{D}} \rangle$, where a set $\Delta_{\mathcal{D}}$ is a *value domain*, and $\Gamma_{\mathcal{D}}$ is a finite set of *predicate symbols*. Examples of data types are: *(i)* **string** $= \langle \mathbb{S}, \{=_s\} \rangle$, strings with the equality predicate; *(ii)* **real** $= \langle \mathbb{R}, \{=_r, <_r\} \rangle$, reals with the usual comparison operators; *(iii)* **int** $= \langle \mathbb{Z}, \{=_{int}, <_{int}, succ\} \rangle$, integers with the usual comparison operators, as well as the successor predicate.

We call $R(\mathcal{D}_1, \ldots, \mathcal{D}_n)$ a $\mathfrak{D}$-*typed relation schema*, where R is a relation name and $\mathcal{D}_i$ indicates the data type associated to an i-th component of R. A $\mathfrak{D}$-*typed database schema* $\mathcal{R}$ is a finite set of $\mathfrak{D}$-typed relation schemas. A $\mathfrak{D}$-*typed database instance* $\mathcal{I}$ over $\mathcal{R}$ is a finite set of facts of the form $R(o_1, \ldots, o_n)$, such that $R(\mathcal{D}_1, \ldots, \mathcal{D}_n) \in \mathcal{R}$ and $o_i \in \Delta_{\mathcal{D}_i}$, for $i \in \{1, \ldots, n\}$. Given a type $\mathcal{D} \in \mathfrak{D}$, the $\mathcal{D}$-*active domain of* $\mathcal{I}$, is the set of $Adom_{\mathcal{D}}(\mathcal{I}) = \{o \in \Delta_{\mathcal{D}} \mid o \text{ occurs in } \mathcal{I}\}$.

Given $\mathfrak{D}$, we fix a countably infinite set $\mathcal{V}_{\mathfrak{D}}$ of typed variables with a *variable typing function* $\mathsf{type} : \mathcal{V}_{\mathfrak{D}} \to \mathfrak{D}$. As query language, we adopt first-order (FO) logic extended with data types under the active-domain semantics [9], that is, the evaluation of quantifiers only depends on the values explicitly appearing in the database instance over which they are applied. This can be seen as the FO representation of SQL queries. A *(well-typed)* $\mathsf{FO}(\mathfrak{D})$ *query* Q over a $\mathfrak{D}$-typed database schema $\mathcal{R}$ has the form $\{\vec{x} \mid \varphi(\vec{x})\}$, where $\vec{x}$ is the tuple of answer variables of Q, and φ is a FO formula, with $\vec{x}$ as free variables, over predicates in $\cup_{\mathcal{D} \in \mathfrak{D}} \Gamma_{\mathcal{D}}$ and relation schemas in $\mathcal{R}$, whose variables and constants are correctly typed. We use $Q(\vec{x})$ to make the answer variables $\vec{x}$ of Q explicit, and denote the set of such variables as $Free(Q)$. When $Free(Q) = \emptyset$, we call Q a *boolean query*.

A *substitution* for a set $X = \{x_1, \ldots, x_n\}$ of typed variables, is a function $\theta : X \to \Delta_{\mathfrak{D}}$, such that $\theta(x) \in \Delta_{\mathsf{type}(x)}$, for every $x \in X$. A *substitution* θ *for*

Fig. 2. The persistence layer for the online shopping scenario. UID and PID respectively represent unique user and product identifiers.

a FO($\mathfrak{D}$) *query* Q is a substitution for the free variables of Q. We denote by $Q\theta$ the boolean query obtained from Q by replacing each occurrence of a free variable $x \in Free(Q)$ with the value $\theta(x)$. Given a $\mathfrak{D}$-typed database schema $\mathcal{R}$, a $\mathfrak{D}$-typed instance $\mathcal{I}$ over $\mathcal{R}$, and a FO($\mathfrak{D}$) query Q over $\mathcal{R}$, the set of *answers* to Q in $\mathcal{I}$ is defined as the set $ans(Q, \mathcal{I}) = \{\theta : Free(Q) \rightarrow Adom_{\mathfrak{D}}(\mathcal{I}) \mid \mathcal{I}, \theta \models Q\}$ of substitutions for Q, where $\models$ denotes standard FO entailment (i.e., we use *active-domain semantics*). We denote by $\text{LIVE}_{\mathcal{D}}(x)$ the unary query returning all the objects of type $\mathcal{D}$ that occur in the active domain (writing such a query is straightforward). When Q is boolean, we write $ans(Q, \mathcal{I}) \equiv \mathsf{true}$ if $ans(Q, \mathcal{I})$ consists only of the empty substitution (denoted $\langle\rangle$), and $ans(Q, \mathcal{I}) \equiv \mathsf{false}$ if $ans(Q, \mathcal{I}) = \emptyset$. Boolean queries are also used to express *constraints* over $\mathcal{R}$. We introduce explicitly two common types of constraints: given relations R/n and S/m, and two index-sets N and M such that $1 \leq i \leq n$ for every $i \in N$, and $1 \leq j \leq m$ for every $j \in M$, we fix the following notation: *(i)* $\text{PK}(R) = N$ expresses that the projection $R[N]$ of R on N is a primary key for R; *(ii)* $R[N] \subseteq S[M]$ expresses that the projection $R[N]$ of R on N refers the projection $S[M]$ of S on M, which has to be a key for S. Both kinds of constraints are obviously expressible as suitable queries [1].

Definition 2. *A $\mathfrak{D}$-typed persistence layer $\mathcal{P}$ is a pair $\langle\mathcal{R}, \mathcal{E}\rangle$ where: (i) $\mathcal{R}$ is a $\mathfrak{D}$-typed database schema; (ii) $\mathcal{E}$ is a finite set $\{\varPhi_1, ..., \varPhi_k\}$ of boolean FO($\mathfrak{D}$) queries over $\mathcal{R}$, modelling constraints over $\mathcal{R}$.*

We say that a $\mathfrak{D}$-typed database instance $\mathcal{I}$ *complies with* $\mathcal{P}$, if $\mathcal{I}$ is defined over $\mathcal{R}$ and satisfies all constraints in $\mathcal{E}$.

Example 1. Let us consider a simplified shopping process used by an e-commerce website. Specifically, we are interested in a simplified scenario in which an already registered user logs in the website and immediately proceeds with selecting products. While products can be selected and added to the shopping cart, the user can occasionally choose a monthly bonus that may be applied when concluding a purchase. We restrict this scenario only by considering cases in which each user ends up buying at least one product.

The persistence layer $\mathcal{P} = \langle\mathcal{R}, \mathcal{E}\rangle$ of this scenario comprises four relation schemas (cf. Fig. 2): *User*(**int**, **string**) lists registered users together with their credit card data, *WithBonus*(**int**, **string**) indicates users that have bonuses, *Product*(**string**) indexes product types offered by the website and *InWarehouse*(**int**, **string**, **real**) indicates which products are stored in the warehouse and with which cost. Note the constraints between these schemas. For example, in order to show that users cannot have more than one bonus at a time, we introduce a foreign key constraint between *WithBonus* and *User* that is denoted as *WithBonus*[{1}] $\subseteq$ *User*[{1}] and formalized in FO logic as: $\forall uid, bt. WithBonus(uid, bt) \rightarrow \exists card. User(uid, card)$. Another constraint limits the bonus type values in *WithBonus* and can be expressed as $\forall uid, bt. WithBonus(uid, bt) \rightarrow bt = 50\% \vee bt = 15\text{eur} \vee bt = \text{extra\_item}$. $\square$

Data Logic Layer. The data logic layer allows one to *extract* data from the database instance using queries as well as to *update* the database instance by adding and deleting possibly multiple facts at once. The updates follow the *transactional* semantics: if a new database instance obtained after some update is still compliant with the persistence layer, the update is *committed*; otherwise it is *rolled back*. Such updates are realized in parametric atomic actions, resembling ADL actions in planning [4], and consist of fact templates – expressions that, once instantiated, assert which facts will be added to and deleted from the database. Specifically, given a typed relation $R(\mathcal{D}_1, \ldots, \mathcal{D}_n) \in \mathcal{R}$, an R-fact template over $\vec{p}$ has the form $R(y_1, \ldots, y_n)$, such that for every $i \in \{1, \ldots, n\}$, y_i is either a value $\text{o} \in \Delta_{\mathcal{D}_i}$, or a variable $x \in \vec{p}$ with $\text{type}(x) = \mathcal{D}_i$.

A *(parameterized) action* over a $\mathfrak{D}$-typed persistence layer $\langle \mathcal{R}, \mathcal{E} \rangle$ is a tuple $\langle \text{n}, \vec{p}, F^+, F^- \rangle$, where: *(i)* n is the *action name*; *(ii)* $\vec{p}$ is a tuple of pairwise distinct variables from $\mathcal{V}_{\mathfrak{D}}$, denoting the *action (formal) parameters*; *(iii)* F^+ and F^- respectively represent a finite set of $\mathcal{R}$-fact templates (i.e., some R-fact templates for some $R \in \mathcal{R}$) over $\vec{p}$, to be *added* to and *deleted* from the current database instance. To access the different components of an action α, we use a dot notation: $\alpha \cdot \text{name} = \text{n}$, $\alpha \cdot \text{params} = \vec{p}$, $\alpha \cdot \text{add} = F^+$, and $\alpha \cdot \text{del} = F^-$. Given an action α and a (parameter) substitution θ for $\alpha \cdot \text{params}$, we call *action instance* $\alpha\theta$ the (ground) action resulting by substituting parameters of α with corresponding values from θ. Then, given a $\mathfrak{D}$-typed database instance $\mathcal{I}$ compliant with $\mathfrak{D}$, the *application* of $\alpha\theta$ on $\mathcal{I}$, written $\text{apply}(\alpha\theta, \mathcal{I})$, is a database instance over $\mathcal{R}$ obtained as $(\mathcal{I} \setminus F^-_{\alpha\theta}) \cup F^+_{\alpha\theta}$, where: *(i)* $F^-_{\alpha\theta} = \bigcup_{R(\vec{y}) \in \alpha \cdot \text{del}} R(\vec{y})\theta$; *(ii)* $F^+_{\alpha\theta} = \bigcup_{R(\vec{y}) \in \alpha \cdot \text{add}} R(\vec{y})\theta$. If $\text{apply}(\alpha\theta, \mathcal{I})$ complies with $\mathcal{P}$, $\alpha\theta$ can be *successfully applied* to $\mathcal{I}$. Note that, in order to avoid situations in which the same fact is asserted to be added and deleted, we prioritize additions over deletions.

Definition 3. *Given a $\mathfrak{D}$-typed persistence layer $\mathcal{P}$, a $\mathfrak{D}$-typed data logic layer over $\mathcal{P}$ is a pair $\langle \mathcal{Q}, \mathcal{A} \rangle$, where: (i) $\mathcal{Q}$ is a finite set of $\text{FO}(\mathfrak{D})$ queries over $\mathcal{P}$; (ii) $\mathcal{A}$ is a finite set of actions over $\mathcal{P}$.*

Example 2. We make the scenario of Example 1 operational, introducing a data logic layer $\mathcal{L}$ over $\mathcal{P}$. To inspect the persistence layer, we use the following queries:

- $\text{Q}_{\text{products}}(pid, n, c)$:- *Product*$(n) \wedge InWarehouse(pid, n, c) \wedge c \neq \text{null}$, to extract products available in the warehouse and whose price is not null (those without prices can be undergoing the stock-taking process);
- $\text{Q}_{\text{users}}(uid)$:- $\exists card. User(id, card)$, to get registered users;
- $\text{Q}_{\text{wbonus}}(uid, bt')$:- *WithBonus*$(uid, bt')$, to inspect all users with bonuses.

In addition, $\mathcal{L}$ provides key functionalities for organizing the shopping process. Such functionalities are realized through four actions (where, for simplicity, we blur the distinction between an action and its name). To manage bonuses we use two actions ADDB and CHANGE. The former is used to assign a bonus of type bt to a user with id uid (ADDB·**params** $= \langle uid, bt \rangle$) and record it into the persistent storage: ADDB·**add** $= \{ WithBonus(uid, bt) \}$, ADDB·**del** $= \emptyset$. Note that, before logging in, the user may have already a bonus assigned during one of the previous sessions. At will, such a bonus can be changed using action CHANGE with CHANGE·**params** $= \langle uid, bt, bt' \rangle$, CHANGE·**add** $= \{ WithBonus(uid, bt) \}$ and CHANGE·**del** $= \{ WithBonus(uid, bt') \}$. In fact, CHANGE realizes an update by first deleting a tuple that is characterized by uid and bt' (the old bonus), and then adding its modified version. We use RESERVE (RESERVE·**params** $= \langle pid, n, c \rangle$) to reserve product pid of price c and stored in cart cid for further processing (e.g., the preparation for shipment) by deleting it from the list of available products: RESERVE·**add** $= \emptyset$, RESERVE·**del** $= \{ InWarehouse(pid, n, c) \}$. At last, the user may utilize her monthly bonus bt (if it has not yet been used) to consider it when paying the order. For that, we use an action called APPLY (with APPLY·**params** $= \langle uid, bt \rangle$) such that: APPLY·**add** $= \emptyset$, APPLY·**del** $= \{ WithBonus(uid, bt) \}$.   □

Control Layer. The control layer employs a fragment of Coloured Petri net to capture the process control flow and a data logic to interact with an underlying persistence layer. We fix some preliminary notions. We consider the standard notion of a *multiset*. Given a set A, the *set of multisets* over A, written $A^{\oplus}$, is the set of mappings of the form $m : A \to \mathbb{N}$. Given a multiset $S \in A^{\oplus}$ and an element $a \in A$, $S(a) \in \mathbb{N}$ denotes the number of times a appears in S. Given $a \in A$ and $n \in \mathbb{N}$, we write $a^n \in S$ if $S(a) = n$. We also consider the usual operations on multisets. Given $S_1, S_2 \in A^{\oplus}$: *(i)* $S_1 \subseteq S_2$ (resp., $S_1 \subset S_2$) if $S_1(a) \leq S_2(a)$ (resp., $S_1(a) < S_2(a)$) for each $a \in A$; *(ii)* $S_1 + S_2 = \{ a^n \mid a \in A$ and $n = S_1(a) + S_2(a) \}$; *(iii)* if $S_1 \subseteq S_2$, $S_2 - S_1 = \{ a^n \mid a \in A$ and $n = S_2(a) - S_1(a) \}$; *(iv)* given a number $k \in \mathbb{N}$, $k \cdot S_1 = \{ a^{kn} \mid a^n \in S_1 \}$.

We shall call *inscription* a tuple of typed variables (and, possibly, values) and denote the set of all possible inscriptions over set $\mathcal{Y}$ as $\Omega_{\mathcal{Y}}$, and the set of variables appearing inside an inscription $\omega \in \Omega_{\mathcal{Y}}$ as $Vars(\omega)$ (such notation naturally extends to sets and multisets of inscriptions). In the spirit of CPNs, the control layer assigns to each place a color type, which in turn combines one or more data types from $\mathfrak{D}$. Formally, a $\mathfrak{D}$-*color* is $\mathcal{D}_1 \times \ldots \times \mathcal{D}_m$, where for each $i \in \{ 1, \ldots, m \}$, we have $\mathcal{D}_i \in \mathfrak{D}$. We denote by Σ the set of all possible $\mathfrak{D}$-colors. To account for fresh external inputs, we employ the well-known mechanism adopted in ν-Petri nets [11,17] and introduce a countably infinite set $\Upsilon_{\mathfrak{D}}$ of $\mathfrak{D}$-typed *fresh variables*. To guarantee an unlimited provisioning of fresh values, we impose that for every variable $\nu \in \Upsilon_{\mathfrak{D}}$, we have that $\Delta_{\text{type}(\nu)}$ is countably infinite. Hereinafter, we shall fix a countably infinite set of $\mathfrak{D}$-typed variable $\mathcal{X}_{\mathfrak{D}} = \mathcal{V}_{\mathfrak{D}} \cup \Upsilon_{\mathfrak{D}}$ as the disjoint union of "normal" variables $\mathcal{V}_{\mathfrak{D}}$ and fresh variables $\Upsilon_{\mathfrak{D}}$.

As we have mentioned before, the control layer can be split into two parts. Let us first define the ν-CPN part that can be seen as an extension of ν-Petri nets with concrete data types, boolean (type-aware) guards and read arcs.

Definition 4. *A $\mathfrak{D}$-typed ν-CPN $\mathcal{N}$ is a tuple $\langle P, T, F_{in}, F_{out}, \texttt{color} \rangle$, where:*

1. *P is a finite set of places.*
2. *$\texttt{color} : P \to \Sigma$ is a color type assignment over P mapping each place $p \in P$ to a corresponding $\mathfrak{D}$-type color.*
3. *T is a finite set of transitions, such that $T \cap P = \emptyset$.*
4. *$F_{in} : P \times T \to \Omega_{\mathcal{V}_\mathfrak{D}}^{\oplus}$ is an input flow from P to T assigning multisets of inscriptions (over variables $\mathcal{V}_\mathfrak{D}$) to input arcs, s.t. that each of such inscriptions $\langle x_1, \dots, x_m \rangle$ is compatible with each of its input places p, i.e., for every $i \in \{1, \dots, m\}$, we have $\texttt{type}(x_i) = \mathcal{D}_i$, where $\texttt{color}(p) = \mathcal{D}_1 \times \dots \times \mathcal{D}_m$.*
5. *$\texttt{guard} : T \to \mathbb{F}_\mathfrak{D}$ is a transition guard assignment over T assigning to each transition $t \in T$ a $\mathfrak{D}$-typed guard φ, s.t.:*
   - *$InVars(t) = \{x \in \mathcal{V}_\mathfrak{D} \mid \text{there exists } p \in P \text{ such that } x \in Vars(F_{in}(\langle p, t \rangle))\}$ is the set of all variables occurring on input arc inscriptions of t;*
   - *a $\mathfrak{D}$-typed guard from is a formula (or a quantifier- and relation-free $\texttt{FO}(\mathfrak{D})$ query) of the form $\varphi ::= \texttt{true} \mid S(\vec{y}) \mid \neg\varphi \mid \varphi_1 \wedge \varphi_2$, where $S/n \in \Gamma_\mathfrak{D}$ and, for $\vec{y} = \langle y_1, \dots, y_n \rangle \subseteq \mathcal{V}_\mathfrak{D}$, we have that y_i is either a value $\texttt{o} \in \Delta_\mathcal{D}$, or a variable $x_i \in \mathcal{V}_\mathfrak{D}$ with $\texttt{type}(x_i) = \mathcal{D}$ ($i \in \{1, \dots, n\}$);*
   - *$\mathbb{F}_\mathfrak{D}$ is the set of all possible $\mathfrak{D}$-typed guards and, with a slight abuse of notation, $Vars(\varphi)$ is the set of variables occurring in φ.*
6. *$F_{out} : T \times P \to \Omega_{\mathcal{X}_\mathfrak{D} \cup \Delta_\mathfrak{D}}^{\oplus}$ is an output flow from transitions T to places P assigning multisets of inscriptions to output arcs, such that all such inscriptions are compatible with their output places.*

According to the diagram in Fig. 1, the DB-net control layer can be obtained on top of ν-CPNs by essentially adding three mechanisms that allow the net to interact with the underlying persistent storage: *(i)* view places, allowing the net to inspect parts of the database using queries; *(ii)* action binding, linking atomic actions and their parameters to transitions and their inscription variables; *(iii)* rollback transition-place arcs, enacted when the action application induced by a transition firing violates some database constraint, so as to explicitly account for "error-handling".

Definition 5. *A $\mathfrak{D}$-typed control layer over a data logic layer $\mathcal{L} = \langle \mathcal{Q}, \mathcal{A} \rangle$ is a tuple $\langle P, T, F_{in}, F_{out}, F_{rb}, \texttt{color}, \texttt{query}, \texttt{guard}, \texttt{act} \rangle$, where:*

1. *$\langle P_c, T, F_{in}, F_{out}, \texttt{color} \rangle$ is a $\mathfrak{D}$-typed ν-CPN, where P_c is a finite set of control places.*
2. *$P = P_c \cup P_v$ is a finite set of places, where P_v are view places (decorated as ⊛ and connected to transitions with special read arcs).*
3. *$\texttt{query} : P_v \to \mathcal{Q}$ is a query assignment mapping each view place $p \in P_v$ with $\texttt{color}(p) = \mathcal{D}_1 \times \dots \times \mathcal{D}_n$ to a query $Q(x_1, \dots, x_n)$ from $\mathcal{Q}$, s.t. the color of p component-wise matches with the types of the free variables in Q: for each $i \in \{1, \dots, n\}$, we have $\mathcal{D}_i = \texttt{type}(x_i)$.*

4. $\texttt{act} : T \rightarrow \mathcal{A} \times \Omega_{\mathcal{X}_{\mathfrak{D}} \cup \Delta_{\mathfrak{D}}}$ *is a partial function assigning transitions in* T
   to actions in $\mathcal{A}$, *where* $\texttt{act}(t)$ *maps* t *to an action* $\alpha \in \mathcal{A}$ *together with a*
   (binding) inscription $\langle y_1, \ldots, y_m \rangle$, *s.t. if* $\alpha \cdot \texttt{params} = \langle z_1, \ldots, z_m \rangle$ *and, for*
   each $i \in \{1, \ldots, m\}$, *we have* $\texttt{type}(y_i) = \texttt{type}(z_i)$ *if* y_i *is a variable from* $\mathcal{X}_{\mathfrak{D}}$,
   or $y_i \in \Delta_{\texttt{type}(z_i)}$ *if* y_i *is a value from* $\Delta_{\mathfrak{D}}$.
5. F_{rb} *is an output flow from* T *to* P_c *called* rollback flow *(we shall refer to* F_{out}
   as normal output flow*).*

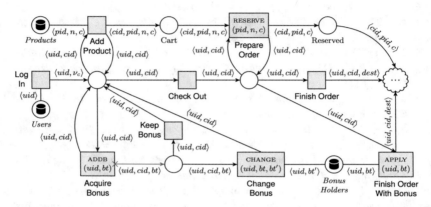

Fig. 3. The control of a DB-net for online shopping. Here, ν_c is a fresh input variable corresponding to a newly created cart, whereas *dest* is an arbitrary input variable representing a destination address. The rollback output arc (corresponds to the rollback flow) is in red and decorated with an "x".

Figure 3 shows the control layer of the shopping cart example. The queries specified in Example 2 are assigned to the corresponding view places: $\texttt{query}(\textit{Products}) := \texttt{Q}_{\texttt{products}}$, $\texttt{query}(\textit{Users}) := \texttt{Q}_{\texttt{users}}$ and $\texttt{query}(\textit{Bonus Holders}) := \texttt{Q}_{\texttt{wbonus}}$. The actions (with their formal parameters) assigned to transitions via $\texttt{act}$ graphically appear in grey transition boxes.

The execution semantics of a DB-net simultaneously accounts for the progression of a database instance compliant with the persistence layer of the net, and for the evolution of a marking over the control layer of the net. Due to space limitations, we refer to the definition of the formal semantics studied in [12]. We thus assume that the execution semantics of both ν-CPNs and DB-nets can be captured with a possibly infinite-state labeled transition system (LTS) that accounts for all possible executions starting from their initial markings. While transitions in such LTSs model the effect of firing nets under given bindings, their state representations slightly differ. Namely, in the case of ν-CPNs we have markings (like, for example, in coloured Petri nets [7]), while in the case of DB-nets one also has to take into account database states. W.l.o.g., we shall use $\Gamma_{\mathcal{N}}^{M_0} = \langle \mathcal{M}, M_0, \rightarrow, L \rangle$ to specify an LTS for a ν-CPN $\mathcal{N}$ with initial marking

M_0 and $\Gamma_{\mathcal{B}}^{s_0} = \langle S, s_0, \rightarrow, L \rangle$ to specify an LTS for a DB-net $\mathcal{B}$ with initial snapshot $s_0 = \langle \mathcal{I}_0, m_0 \rangle$, where $\mathcal{I}_0$ is the initial database instance and m_0 is the initial marking of the control layer.

3 Translation

We are now ready to describe the translation from DB-nets to ν-CPNs with priorities (we assume the reader is familiar with transition priorities). Recall that this is not just of theoretical interest, but has also practical implications. In [16], we have presented a prototypical implementation of DB-nets in CPN Tools that, using Access/CPN and Comms/CPN, allow to model and simulate DB-nets. However, we realized that CPN Tools would not correctly generate the state space of the DB-net at hand. This is due to the fact that the CPN Tools state space construction module does not consider third-party extensions which, in our setting, implies that the content of the view places is not properly recomputed after each transition firing.

The first challenge to overcome is how the database schema is represented in the target net. To this aim, we introduce special *relation places* that copy corresponding database relations by mirroring their signature to the type definitions of places.[1] In this light, database instances will correspond to relation place markings, where tokens are nothing but tuples. All other DB-net elements (for example, bindings for fresh variables, action execution) require actual computation that happens when a transition fires. Intuitively, every DB-net transition T is represented using the following four phases:

1. Collect bindings and compute the content of view places adjacent to T.
2. If there is an action assigned to T, execute it. We employ auxiliary boolean places that control whether an update has actually happened (that is, a token representing a tuple has been removed from or added to a relation place).
3. Check the satisfaction of integrity constraints.
4. Finish the computation and generate a new marking.
   (a) If all constraints are satisfied, empty the auxiliary boolean places used in (2), release the lock, and populate the postset of T.
   (b) If some constraint is violated, roll-back the effects. This is done in reverse order w.r.t. phase (2), applying or skipping a reverse update depending on how the values in the special places. After this, the relation places have the content they had before the action was applied. Then, one releases the lock and pushses the special postset corresponding to the roll-back arc (if any) attached to T.

To realize the execution of an original DB-net transition, all the four phases are executed uninterruptedly (under lock). In the reminder of the section we formalize the phases discussed above.

---

[1] Relation places do not differ from the normal ν-CPN places. We use the different name in order to conceptually distinguish their origin.

A generic DB-net $\mathcal{B}_\tau$ that we use to demonstrate the translation is represented in Fig. 4(a). Here, we assume that T contains enough of tokens assigned by its input flow and its eventual firing is subject to the $G(\vec{y})$ guard evaluation. $\vec{y}$, in turn, is bound to values from $\vec{z}$ and from $m \in \mathbb{N}$ ordered view places, where each view place V_i has a query $\mathtt{Q}_{V_i}$ assigned to it. The ν-CPN $\mathcal{N}_\tau$ representing

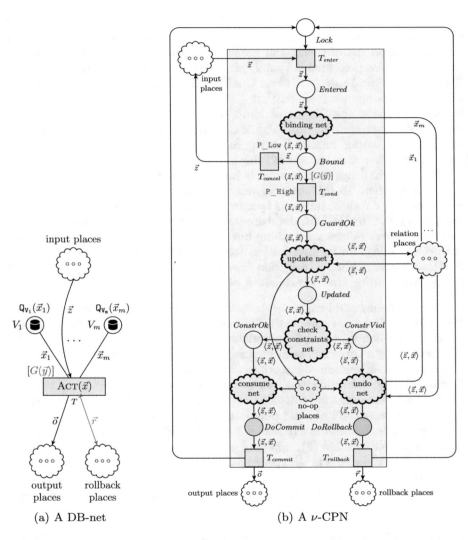

(a) A DB-net   (b) A ν-CPN

Fig. 4. A generic DB-net transition accessing multiple view places (left) and its overall ν-CPN encoding (right). Blue clouds stand for subnets that are expanded next, and $\vec{x}$ is a shortcut for $\vec{x}_1, \ldots, \vec{x}_m$. Elements within the gray rectangle are local to the transition, whereas external elements are shared at the level of the whole net. (Color figure online)

$\mathcal{B}_\tau$ is depicted in Fig. 4(b). To facilitate the translation, we make three working hypothesis. First, we assume that the relational schema is equipped only with three types of constraints: primary keys, foreign keys and domain constraints. Second, for ease of presentation, we consider that the resulting ν-CPN model can deal with DB-net external inputs. For each $t \in T$, such inputs are in general modeled using variables that do not appear in $InVars(t)$ (including those from $\Upsilon_{\mathfrak{D}}$). The preliminary implementation of DB-nets in CPN Tools [16] has provided functionality demonstrating the feasibility of the binding computation for such variables. Third, we naturally extend the notion of ν-CPN with read arcs.

3.1   Computing Views Using CPN Places

We start by describing how the view computation should work using only ν-CPN places. Let us consider as an example a subnet $\mathcal{B}_{tr}$ of the DB-net present in Fig. 3 that models only the selection of available products. To access products that are available in the warehouse and that have prices assigned to them, we need to run a query $\mathsf{Q_{products}}(pid, n, c) = Product(n) \wedge InWarehouse(pid, n, c) \wedge c \neq \mathtt{null}$. Interestingly, such a query can be formulated directly using standard elements of ν-CPNs. Indeed, we may transfer the DB-net in Fig. 5(a) into a ν-CPN $\mathcal{N}_{tr}$ in Fig. 5(b) representing the project selection step. As one can see, the relations of $\mathcal{B}_{tr}$ have been copied to the same-named relation places, when $\mathsf{Q_{products}}$ is treated as follows: $\mathcal{N}_{tr}$ accesses relation places with read-arcs (that have relation attributes as their inscriptions) so as to realize the projection, while the filter (i.e., $c \neq \mathtt{null}$) is basically plugged into the guard of **Add Product**. The result of the query is then propagated into the post-set of **Add Product** using the free variables of $\mathsf{Q_{products}}$ (i.e., pid, n and c) in the arc inscriptions.

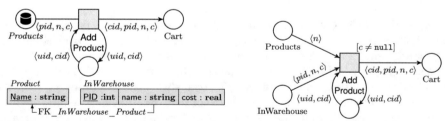

(a) Fragments of the process and persistence layers of the DB-net in Figure 3

(b) A ν-CPN representation of the DB-net in Figure 5(a)

Fig. 5. Example of a view computation in ν-CPNs

However, one may see that not every query can be handled when only using standard ν-CPN elements. Assume a query $\mathsf{Q_{\neg available}}(n) = Product(n) \wedge \neg \exists pid, c. InWarehouse(pid, n, c)$ that lists products not available in the warehouse. In order to represent $\mathsf{Q_{\neg available}}$ in a ν-CPN, one would need to extend the net with constructs allowing to fire a transition only if a certain element does not exists in a place incident to it. Thus, we restrict ourselves to the union of conjunctive queries with negative filters (or atomic negations) (UCQFs$^{\neq}$), that is $FO(\mathfrak{D})\mathfrak{D}$ queries of the form $\bigvee_{i=1}^{n} \exists \vec{y_i}.conj_i(\vec{x})$, where $conj_i(\vec{x})$ is also a $FO(\mathfrak{D})\mathfrak{D}$ query that is a conjunction of relations $R(\vec{z})$, predicates $P(\vec{y})$ and their negations $\neg P(\vec{y})$. Henceforth, we use $\mathcal{Q}^{UCQF^{\neq}}$ to define a UCQF$^{\neq}$ subset of $\mathcal{Q}$. In SQL, a conjunctive query is a query representable with a SELECT-FROM-WHERE expression. As it has been already shown, the filter conditions (of the UCQFs$^{\neq}$ attached to view places) can be modeled using transition guards.

In case of multiple view places attached to one transition, we construct a net that computes them in a sequential manner. One may see the computation process as a pipeline. Whenever a transition that corresponds to a certain view place is enabled, it fires and generated tokens that represent one of the tuples of the view. Then, acquired tokens are transferred to the next transition using variables in the arc inscriptions. The computation continues until the last view. After that, the results of all the computations are transferred to the corresponding places, following the topology (i.e., the organization of arcs defined by the flow relations) of the original DB-net. Note that the order in which views are computed has to be the same as the one defined for $\mathcal{B}_\tau$.

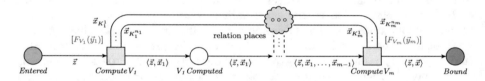

Fig. 6. Expansion of the binding net from Fig. 4(b)

A ν-CPN in Fig. 6 shows how bindings and view places are computed in the case of the generic DB-net $\mathcal{B}_\tau$. The computation process per view V_i is realized by a transition called $Compute\,V_i$ and analogous to the one explained before: we read necessary data from relation places, representing relations used in Q_{V_i}, and filter these data by means of $F_{V_i}(\vec{y})$. Note that variables on every read-arc adjacent to $Compute\,V_i$ represent attributes of some relation R. The intermediate result of the view computation is then stored in a place called $V_i\,Computed$. As one can see from Fig. 6, all the intermediate results are accumulated along the computation cycle. Moreover, we carry data provided with input variables of T so as to check the validity of the guard G (see Fig. 4(b)). This is done using prioritized transition T_{cond}. If the guard is not satisfied, one has to reset the computation process by returning tokens that have been consumed at the beginning of the view computation (that is, tokens that have been assigned to z). We resolve this issue by introducing an auxiliary transition called T_{cancel} that may fire only when the guard has been evaluated to false. Scheduling between T_{cond} and T_{cancel} is managed by means of two priority labels P_High and P_Low (where P_High > P_Low) respectively assigned to them.

3.2   Modelling RDBMS Updates in CPNs

We now show how database updates exploited by DB-nets could be represented using regular coloured Petri nets. We recall, that actions assigned to DB-net transitions support addition and deletion of $\mathcal{R}$-fact, which should preserve the set semantics adopted by the persistence layer.

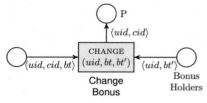

In Fig. 7 we consider a DB-net describing the bonus change step of he online shopping process. Here, for ease of presentation, instead of considering a view place for bonus holders, we use a regular (control) place that stores the same kind of data.

The translation of DB-net-like database updates into ν-CPNs is conceptually similar to the representation of the view computation process: DB-net actions must be performed sequentially within a critical section that can be entered whenever a special write lock is available (cf. place *Lock* in Fig. 4(b)). For preserving the set semantics over every relation place, we use prioritized transitions so as to check whether a tuple to be added or deleted already exists in the relation place. Specifically, for each tuple we would introduce two transitions, one with a higher priority and another with a lower priority, and an auxiliary (*no-op*) boolean place. The first transition can fire if the tuple is in the corresponding relation place, while the second one would fire otherwise. Both transitions are adopted to deal with additions and deletions. In case of additions, the highly prioritized transition would not add the tuple, while the one with the lower priority would do otherwise. To deal with deletions, we mirror the previous case: if the tuple exists, then one can safely remove it; otherwise, one proceeds without changes. Upon firing of any of these transitions, the auxiliary place receives a boolean token. If the value of the token is true, then it means that the tuple has been successfully added or deleted. In case the database update has not taken

Fig. 7. A subnet of the DB-net in Fig. 3 describing the bonus change step

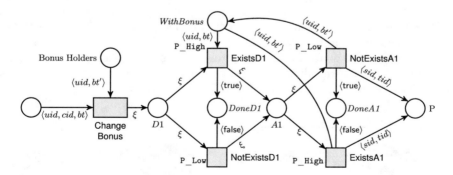

Fig. 8. The CPN representation of the DB-net in Fig. 7, where ξ, for ease of reading, denotes the tuple $\langle uid, cid, bt, bt' \rangle$

place, the token value is going to be false. It is important to note that the update execution order of DB-net actions must be also preserved in their ν-CPN representation. That is, since additions are prioritized over deletions, for every action α we first delete all the tuples from $\alpha\cdot\mathtt{del}$, and only then add those from $\alpha\cdot\mathtt{add}$.

We incorporate aforementioned modelling guidelines in the ν-CPN depicted in Fig. 8. Since CHANGE in $\mathcal{B}_{tr}^{\alpha}$ contains multiple database updates, the model starts with deleting $WithBonus(uid, bt)$ from $WithBonus$. To do so, at first one checks whether the relation place $WithBonus$ contains the tuple we would like to remove. This is done using **Exists D1** that performs conditional removal of $WithBonus(uid, bt)$, that is, if there is a token in $WithBonus$ such that bindings of inscriptions on $(D1, ExistsD1)$ and $(WithBonus, ExistsD1)$ coincide, then **ExistsD1** is enabled, and upon firing consumes the selected token from $With$-$Bonus$ and populates one token with value $\langle true \rangle$ (the value true means that the update has been successfully accomplished) in $DoneD1$. Note that **Exists D1** is always checked first given the higher priority label assigned to it. If the tuple does not exist, then one proceeds with firing **NotExistsD1** and populating one token with value false in $DoneD1$. Now, when we reach the first control place allowing to perform the add operation over $WithBonus$, we start by checking whether $WithBonus$ already contains the $WithBonus(uid, bt')$ tuple. Specifically, we use the read arc $(ExistsA1, WithBonus)$ that has the only purpose of checking whether the token is present in the place. In case there is no token that matches values assigned to ξ, we proceed with adding $WithBonus(pid, bt')$ with **NotExistsA1** that has the lower priority label assigned to it and consequently populate a $\langle true \rangle$-valued token in $DoneA1$. Note that the whole computation process is "guarded" with the global lock variable (needed for the consequent execution of all the steps defined in Fig. 4(b)): whenever started, the token is removed from it and can be returned only after the last operation of the action has been carried out.

Next we show how an action is encoded considering the general DB-net $\mathcal{B}$ in Fig. 4(a). Note that T is equipped with action ACT, where some of the action parameters $\vec{x}$ coincide with external inputs. ACT is defined on top of $\mathcal{P}$ with $\text{ACT}\cdot\mathtt{params} = \langle x_1, \ldots, x_n \rangle$, $\text{ACT}\cdot\mathtt{del} = F^-$ and $\text{ACT}\cdot\mathtt{add} = F^+$, where F^- and F^+ are two sets of $\mathcal{R}$-facts that should be respectively deleted and added. The CPN representing that expansion of the update net from Fig. 4(b) is depicted in Fig. 9. The computation starts by checking the guard of T with transition T_{cond} (cf. Fig. 4(b)). If the guard evaluates to true, T_{cond} puts a token in a place called $GuardOk$ that, in turn, allows to initiate the action execution process that is sequentially realized for all R-facts from F^- or F^+ in ACT.

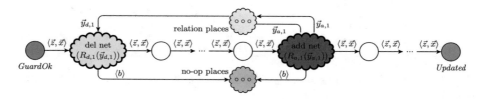

Fig. 9. Expansion of the update net from Fig. 4(b)

We first proceed with deleting all the $R_{d,i}$-facts from F^- (i.e., facts of the form $R_{d,i}(\vec{y}_{d,i})$). This process is sequntialized and at each of its step the net models the deletion of only one $R_{d,i}$-fact. Specifically, the deletion of each $R_{d,i}$-fact (see Fig. 10(a)) is realized by a pair of prioritized transitions **ExistsD**$_i$ and **NotExistsD**$_i$ and one auxiliary place $DoneD_i$, and is analogous to the example in Fig. 8. After all the R-facts from F^- have been deleted, the net switches to performing the insertion of R-facts from F^+. We omit the details of the addition process as it can be defined analogously to the one from the bonus change example and refer to Fig. 10(b). As soon as all R-facts are added, the update net completes its work by putting a token into a place called *Updated*.

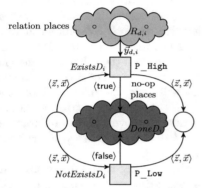
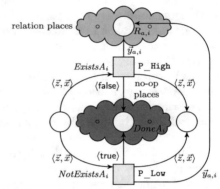

(a) Expansion of the i-th deletion component in the net of Figure 9

(b) Expansion of the i-th addition component in the net of Figure 9

Fig. 10. Expansion of deletion and addition nets from Fig. 9

3.3   Checking Integrity Constraints and Generating a New Marking

Let us now remind that the relational schema of $\mathcal{B}_\tau$ is equipped with three types of integrity constraints: primary keys, foreign keys and domain constraints. When the first and the last one could be relatively easy to check during the update phase, assuming that the computation results are accumulated in arc inscriptions analogously to the binding net in Fig. 6[2], the process of managing updates in the presence of arbitrary many foreign key dependencies is quite involved. To manage it correctly we first perform the updates and only then check whether the

---

[2] Both primary keys and domain constraints can be violated when a tuple is about to be inserted into a table. Specifically, to guarantee that former are respected, it is enough to check with $ExistsA_i$ whether there is a token in $R_{a,i}$ that has the same primary key value, and, if so, cancel the computation process. In the case of domain constraints, one may insert a third transition t' that has a normal priority and that will be fired whenever one of the values we want to insert is not in the allowed range. Firing of t' will have the same consequences as in the case of primary keys.

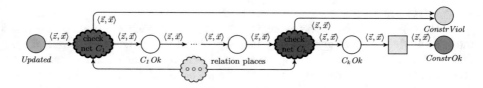

Fig. 11. Expansion of the check constraint net from Fig. 4(b)

generated marking represents a database instance that satisfies all the integrity constraints contained in the persistence layer of $\mathcal{B}_\tau$. A ν-CPN representing the check constraint phase is depicted in Fig. 11. The net works as follows: it consequently runs small nets for verifying the integrity of constraints and, in case of violation, puts a token in a special place called *ConstrViol*. As soon as there is at least one token in *ConstrViol* place, the big net in Fig. 4(b) terminates the constraint checking process and switches to the phase 4.(b) (that is, runs the undo net) explained in the beginning of this section. For ease of presentation we assume that every relation R/n_r from $\mathcal{R}$ of $\mathcal{B}_\tau$ will have the following look: the first k attributes form a primary key, while the rest of $n_r - k + 1$ attributes can be unconstrained or bounded by domain constraints. Moreover, if R is referencing some other relation S, then among these $n_r - k + 1$ attributes we reserve the last m such that $S[\{1, \ldots, m\}] \subseteq R[\{n_r - m, \ldots, n_r\}]$.

The constraint checking process starts with verifying that all the updates performed using the net in Fig. 9 are satisfying primary key constraints C_i. This is done by sequentially running small check nets in Fig. 12(a), where the constraint integrity is verified for some relation R by a pair of prioritized transitions *RepeatedKey* (high priority) and *NoRepeatedKey* (low priority). Note that *RepeatedKey* accesses the content of R with two read arcs and using the guard assigned to it verifies whether there exist two tokens, such that their first

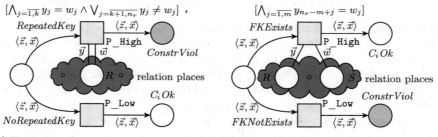

(a) Expansion for a primary key constraint C_i indicating that the first k components of relation R with arity $n_r \geq k$ form a key for R. In the figure, $\vec{y}$ and $\vec{w}$ are tuples containing n_r variables.

(b) Expansion for a foreign key constraint C_i indicating that the last m components of relation R with arity $n_r \geq m$ reference the last m components of relation S with arity $n_s \geq m$. In the figure, $\vec{y}$ and $\vec{w}$ are tuples containing n_r and n_s variables.

Fig. 12. Expansion of the check net from Fig. 11 in the case of key constraints

k values coincide and in the rest of $n_r - k + 1$ values there is at least one distinct pair of values. The satisfaction of such guard would mean that, essentially, we have inserted a token in R whose primary key values were not unique. Firing of *RepeatedKey* will produce one token in *ConstViol* and terminate the run of the check constraint net.

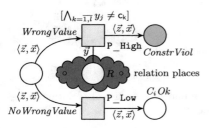

Fig. 13. Expansion of the check net from Fig. 11 in the case of a domain constraint C_i indicating that the j-th component of relation R with arity $n_r \geq j$ must contain a value that belongs to $\{c_1, \ldots, c_l\}$; In the figure, $\vec{y}$ is a tuple containing n_r variables.

The next type of constraints to verify is the foreign key dependency. Analogously to the previous case, we successively run small check nets like the one in Fig. 12(b) and in each of them control that R correctly references S (that is, there are no tuples in R that do not depend on any tuple in S). This is realized with two prioritized transitions *FKExists* (high priority) and *FKNotExists* (low priority). The first one, as the name suggests, checks whether the dependency between R and S is preserved for all the tokens in the corresponding relation places. *FKExists* makes use of the guard attached to it that performs pairwise comparison of m last values of a token from R to m first values of a token from S. If the guard is not satisfied, then the dependency relation between R and S has been violated and one fires *FKNotExists* so as to terminate the constraint checking process.

The last series of constraints to be checked is the one of domain constraints. The net in Fig. 13 employs two prioritized transitions, *WrongValue* and *NoWrongValue*, to verify whether all the tuples inserted into R had correct values. First, using the guard of *WrongValue*, we check whether there is at least one value that breaks the integrity of the domain constraint C_i of R. If *WrongValue* fires, the process is terminated by putting a token into *ConstViol*. Otherwise, *NoWrongValue* is executed and the constraint checking process continues.

Now let us show how the computation of all the effects of T is finished and a new marking is generated. If one of the constraints has been violated, we have to roll back all the effects pushed using the net in Fig. 9. To do so, we employ the net in Fig. 14 that reverts the update process by first canceling all the additions, and

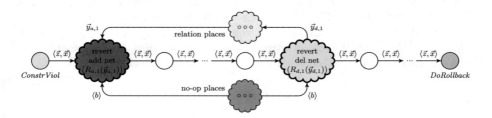

Fig. 14. Expansion of the undo net from Fig. 4(b)

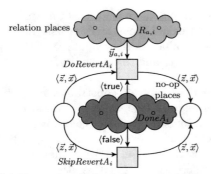
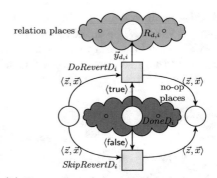

(a) Expansion of the i-th revert addition component

(b) Expansion of the i-th deletion component

Fig. 15. Expansion of revert deletion and addition nets from Fig. 14

then canceling all the deletions. Let us briefly explain how the rollback process is performed for each component net.

We start by removing all the tuples that have been successfully added to relation places following the definition of ACT. The net in Fig. 15(a) shows how to revert the result of inserting $R_{a,i}$-fact from F^+ (i.e., a fact of the form $R_{a,i}(\vec{y}_{a,i})$). If a fact has been added, that is, there is a token with value $\langle \text{true} \rangle$ in $DoneA_i$, then the net removes it by firing $DoRevertA_i$. Otherwise, if the fact has not been added, that is, there is a token with value $\langle \text{false} \rangle$ in $DoneA_i$, then the net proceeds without reverting by firing $SkipRevertA_i$. Then, for each $R_{d,i}$-fact, we go on with adding all the tuples that have been deleted by using the net depicted in Fig. 15(b). The update reverting process is analogous to the one dealing with reverted additions, but with only one exception: whenever $DoneD_i$ has a $\langle \text{true} \rangle$ token, then we put the deleted tuple (specified in F^-) back into $R_{d,i}$. Note that every revert deletion or addition net removes a token from a corresponding auxiliary no-op place.

As soon as all the operations of ACT have been undone and all the corresponding tokens have been withdrawn from relation places, the net places a token in $DoRollBack$ (cf. Figure 4(b)) and allows us to fire a transition called $T_{rollback}$ that implements the generation of the tokens in the postset corresponding to the rollback flow of T.

If the check constraint net's work has not been interrupted an the token was placed in $ConstrOk$ (cf. Fig. 4(b)), then we proceed with the consume net (cf. Fig. 16) that removes all tokens from the auxiliary no-op places and places a token into $DoCommit$. This, in turn, allows $\mathcal{N}_\tau$ to execute T_{commit} that populates tokens in the postset corresponding to the normal flow of T.

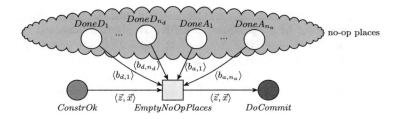

Fig. 16. Expansion of the consume net from Fig. 4(b)

3.4 The General Translation

In this section we bring together the modelling approaches described in the previous three sections and quickly summarize the translation from DB-nets to ν-CPNs with priorities. Specifically, we show that, given a DB-net, it is possible to build a ν-CPN that is weakly bisimilar to it.

Intuitively, $\mathcal{N}_\tau$ from Fig. 4(b) behaves just like the $\mathcal{B}_\tau$ in Fig. 4(a) and hence LTSs of these two nets are weakly bisimilar [10]. Notice that, in order to correctly represent the behavior of $\mathcal{B}_\tau$, $\mathcal{N}_\tau$ includes many intermediate steps that are, however, not relevant for comparing content of the states and behavior of the nets. For this we are going to resort to a form of bisimulation that allows to "skip" transitions irrelevant for the behavioral comparison [10]. Specifically, given two transition systems $\Gamma_1 = \langle S_1, s_{01}, \rightarrow_1, L \rangle$ and $\Gamma_2 = \langle S_2, s_{02}, \rightarrow_2, L \rangle$ defined over a set of labels L, relation $wb \subseteq S_1 \times S_2$ is called a *weak bisimulation* between Γ_1 and Γ_2 iff for every pair $\langle p, q \rangle \in wb$ and $a \in L \cup \{\epsilon\}$ the following holds: (1) if $p \xrightarrow{a}_1 p'$, then there exists $q' \in S_2$ such that $q \text{ -}^a\!\!\twoheadrightarrow_2 q'$ and $\langle p', q' \rangle \in wb$; (2) if $q \xrightarrow{a}_2 q'$, then there exists $p' \in S_1$ such that $p \text{ -}^a\!\!\twoheadrightarrow_1 p'$ and $\langle p', q' \rangle \in wb$. Here, $\epsilon \neq a$ is a special silent label and $p \text{ -}^a\!\!\twoheadrightarrow q$ is a weak transition that is defined as follows: *(i)* $p \text{ -}^a\!\!\twoheadrightarrow q$ iff $p(\xrightarrow{\epsilon})^* q_1 \xrightarrow{a} q_2 (\xrightarrow{\epsilon})^* q$; *(ii)* $p \text{ -}^\epsilon\!\!\twoheadrightarrow q$ iff $p(\xrightarrow{\epsilon})^* q$. We use $(\xrightarrow{\epsilon})^*$ to define the reflexive and transitive closure of $\xrightarrow{\epsilon}$. We say that a state $p \in S_1$ is weakly bisimilar to $q \in S_2$, written $p \approx^{wb} q$, if there exists a weak bisimulation wb between Γ_1 and Γ_2 such that $\langle p, q \rangle \in wb$. Finally, Γ_1 is said to be weakly bisimilar to Γ_2, written $\Gamma_1 \approx^{wb} \Gamma_2$, if $s_{01} \approx^{wb} s_{02}$. Let us now define a theorem that sets up the behavioral correspondence between DB-nets and ν-CPNs.

Theorem 1. *Let $\mathcal{B} = \langle \mathfrak{D}, \mathcal{P}, \mathcal{L}, \mathcal{N} \rangle$ be a DB-net with $\mathcal{P} = \langle \mathcal{R}, \emptyset \rangle$, $\mathcal{L} = \langle \mathcal{Q}^{UCQF^{\neq}}, \mathcal{A} \rangle$ and $\mathcal{N} = \langle P, T, F_{in}, F_{out}, F_{rb}, \textsf{color}, \textsf{query}, \textsf{guard}, \textsf{act} \rangle$, and s_0 is the initial snapshot. Then, there exists a ν-CPN $\mathcal{N} = \langle P \cup P_{rel} \cup P_{aux}, T \cup T_{aux}, F_{in}, F_{out}, \textsf{color}, M_0 \rangle$ with (1) a set of relation places P_{rel} acquired from $\mathcal{R}$, (2) two sets P_{aux} and T_{aux} of auxiliary places and transitions (required by the encoding algorithm), and such that $\Gamma_{\mathcal{B}}^{s_0} \approx_{flat}^{wb} \Gamma_{\mathcal{N}}^{M_0}$.*

The proof of the theorem (see [13] for the full version) can be obtained by induction on the construction of $\Gamma_{\mathcal{B}}^{s_0}$ and $\Gamma_{\mathcal{N}}^{M_0}$, where the latter is induced by $\mathcal{N}$ that has been modularly generated using the encoding defined in Sects. 3.1, 3.2 and

3.3. Intuitively, such encoding lifts the persistence and data logic layers to the control layer, resulting in a "pristine" ν-CPN. To show behavioral correspondence, one should make sure that states of $\Gamma_{\mathcal{B}}^{s_0}$ and $\Gamma_{\mathcal{N}}^{M_0}$ are comparable. This can be achieved by slightly modifying the notion of weak bisimulation in such a way that, for each $\langle \langle \mathcal{I}, m \rangle, M \rangle \in wb$, we compare elements stored in $\mathcal{I}$ only with their "control counterparts" in P_{rel} of M, whereas $m \subseteq M$. Moreover, we assume that states of $\Gamma_{\mathcal{N}}^{M_0}$ are restricted only to places in $P \cup P_{rel}$, that is, each marking M shall reveal tokens stored only in P and P_{rel}, and that when constructing $\Gamma_{\mathcal{N}}^{M_0}$ all the auxiliary transitions of $\mathcal{N}$ (i.e., all the transitions within the grey lane in Fig. 4(b)) are going to be labeled with ϵ. Note that such an extended definition allows to establish equivalence not only in terms of behaviors of two systems, but also in terms of their (data) content.

4   Conclusions

We have shown that the large and relevant fragment of DB-nets employing unions of conjunctive queries with negative filters as a database query language, can be faithfully encoded into a special class of Coloured Petri nets with transition priorities. This result is of particular interest as it demonstrates how to represent full-fledged databases with corresponding data manipulation operations in a conventional Petri net class. Since the encoding is based on a constructive technique that can be readily implemented, the next step is to incorporate the encoding into the DB-net extension of CPN Tools [16], in turn making it possible to make the state-space construction mechanisms available in CPN Tools also applicable to DB-nets. It must be noted that, due to the presence of data ranging over infinite colour domains, the resulting state-space is infinite in general. However, in the case of state-bounded DB-nets [12], that is, DB-nets for which each marking contains boundedly many tokens and boundedly many database tuples, a faithful abstract state space can be actually constructed using the same approach presented in [2]. Interestingly, this can be readily implemented by replacing the ML code snippet dealing with fresh value injection with a slight variant that recycles, when possible, old data values that were mentioned in a previous marking but are currently not present anymore.

Acknowledgments. This work has been partially supported by the UNIBZ projects PWORM and REKAP.

References

1. Abiteboul, S., Hull, R., Vianu, V.: Foundations of Databases. Addison Wesley, Boston (1995)
2. Calvanese, D., De Giacomo, G., Montali, M., Patrizi, F.: First-order mu-calculus over generic transition systems and applications to the situation calculus. Inf. Comput. **259**(3), 328–347 (2018)
3. Calvanese, D., De Giacomo, G., Montali, M.: Foundations of data aware process analysis: a database theory perspective. In: Proceedings of PODS (2013)

4. Drescher, C., Thielscher, M.: A fluent calculus semantics for ADL with plan constraints. In: Hölldobler, S., Lutz, C., Wansing, H. (eds.) JELIA 2008. LNCS (LNAI), vol. 5293, pp. 140–152. Springer, Heidelberg (2008). https://doi.org/10.1007/978-3-540-87803-2_13

5. Emmerich, W., Gruhn, V.: Funsoft nets: a petri-net based software process modeling language. In: Proceedings of IWSSD, pp. 175–184 (1991)

6. Hull, R.: Artifact-centric business process models: brief survey of research results and challenges. In: Meersman, R., Tari, Z. (eds.) OTM 2008. LNCS, vol. 5332, pp. 1152–1163. Springer, Heidelberg (2008). https://doi.org/10.1007/978-3-540-88873-4_17

7. Jensen, K., Kristensen, L.M.: Coloured Petri Nets - Modelling and Validation of Concurrent Systems. Springer, Heidelberg (2009). https://doi.org/10.1007/b95112

8. de Leoni, M., Felli, P., Montali, M.: A holistic approach for soundness verification of decision-aware process models. In: Trujillo, J.C., et al. (eds.) ER 2018. LNCS, vol. 11157, pp. 219–235. Springer, Cham (2018). https://doi.org/10.1007/978-3-030-00847-5_17

9. Libkin, L.: Fixed point logics and complexity classes. Elements of Finite Model Theory. LNCS, vol. 7360. Springer, Heidelberg (2004). https://doi.org/10.1007/978-3-662-07003-1

10. Milner, R.: Communication and Concurrency. Prentice-Hall Inc., Upper Saddle River (1989)

11. Montali, M., Rivkin, A.: Model checking Petri nets with names using data-centric dynamic systems. Formal Asp. Comput. 28(4), 615–641 (2016)

12. Montali, M., Rivkin, A.: DB-nets: on the marriage of colored petri nets and relational databases. In: Koutny, M., Kleijn, J., Penczek, W. (eds.) Transactions on Petri Nets and Other Models of Concurrency XII. LNCS, vol. 10470, pp. 91–118. Springer, Heidelberg (2017). https://doi.org/10.1007/978-3-662-55862-1_5

13. Montali, M., Rivkin, A.: From DB-nets to coloured petri nets with priorities (extended version). Technical Report arXiv:1904.00058, arXiv.org (2019)

14. Oberweis, A., Sander, P.: Information system behavior specification by high level petri nets. ACM Trans. Inf. Syst. 14(4), 380–420 (1996)

15. Reichert, M.: Process and data: two sides of the same coin? In: Meersman, R., Panetto, H., Dillon, T., Rinderle-Ma, S., Dadam, P., Zhou, X., Pearson, S., Ferscha, A., Bergamaschi, S., Cruz, I.F. (eds.) OTM 2012. LNCS, vol. 7565, pp. 2–19. Springer, Heidelberg (2012). https://doi.org/10.1007/978-3-642-33606-5_2

16. Ritter, D., Rinderle-Ma, S., Montali, M., Rivkin, A., Sinha, A.: Formalizing application integration patterns. pp. 11–20 (2018)

17. Rosa-Velardo, F., de Frutos-Escrig, D.: Decidability and complexity of petri nets with unordered data. Theor. Comput. Sci. 412(34), 4439–4451 (2011)

18. Triebel, M., Sürmeli, J.: Homogeneous equations of algebraic petri nets. In: Proceedings of CONCUR, LNCS, pp. 1–14. Springer (2016)

19. Weitz, W.: SGML nets: integrating document and workflow modeling. In: Proceedings of HICSS, vol. 2, pp. 185–194 (1998)

20. Westergaard, M., Verbeek, H.M.W.E.: Efficient implementation of prioritized transitions for high-level Petri nets. In: Proceedings of PNSE, pp. 27–41 (2011)

Author Index

Printed in the United States
By Bookmasters